Micro- and Opto-Electronic Materials, Structures, and Systems

Series Editor
E. Suhir
University of California, Santa Cruz, CA, USA

For further volumes:
http://www.springer.com/series/7493

Xingsheng Liu • Wei Zhao • Lingling Xiong
Hui Liu

Packaging of High Power Semiconductor Lasers

Springer

Xingsheng Liu
Focuslight Technologies Co., Ltd
Xi'an Institute of Optics
& Precision Mechanics
Chinese Academy of Sciences
Shaanxi, People's Republic of China

Wei Zhao
Xi'an Institute of Optics
& Precision Mechanics
Chinese Academy of Sciences
Shaanxi, People's Republic of China

Lingling Xiong
Xi'an Institute of Optics
& Precision Mechanics
Chinese Academy of Sciences
Shaanxi, People's Republic of China

Hui Liu
Xi'an Institute of Optics
& Precision Mechanics
Chinese Academy of Sciences
Shaanxi, People's Republic of China

ISBN 978-1-4939-5590-9 ISBN 978-1-4614-9263-4 (eBook)
DOI 10.1007/978-1-4614-9263-4
Springer New York Heidelberg Dordrecht London

Softcover reprint of the hardcover 1st edition 2015

Printed on acid-free paper

Springer is part of Springer Science+Business Media (www.springer.com)

Preface

Over the past decades, laser technologies have developed tremendously and are playing an increasingly important role in economic development in highly industrialized countries. Lasers are now used in a wide range of applications such as advanced materials processing, communication, information technology, and medical and military fields. As one of the major types of laser technologies, semiconductor lasers are by far the most widespread and economically significant laser technology ever developed. The semiconductor lasers exhibit many advantages such as compact size, wide range of output power, wavelength versatility, and low cost per watt, etc. Traditionally, semiconductor lasers have been mainly used in fiber-optic communications and optical storage. Today, semiconductor lasers are moving towards the direction of high power and high brightness and extending their applications from "signal" to "energy". High power semiconductor lasers have found increased applications in pumping of solid-state lasers and fiber lasers as well as direct applications. They are widely used in materials processing, medical and cosmetic applications, military and defense, displays, printing, and scientific research, etc.

To realize the potential of high power semiconductor lasers, a convergence of many technologies must happen. Over recent years, high power semiconductor lasers have seen a tremendous evolution based on material epitaxial growth technology, advanced epi-structure design and optimization techniques, and facet passivation technologies. The successful use of high power semiconductor lasers depends on their high power, high reliability, acceptable lifetime, and high durability. For some of the direct applications, high brightness and good beam quality are also required. All these have to do with packaging of high power semiconductor lasers or sometimes even largely rely on packaging technologies. Higher output powers lead to an increased thermal power density in the high power semiconductor laser, and the package has to carry a higher thermal load. The reliability, lifetime, and durability of high power semiconductor lasers are related to thermal management, packaging material properties, bonding processing, and testing and burn-in of the devices. Achieving high brightness and good beam quality requires not only high brightness diode laser chips or bars, but also micro-optics, optical beam shaping techniques, and beam combining technologies. These techniques and technologies require sophisticated optical systems and increased precision of the semiconductor laser packaging in terms of positioning tolerances relative to the mounting substrate and smile of the semiconductor laser. Packaging technology has become an important part of high power semiconductor laser development. With the vast advance of semiconductor laser chip and bar technology in recent years, packaging has become one of the bottlenecks of the advancement of high power semiconductor lasers.

Many books on semiconductor lasers have been published. There are also a few books on high power semiconductor lasers. These books mainly focus on semiconductor physics and diode laser physics, epitaxial structures and growth, chip fabrication process, and device performance. As mentioned earlier, packaging is a significant part of high power semiconductor laser development and it becomes one of the limiting factors of the progression of high power semiconductor lasers. Packaging is truly a multi-discipline subject involving physics, mechanical design, thermal management, mechanics, optics, materials science, processing, manufacturing engineering, electrical and reliability engineering, and quality engineering. To the authors' best knowledge, there is still no published book dedicated to packaging of high power semiconductor lasers. The aim of this book is to bridge this gap and to devote it to the packaging aspects of high power semiconductor lasers. This book covers the basic essentials and fundamentals of high power semiconductor laser packaging technology and is intended to provide a practical guidance and reference for engineers who work on semiconductor lasers and optoelectronics. This book could also serve as a text book for undergraduate and graduate students who study semiconductor lasers, optoelectronics, and packaging technology.

This book presents systematic information on the packaging technology of high power semiconductor lasers. There are 11 chapters in this book, with each chapter focusing on one technical aspect of high power semiconductor lasers. Chapter 1 gives a general description on the basics of semiconductor lasers and some important specific designs and considerations for high power semiconductor lasers. Chapter 2 introduces the packaging types of high power semiconductor lasers. Chapter 3 discusses thermal design and management of high power semiconductor lasers. Chapter 4 focuses on the effects of thermal stress on performances of high power semiconductor lasers, thermal stress analysis, and thermal stress minimization. Chapter 5 describes the optical design and beam shaping, followed by the study of the packaging materials in Chap. 6. Packaging process and the testing technology for high power semiconductor lasers are discussed in detail in Chaps. 7 and 8, respectively. Chapter 9 focuses on the failure analysis and the reliability assessment of high power semiconductor lasers. In Chap. 10, examples of high power semiconductor lasers in pumping, material surface treatment, and hair removal applications are presented. As a conclusion of this book, the technology trend and challenges in high power semiconductor laser packaging are discussed in Chap. 11.

We express our sincere thanks to our colleagues Dr. Guodong Xu, Dr. Jingwei Wang, Dr. Shuangyan Xu, Mr. Dong Hou, Dr. Pu Zhang, Ms. Bei Hao, Mr. Tao Song, Ms. Xia Yin, Mr. Yao Sun, Mr. Steven Chen, and Dr. Guowen Yang for their contributions. Dr. Guodong Xu and Dr. Shuangyan Xu reviewed most of our manuscript and provided many valuable and supportive discussions and modifications. Dr. Jingwei Wang contributed significantly to Chaps. 6–8. Mr. Dong Hou and Ms. Bei Hao made great contribution to Chap. 7. Dr. Pu Zhang contributed to Chap. 6. Mr. Tao Song, Ms. Xia Yin and Mr. Yao Sun helped on Chap. 10. Mr. Steven Chen and Dr. Guowen Yang proof read some chapters and provided helpful suggestions. We are also grateful to our colleagues Ms. Di Wu, Ms. Min Wang, and Mr. Ye Dai for providing some materials and suggestions. We also extend our thanks to our graduate students Mr. Zhihuang Huang, Mr. Dihai Wu, and Ms. Shuna Wang for their work on the drawings and format. We express our gratitude to Ms. Ania Levinson and Mr. Brian Halm of Springer for their support, patience, and understanding. Furthermore, we owe great thanks to Prof. Ephraim Suhir for his support and guidance.

We acknowledge Focuslight Technologies Co., Ltd. and Xi'an Institute of Optics and Precision Mechanics of CAS for tremendous supports on allowing us to use invaluable technical and production information and many data resources in this book.

Last but not the least, we are indebted to our family members for their invaluable support and understanding during the process of writing this book.

Xi'an, P.R. China
January, 2014

Xingsheng Liu
Wei Zhao
Lingling Xiong
Hui Liu

Contents

Chapter 1
Introduction of High Power Semiconductor Lasers

In this chapter, we will give a general description on the basics of semiconductor lasers. Starting with a general introduction about semiconductor lasers, other important specific basics for high power semiconductor lasers are also introduced, covering the fundamental principles, device structures, and basic characteristic parameters.

1.1 The Fundamental Principle of Semiconductor Lasers

1.1.1 Energy Band Theory

Laser operation relies on three conditions, stimulated emission of the amplifying medium, feedback by an optical resonator, and the population inversion. The threshold of laser operation is obtained if the gain in the resonator compensates for the overall losses, i.e., the propagation losses and the apparent losses due to the extraction of light [1–4].

Figure 1.1 shows the energy band structure with nearly free carrier approximation [2]. In Fig. 1.1, $E_c(k)$ and $E_v(k)$ are the electron energies in the conduction and valence bands, respectively. They behave similarly for small wave numbers k and are expressed as

$$E_c(k) = E_g + \hbar^2 k^2/2m_e, \tag{1.1}$$

$$E_v(k) = -\hbar^2 k^2/2m_h, \tag{1.2}$$

where $\hbar$ is Planck constant of $\hbar = h/2\pi$ and k is the wave number $k = 2\pi/\lambda$; m_e and m_h are the effective masses of the electrons and the holes, which are in general different from the free-electrons rest mass. If the parameter k is continuous variable in Eqs. (1.1) and (1.2), $E_c(k)$ and $E_v(k)$ are continuous curves as shown in Fig. 1.1.

X. Liu et al., *Packaging of High Power Semiconductor Lasers*, Micro- and Opto-Electronic Materials, Structures, and Systems, DOI 10.1007/978-1-4614-9263-4_1,

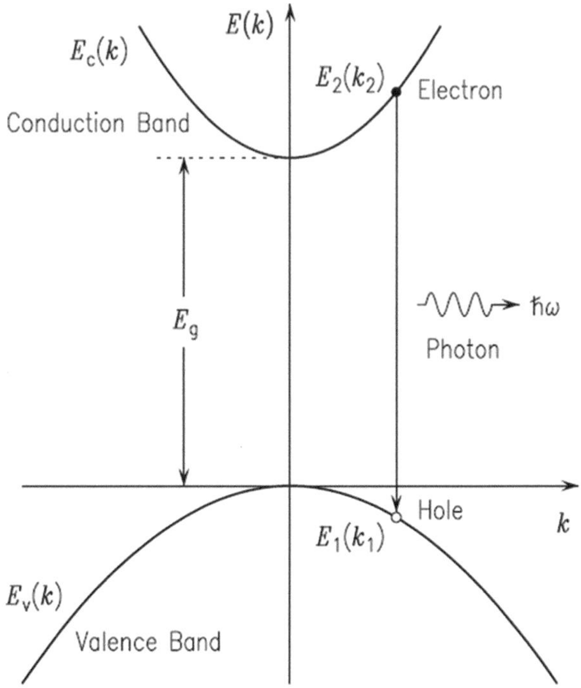

Fig. 1.1 The energy band structure with nearly free carrier approximation [2]

Radiative band-to-band transitions are generation and recombination of electron–hole pairs associated with absorption or emission of photons. E_g is the value of the forbidden band which is to separate the conduction and valence bands. E_g is the energy difference between the minimum electron energy of the conduction band and the maximum electron energy of the valence band. $E_{c2}(k_2)$ and $E_{v1}(k_1)$ are one recombination of an electron in the conduction band and a hole in the valence band, which generates a photon with energy ħω. Since the momentum of the photon ħω is negligibly small, radiative electronic transitions between conduction and valence bands only occur at the same wave number k.

$$E_{ph} = \hbar w = E_2 - E_1, k_2 = k_1. \tag{1.3}$$

As shown in Fig. 1.1, these transitions can be illustrated by vertical arrows with the length of the photon energy ħω pointing upwards for generation and downwards for recombination of an electron–hole pair. In thermal equilibrium, the carriers tend to occupy the states with lowest energy. For electrons, these are states at the minimum of the conduction band. On the ordinate axis of a band diagram, the electron energy is plotted; therefore, the minimum energy of the positively charged holes is in the maximum of the valence band. The valence-band maximum and the conduction-band minimum of direct semiconductors are both located at the point $k = 0$. In indirect semiconductors like silicon and germanium, minimum and maximum have different k-values; therefore, band-to-band recombination can only occur with the contribution of phonons or traps. These transitions are

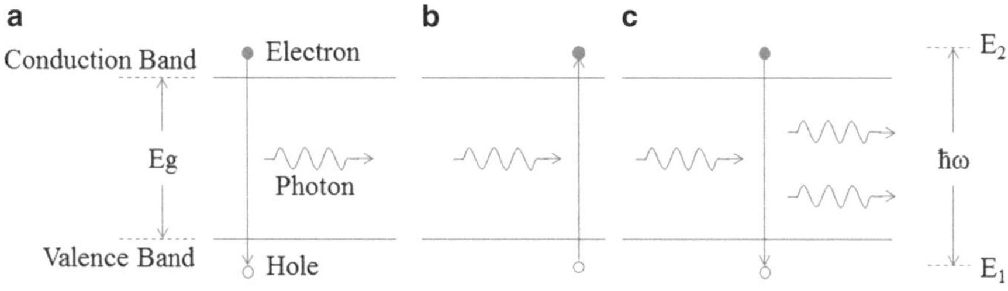

Fig. 1.2 Three types of radiative band-to-band transition in semiconductor, (**a**) spontaneous emission, (**b**) absorption, and (**c**) stimulated emission [2]

unsuitable for laser activity, because the spatial density of phonons and traps is very low. Furthermore, they are mostly non-radiative and rather unlikely since more partners are involved.

1.1.2 Principle of Semiconductor Lasers

For semiconductor lasers, the stable laser beams are generated when three conditions are satisfied: the population inversion, the optical gain, and the stable laser oscillation.

There are three types of radiative band-to-band transition in semiconductor as shown in Fig. 1.2, which are spontaneous emission, stimulated absorption, and stimulated emission [2]. Figure 1.2a shows the spontaneous emission where the emission of photons is generated by recombination of electro-hole pairs and the photons are random in direction, phase, and time resulting in incoherent light. This emission is the process in light-emission diodes (LED).

The stimulated absorption is shown in Fig. 1.2b. The transition rate R_{12} is proportional to three densities: first, the density of non-occupied states $D(E_2)[1-f(E_2, T)]$ in the conduction band at the energy E_2; second, the density of states occupied by electrons $D(E_1)f(E_1, T)$ in the valence band at E_1; and third, the density of the photons $\rho(\hbar\omega)$ with energy $\hbar\omega = E_2 - E_1$.

$$R_{12} = B_{12}\rho(\hbar w)D(E_1)f(E_1, T)D(E_2)[1 - f(E_2, T)]. \tag{1.4}$$

B_{12} is proportionality constant for stimulated absorption.

The third process is stimulated emission shown in Fig. 1.2c. The stimulated absorption and emission are inverse processes to each other.

The combination of an electron hole pair is stimulated by a photon and a second photon is generated simultaneously which has the same direction and phase as the first photon. This process can be used to amplify optical radiation, since the photons are emitted into the optical mode of the stimulating photon resulting in coherent radiation.

Light sources based on this emission process are called lasers, which is an abbreviation of light amplification by stimulated emission of radiation. Analogous to the stimulated absorption Eq. (1.4), the transition rate R_{21} for stimulated emission can be described as

$$R_{21} = B_{21}\rho(\hbar w)D(E_2)f(E_2, T)D(E_1)[1 - f(E_1, T)], \tag{1.5}$$

where B_{21} is the proportionality constant for stimulated emission.

In order to achieve semiconductor laser beam, stimulated absorption and emission are both needed. To determine whether an optical wave with quantum energy $\hbar\omega$ is absorbed or amplified by stimulated emission, the quotient of the corresponding rates R_{12} and R_{21} is calculated. The proportion of R_{12} to R_{21} can be expressed as [1]:

$$\frac{R_{12}}{R_{21}} = \exp\left[\frac{\hbar\omega - (E_{Fc} - E_{Fv})}{K_B T}\right], \tag{1.6}$$

where, E_{Fc} and E_{Fv} are the equilibrium Fermi function with different Fermi-level energies in the conduction and valence bands, respectively. In thermal equilibrium there is $E_{Fc} = E_{Fv} = E_F$, where E_F is Fermi-level energy at $T = 0$ K, and the transition rate R_{12} is always larger than R_{21}, The operation of the semiconductor lasers can only be achieved only if the transition rate R_{21} is larger than R_{12} in Eq. (1.6). When $R_{12}/R_{21} < 1$, it is called population inversion and the relationship can be written as

$$E_{Fc} - E_{Fv} > \hbar w > E_2 - E_1, \tag{1.7}$$

when the Eq. (1.7) is satisfied, the population inversion in semiconductor is built. Then the laser beam generated during stimulated emission is reflected in the optical resonant cavity, and it is amplified after being reflected repeatedly by the optical resonators which is called optical oscillation.

The optical loss of the laser beam also occurs during the multiple oscillations. So only when the optical amplifying of the laser beam is larger than the optical loss during the oscillations, the laser beam can be increased and the stable laser operation can form. The beam amplifying is larger than the optical loss during the oscillations is called optical gain.

The value is called the threshold gain (denoted by g_{th}), when the optical gain is equal to the loss. Only if the optical gain is greater than the threshold gain g_{th}, the laser operation can be established. The threshold gain g_{th} is calculated and expressed as [2, 3]

$$g_{th} = a_i + \frac{1}{2L}\ln\frac{1}{R_1 R_2}, \tag{1.8}$$

where a_i is the inner loss coefficient. L, R_1, and R_2 in Eq. (1.8) are shown in Fig. 1.3 in which L is the length of the optical resonators, and R_1 and R_2 are the two mirrors

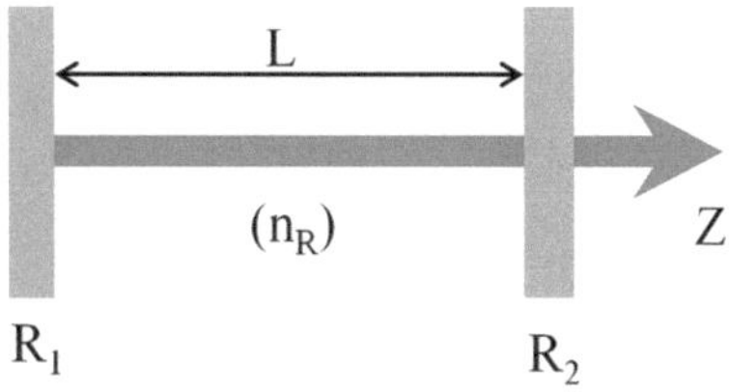

Fig. 1.3 A Fabry-Perot resonator with cavity length L and the reflectivity of R_1 and R_2 [5, 6]

to compose optical resonator, and the R_1 is the holophote and the mirror of R_2 has low reflectivity [5].

Equation (1.8) indicates that when the optical gain arrives at the threshold gain g_{th}, the optical gain can overcome the inner losses such as the absorption of the optical medium and optical scattering, and the loss caused by the output. When the optical gain is above the g_{th}, the positive optical gain is obtained and the laser operation is realized.

1.2 Device Structure

1.2.1 Gain Medium Structures

Double heterojunction structure: The homojunction semiconductor laser was initially used in the laser industry but it could only operate under very low environment temperature with high threshold current. As a replacement of the homojunction laser, the single heterojunction laser was developed, but it still could not work efficiently in room temperature and the threshold current was also too high. With the development of semiconductor lasers, the double heterojunction semiconductor laser has been proposed and produced, and it is successful to overcome the drawbacks in the former two semiconductor lasers. The typical structure of double heterojunction is shown in Fig. 1.4 [6]. In the figure, the junction of p-GaAs is between the layers of N-$Al_xGa_{1-x}As$ and P-$Al_xGa_{1-x}As$, and two heterojunction barriers are formed. The current carrier and optical gain are restricted in active region of p-GaAs by the heterojunction barriers. It has been formally used in the products of semiconductor lasers due to the low threshold current at room temperature. This makes it possible to achieve the high power semiconductor laser.

The double heterojunction semiconductor lasers have high threshold current, low photoelectric conversion efficiency and output power, and hence, it should be developed by new structures to improve the performances and obtain the high efficiency and output power.

Quantum well structures: As the generation of the new semiconductor theory and the progress of the manufacturing process of the crystal epitaxy growth process, the quantum well semiconductor lasers were born in the 1980s of last century.

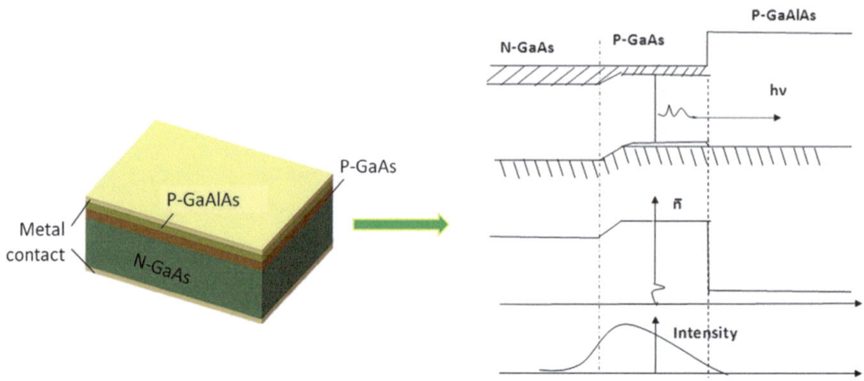

Fig. 1.4 The structure of a double heterojunction [5, 6]

The difference between the quantum well and the double heterojunction laser is the size of the active region. Typical thickness of the active layer for the double heterojunction is 50–300 nm while the thickness of the quantum well is just 5–10 nm. Due to the small active volume of the quantum well lasers, the low threshold current can be achieved. As the thickness of the active layer is 5–10 nm, and the electronic wave functions in quantum well show quantization which results in discrete energy levels in the vertical direction. In this case, the electronic state density, which are located at the electronic energy levels of the quantum well, increases in steps. Thus, the state density which is close to the lowest energy level in quantum well is much higher than that at the band edge in bulk material. The density of carriers at a given energy is determined by the state density and the occupied probability of electron or hole, which is exponentially decreasing function. So the carrier distribution for a quantum well laser structure has a high maximum value.

Figure 1.5a shows the structure of a single quantum well [5, 6]. In the quantum well, the ultra-thin layer with narrow band gap is inlayed into the middle of the two kinds of material with wide band gap. If the ultra-thin layers with narrow band gap and broad band gap are deposited alternately on the chip, the multi-quantum wells can be formed as shown in Fig. 1.5b [5, 6].

Quantum wire and dot structures: The high performances of semiconductor lasers are realized by the strained quantum well structures based on the development in epitaxial growth technologies. A further improvement in semiconductor laser performance is made by the introduction of quantum confinement in more than one dimension. In low dimensional quantum well structures, such as quantum wire and quantum dot (or quantum box), carriers are more strongly confined than that in quantum film due to further modification of band structures and density of states DOS distributions [4, 7–9]. The low dimensional quantum well laser, such as quantum wire and dot, are more complicated than that of the quantum well due to the multidimensional structures as shown in Fig. 1.6 [10]. The cube on the left

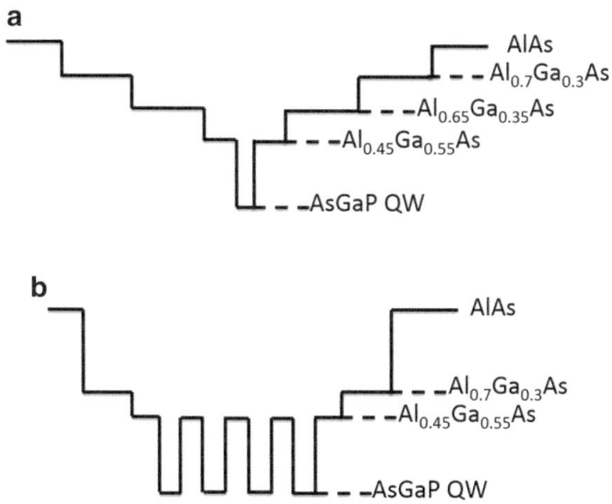

Fig. 1.5 The structure of a single quantum well (**a**) and the multi-quantum wells (**b**) [5, 6]

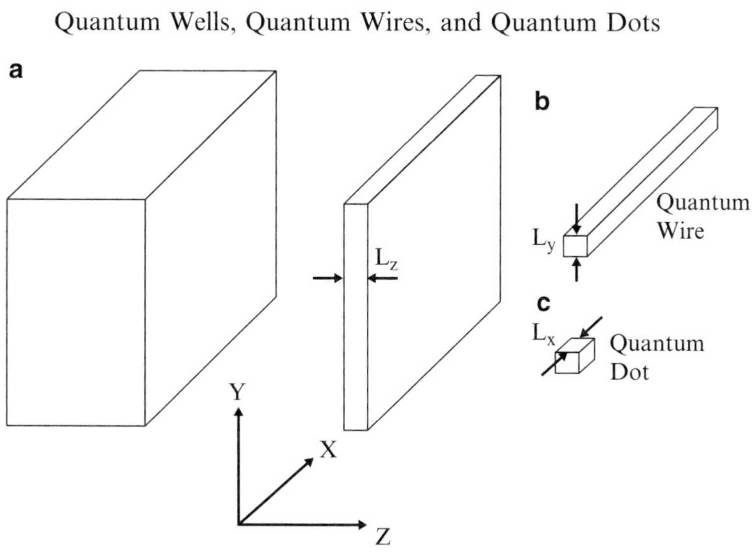

Fig. 1.6 Quantum wells, quantum wires, and quantum dots. The cube on the left represents bulk semiconductor. The drawings labeled (**a**), (**b**), and (**c**) represent quantum well, quantum wire, and quantum dot structures, respectively, with planar boundaries [10]

represents bulk semiconductor in the figure. The drawings labeled (a), (b), and (c) represent quantum well, quantum wire, and quantum dot structures with planar boundaries, respectively.

Owing to the strong confinement of carriers into low dimensional quantum well structures as shaper DOTS features, low dimensional quantum structure lasers give higher optical gain and a narrower gain spectrum at the same injection current

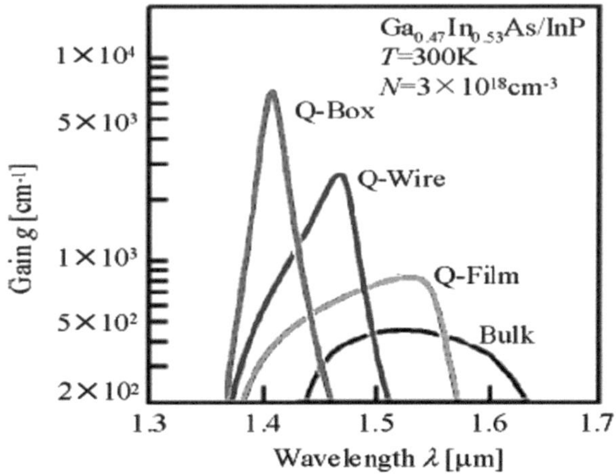

Fig. 1.7 The gain spectra of different quantized dimensions in the case of lattice-matched GaInAs/InP under the same injection carrier density [11]

density and intraband relaxation time (0.1 ps). Figure 1.7 shows the calculated gain spectra of different quantized dimensions in the case of lattice-matched GaInAs/InP under the same injection carrier density [11]. Not only the sharpness of the gain spectrum but also its symmetric shape of lower dimensional structures was found to be very promising for narrow spectral chirp under a high-speed direct modulation and a narrow-linewidth operation due to reduced linewidth enhancement factor [11, 12]. Therefore, the threshold current, differential quantum efficiency, and linewidth of Q-wire and Q-box lasers have been expected to be superior to those of Q-film lasers [13]. Further improvements by a combination of strain and low dimensional QW structures have also been expected [14].

1.2.2 *Physical Structure*

The scheme of a semiconductor laser is shown in Fig. 1.8. As shown in Fig. 1.8, an optical gain with yellow color is between the epitaxial layer, and the whole thickness of the epitaxial layer is ~5 μm. The p and n contacts are packaged up and down side of the epitaxial layer. This section will present the physical structure of high power semiconductor lasers by examining the current injection of carriers, the optical gain, and optical resonator structures [4].

Current injection: The current injection of carriers is accomplished through the p-contact and n-contact which connect to the anode and cathode of a power, respectively. The optical gain medium of semiconductor lasers is doped semiconductor material, which is either n-doped or p-doped. The package structure with p-contact on the upper side is called as epi-up package structure and has been used for many years. Since the epitaxial layers containing the active region are typically much thinner than the substrate, it is obviously advantageous to mount the laser in the epi-down configuration to minimize the thermal resistance [15–18].

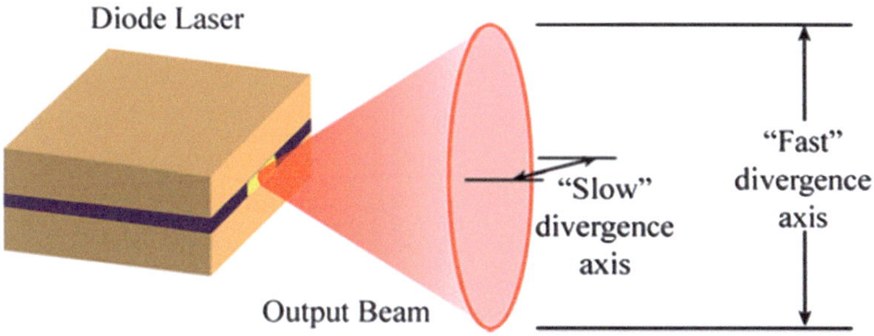

Fig. 1.8 The structure of a semiconductor laser emitter [4]

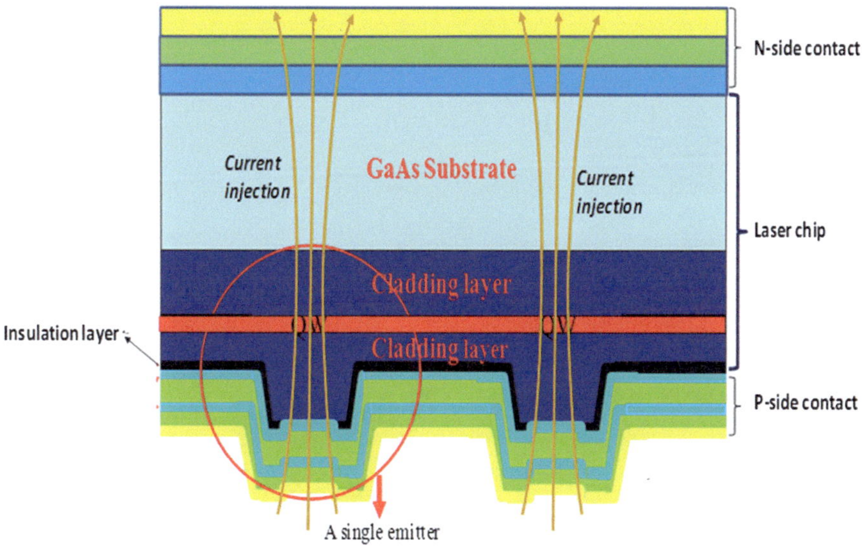

Fig. 1.9 Schematic structure of an epi-down bonded semiconductor laser [4, 17]

The epi-down structure is shown in Fig. 1.9, and it replaced the epi-up structure and has been applied in the packaging of high power semiconductor lasers [17]. The current injection flows into laser chip from N-side metallization to P-side metallization as shown in the figure.

P-side and N-side contacts are composed by multiple layers of metal materials. In order to obtain good ohmic contact, the different structures of P-side and N-side contacts have been designed. Figures 1.10 and 1.11 show three typical structures of P-side and N-side contacts, respectively [5].

Optical resonator: The optical resonator of a semiconductor diode laser consists of a waveguide structure between the mirrors build by crystal facet, and the structure is shown in Fig. 1.12 [19]. These facets are coated to achieve the optimum reflectivity.

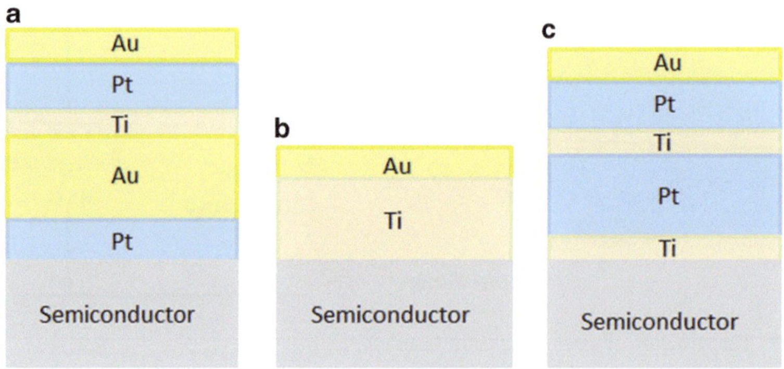

Fig. 1.10 Some examples of the structure of P-side contact [5]

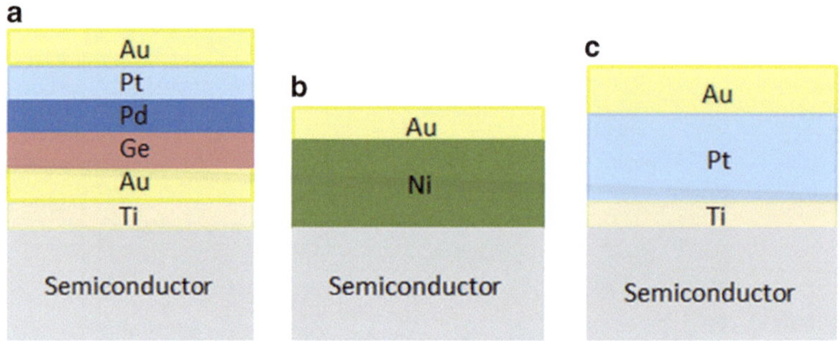

Fig. 1.11 Some examples of the structure of N-side contact [5]

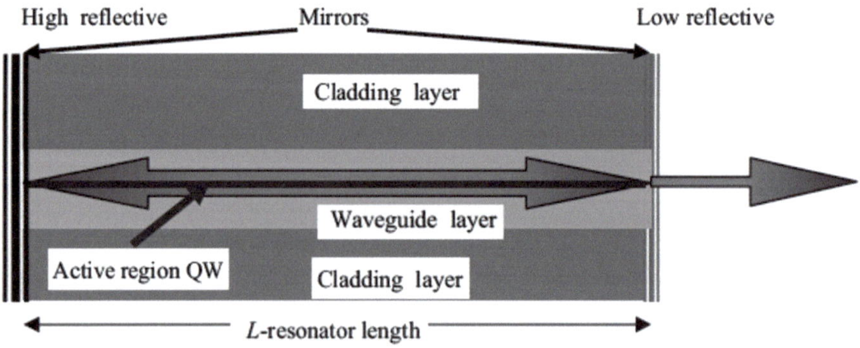

Fig. 1.12 The structure of the Fabry-Perot resonator of a diode laser [19]

In the vertical dimension, perpendicular to the pn-junction, the mode intensity distribution and the number of modes are determined by the thickness and composition of the grown layers. Wave guiding is supported for modes with two polarizations, one nearly transverse electric (TE) and one nearly transverse magnetic (TM).

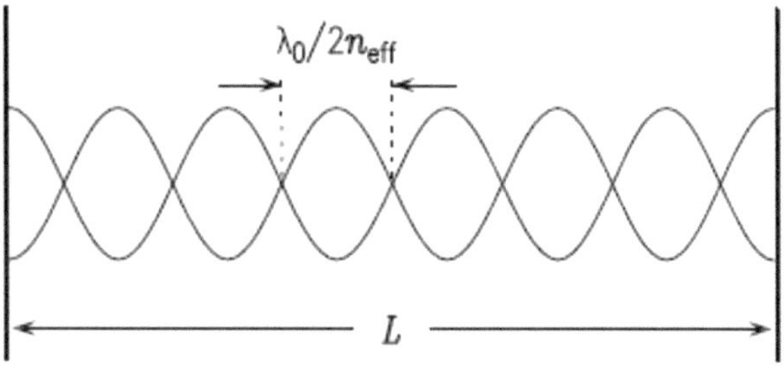

Fig. 1.13 The beam propagates in the form of the standing wave in the resonator [2]

For the TE case, the electrical field vector oscillates parallel to the epitaxial layers; for the TM case the magnetic field vector oscillates vertical to the epitaxial layers. The polarization of the diode laser beam is determined by the kind of QW. Tensile-strained QWs yield more gain for TM modes, compressively strained QWs for TE-modes.

In the lateral direction the mode distribution is determined by geometrical aspects of the current injection and/or by lateral waveguide, for example, due to the etched waveguide structure. A broad contact stripe represents the most elementary structure from a fabrication point of view. "Broad" means that the lateral dimensions are large compared to both wavelengths and carrier diffusion length. Broad-Area (BA) diode lasers exhibit widths of around 100 μm corresponding to about 400 wavelengths and about 50 diffusion lengths, respectively. This broad lateral waveguide supports many guided modes resulting in the typical multimode beam characteristics of semiconductor diode lasers. Since nearly all recombination processes can contribute to the mode gain, the efficiency of such devices is very high.

For a semiconductor laser, the beam intensity distribution and beam characteristics are determined by the thickness and composition of the grown layer. The semiconductor laser beam propagates in the form of the standing wave in the resonator as shown in Fig. 1.13 [2]. The distance between two mirrors is L and the effective refractive index of the optical waveguide is denoted by n_{eff} in the propagating direction. The resonator provides feedback, when a standing wave develops between the mirrors.

$$L = m\,\lambda_0/2n_{eff}, \quad m = 1, 2, 3\ldots, \tag{1.9}$$

where m presents the number of nodes of the standing wave and the order number of the longitudinal mode, and λ_0 is the wavelength in vacuum.

1.3 High Power Designs and Considerations

Generally, high output power is related to high operating current, high optical power load at the chip end facets, and high heat load. In order to achieve high power output, a number of design aspects are considered [19]:

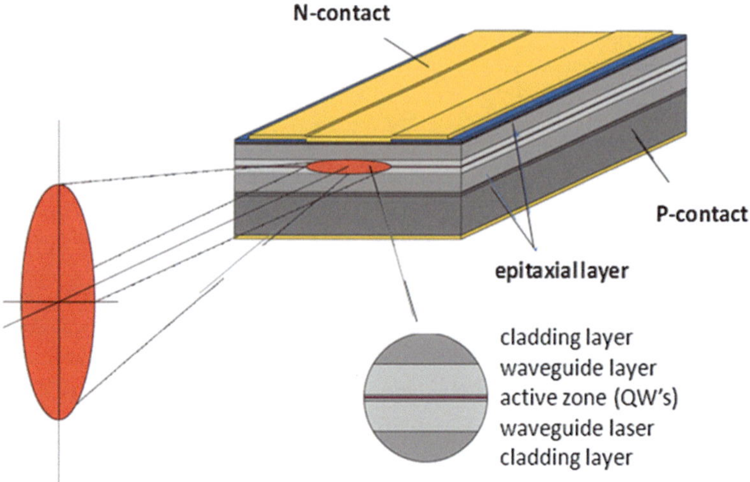

Fig. 1.14 Typical structure of the single emitter [19]

1. Minimizing the local current densities, e.g., by using large contact area and a large pn-junction area for the optical emission.
2. Dissipating the generated heat through the contact area more efficiently, e.g., by using larger heat transfer areas.
3. Minimizing electrical losses, especially at high current densities, requires low series resistances, which can be reduced by enlarging the contacts, increasing the doping levels, and improving the electron and hole mobility inside the epitaxial layers.
4. Minimizing scattering or absorption losses, e.g., by optimizing the quality of the epitaxial layers and by improving the waveguide and quantum well design.
5. Increasing the threshold of catastrophic optical mirror damage (COMD), e.g., by using a large optical cavity (LOC) in order to reduce the local optical power density at the end facets.

Current high power semiconductor lasers consider most of these aspects, and many semiconductor lasers have been designed and fabricated. A typical single emitter semiconductor laser is normally of 500–600 μm wide and 100–150 μm thick, the typical structure is shown in Fig. 1.14 [19]. The cavity length is generally 1–4 mm depending on the chip design and the stripe width typically ranges from 100 to 200 μm. A semiconductor laser bar is composed by multiple emitters. A standard laser bar is of 10 mm wide and generally has 19–75 emitters depending on the fill factor and stripe width. Each emitter in the laser bar is arranged in a row, and a typical structure of a laser bar is shown in Fig. 1.15 [19]. Each emitter is of equal interval and parallel connection. Presently, the reliable power of a state-of-the-art single emitter can reach ~15 W in continuous wave (CW) operation mode, while the laser bar with 19 emitters can be up to ~150 W in CW operation mode.

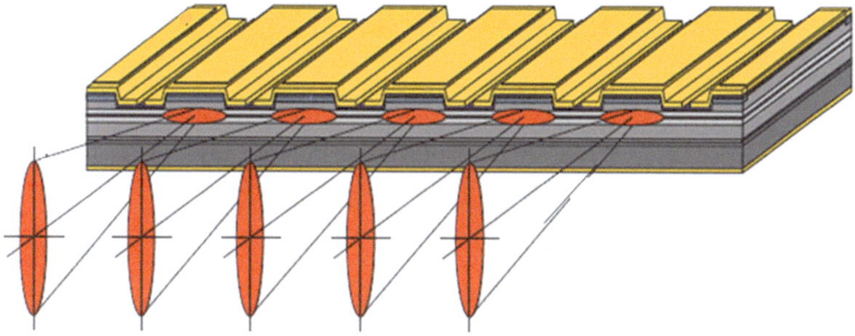

Fig. 1.15 Typical structure of the semiconductor laser bar [19]

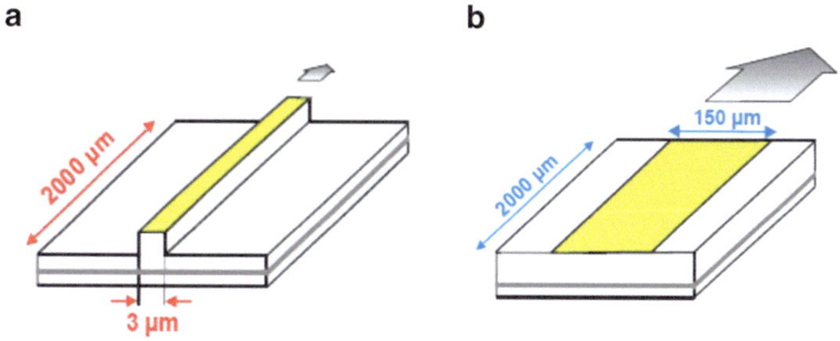

Fig. 1.16 Scheme of the (a) ridge laser and (b) broad area laser [21]

1.3.1 Broad Emitting Area

The maximum power from a diode laser P_{max} is proportional to width of the emitter according to the following expression [20]:

$$P_{\max} = \left(\frac{d}{\Gamma}\right) W \left(\frac{1-R}{1+R}\right) P_{COMD}, \tag{1.10}$$

where W is the stripe width, R is the front facet reflectivity, d is the quantum well thickness, and Γ is the transverse optical confinement factor such that d/Γ is the equivalent spot size. Broad-area lasers are able to achieve high output powers due to a wide stripe width W. Figure 1.16 shows the scheme of the broad area semiconductor laser and compared with a single mode ridge laser [21].

1.3.2 Long Cavity

According to Eqs. (1.11), (1.12), and (1.13), the output power from a semiconductor laser P_{opt} is inversely proportional to thermal resistance (R_{th}) and series resistance (R_s) [22]. From Eq. (1.13), the R_{th} and R_s are inversely proportional to cavity length (L) [23]. So the use of long cavity (L) can reduce not only R_{th} but

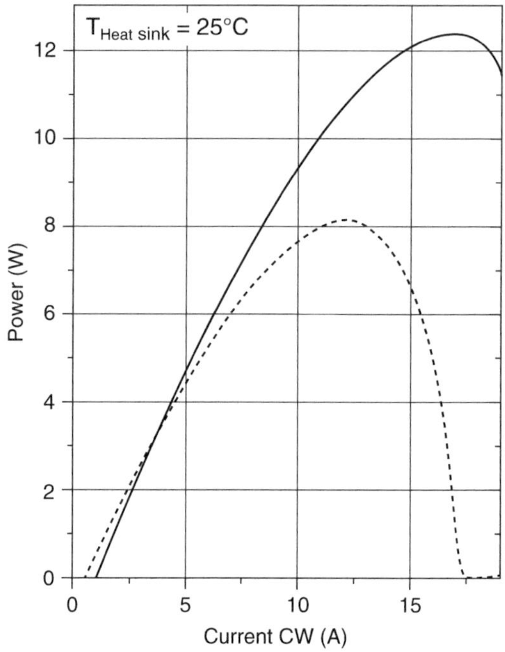

Fig. 1.17 Output power of semiconductor laser at 808 nm with cavity lengths 4.5 mm (*solid line*) and 3.0 mm (*dashed line*) [24]

also R_s in semiconductor laser. Also, long cavity can increase the volume of the gain medium. As a result of these two reasons, the use of long cavity can increase the device maximum CW output power.

$$P_{opt}(T) \propto \eta_d(T_h)\exp\left(-\frac{\Delta T_j}{T_1}\right) \tag{1.11}$$

$$\Delta T_j = R_{th}\left[I(V_0 + IR_s) - P_{opt}\right] \tag{1.12}$$

$$R_{th}, R_s \propto \frac{1}{L} \tag{1.13}$$

where η_d is slope efficiency, T_h is the heatsink temperature, ΔT_j is the junction temperature, T_1 is the characteristic temperature for η_d, I is injection current, V_0 is built-in voltage, P_{opt} is output power, R_{th} is thermal resistance, R_s is series resistance, and L is cavity length. Figure 1.17 shows the higher power can be achieved by using longer cavity in semiconductor laser [24].

1.3.3 Broadened Waveguide Design

The internal losses are related to free carrier absorption. The main contribution into the net internal loss comes from absorption within highly doped cladding regions. Confinement factor Γ for broadened waveguide device is slightly reduced but the

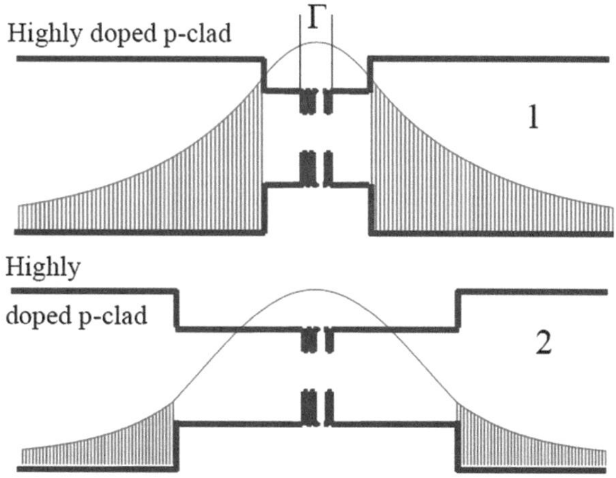

Fig. 1.18 Scheme of the broadened waveguide design [23]

losses decrease dramatically resulting in net decrease of the threshold current density. The efficiency of the broadened waveguide lasers are less sensitive to cavity length increase and long cavity lasers can be used for high power applications. Figure 1.18 shows the scheme of the broadened waveguide design [23].

1.3.4 Fill Factor

An important factor in bar design is its fill factor. This describes the percentage of the bar that is occupied by emitters, and equals the width of one emitter divided by the center-to-center spacing between emitters. A typical bar has 19 emitters with each emitter 150 μm wide. The pitch is 500 μm, resulting in a 30 % fill factor. Commercially available conduction-cooled bars with a 30 % fill factor and 2 mm-long cavity can produce up to 80 W of total CW power, while water-cooled versions with an 50 % fill factor can provide up to 150 W CW. For quasi-CW (QCW) operation, much higher peak-power levels can be achieved. This comes from the use of diode bars with much higher fill factors such as 80–90 % instead of the 30–50 % fill factors used for CW bars. A higher fill factor bar roughly doubles the peak power compared with a CW bar which uses the same design [25].

1.3.5 Facet Passivation

By proper design, the maximum output power P_{max} of a 100 μm aperture emitter can be ~20 W. This corresponds to optical power density at the output mirror of about 20 MW per cm^2. At such high optical power density, the facet of the diode

Table 1.1 Conventional facet-passivated lasers [22]

Active-region material	P_{COMD}
InGaAs (λ ~1.12 μm)	23 MW/cm^2
InGaAs (λ ~0.92–0.98 μm)	~19 MW/cm^2
InGaAsP (λ ~0.81 μm)	17–18 MW/cm^2
InAlGaAs (λ ~0.81 μm)	13–14 MW/cm^2
GaAs (λ ~0.81–0.87 μm)	10–12 MW/cm^2
Al0.07Ga0.93As (λ ~0.81 μm)	8–9 MW/cm^2
Al0.13Ga0.87As (λ ~0.78 μm)	~5 MW/cm^2

laser is susceptible to COMD. According to Eq. (1.10), the P_{max} can be enhanced by increasing the threshold of the COMD power (PCOMD). For the conventionally facet-passivated devices, PCOMD is a function of the active-region material and the corresponding PCOMD are listed in Table 1.1 [22]. One can select certain material for increasing the PCOMD. From Table 1.1, we can see PCOMD(InGaAs) is higher than PCOMD(GaAs) and PCOMD(AlGaAs) is the smallest.

P_{COMD} could be further increased by reducing absorption coefficient of the mirror coatings [23]. Completely Facet-Passivated technique or Nonabsorbing-Mirror (NAM) technique is usually used to increase the P_{COMD} [22]. The facet passivation technique is based on facet cleaving and coating in vacuum to reduce optical absorption and non-radiative recombination at the front facet. The NAM technique is to create high bandgap nonabsorbing regions at the mirror facets.

1.3.6 Tapered Lateral Structure

Today broad-area diode lasers are used to achieve high outputs [26]. But standard broad-area waveguide designs are susceptible to modal instabilities, filamentation, and COMD failure [27]. This results in low beam qualities and value for the brightness is limited to around 1×10^7 Wcm^{-2} sr^{-1} [28]. A lot of different solutions have been proposed in the last few years to overcome these problems and to achieve high output power together with high beam quality. The main effort has been directed to develop broad-area structures that support only one lateral mode. Tapered devices [29, 30], Distributed-Feed-Back (DFB) lasers [31, 32], and monolithically integrated Master-Oscillator Power-Amplifiers (MOPAs) [33] have been demonstrated, and all of them are able to produce output powers well above 1 W together with a high beam quality. Among these designs, tapered lateral structure semiconductor lasers show the promise to achieve high power with a good beam quality.

A tapered diode laser bar consisting of many tapered emitters could scale up the power levels even up to tens of Watts [33–35]. The vertical structure (fast axis) of tapered diode laser bars is similar to that of common broad area diode laser bars. The significant difference between tapered diode lasers and broad area diode lasers becomes apparent when comparing the lateral structure shown in Fig. 1.19 [36].

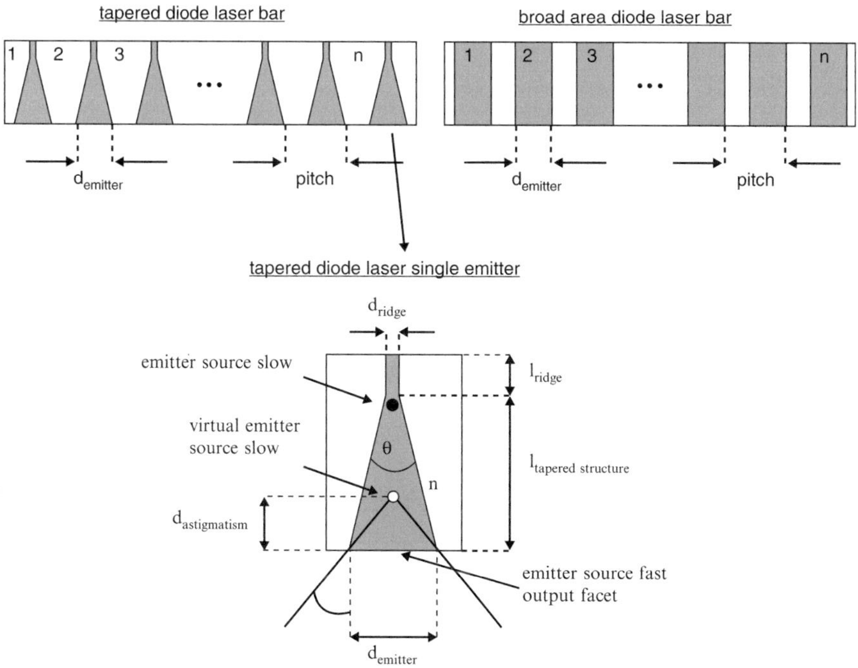

Fig. 1.19 Lateral structure of a tapered diode laser bar (*top left*) compared to a broad area diode laser bar (*top right*) [36]

In contrast to a broad area multimode emitter, the single emitter of a tapered diode laser bar consists of a small ridge waveguide followed by a tapered amplifier section. The high beam quality is defined by the ridge waveguide and the high output power is provided by the tapered section, while maintaining the beam quality of the ridge waveguide. Typical data for a single emitter (bottom) of a tapered diode laser bar are summarized in Table 1.2 [36].

One important consequence of the tapered structure is the difference of source position for the fast- and the slow axis, as shown in Fig. 1.19. Whereas the source position of the fast axis is on the output facet of the diode bar, the source position of the slow axis is located inside the diode bar at the transition between the ridge waveguide and the tapered part.

High beam quality at the output facet can be achieved using a tapered laser diode in which a short and narrow straight ridge waveguide provides mode selection and a tapered amplifier performs large-signal amplification and avoids damage to the output facet [35–40]. Such diodes yield high output beam quality. When the fast axis is collimated using a micro-lens, its divergence is lower than 1°, and the slow axis typically exhibits divergences of around 3°. Due to the good beam quality, tapered semiconductor laser could make them ideal candidates for industrial applications such as cutting or marking of certain artificial materials and metals [36].

Table 1.2 Typical data for the single emitter of a tapered diode laser bar (CF. Fig. 1.19) [36]

Power (W)	Output fact $d_{emitter}$ (μm)	Source width d_{ridge} (μm)	I_{ridge} (μm)	I_{taper} (μm)	Refractive index	$D_{astigmation}$ (I_{taper}/n) (μm)	Taper angle (°)	M^2 single emitter slow	BPP single emitter slow (mm*mrad)
2.5	200	5–20	500	2,000	3.5	600	6	4	1.2

1.4 Basic Characteristics and Parameters

The parameters, which include wavelength, output power, slope efficiency, threshold current, spectrum, near field, far field, beam quality, and life time, are used to describe the performance of a semiconductor laser. Currently, the development trends of the semiconductor laser are towards to high power, narrow width spectrum, and high reliability. With the development of the packaging and processing technologies, the performances of the semiconductor lasers have been improved continually. On the other hand, the packaging technology is still one of the bottlenecks for further advancement of high power semiconductor lasers. In this section, the important parameters characterizing the performance of the semiconductor laser in applications are presented.

1.4.1 Wavelength

Peak wavelength is the wavelength corresponding to maximum of luminous intensity or radiation power of laser radiation spectrum. Central wavelength is the middle-point wavelength between two points whose intensities are half of peak wavelength intensity. Semiconductor lasers have broad wavelength which extends from near-ultraviolet to far-infrared depending on different semiconductor material (materials). The relationships between wavelength and some semiconductor materials are shown in Fig. 1.20 [41].

The active-region materials of visible light laser in red region are usually InGaP/GaAsP, GaAlAs/GaAs, and InGaAlP. When $Al_xGa_{1-x}As/GaAs$ heterojunction is applied as visible light laser material and $x > 0.47$, the wavelength of 670 nm can be achieved at room temperature, which approaches the theoretical limit. For blue light laser, the active-region materials can be group three elements nitride compounds, wide band II–VI compound semiconductor material, and SiC. As SiC is indirect band gaps semiconductor material, the emitting efficiencies of group three elements nitride compounds and wide band II–VI compound semiconductor material are significantly higher. In recent years, progress has been made in short wavelength laser diode made by GaN-type semiconductor and ZnSe-type semiconductor materials, which are believed to be the most promising laser diode materials. The semiconductor lasers with radiation wavelength longer than 2 μm are collectively called mid and far-infrared semiconductor laser. The active region of this

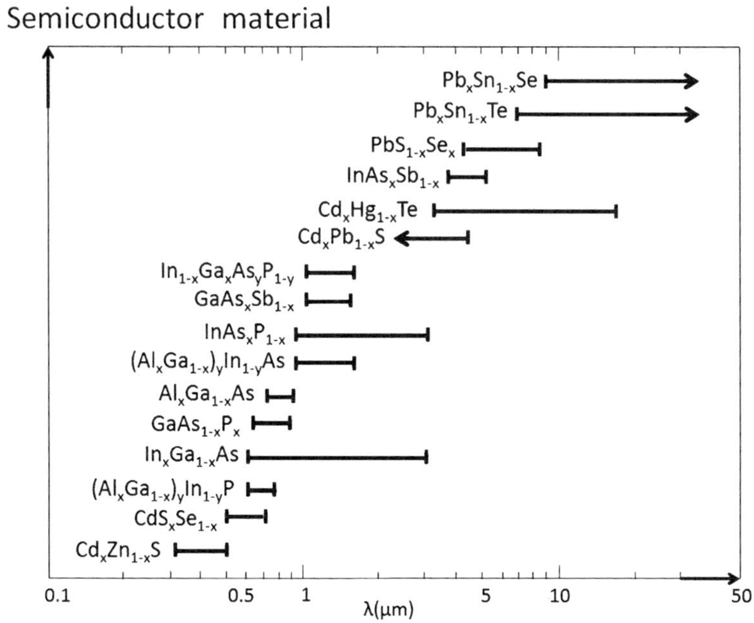

Fig. 1.20 The relationships between wavelength and some semiconductor materials [41]

type laser is direct transition semiconductor material with narrow band gap, such as group III–V, IV–VI, and II–VI compound semiconductor materials.

The semiconductor lasers with different wavelengths are applied in different realms, and the application of semiconductor lasers are shown in Fig. 1.21 [42]. However, the applications of high power semiconductor lasers, such as in the pumping of solid state and fiber lasers, laser medical, laser cosmetic, and laser processing, have been mainly developed in recent years.

1.4.2 Output Power

Output power is the most important electric-optical characteristics for high power semiconductor laser. It is defined as the nominal output power under the precondition that laser runs well with long-term stabilization. The output power of a semiconductor laser is limited either by COMD or by thermal rollover. Taken a single emitter as an example, Fig. 1.22 is an illustration of the output optical power as a function of driving current [16, 37]. The maximum power of P_{max} is defined as the value of the output power before declining. The operating power is from 30 to 80 % of P_{max} and cannot be over the power of P_{COMD} which is shown in Fig. 1.22.

High power semiconductor lasers can be a single emitter, a laser bar, horizontal bar arrays, or a vertical bar stack. The output power of the single emitter laser has

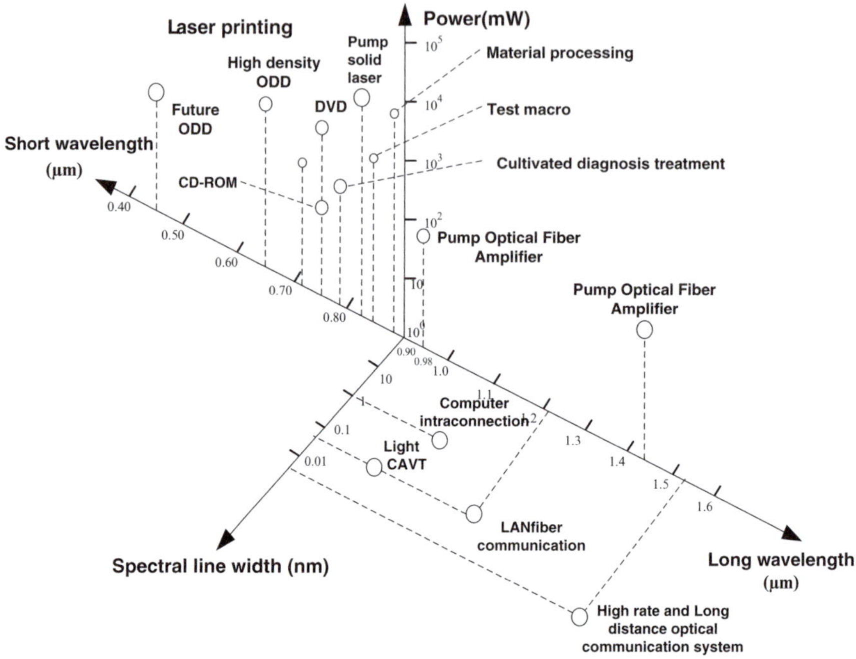

Fig. 1.21 The applications distribution of diode laser along the wavelengths [42]

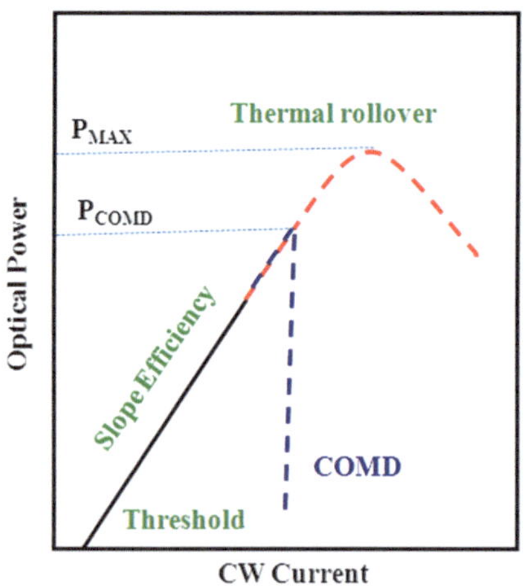

Fig. 1.22 An illustration of the output optical power as a function of driving current (*red dot line* and *blue dot line* present the laser with thermal-over and COMD, respectively) [16, 37]

been significantly improved with advancement of laser chip technology. A laser bar provides the output power with one order of magnitude higher than single emitter by integrating the single emitters at the wafer level. To further increase the output power, several packaging technologies have been developed including horizontal

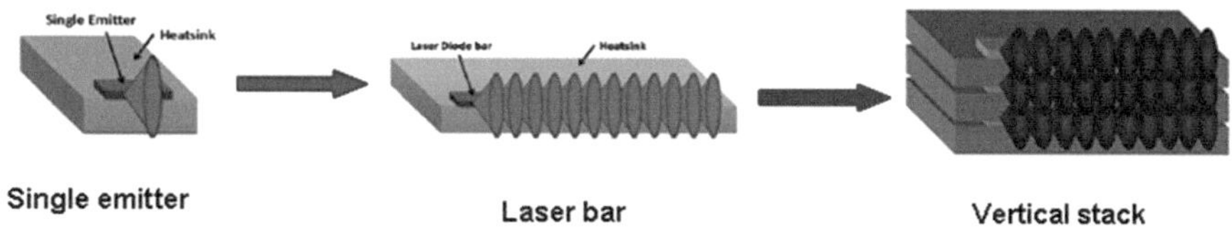

Fig. 1.23 The output power scaling [37]

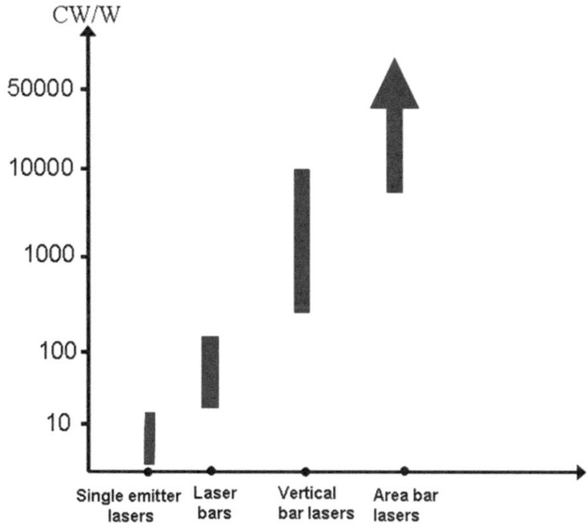

Fig. 1.24 The power scaling with packaging structures [37]

bar arrays, vertical bar stacks, and area bar arrays. The single emitter is unit in semiconductor lasers and the laser bars and stacks are made by multiple emitters as shown in Fig. 1.23 [37].

15 W 9xx nm single emitter lasers are commonly available commercially. Semiconductor laser bars integrated with single emitters can improve the output power to tens watts. Currently, the output power has been enhanced to 150 W per bar. With the demand of high power, vertical bar stack and area bar arrays become the choices. The output power of semiconductor laser stacks can reach hundreds and thousands watts, and the more high output power, such as ten thousands watt power, can be achieved by semiconductor laser stack arrays composed by multiple laser stacks. With the development of the packaging technology, the output power has been further increased. The ranges of the output power with different structures are shown in Fig. 1.24 [37].

In addition, operating current, operating voltage, threshold current, slope efficiency, electric-optical converse efficiency, the max power converse efficiency, and series resistance are also the important electric-optical characteristics for high power semiconductor laser.

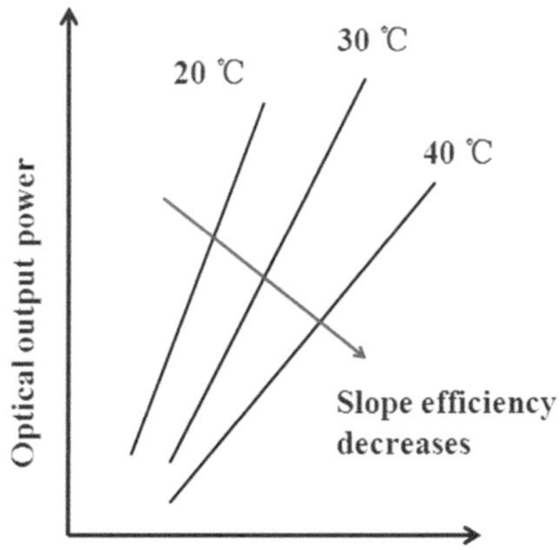

Fig. 1.25 The slope efficiency changes with operation temperature [38]

1.4.3 Slope Efficiency

The slope efficiency is defined as the ratio of the optical power to the current above the threshold current, which is shown in Fig. 1.22 and measured in W/A. Above the threshold, the output power increases linearly with the driving current. The slope efficiency in typical devices is of the order of 1 W/A, and the higher slope efficiency the better the laser. Generally, the slope efficiency depends on internal quantum efficiency, the reflectivity of rear and front facets and absorption within the semiconductor material and on the frequency of the emitted photons. Also, the slope efficiency changes with operation temperature as shown in Fig. 1.25 [38].

1.4.4 Threshold Current

To achieve lasing, the gain coefficient must be larger than the effective loss coefficient. This occurs at the threshold current. Above the threshold current, the gain will be larger than the loss. The experimentally measured light output versus current graph shows a distinct break at I_{thres}, as can be seen in Fig. 1.26 below [38]. At currents larger than I_{thres} the gain of the device is higher than the total loss, and the device is an amplifier. Below the threshold current the diode is basically a spontaneous radiation. I_{thres} is an important parameter. It is often used to characterize the quality of the laser: the smaller I_{thres} the better the laser.

1.4.5 Spectrum

Spectrum is one of the key specifications of semiconductor laser products. The spectrum width and the peak wavelength are two important parameters describing a spectrum. In many applications, the narrow spectrum is required, for example,

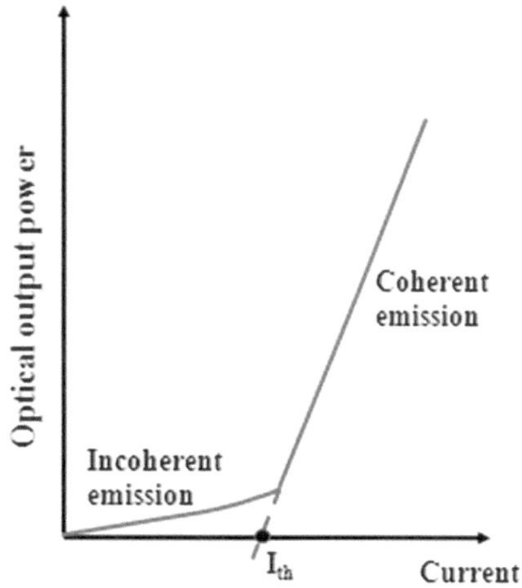

Fig. 1.26 The experimentally measured light output versus current [38]

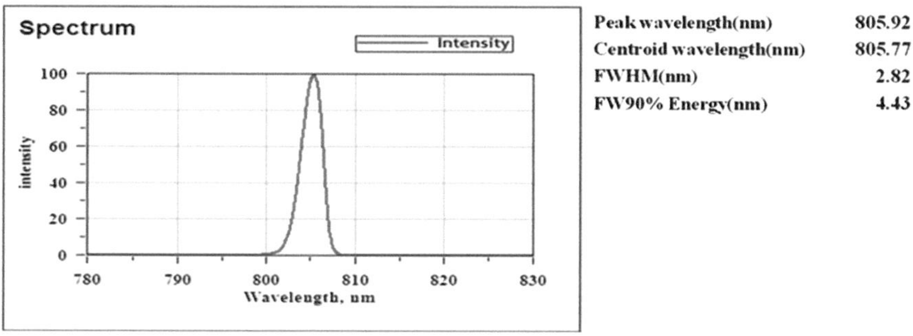

Fig. 1.27 A prototype spectrum curve of a semiconductor laser bar [16]

as the pumping source for the pumping solid state lasers. Figure 1.27 depicts a prototype spectrum curve of a semiconductor laser bar, in which the spectral width (the full width at half maximum (FWHM) of the spectral width) is below 3 nm [16].

1.4.6 Near Field

Near field is defined as the laser irradiation of near to laser front facet. Near field is the image of a semiconductor laser at active region when it works. Optical imaging system is used to collect all the beams from the light source and image the light source on the screen clearly. The image is processed to obtain the emitting information of each emitter of a semiconductor laser. Figure 1.28 [37] shows a typical near field image of a semiconductor laser bar with 19 emitters. Furthermore, the beam profile of each emitter of a semiconductor laser is also a characterization

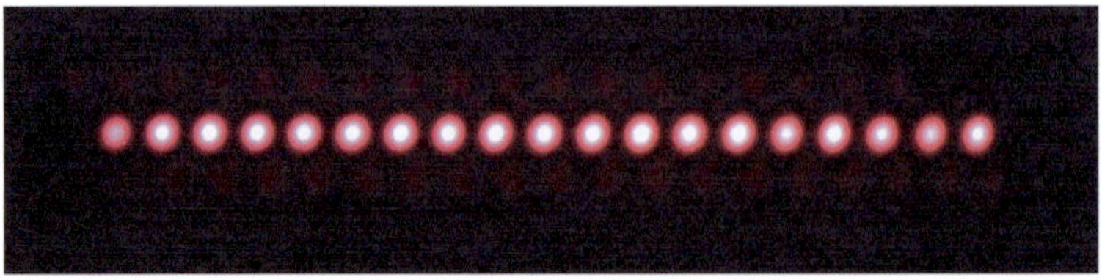

Fig. 1.28 A typical near field image of a semiconductor laser bar [37]

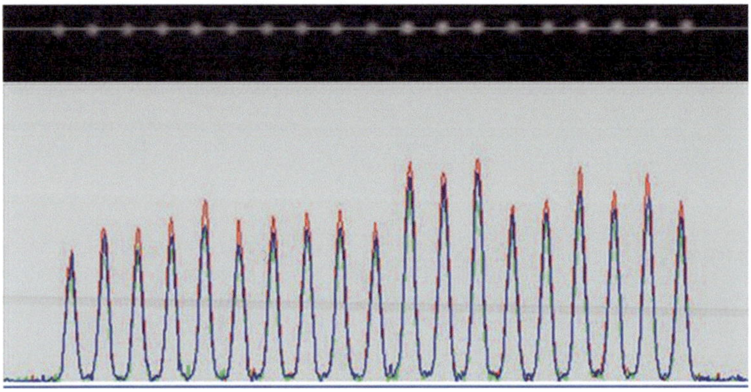

Fig. 1.29 Near field intensity profile of each emitter of a semiconductor laser bar [37]

of near field. Figure 1.29 [37] shows the near field intensity profile of each emitter of a semiconductor laser bar with 19 emitters.

1.4.7 Far Field

Far field indicates the radiation field of a laser at a distance from the beam waist which is much greater than the Raleigh length. A laser diode with an active region of dimensions (L μm × W μm) emits light with a far-field angular divergence. The radiation is produced in the active layer, which is a small fraction of the thickness of the chip; hence the radiation is diffracted when it emerges from the active region analogously to the diffraction of radiation passing through a narrow slit. Because of the small dimensions, the beam diverges with angles of the order of a few degrees, and because of the unequal dimensions, the beam spreads unequally in two directions.

A diode laser has large divergence angles, and especially on fast axis the value of the divergence angle (FWHM) is normally about 45°. The sizes of active layer of a single emitter are asymmetric, where on fast axis the width is 1 μm and is 100–200 μm on slow axis. Moreover, the far-field pattern of laser beam is also quite

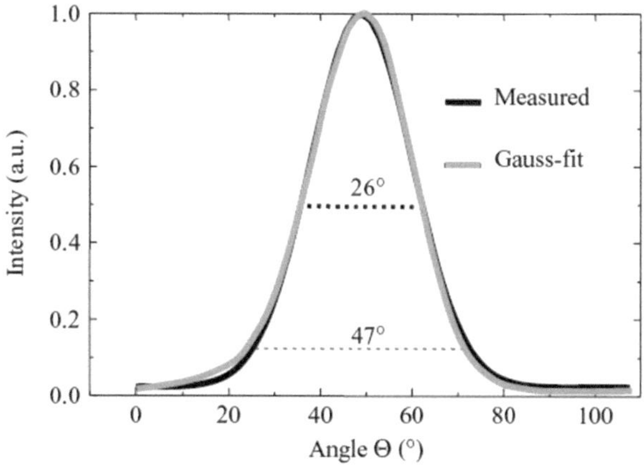

Fig. 1.30 The intensity distribution for the far field along fast axis [19]

asymmetric on the two axes. The distribution of the lowest-order mode inside the waveguide and the resulting intensity distribution for the far field are depicted in Fig. 1.30 [19]. The thickness of the waveguide layers is small enough to support only the fundamental mode. Generally, in fast axis it can be well approximated as a Gauss distribution.

In slow axis direction, the beam profile is much complex than that in fast axis direction. Generally, the beam is dominated by the presence of higher order modes in slow axis, and the detailed profile depends on the structure and type of the laser. The far-field intensity of a semiconductor bar or stack is considered as the incoherence superposition of intensity of the emitters, approximately.

1.4.8 Beam Quality

The beam quality of a laser beam can be defined in different ways, but is essentially a measure of how tightly a laser beam can be focused under certain conditions (e.g., with a limited beam divergence). The most common ways to quantify the beam quality are

1. The beam parameter product (BPP) is the product of the waist radius w_0 and the far-field divergence θ_0 of the beam: $BPP = w_0 \times \theta_0$.
2. The M^2 factor is defined as the beam parameter product divided by the corresponding product for a diffraction-limited Gaussian beam with the same wavelength.

A high beam quality implies smooth wave fronts, such that focusing the beam with a lens allows one to obtain a focus where the wave fronts are plane. Scrambled wave fronts make beam focusing more difficult, i.e., the beam divergence for a given spot size is increased.

The best possible beam quality is achieved for a diffraction-limited Gaussian beam, having $M^2 = 1$. For high power semiconductor laser, it can have a very large M^2 of more than 100 or even well above 1,000. In high power semiconductor lasers the poor beam quality results from the broad area radiation and large divergence angle.

1.4.9 Life Time

There are two ways used to define operating lifetime: (1) the time that takes for output power going down to 80 % of the original value, as the laser continuously works at a rated operating current; (2) the time that takes for operating current going up to 120 % of the original value as the laser continuously works at a rated power. Constant current control is used in the former way while constant power control for the later.

There are a lot of factors which could result in the degeneration and invalidation of semiconductor laser. Some factors are avoidable, e.g., laser usage at improper temperature or current control and optical fiber head impinging cleavage face. However, some are unavoidable since they come from laser itself, e.g., the degeneration of cavity surface. In order to guard long operating lifetime, semiconductor lasers must be tested based on a series of standards on vibration, impact, moisture, static electricity, high and low temperatures. Among all the reliability tests, the most important and time consuming one is the lifetime testing. Mass production of a semiconductor laser in the market cannot be implemented only after it can pass all the reliability qualification testing.

References

1. W. Koechner, *Solid State Laser Engineering* (Springer, Berlin, 1966)
2. R. Diehl, *High-Power Diode Laser: Fundamental, Technology, Applications* (Springer, Berlin, 2000)
3. B. Mroziewicz, M. Buqajski, W. Nakwaski, *Physics of Semiconductor Laser* (North-Holland, Amsterdam, 1991)
4. S.W. Koch, F. Jahnke, W.W. Chow, Physics of semiconductor micro-cavity lasers. Semicond. Sci. Technol. **10**(6), 739–751 (1995)
5. L. Guo, G.D. Xu, X.S. Liu, *The Development of Semiconductor Laser Chips*, Internal Talk from Focuslight Technologies Co., Ltd. (2012), pp. 56–60.
6. X.D. Huang, X.F. Liu, *Technology and Applications of Diode Lasers* (Science, Chen du, 1999)
7. I. Vurgaftman, J.M. Hinckley, J. Singh, A comparison of optoelectronic properties of lattice-matched and strained quantum-well and quantum-wire structures. IEEE J. Quantum Electron. **30**(1), 75–84 (1994)
8. J.P. Reithmaier, G. Sek, A. Loffler, C. Hofmann, S. Kuhn, S. Reitzenstein, L.V. Keldysh, V.D. Kulakovskii, T.L. Reinecke, A. Forchel, Strong coupling in a single quantum dot semiconductor micro cavity system. Lett. Nat. **432**, 197–200 (2004)

9. W.W. Chow, S.W. Koch, Theory of semiconductor quantum-dot laser dynamics. IEEE J. Quantum Electron. **41**(4), 495–506 (2005)
10. P.C. Sercel, K.J. Vahala, Analytical formalism for determining quantum- wire and quantum-dot band structure in the multiband envelope-function approximation. Phys. Rev. B **42**(6), 3690–3712 (1990)
11. S. Arai, GaInAsP/InP Quantum wire lasers. IEEE J. Sel. Top. Quantum Electron. **15**(3), 731–743 (2009)
12. C. Henry, Theory of the linewidth of semiconductor lasers. IEEE J. Quantum Electron. **QE-18**(2), 259–264 (1982)
13. Y. Miyake, M. Asada, Spectral characteristics of linewidth enhancement factor α of multidimensional quantum wells. Jpn. J. Appl. Phys. **28**(7), 1280–1281 (1989)
14. Y. Arakawa, A. Yariv, Quantum well lasers-gain, spectra, dynamics. IEEE J. Quantum Electron. **22**(9), 1887–1899 (1986)
15. Y. Qiu, P. Gogna, S. Forouhar, A. Stintz, L.F. Lester, High-performance InAs quantum-dot lasers near 1.3 mm. Appl. Phys. Lett. **79**(22), 3570–3572 (2001)
16. X.S. Liu, W. Zhao, in *Technology Trend and Challenges in High Power Semiconductor Laser Packaging*. 2009 Electronic Components and Technology Conference (2009), pp. 2106–2114.
17. X.S. Liu, M.H. Hu, R.W. Davis, C.E. Zah, Comparison between epi-down and epi-up bonded high-power single-mode 980-nm semiconductor lasers. IEEE Trans. Adv. Packag. **27**(4), 640–646 (2004)
18. X.S. Liu, L.C. Hughes, M.H. Rasmussen, M.H. Hu, V.A. Bhagavatula, R.W. Davis, S. Coleman, R. Bhat, C.E. Zah, Packging and performance of 980 nm broad area semiconductor laser. Electron. Packag. **6**(3), 6–14 (2006)
19. F. Bachmann, P. Loosen, R. Poprawe, *High Power Diode Lasers Technology and Applications* (Springer Science Business Media, LLC, New York, 2007)
20. H. Injeyan, G.D. Goodno, *High-Power Laser Handbook* (McGraw-Hill, New York, 2011)
21. http://www.ist-brighter.eu
22. D. Botez, in *Advances in High-Power Monolithic Semiconductor Lasers*. IEEE ISLC'04: SC3 Matsue-shi, Japan (2004)
23. www.ece.sunysb.edu/~oe/ESE519/lecture10.pdf
24. V. Gapontsev, N. Mozhegov, P. Trubenko, A. Komissarov, I. Berishev, O. Raisky, N. Strouglov, V. Chuyanov, G. Kuang, O. Maksimov, A. Ovtchinnikov, High-brightness fiber coupled pumps. Proc. SPIE **7198**, 71980O(1–9) (2009)
25. www.nlight.net/nlight-files/file/articles/HP_May05_OLE.pdf
26. http://www.optosolutions.com/doc/TaperedLaser_intro_070209.pdf
27. A. Moser, E.E. Latta, Arrhenius parameters for the rate process leading to catastrophic damage of AlGaAs-GaAs laser facets. J. Appl. Phys. **71**(10), 4848–4853 (1992)
28. S. Pawlik, S. Traut, A. Thies, B. Sverdlov, B. Schmidt, in *Ultra-High Power RWG Laser Diodes With Lateral Absorber Region*. 18th IEEE international semiconductor laser conference, Garmisch, Germany (2002), pp. 163–164
29. E.S. Kintzer, J.N. Walpole, S.R. Chinn, C.A. Wang, L.J. Missaggia, High power strained-layer amplifiers and lasers with tapered gain regions. IEEE Photon. Technol. Lett. **5**(6), 605–608 (1993)
30. S. O'Brien, A. Schönfelder, R.J. Lang, 5W CW diffraction-limited InGaAs broad-area flared amplifier at 970 nm. IEEE Photon. Technol. Lett. **9**(9), 1217–1219 (1997)
31. S.D. DeMars, K.M. Dzurko, R.J. Lang, D.F. Welch, D.R. Scifres, A. Hardy, *Angled-Grating Distributed Feedback Laser With 1 W Single-Mode Diffraction-Limited Output at 980 nm*. Technical Digest of CLEO'96 (1996), pp. 77–78.
32. K. Paschke, A. Bogatov, F. Bugge, A.E. Drakin, J. Fricke, R. Guther, A.A. Stratonikov, H. Wenzel, G. Erbert, G. Tränkle, Properties of ion-implated high-power angled-grating distributed-feedback lasers. IEEE J. Sel. Top. **9**(5), 1172–1178 (2003)
33. S. O'Brien, D.F. Welch, R.A. Parke, D. Mehuys, K. Dzurko, R.J. Lang, R. Waarts, D. Scifres, Operating characteristics of a high-power monolithically integrated flared amplifier master oscillator power amplifier. IEEE J. Quantum Electr. **29**(6), 2052–2057 (1993)

34. M.T. Kelemen, J. Weber, G. Kaufel, G. Bihlmann, R. Moritz, M. Mikulla, G. Weimann, Tapered diode lasers at 976 nm with 8W nearly diffraction limited output power. Electron. Lett. **41**(18), 1011–1013 (2005)
35. C. Scholz, K. Boucke, R. Poprawe, M.T. Kelemen, J. Weber, M. Mikulla, G. Weimann, Comparison between 50 W tapered laser arrays and tapered single emitters. High Power Diode Laser Technology and Applications IV, Proc. of SPIE **6104**, 61040G(1–8) (2006)
36. B. Köhler, S. Ahlert, T. Brand, M. Haag, H. Kissel, G. Seibold, M. Stoiber, J. Biesenbach, W. Reill, G. Grönninger, M. Reufer, H. König, U. Strauss, Diode laser modules based on new developments in tapered and broad area diode laser bars. Proc. of SPIE **6876** 6876(1–11) (2008)
37. D. Wu, L.L. Zhu, M.N. Yan, X.S. Liu, Measurement method of the high power semiconductor laser, Internal report, Xi'an Focuslight Co. Ltd. (2012)
38. http://course.ee.ust.hk/elec342/notes/Lecture%2013_laser%20diodes(2).pdf
39. G.L. Bourdet, I. Hassiaoui, R. McBride, J.F. Monjardin, H. Baker, N. Michel, M. Krakowski, High-power, low-divergence, linear array of quasi-diffraction-limited beams supplied by tapered diodes. Appl. Opt. **46**(25), 6297–6302 (2007)
40. B. Kőhler, J. Biesenbach, T. Brand, M. Haag, S. Huke, A. Noeske, G. Seibold, M. Behringer, J. Luft, High-brightness high-power kW-system with tapered diode laser bars. Proc. SPIE **5711**, 5711(1–12) (2005)
41. B.E.A. Saleh, M.C. Teich, *Fundamentals of Photonics* (Wiley, New York, 1991)
42. J.P. Jiang, *Semiconductor Laser* (Publishing House of Electronics Industry, Beijing, 2000)

Chapter 2
Overview of High Power Semiconductor Laser Packages

Semiconductor lasers can be divided into many types according to different classification criteria, such as cooling types, die bonding techniques, and the arrangement of the chips or bars in the assembly. In this chapter, the packaging structures of semiconductor lasers are divided into two categories which are open packages and fiber-coupled modules. The basic and typical structures of the two categories are discussed in detail according to single emitters, single bars, and laser stacks.

The basic structure of a semiconductor laser, shown in Fig. 2.1, consists of four parts: a laser chip, a bonding solder, a mounting substrate, and a heat sink [1]. A laser chip can be a single emitter, or multiple emitters such as a bar. Bonding solder is under the laser chip, and it is used to bond the laser chip to the mounting substrate. Bonding solder can be divided as soft and hard solders. The mounting substrate under laser chips is used to mount the laser chip. The heat sink is attached under the mounting substrate, and the function is to dissipate heat from laser chips and serve as a mechanical basis. And hence, the heat sink is sometimes named cooler due to its function. Sometimes the mounting substrate and heat sink are combined and of the same material. In this case, it is named a mounting substrate or cooler.

2.1 Open Packages

2.1.1 Packaging Structure of Single Emitter Semiconductor Lasers

There are two cooling approaches for the semiconductor lasers, which are the conduction cooling and the liquid cooling. Generally, the output power of a single emitter is not high and less than 15 W, and the heat is also low. Hence, conduct cooling is adopted for most of the single emitter lasers. There are mainly

X. Liu et al., *Packaging of High Power Semiconductor Lasers*, Micro- and Opto-Electronic Materials, Structures, and Systems, DOI 10.1007/978-1-4614-9263-4_2,

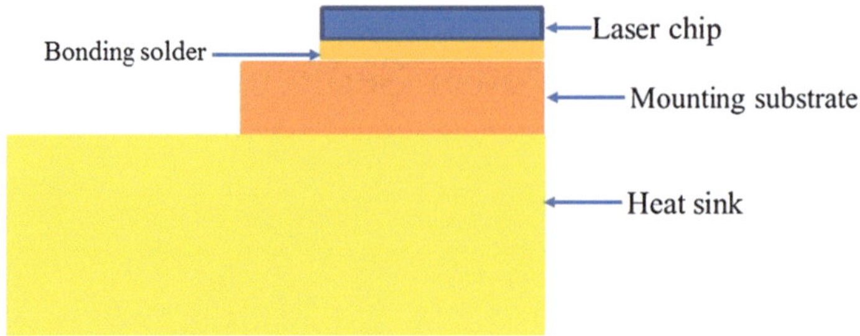

Fig. 2.1 Basic structure of a semiconductor laser [1]

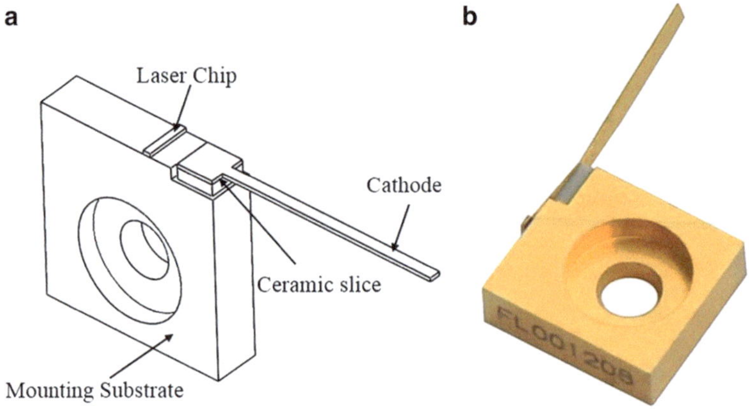

Fig. 2.2 The typical C-mount packaging structure and a picture of a C-mount laser device [1, 3]

four packaging structures of single emitter semiconductor lasers: C-mount, B-mount, CT-mount, and F-mount. The packaging structures are presented and discussed below.

1. C-mount

   The package structure schemes of C-mount and C-mount laser are shown in Fig. 2.2a, b, respectively [1, 3]. C-mount also contains four main parts: cathode, a ceramic slice, a laser chip, and a mounting substrate which also acts as the anode. The hole in the mounting substrate is designed for mounting the C-mount to an installation platform. The ceramic slice is used to insulate the cathode from the anode. The fly-lead of the cathode is wire-bonded to the n-side of the laser chip.

   As shown in Fig. 2.3a, in applications, a C-mount is installed on an installation platform at the side of the mounting substrate [2]. The heat conduction path is shown in Fig. 2.3b. The heat conduction path is along vertical direction first, and then along horizontal direction towards the platform. This means that when it works, the heat is first conducted to the mounting substrate vertically and then horizontally conducted to the installation platform. However, the two paths

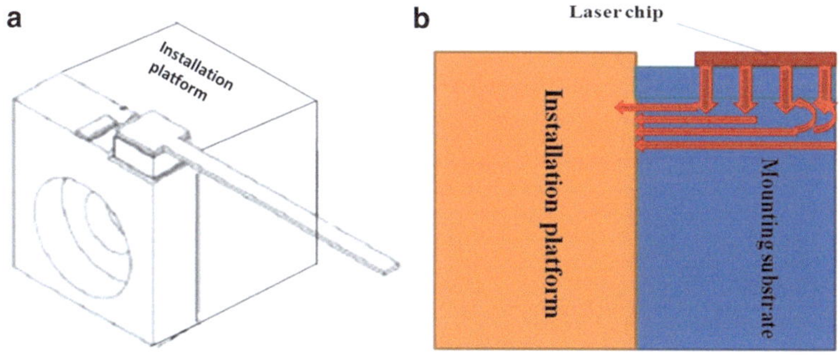

Fig. 2.3 The installation structure of C-mount and its heat conduction paths [2]. (**a**) The installation of C-mount. (**b**) The heat conduction paths

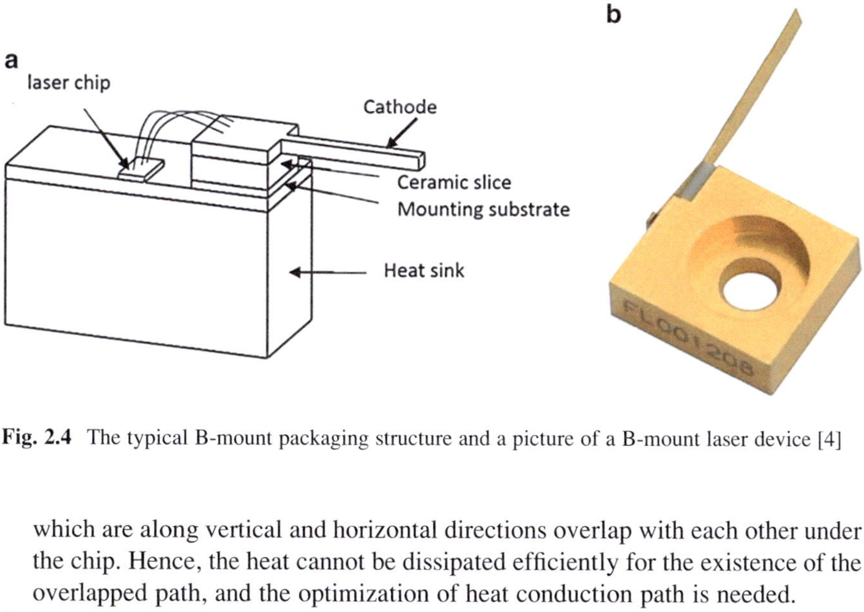

Fig. 2.4 The typical B-mount packaging structure and a picture of a B-mount laser device [4]

which are along vertical and horizontal directions overlap with each other under the chip. Hence, the heat cannot be dissipated efficiently for the existence of the overlapped path, and the optimization of heat conduction path is needed.

2. B-mount

   The package structure scheme of B-mount and a B-mount laser is shown in Fig. 2.4a, b [4], respectively. It contains four parts: a mounting substrate, a laser chip, a ceramic slice, and the cathode. Same as the C-mount, the mounting substrate of B-mount has two functions in the single emitter laser: one acts as the anode, and another is to conduct the heat. As shown in the figure, the fly-lead of the cathode is placed on the top of a ceramic slice. The ceramic slice is used to insulate the fly-lead from the mounting substrate. The cathode is wire-bonded to the n-side of the laser chip.

   The heat conduction path of B-mount is shown in Fig. 2.5 [5]. The heat conduction path is along vertical direction. As shown in the figure, the width

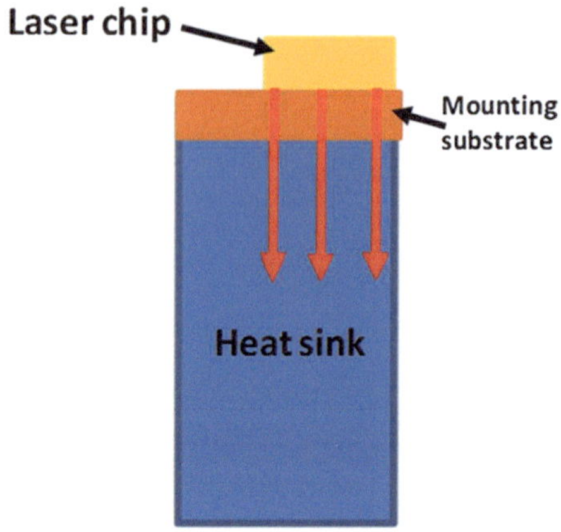

Fig. 2.5 The heat conduction paths [5]

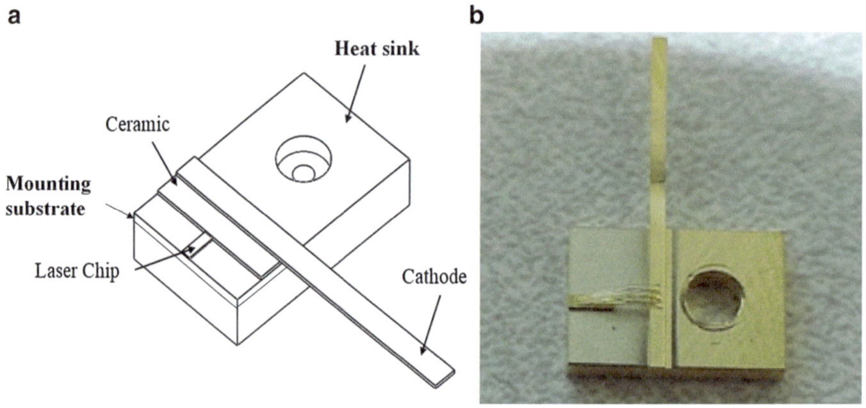

Fig. 2.6 The typical CT-mount packaging structure and a picture of a CT-mount laser device [3, 6]

is much larger than the size of chip, and hence heat dissipation efficiency is improved than that of C-mount.

3. CT-mount

   The package structure scheme of CT-mount and a finished CT-mount laser are shown in Fig. 2.6a [3] and b [6], respectively. Compared with C-mount laser, the difference of these two lasers is that a heat sink can be directly under the mounting substrate in CT-mount laser.

   The heat dissipation path of CT-mount is also improved, and the heat conduction path is similar to that of B-mount which is shown in Fig. 2.6.

   Similar to that of the C-mount the cathode of CT-mount is also wire-bonded to the n-side of laser chip, and the mounting substrate is the anode. This implies that the cooler of the CT-mount still have charges when the laser works, and thus the CT-mount laser is also limited in some applications.

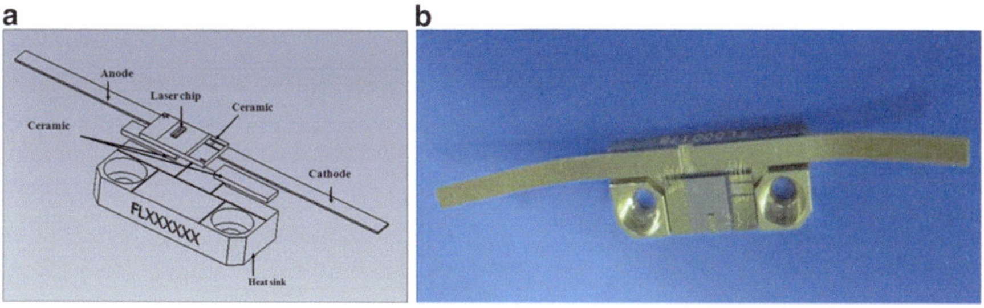

Fig. 2.7 The typical F-mount packaging structure and a picture of a F-mount laser device [7]. (**a**) The structure scheme of F-mount. (**b**) A F-mount laser

Fig. 2.8 The structure of mounting substrate [7]

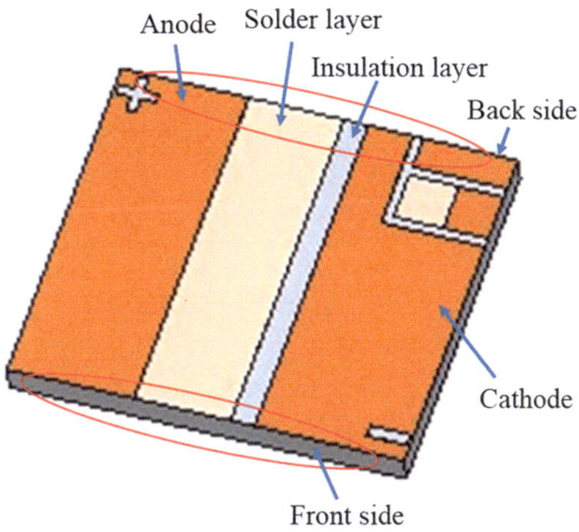

4. F-mount

   F-mount is a new packaging structure developed in recent years [1, 3, 7]. The package structure scheme of F-mount and an F-mount laser is shown in Fig. 2.7a, b, respectively [7]. As marked in Fig. 2.7a, six parts constitute the F-mount: the laser chip, the ceramic mounting substrate, the cathode, the anode, the ceramic slices, and the heat sink. The ceramic slices are used to isolate the anode and cathode from the heat sink. Thus the heat sink in the F-mount is not charged when the laser is in operation, that is, the heat sink is insulated from the laser diode.

   A schematic drawing of the ceramic mounting substrate is shown in Fig. 2.8 [7]. The surface of the ceramic mounting substrate is metallized with gold layer finish. The ceramic mounting substrate is divided into two parts by the insulation grove as shown in the figure, which are marked "+"and "−" respectively, as the anode and cathode. The laser chip is bonded onto the ceramic mounting substrate by the bonding solder layer. The cathode is connected to the n-side of the laser chip by wire bonding. Furthermore, the anode and cathode of the ceramic mounting substrate is connected to the two fly-leads which are the anode and cathode of the F-mount by wire bonding [7]. A thermistor pad is

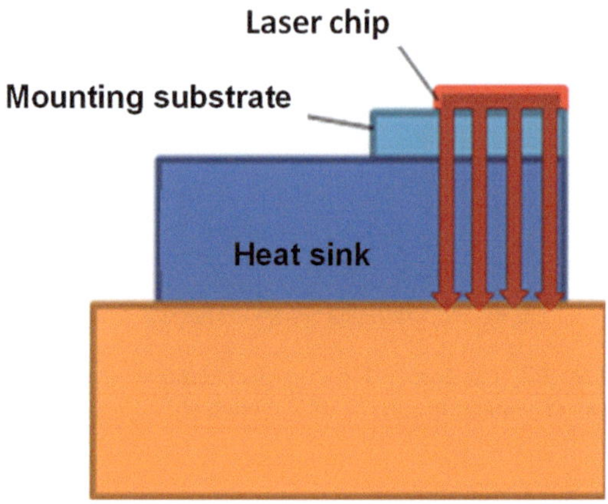

Fig. 2.9 The heat conduct path of F-mount [2]

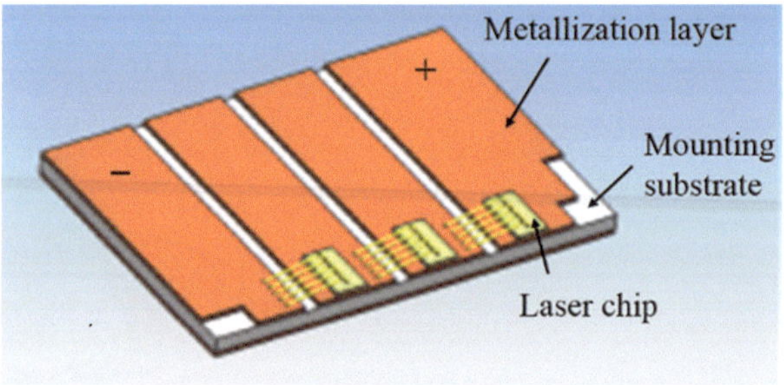

Fig. 2.10 A carrier design of a multiple single emitter laser [5]

designed on the ceramic mounting substrate which could be used to monitor and control the temperature of the laser diode.

The heat conduction path of F-mount is shown in Fig. 2.9 [2]. As shown in the figure, the heat is conducted to the mounting substrate firstly and then directly onto the heat sink both along vertical direction.

2.1.2 Packaging Structure of Multi-emitter Semiconductor Lasers

A basic carrier of a multiple single emitter laser is shown in Fig. 2.10 [5]. It contains three parts: laser chips, metallization layer, and a mounting substrate. As shown in the figure, the metallization layer is divided by insulation groves, and the cathode and anode are designed on this layer. The mounting substrate is also used to isolate the heat sink.

Each of the laser chips is serially wire-bonded to the carrier. A packaging structure based on the carrier has been designed and is shown in Fig. 2.11 [5]. The packaging structure which is similar to that of F-mount has an insulation sheet between the heat sink and electrodes, and the heat dissipation path is along the vertical direction. Figure 2.12 shows a finished three single emitter semiconductor laser [5].

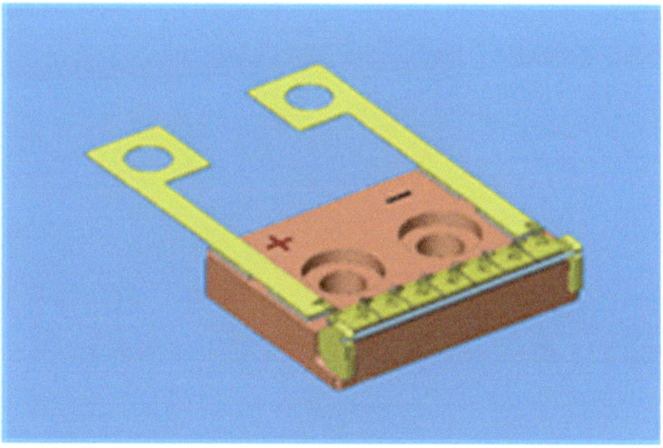

Fig. 2.11 Packaging structures of a laser device with seven emitters [5]

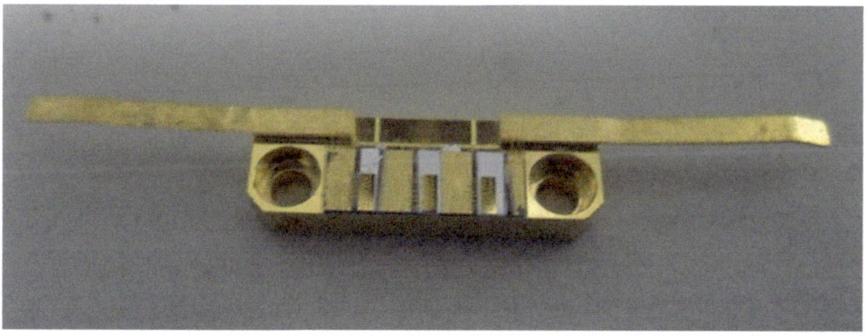

Fig. 2.12 A multiple single emitter semiconductor laser device [5]

2.1.3 Packaging Structure of Single Bar Semiconductor Lasers

The entire heat generation of a single emitter semiconductor laser is several watts and it can be easily dissipated by the copper conduction cooler, even in CW operation. But for diode laser bars, tens watts or hundred watts of heat will be generated during operation. The structure of the coolers for diode laser bars must be optimized, and in some cases, such as in the high duty cycle and CW mode, liquid coolers are needed [8, 9]. There are two types of liquid coolers: the micro-channel cooler (MCC) and the macro-channel cooler (MaCC) [10–12]. Based on the liquid cooler, bar-based semiconductor lasers can be constructed with four important parts: A mounting substrate or cooler, a laser bar, a cathode plate, and an insulator, as shown in Fig. 2.13 [8]. Mounting substrate or cooler shown in the drawing can be a Cu/CuW heat sink, a MCC, or MaCC. The laser bar with conduction heat sink such as Cu or CuW is called conduction-cooled semiconductor laser bar (CS).

1. Conduction-cooled single bar semiconductor lasers
   A three-dimensional drawing, the main constituent parts, and a conventional laser device of a CS are, respectively, shown in Fig. 2.14a–c [8]. In addition to

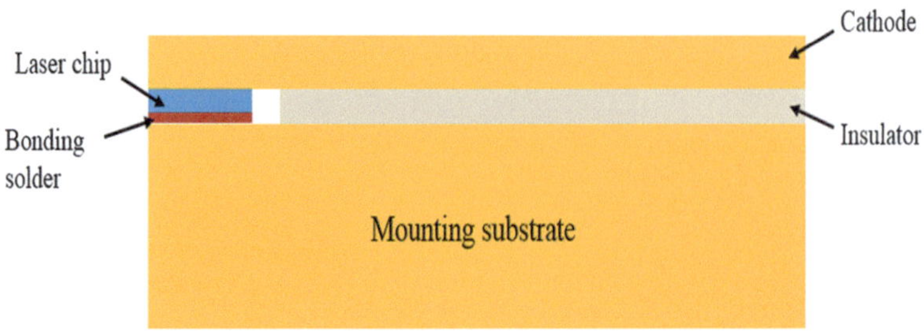

Fig. 2.13 The schematic structure of a semiconductor laser bar [8]

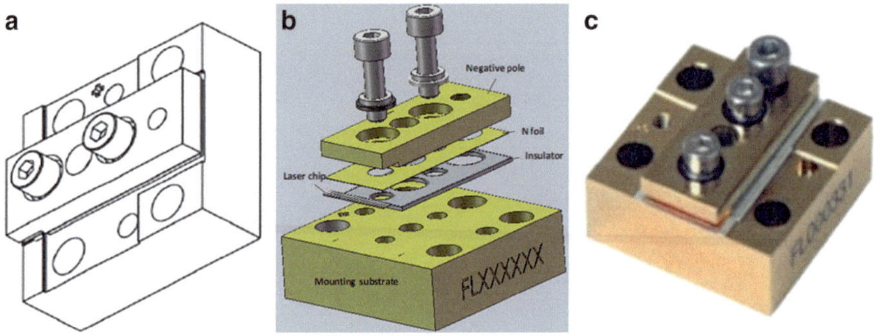

Fig. 2.14 The packaging structure of a conduction-cooled semiconductor laser bar [8]

the name of CS, CCP is also commonly called in the arena for the conduction-cooled single bar package. It should be noted that the solder layer between the laser bar and heat sink is very thin and not easy to be seen, so it is not shown in Fig. 2.14b. The laser chip is epi-down mounted on the heat sink. The heat sink is a thick copper block and has two functions. One function is as a cooler to dissipate heat and the other is as the anode of the laser bar. The four big holes in the heat sink are the mounting holes which are used to install the device on a platform. On the left side of the heat sink the small hole among the big holes is the anode thread and the small hole on the right is used to fix a thermistor.

In order to obtain good heat dissipation, the size of the CS cooler is much larger than the laser bar. For conduction cooling laser devices with low duty cycle, a narrower cooler size was developed, called NCCP which is shown in Fig. 2.15a, b [9].

2. Liquid-cooled single semiconductor laser bars
The MCC and an MaCC laser bar products are shown in Fig. 2.16a, b, respectively [10]. The main components, such as the laser bar, FAC, a cathode, and anode are marked in the figure. The sizes of the MCC and MaCC can be the same and are normally several millimeters in height and 10 mm in width. Except for acting as a cooler, it is also used as the anode of the laser bar. The main structural distinction between the MCC and MaCC lasers is the designs of the cooler channels.

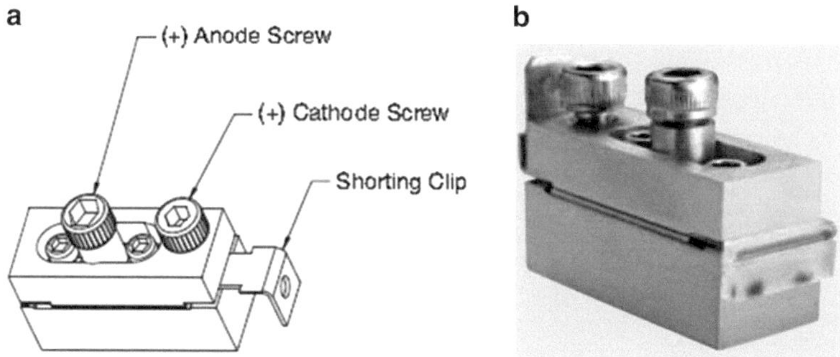

Fig. 2.15 The packaging structure of a narrow conduction-cooled semiconductor laser bar [9]

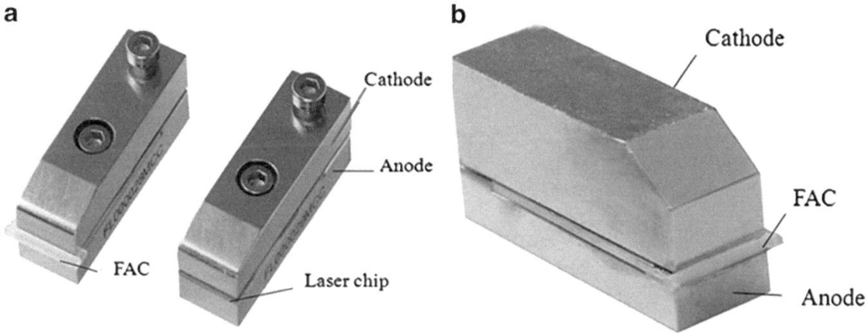

Fig. 2.16 MCC (**a**) and MaCC (**b**) semiconductor lasers [10]

The MCC semiconductor laser: Micro-channel cooler has high efficiency of heat dissipation, and the high power semiconductor laser bar with MCC can operate under CW and QCW with high duty cycle. Figure 2.17a shows the packaging structure of MCC laser [13]. The coolant inlet of the MCC is located near the laser bar. Between the coolant inlet and outlet there is a small hole and the function is for assembling or installation. The MCC laser is composed with three layers as shown in Fig. 2.17b. The upper layer is the cathode. The down layer is the micro channel which also acts as the anode of the laser bar. The middle layer between the cathode and micro-channel cooler is the insulator layer, which separates the cathode from the anode. The laser chip is packaged on the micro-channel cooler with a thin layer of bonding solder as shown in Fig. 2.17b.

The structure of MCC itself is complex and a typical one can consists of five copper layers as shown in Fig. 2.18 [14]. In Fig. 2.18 the copper layers are arranged as "1, 2, 3, 4, 5", and the first, third, and last copper layers are coolant passageway. The second and fourth layers are the micro-channel layers, and the size of channel is about hundreds micrometer wide. MCC is fabricated with very thin copper foils with micro-channel cut outs. Note: The micro channel can easily be blocked if there

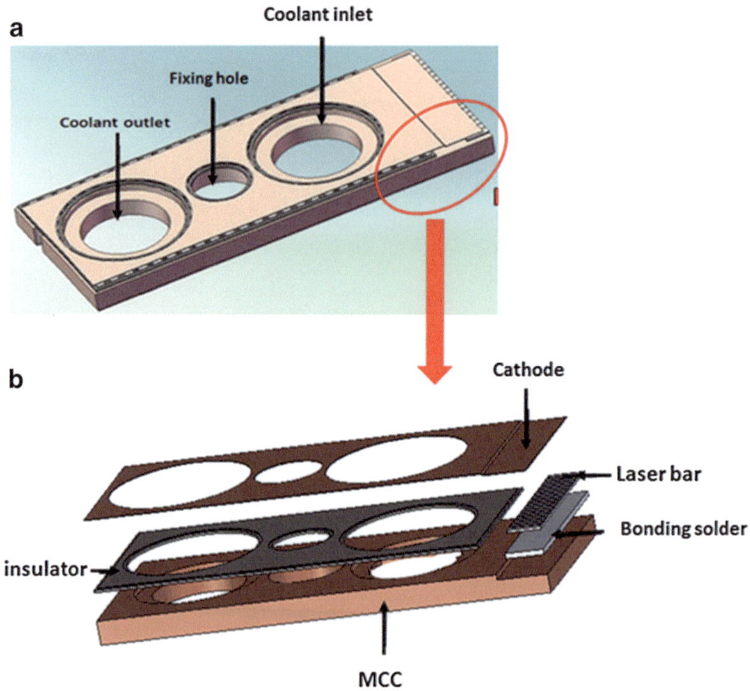

Fig. 2.17 The packaging structure of a MCC laser bar [13]

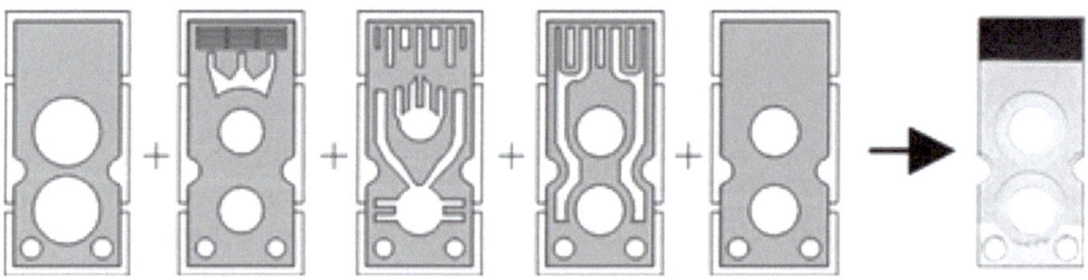

Fig. 2.18 The inner structure of a MCC [14]

is excessive debris in coolant liquid. Additionally, the channels are susceptible to electrochemical erosion and corrosion if there is too much electrolyte in the liquid. Hence, industrial water or tap water cannot be used and only deionized water should be used.

Figure 2.19 shows the three-dimensional profile of an assembled MCC laser [13]. In the figure, the upper and lower copper blocks act as the cathode and anode respectively. The hole in the back of the cathode is the cathode contact, and the hole in the back of the anode is the anode contact. It is seen from the figure that the coolant inlet and output channels are designed in the anode of the device. In order to separate the cathode from anode, there is an air gap or insulator between the cathode and anode, this area is marked in Fig. 2.19 with red circle ring.

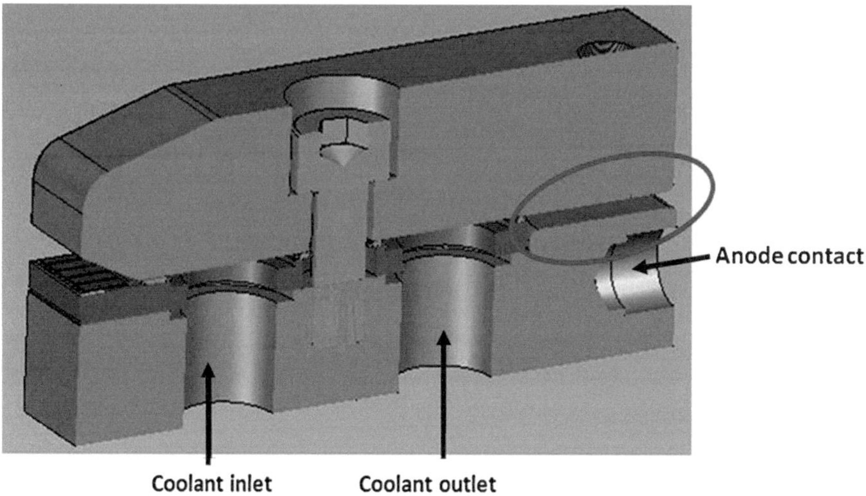

Fig. 2.19 The three-dimensional profile of the MCC laser bar [13]

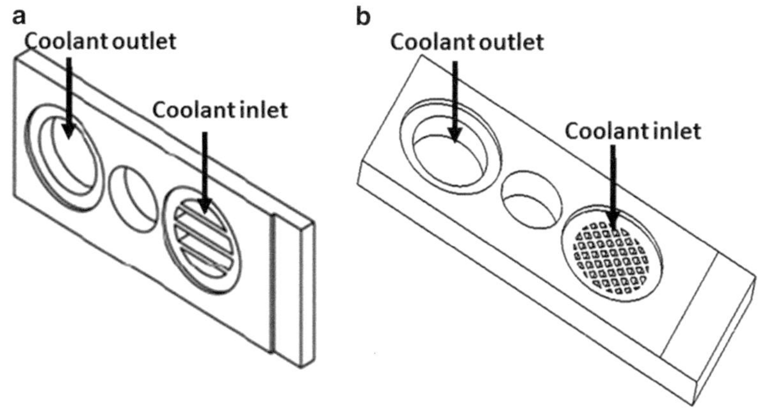

Fig. 2.20 Two typical MaCC [12]. (**a**) The strip type; (**b**) The mesh type

The MaCC semiconductor laser: MaCC also uses copper as the cooler material and two typical package structures are shown in Fig. 2.20a, b [12]. The two types shown in the figure have similar structural arrangement: both of them have coolant outlet and inlet with a fixing hole between them; and both of them have the laser chip attached to the front side of the cooler. One obvious difference is however in the coolant inlet structure for the two types of MaCC: one strip format and the other small mesh channel format, as illustrated in Fig. 2.20a, b, respectively. The size of the strip or mesh is about one millimeter, much larger than the channel size of MCC.

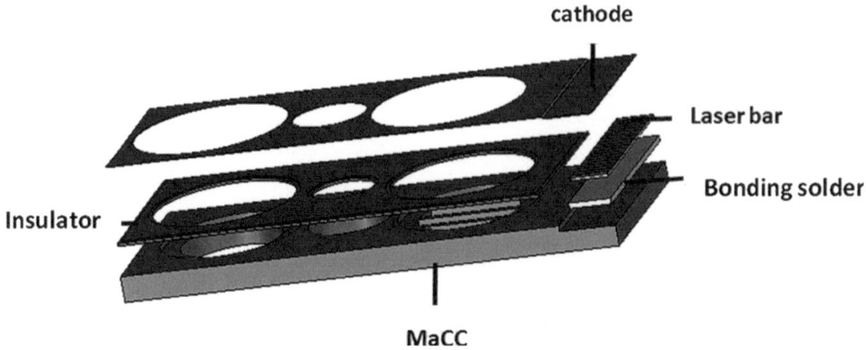

Fig. 2.21 The inner structure of MaCC [13]

An MaCC semiconductor laser consists of a cathode, an MaCC cooler, a laser chip, a solder layer, and an insulator. They are shown in Fig. 2.21 [13]. From Fig. 2.21, a typical MCC cooler is made up of five thin copper foils. MaCC cooler is however manufactured with one sheet of copper only with the thickness of 1.5–1.6 mm. As shown in Fig. 2.21, the coolant inlet is divided into small channels, and the size is much larger than the channel of MCC. So debris blockage is not as much a concern from electrochemical erosion and corrosion. Industrial tap water may be used and deionized water is not needed. The benefit of MaCC cooling cost is much lower than MCC. In addition, due to the simple structure of MaCC, it is easier to produce and modify than MCC. However, the efficiency of heat dissipation of MaCC is much lower than that of MCC, especially under high power, and hence, high power MaCC semiconductor lasers operate under QCW with low duty cycle.

Figure 2.22 shows the three-dimensional profile of the MaCC laser. In the figure, the coolant inlet and output channels are designed in both of the cathode and anode of the bar. The cooling water flows along the path with blue arrows. As shown in Fig. 2.22, the cathode and anode have been separated by the air gap marked by red circle [13].

2.1.4 *Packaging Structure of Multi-bar Semiconductor Lasers*

In order to obtain high power, the package structures of semiconductor laser bar arrays have been developed. Semiconductor laser bar arrays are composed of multiple semiconductor laser bars, and these bars have two arranged styles which are shown in Fig. 2.23a, b [13]. Figure 2.23a shows that the bars are arranged along horizontal direction and this laser array is called as the horizontal array (H-array). Figure 2.23b shows that the bars are arranged along vertical direction and this laser array is called as the vertical stack (V-stack). Generally, the commercial H-array and V-stack semiconductor lasers use liquid coolers. The semiconductor laser bar

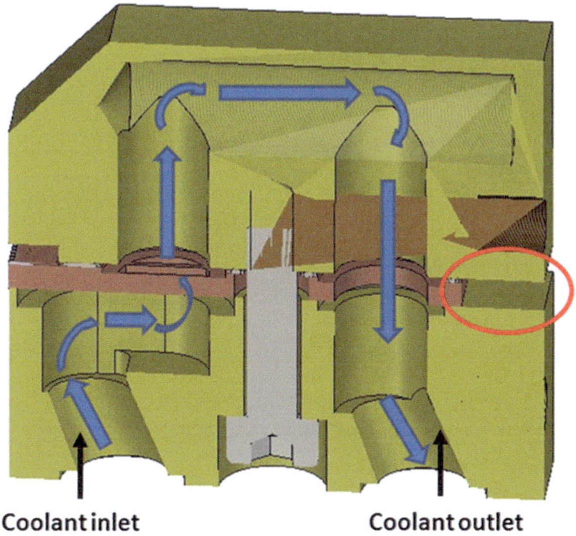

Fig. 2.22 The profile of the packaged MaCC laser bar [13]

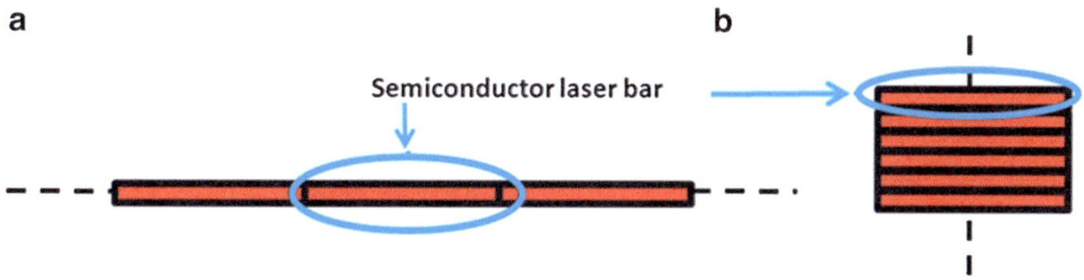

Fig. 2.23 The structures of diode laser arrays: (**a**) horizontal array; (**b**) vertical array [13]

array which use conduction cooler is called conduction-cooled semiconductor laser stack (G-stack). The H-array and V-stack semiconductor laser can be operated at CW and QCW, but the G-stack semiconductor laser can only work at QCW operation with low duty cycle. In this section, the package structures, the thermal designs, and the spectrum controlling technology will be discussed. The performances of semiconductor laser arrays with different package structures will be introduced as well.

1. Packaging structure of H-array
   The typical packaging structures of H-array lasers with four and three laser bars are shown in Fig. 2.24a, b [10, 15], respectively. We can see that the laser bars are arranged along horizontal direction, so the output beam is a line beam. The beam product parameter (BPP) of H-array lasers becomes worse with the increasing number of semiconductor laser bars. However, in some special applications where line beam is needed, such as the side pumping of solid

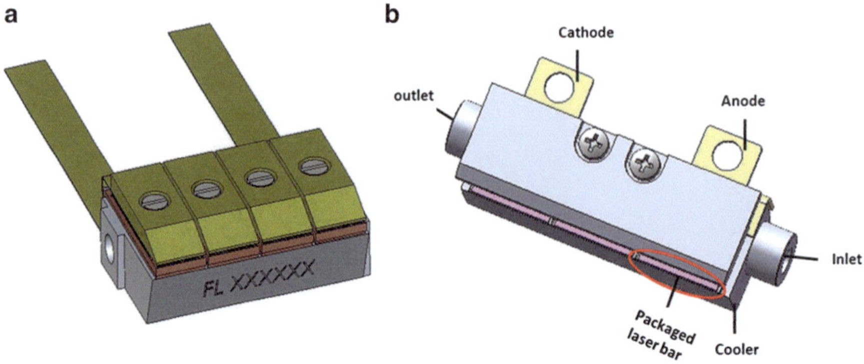

Fig. 2.24 The packaging structures of the H-array laser [1]. (**a**) H-array laser with four bars. (**b**) H-array laser with three bars

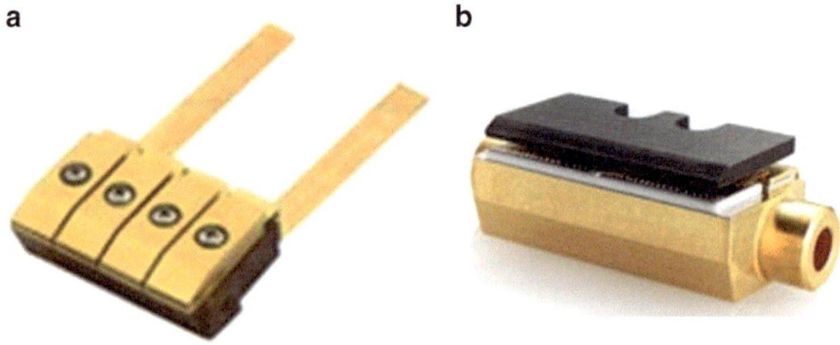

Fig. 2.25 H-array semiconductor lasers [10, 15]

state laser, H-array lasers are preferred and the numbers of the laser bars in the array are not limited theoretically. The packaged structures of these two H-array lasers in Fig. 2.24 are different, but both of them are composed with coolant inlet and outlet, heat sink, cathode, anode, and semiconductor laser bars. In addition, the coolant inlet of the cooler is near the anode of the laser array while the coolant outlet is beside the cathode. Figure 2.25 shows the pictures of finished H-array semiconductor lasers with four and three laser bars [1].

2. Packaging structure of V-stack

   Due to the decreasing beam quality along slow axis, H-arrays are not preferred for applications when high beam quality is required. The vertical stack semiconductor laser overcomes this shortcoming of the H-array laser. The beam quality of V-stack laser along slow axis keeps the same as a single laser bar. With the improvement of the packaging technology, the numbers of laser bars in a V-stack laser are also increased from several bars to 70 bars, and the maximum output power of a V-stack laser is from several hundred watts to up to twenty thousand watts.

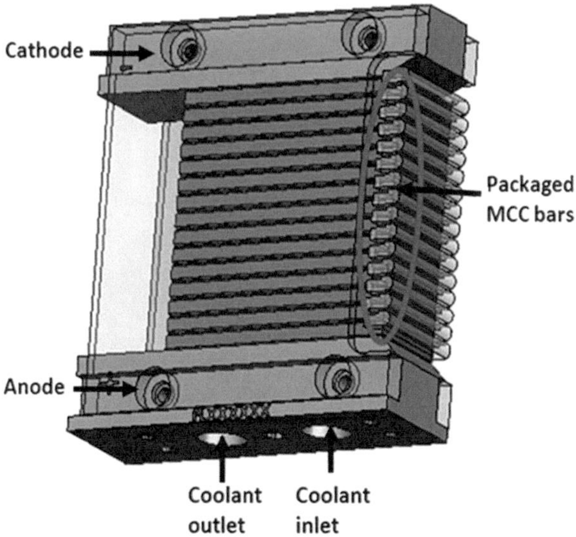

Fig. 2.26 The V-stack semiconductor laser packaged by MCC laser bars [16]

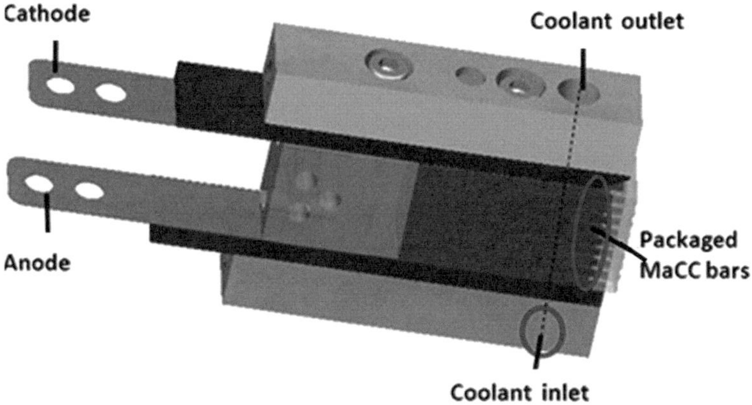

Fig. 2.27 The V-stack semiconductor laser packaged by MaCC laser bars [16]

The V-stack semiconductor laser can be packaged with either MCC or MaCC laser bars. Figures 2.26 and 2.27 show the packaging structures of the V-stack semiconductor lasers with MCC and MaCC laser bars, respectively [16]. The important constituent parts, i.e., the cathode, the anode, the coolant inlet and outlet, and the packaged MaCC (or MCC) bars, are noted in both figures.

The V-stack lasers shown in Figs. 2.26 and 2.27 are assembled with examples of 15 MCC and 9 MaCC laser bars, respectively. To compare the two V-stack lasers, the distinction between the two V-stack lasers is the locations of the

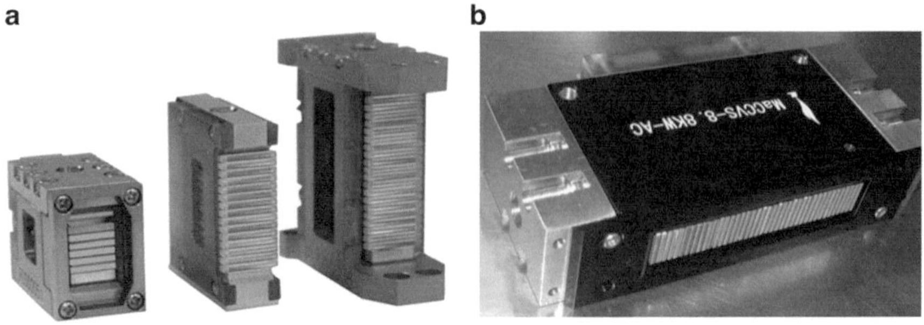

Fig. 2.28 The vertical stack semiconductor laser [10]. (**a**) MCC-based V-stack. (**b**) MaCC-based V-stack

coolant outlet and inlet. As seen in Fig. 2.26, the coolant outlet and inlet are in the cathode and anode of the MCC V-stack. Both of the coolant outlet and inlet are in the anode of the MaCC V-stack laser in Fig. 2.27, and they are arranged in a line. Same to the MCC and MaCC lasers, the MCC V-stack lasers can work at CW and QCW operation, and the MaCC V-stack semiconductor laser can only work at QCW mode with low duty cycle.

Based on the package structure illustrated in Figs. 2.26 and 2.27, Fig. 2.28a, b represents the commercially manufactured MCC V-stack laser products with different numbers of MCC laser bars and the MaCC V-stack laser with 40 bars [10].

3. Packaging structure of G-stack

The packaging structure of a G-stack semiconductor laser is shown in Fig. 2.29 [16]. In the figure, three semiconductor laser bars are packaged in one G-stack laser, and the cathode and anode of the laser stack are on the left and right side, respectively. The heat conduction cooler is under the laser chips, and the layer between laser chips and conduction cooler is the insulator layer. There are two insulator layers used for separating cathode and cooler, anode and cooler, respectively. There are four fixing holes in the stack, and the functions of them are to fix the stack and to connect the anode and cathode of the stack laser to the power driver, respectively.

To simplify the package structure, many other structures of G-stack lasers have been developed and another typical example is shown in Fig. 2.30 [16]. Compared with H-array and V-stack lasers, the G-stack laser has merits of high peak power, small volume, and ease for integration. They, however, are limited to QCW mode and low duty cycle operation. Figure 2.31a, b shows the pictures of finished G-stack semiconductor lasers with three and five laser bars, respectively [10].

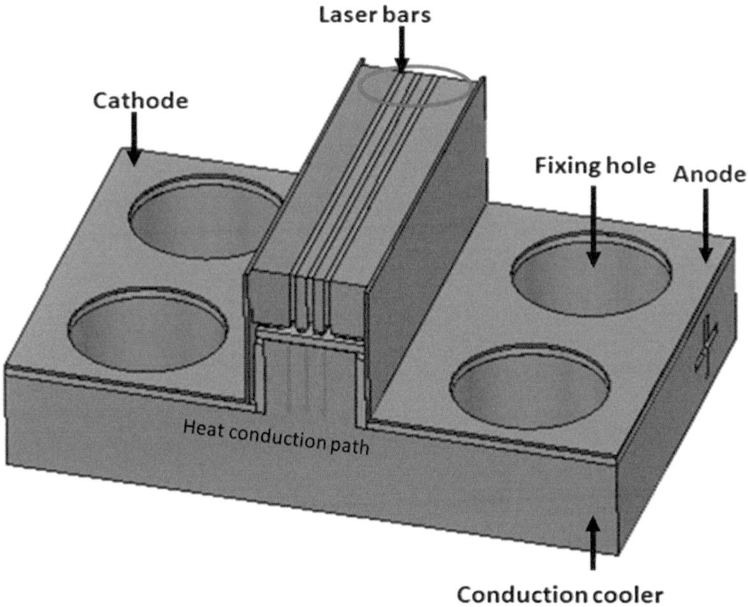

Fig. 2.29 The packaging structure of the G-stack [16]

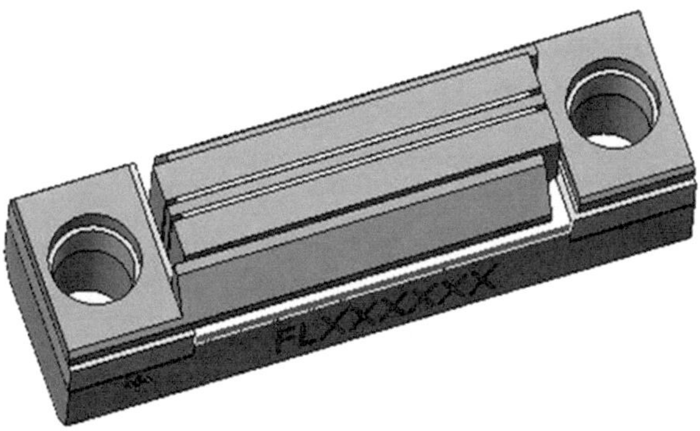

Fig. 2.30 The packaging structure of a two bar G-stack [16]

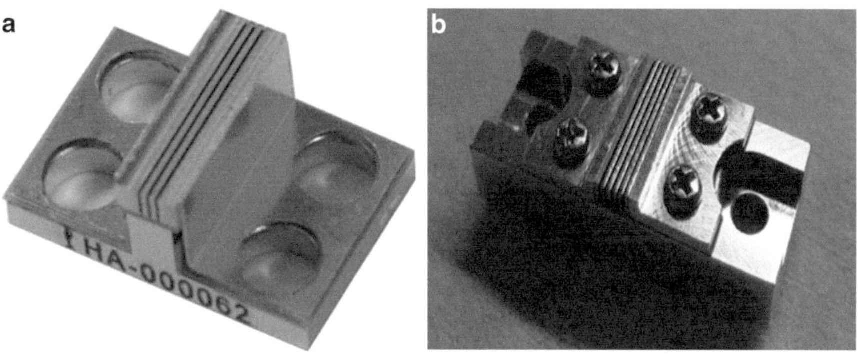

Fig. 2.31 Two typical structures of G-stack lasers [10]

2.2 Fiber-Coupled Packages

The fiber-coupled modules have relatively complex packaging structures. In addition to the laser chips or laser bars, they contain optical beam shaping and fiber-coupled optics, cooler, module house, and fiber ferrule assembly. In some applications, a feedback system is also needed.

2.2.1 Packaging Structure of Single Emitter Modules

The package structure scheme of fiber-coupled modules of single emitter is shown in Fig. 2.32a [5]. The major parts forming the module are illustrated in Fig. 2.32b. In addition to the single emitter, here the F-mount, the module contains components such as optical system (OS), thermoelectric cooler (TEC), module body, fiber ferrule assembly, and solder perform. Due to its small beam size, the beam of a single emitter module can be coupled into 200 μm fiber just with beam collimator at fast axis. In a fiber-coupled module, there are various types of feedback systems. The feedback systems include temperature sensor, power photoelectric detector (PD), fiber detection, red indicator, and so on.

The packaging structure of single emitter semiconductor lasers can be simplified. For a given application the fiber module can be made even without feedback systems and TEC. Figure 2.33 shows the design and structure of a typical fiber-coupled module of such type. Figure 2.34 shows a typical fiber-coupled single emitter modules [1].

2.2.2 Packaging Structure of Multi-emitter Modules

Generally, the structure of multi-emitter fiber-coupled modules mainly consists of multiple single emitters, beam shaping optical system, pigtail, and module body. The beam shaping optical system in the modules is to improve the beam quality and couple laser beam into optical fiber.

a b

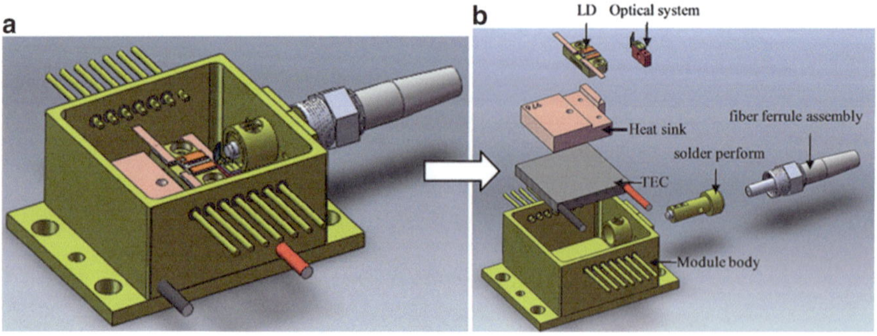

Fig. 2.32 Structure of a fiber-coupled single emitter module [5]

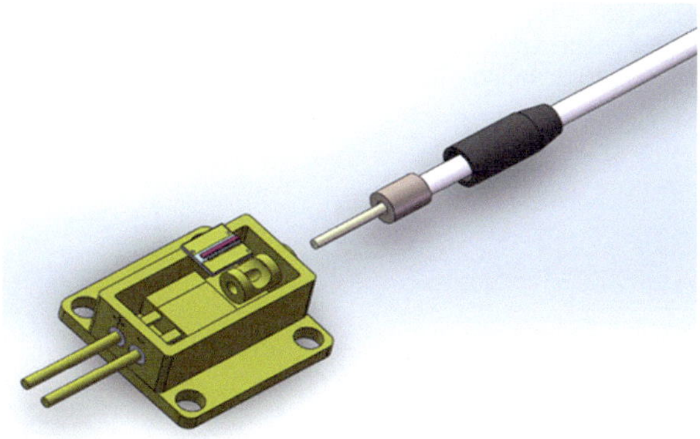

Fig. 2.33 The design and structure of a typical fiber-coupled module [1]

Fig. 2.34 Typical fiber-coupled single emitter module [10]

Figure 2.35 shows an example of multi-emitter fiber-coupled modules [1]. As shown in Fig. 2.35, the module body and pigtail have simple structure, and the module contains three emitters which are spatially divided by the steps. The beam shaping optical system is composed by three collimator lenses along slow axis, three mirrors, and two cylindrical lenses to focus laser beam. The example of multi-emitter fiber-coupled module is shown in Fig. 2.36 [10].

2.2.3 *Packaging Structure of Single Bar Modules*

The typical package structure scheme of fiber-coupled modules of single bar is shown in Fig. 2.37 and an open device example is shown in Fig. 2.38 [10]. Usually, a conduction cooling packaged semiconductor laser bar with 19 emitters is set in

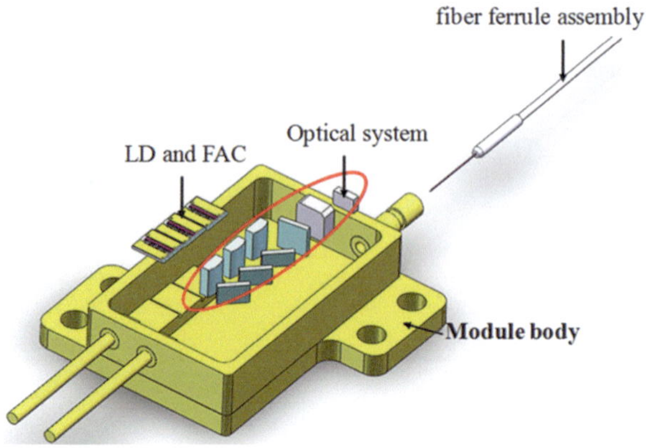

Fig. 2.35 Structure of a fiber-coupled multiple emitters-based module [1]

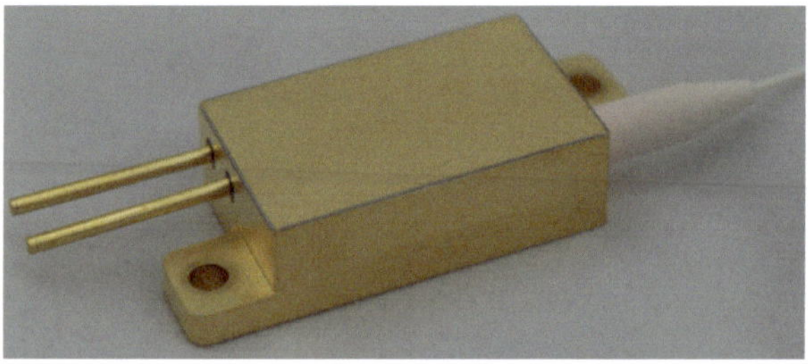

Fig. 2.36 Fiber-coupled multiple emitters-based module [10]

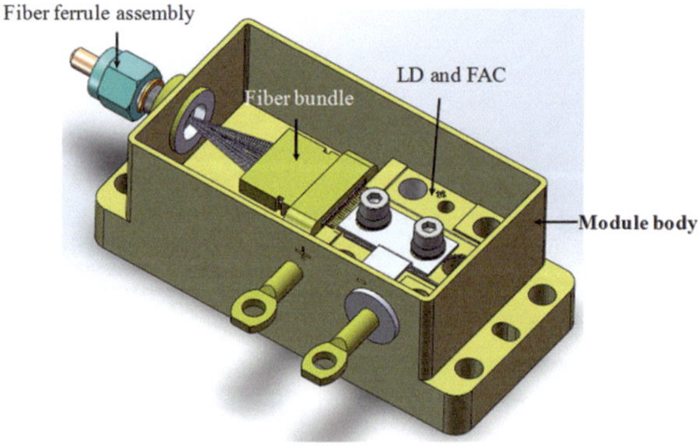

Fig. 2.37 Structure of a fiber bundle-coupled module [1]

this module. After fast-axis collimation for the 19 emitters, the 19 laser lights are coupled into an array of 19 fibers. The output power is the superposition of the 19 beam coupled in the fiber. The difficult in fabricating this module is the alignment accuracy between semiconductor laser and array fiber.

Figure 2.39 shows another packaging structure of fiber-coupled module of single semiconductor laser bar [17]. The main difference of the two kinds in Figs. 2.38

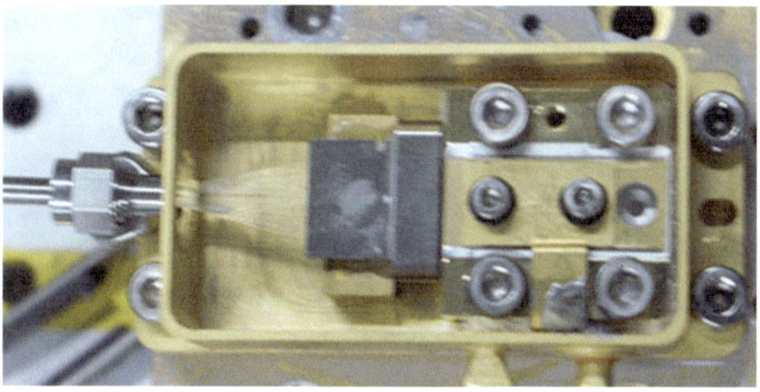

Fig. 2.38 Fiber bundle module of single bar [10]

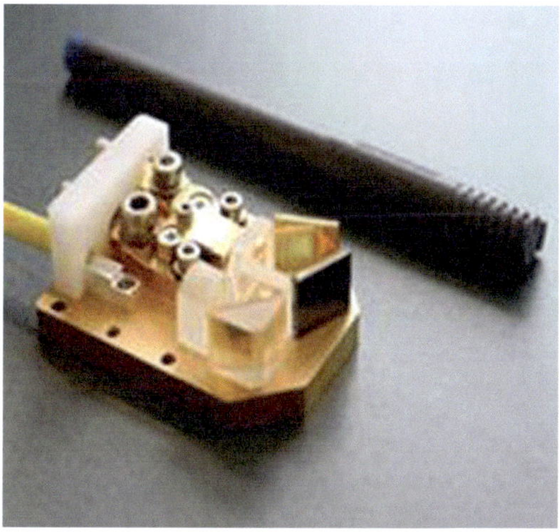

Fig. 2.39 Beam shaping for a laser bar [17]

and 2.39 is the beam shaping system in module body. In fiber bundle mode of Fig. 2.38, the beam doesn't need complex beam shaping and solely beam collimation along the fast axis is needed before fiber coupling. However, in Fig. 2.39 in addition to beam collimation a complex beam shaping system is needed, and based on this style it is called a free space fiber-coupled structure [18]. The module shown in Fig. 2.39 has higher brightness than that of fiber bundle module shown in Fig. 2.38. A typical optical structure of an optical system for a free space fiber-coupled device is shown in Fig. 2.40 [19].

2.2.4 *Packaging Structure of Multi-bar Modules*

The typical structure scheme of fiber-coupled modules of multiple bars is shown in Fig. 2.41 [13]. The module contains three conduction-cooled semiconductor lasers (CS) that are fixed on the ladder platform. To insulate, a piece of ceramic is

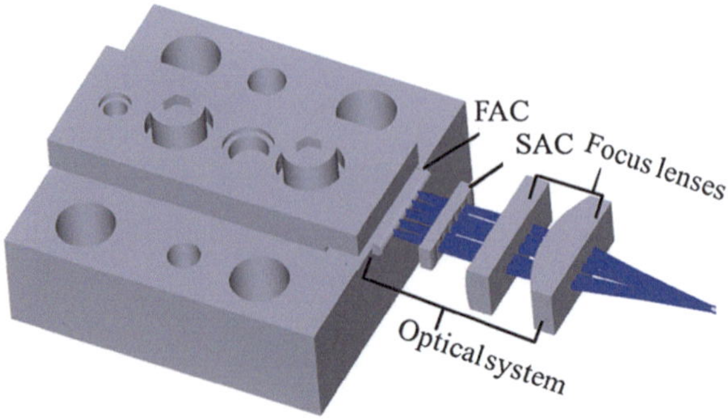

Fig. 2.40 Fiber-coupled module of a single bar [19]

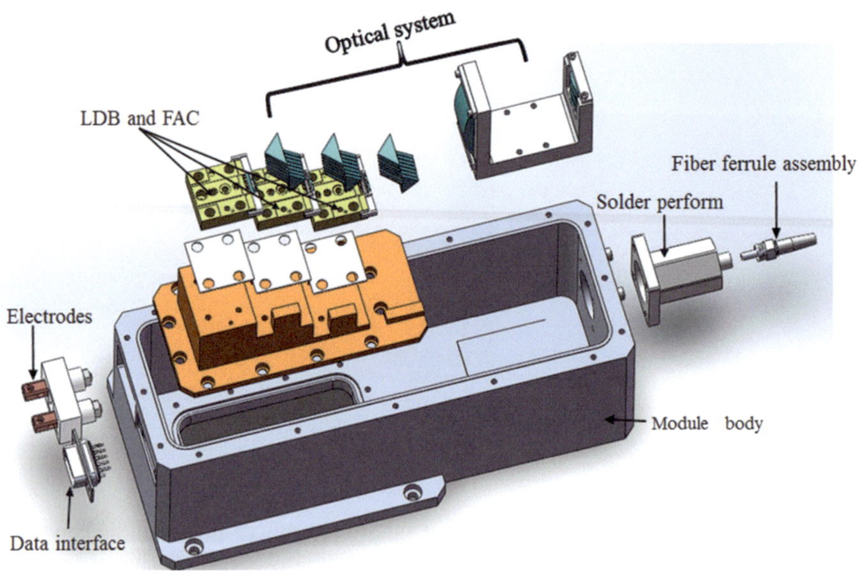

Fig. 2.41 Scheme of fiber-coupled modules of multiple bars [13]

arranged underneath each of the CS devices. Fast-axis and slow-axis collimation lens must be added to collimate the beam and to reduce the divergence angles.

In the multiple bar-coupled module, optical coupling and beam shaping are the key techniques employed in the beam quality improvement. Generally speaking, the beam combining technologies are wavelength combination, polarization combination, and spatial combination (also can be the mostly used beam treatment methods). Figure 2.42 is an example of spatial combination for three CS [13]. Based on these combination approaches, higher output power can be obtained for fiber coupling. Finally, two cylindrical lenses also are used to shape the beam into the fiber.

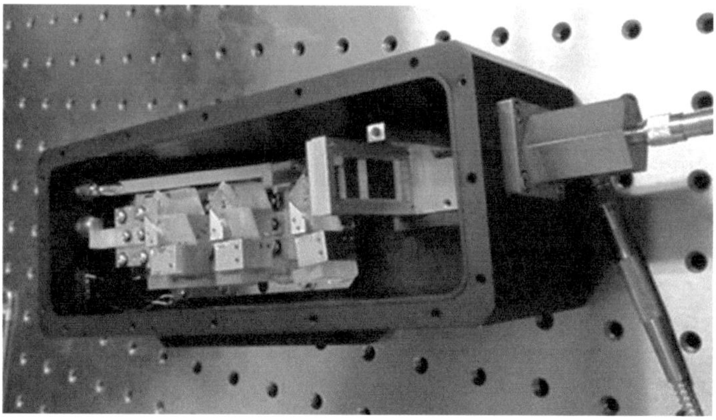

Fig. 2.42 Fiber-coupled modules of multiple bars: an example of spatial combination [13]

References

1. J.W. Wang, Y.X. Zhang, Packaging design method of high power semiconductor lasers. Internal Report **4**, 21–27 (2010)
2. X.N. Li, Y.X. Zhang, J.W. Wang, L.L. Xiong, P. Zhang, Z.Q. Nie, Z.F. Wang, H. Liu, X.S. Liu, Influence of package structure on the performance of the single emitter diode laser. IEEE Trans. Compon. Packag. Manuf. Technol. **2**(10), 1592–1599 (2012)
3. Y.X. Zhang, J.W. Wang, C.H. Peng, X.N. Li, L.L. Xiong, X.S. Liu, *A New Package Structure for High Power Single Emitter Semiconductor Lasers*. 2010 11th International Conference on Electronic Packaging Technology & High Density Packaging (ICEPT-HDP), 2010, pp. 1346–1349
4. http://laserpointerforums.com/f44/demystifying-808nm-laser-diode-81627.html
5. L. Guo, J.W. Wang, Z.B. Yuan, Y.X. Zhang, Packaging of multiple single emitter semiconductor laser. Internal Report **12**, 12–17 (2011)
6. X.S. Liu, L.C. Hughes, M.H. Rasmussen, M.H. Hu, V.A. Bhagavatula, R.W. Davis, S. Coleman, R. Bhat, C.E. Zah, Packaging and performance of 980 nm broad area semiconductor lasers. Electron. Packag. **6**(3), 6–14 (2006)
7. X.S. Liu et al., A novel single emitter high power semiconductor laser and method of preparation, China patent, No. 200910020854.5, 2009
8. J.W. Wang, Z.B. Yuan, Y.X. Zhang, E.T. Zhang, D. Wu, X.S. Liu, *250W QCW Conduction Cooled High Power Semiconductor Laser*. 2009 International Conference on Electronic Packaging Technology & High Density Packaging, IEEE (2009), pp. 451–455
9. http://www.dilas-inc.com/
10. http://www.focuslight.com.cn/products.asp
11. J.W. Wang, Z.B. Yuan, L. Guo, L.L. Xiong, Y.X. Zhang, *Packaging of High Power Semiconductor Laser Arrays Using a Novel Macro-Channel Cooler*. 2010 11th International Conference on Electronic Packaging Technology & High Density Packaging (ICEPT-HDP), 2010, pp. 92–97
12. X.S. Liu, A design of liquid cooled plate and method of preparation for semiconductor laser array, China patent, No. ZL200910023753.3, 2009, International PCT patent is pending
13. D. Wu, J.W. Wang, Z.B. Yuan, Y.X. Zhang, Packaging of high power semiconductor laser bars. Inner Report **1**, 34–42 (2010)
14. F. Bachmann, *High-Power Diode Lasers Technology and Applications* (Aachen, Germany, 2006)
15. http://www.coherent.com/products/?1534/Diode-Lasers

16. H.J. Zhong, J.W. Wang, D. Wu, Y.X. Zhang, Packaging of high power semiconductor laser arrays. Inner Report **1**, 34–42 (2010)
17. http://www.iof.fraunhofer.de/en/buiness-fields/laser/faserlaser/fasereinkopplung/fiber_coupling_forlaserdiodebarsandstacks.html
18. http://ir.ciomp.ac.cn/bitstream/181722/4286/1/Mini-bar%20based%20diode%20lasers%20with%20high%20brightness%20high%20power.pdf
19. http://an.hitchcock.org/repairfaq/sam/laserfil.htm

Chapter 3
Thermal Design and Management in High Power Semiconductor Laser Packaging

Thermal management of high power lasers is critical since the junction temperature rise originating from large heat fluxes strongly affects the device characteristics, such as wavelength, power, threshold current, efficiency, and reliability. In this chapter, the temperature effect on the performances of high power semiconductor lasers is introduced in Sect. 3.1; the heat generation sources in semiconductor lasers are analyzed in Sect. 3.2; thermal modeling, design, and analysis are discussed in Sect. 3.3. Furthermore, some thermal management techniques for reducing the junction temperature of the diode laser devices are proposed, including optimization of the cooling structure, cooler design, and packaging materials.

3.1 Temperature Effect on Performances of High Power Semiconductor Lasers

The conversion efficiency of high power diode lasers is much higher than other types of lasers. The conversion efficiency is approaching 75–80 % in laboratory and the conversion efficiency of commercial product is around or slightly above 55 % for 808 nm and 65 % for 980 nm. Nevertheless, due to its high output power, a considerable amount of heat is still generated in high power diode lasers, causing a high temperature rise at the active region of the diode laser. Junction temperature is related to temperature rise and heat sink temperature by the following relationship:

$$T_J = T_{heat\,\sin k} + \Delta T, \tag{3.1}$$

where ΔT is the temperature rise between active region temperature and the ambient environment temperature, T_J and $T_{heatsink}$ are the temperatures of the junction and heat sink of the laser device, respectively. By approximation,

X. Liu et al., *Packaging of High Power Semiconductor Lasers*, Micro- and Opto-Electronic Materials, Structures, and Systems, DOI 10.1007/978-1-4614-9263-4_3,

the temperature rise ΔT equals to the product of the thermal resistance and ($P_{in}-P_{out}$) of the laser, which is calculated as follows [1]:

$$\Delta T = R_{th}(P_{in} - P_{out}) \tag{3.2}$$

where R_{th} is thermal resistance, P_{in} is input power which is the product of input current I_0 and input voltage V_0. P_{out} is output optical power.

By substituting Eq. (3.2) into Eq. (3.1), the diode laser junction temperature can be expressed as

$$T_J = T_{heat\,\sin k} + R_{th}\left(I_0 V_0 - P_{output}\right). \tag{3.3}$$

According to Eq. (3.3), three approaches may be proposed to reduce the junction temperature T_J. The first approach is to reduce the heat sink temperature. The second approach is to reduce the thermal resistance of the laser device. The third approach is to increase the optical output power.

The temperature increase at the active region of diode lasers reduces the carrier confinement and increases the non-radiative recombination probability. Both effects lead to a higher threshold current and a lower conversion efficiency. The maximum output power is thus limited by the temperature rise of the active region. In addition, other important parameters, such as wavelength, spectrum, and lifetime, are also negatively affected by the elevated temperature of the active region of diode lasers.

3.1.1 *Threshold*

The impact of temperature rise on threshold depends on the given structure and resonator quality of the diode laser considered [1]. The temperature dependence of the threshold current can be expressed empirically as [2]

$$I_{th}(T) = I_{\mathrm{Ref}}e^{\frac{T-T_{\mathrm{Ref}}}{T_0}} \tag{3.4}$$

where T_0 is the characteristic temperature given in Kelvin and I_{Ref} is the threshold current at the reference temperature T_{Ref}. The characteristic temperature T_0 depends on the barrier height and the necessary threshold-current density, which depends on the resonator losses and the optical confinement factor for a given active region. The values for T_0 are around 200 K for wavelengths at 980 nm, come down to ~140 K at 800 nm and further get below 100 K for shorter wavelength (630–760 nm) GaAs diode lasers [3].

Figure 3.1 shows the threshold changes with operation temperature of a typical single emitter diode laser [4]. As shown in the figure, the threshold increases with the increasing of temperature.

Fig. 3.1 The threshold of a typical single emitter laser as a function of operation temperature [4]

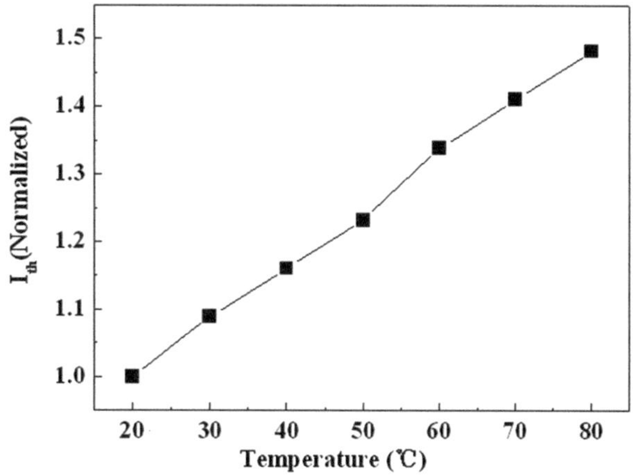

3.1.2 Slope Efficiency

Above the threshold, the output power increases linearly with the driving current. The slope efficiency of a typical 800–1,000 nm diode laser device is on the order of 1 W/A. The higher the slope efficiency, the better the laser is. The change of η_{slope} with temperature can be expressed as [2]

$$\eta_{slope}(T) = \eta_{\mathrm{Ref}} e^{-\left(\frac{T - T_{\mathrm{Ref}}}{T_1}\right)} \tag{3.5}$$

where T_1 is the characteristic temperature given in Kelvin and η_{Ref} is the slope efficiency at the reference temperature T_{Ref}.

Generally, the slope efficiency depends on internal quantum efficiency, the reflectivity of rear and front facets and absorption within the semiconductor material, and the frequency of the emitted photons. Also, the slope efficiency changes with operation temperature as illustrated in Fig. 3.2 [5].

3.1.3 Output Power

Figure 3.3 shows the output power of a single emitter diode laser at different temperatures [3]. It clearly shows the temperature effect on output power of semiconductor lasers. The output power is decreased obviously with the increasing temperature. The expression of the output power is [6]

$$P_{output} = \eta_{slope}(T)(I - I_{th}(T)) \tag{3.6}$$

Fig. 3.2 Illustration of the slope efficiency change with operation temperature [5]

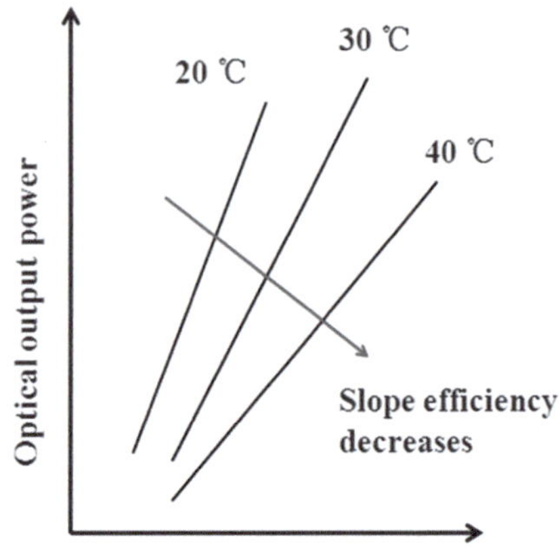

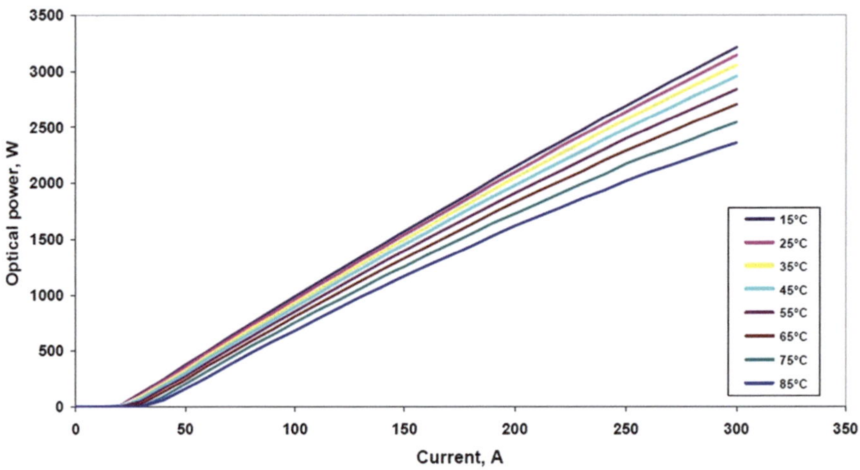

Fig. 3.3 Output power of a single emitter diode laser at different temperatures [3]

Using the temperature dependence of threshold and slope efficiency Eqs. (3.4) and (3.5), the power-current characteristic Eq. (3.6) can be described approximately by the following expression:

$$P_{output} = \eta_{\mathrm{Ref}} e^{-\left(\frac{T-T_{\mathrm{Ref}}}{T_1}\right)} \left(I - I_{\mathrm{Ref}} e^{\frac{T-T_{\mathrm{Ref}}}{T_0}}\right) \quad (3.7)$$

In Eq. (3.7), the higher the characteristic temperatures T_0 and T_1 the more stable the operation in the CW regime at elevated temperatures [1]. For a given diode laser, which means its T_0 and T_1 are given, the output power is significantly influenced by its junction temperature.

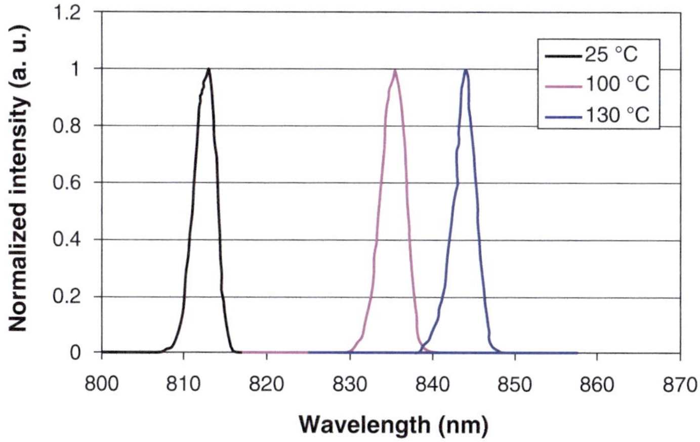

Fig. 3.4 Output wavelength at different temperature for 808 nm semiconductor laser [8]

3.1.4 Wavelength

When the junction temperature of the emitter increases, the output wavelength has red shift. The wavelength of a semiconductor laser will shift at a certain rate with junction temperature. The relationship between bandgap and temperature can be described by Varshni equation [7]:

$$E_g(T) = E_g(0) - \frac{\alpha T^2}{\beta + T} \tag{3.8}$$

where $E_g(0)$ is the bandgap of semiconductor material at $T = 0$ K, α is an empirical constant, and β is associated with the Debye temperature.

Figure 3.4 shows the output wavelength at different temperature for 808 nm semiconductor laser [8]. The device exhibits lasing spectral red shift with increase of temperature at the rate of ~0.28 nm/°C for 808 nm laser [8]. For the 980 nm semiconductor laser, the coefficient of the wavelength-temperature shift is ~0.32 nm/°C [9].

3.1.5 Lifetime

The temperature rise at the active region of diode lasers can reduce the lifetime of the lasers. Generally, the lifetime of the high power semiconductor laser can be expressed by the relationship described below [10]:

$$MTTF \propto AJ^{-N} Exp(E_a/kT) \tag{3.9}$$

where MTTF is the mean time to failure, A and N are constants, J is current density, Ea is active energy, k is Boltzmann constant, and T is junction temperature.

The lifetime of the semiconductor laser can be analyzed by the lifetime testing data. According to the Eq. (3.9), the most direct and effective approach to increase the lifetime and improve the reliability of devices is to decrease the junction temperature.

3.2 Heat Generation Sources

As shown in Chap. 1, the structure of a diode laser chip consists of multiple layers. At each layer, heat may be generated when the laser is working. When the diode laser chip is packaged and becomes a device, heat may be generated at the packaging interfaces or materials. According to the heat generation principle, there are five heat sources in a semiconductor laser device, as listed below [11–15]:

1. Non-radiative recombination in the active region
2. Absorption of radiation in an optical cavity
3. Absorption of radiation outside an optical cavity
4. Joule heating in all layers of a laser structure
5. Surface Joule heating at semiconductor metal interfaces.

Among the five heat sources, non-radiative recombination in the active region, absorption of radiation in an optical cavity, absorption of radiation outside an optical cavity, and surface Joule heating at semiconductor metal interfaces are from diode laser chip. The heat generation from these heat sources depends on the laser chip design and structure and the calculation is complex [11–13, 15]. The analysis and calculation of these heat sources will not be discussed in this book. This chapter only discusses the heat sources related to packaging. Characteristics of the laser device are greatly influenced by the package structure of the device. The following will use C-mount and F-mount-packaged semiconductor lasers as examples to analyze and compare the heat sources of different packaging structures.

3.2.1 Heat Sources of Single Emitter Semiconductor Lasers

1. C-mount
   The schematic diagram of the structure of C-mount lasers is shown in Fig. 3.5 [4]. C-mount-packaged device has four main parts: bonding wires, chip, solder, and mounting substrate. The heat generated by these parts is denoted Q1, Q2, Q3, and Q4, respectively. Besides heat generated from these bulk materials, the interfaces between these parts can also produce heat. If the solder between the neighboring layers is not bonded well, voids would exist at the interfaces and the electrical contact resistance could be high, which would generate a lot of heat when laser device is operated.

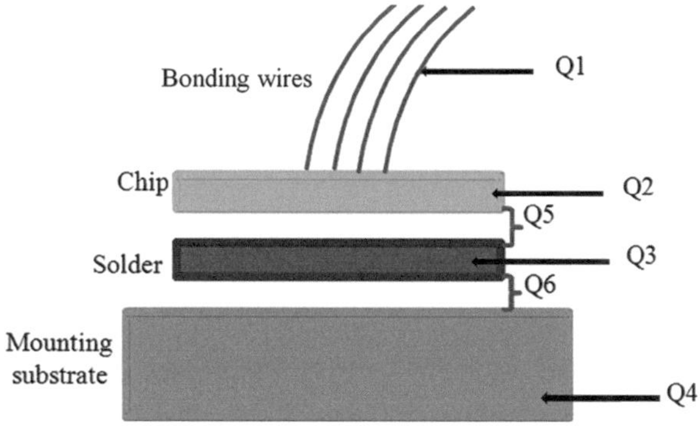

Fig. 3.5 The schematic diagram of the structures of C-mount [4]

As shown in Fig. 3.5, Q5 and Q6 represent the heat produced from the interfaces between chip and solder, and solder and mounting substrate, respectively. The total heat Q of the laser device can be expressed by the following equation [14, 16]:

$$Q = Q1 + Q2 + Q3 + Q4 + Q5 + Q6. \tag{3.10}$$

where Q1, Q3, and Q4 are the Joule heating by wire, solder, and mounting substrate, respectively. Q2 is generated by laser chip includes four kinds of heat induced by non-radiative recombination in the active region, absorption of radiation in an optical cavity, absorption of radiation outside an optical cavity, and Joule heating from the chip.

The Joule heating Q1, Q3, and Q4 can be obtained by the equation,

$$Q = I^2\rho, \tag{3.11}$$

where Q is Joule heat, I is the inject current, and ρ is the local electrical resistivity. In the equation the local resistance ρ is calculated by the following equation:

$$\rho = \rho_0 l/A \tag{3.12}$$

where ρ_0 is the electrical resistivity of material; l and A are the length and cross-sectional area of the material, respectively.

The surface Joule heating Q5 and Q6 is generated by the imperfect contact of two surfaces. Generally, the contact of two surfaces can be strengthened and improved by material selection and optimizing the soldering process. If solder interface and the soldering process are optimized, the surface resistance of the interface would be very low and the surface Joule heating can be ignored. For semiconductor lasers, the contact between laser chip and solder layer, and between solder layer and Cu heat sink is done by metallization and optimized soldering process. Generally the surface Joule heating is very small and thus Q5 and Q6 can be ignored in this case.

For heat source of Q2, as discussed previously, it consists of several heat generation mechanisms. Here we only discuss Joule heating of the laser chip.

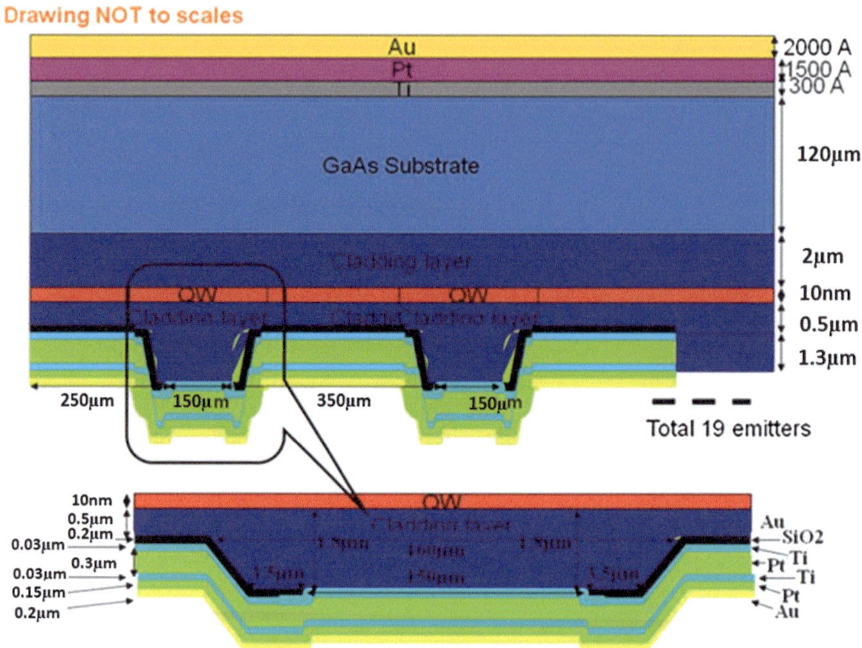

Fig. 3.6 The typical structure of a diode laser chip [4]

As shown in Fig. 3.6, the laser chip typically consists of seven or more layers [4]. The local resistance of a laser chip is the sum of the resistance value of all layers

$$\rho_{chip} = \sum_{i=1}^{N} \rho_i, \tag{3.13}$$

where N is the layer number of laser chip and ρ_i is the resistance of the i_{th} layer. Then the Joule heating from the chip can be calculated by $I^2\rho_{chip}$.

The conversion efficiency of a laser device, denoted by η, can be known by the measurement of LIV characteristics. The total heat of the laser device can be obtained by the following equation [14]:

$$Q = IV(1 - \eta) \tag{3.14}$$

where I and V are the injection current and voltage, respectively.

2. F-mount

   Figure 3.7 illustrates the F-mount structure, which includes five main parts: bonding wires, laser chip, solder layer, metallization layer, ceramic layer, and copper heat sink [4]. The top view of F-mount is shown in Fig. 3.8 [4]. The current flow path in the metallization layer can be fairly long.

   In the F-mount, as no current goes through the ceramic and heat sink, no heat is generated in the ceramic and heat sink layer when the laser works. Hence, heat

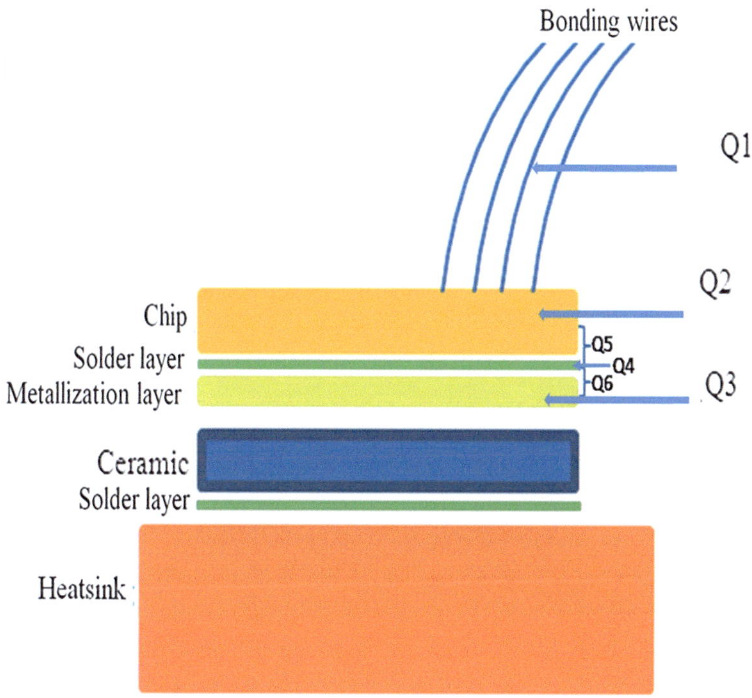

Fig. 3.7 The schematic cross-section view of F-mount [4]

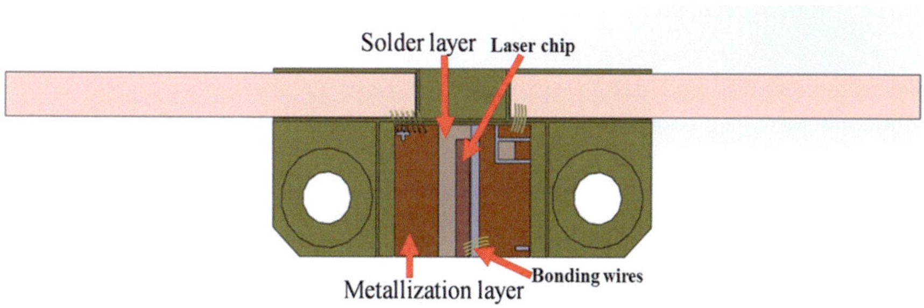

Fig. 3.8 The top view of F-mount [4]

sources in the F-mount laser include six parts: bonding wires, laser chip, metallization layer, and solder layer, which are denoted as Q1, Q2, Q3, and Q4, respectively; Similar to the C-mount structure, the heat of Q5 and Q6, which are generated between each joint interface as marked in Fig. 3.7, can be ignored in this case. The total heat Q of the F-mount laser device can be expressed by the following equation:

$$Q = Q1 + Q2 + Q3 + Q4. \tag{3.15}$$

In Eq. (3.15), Q1, Q2, Q3, and Q4 can be calculated by the similar methods as in the C-mount structure, so no further discussions are made here.

3.2.2 *Heat Sources of Semiconductor Laser Bars*

A semiconductor laser bar is composed by multiple emitters as described in Chap. 1, with each emitter in the laser bar being arranged in a row. The structure of a diode laser emitter was discussed in Chap. 1.

The total heat of a laser bar consists of the heat generated from each emitter of the laser bar, and hence, it can be expressed as [14]

$$Q = \sum_{i=1}^{N} Q_i \tag{3.16}$$

where Q is the total heat of the laser bar, Q_i is the heat generated from the i-th emitter, and N is the number of the emitters in the laser bar. Thermal simulation with the three-dimension finite element method (FEM) is used to analyze the heat in the laser bar as described in Sect. 3.3, along with analysis on several other major package structures.

3.3 Thermal Modeling, Design and Analysis

3.3.1 *Finite Element Thermal Modeling and Design*

Thermal modeling and design are important steps in developing a high power semiconductor laser package. There are four major purposes in conducting thermal modeling and design:

1. To optimize package structure from thermal management perspective
2. To calculate the thermal resistance
3. To calculate the max temperature of the active region
4. To stimulate the steady state and transient thermal behavior, analyze temperature field distribution and heat flow.

With the development of the computer technology, the finite element modeling (FEM) has been a very effective tool for thermal design and analysis of semiconductor laser packages. The main principle of FEM is that the laser device is divided into many small units, and thermal properties of every unit are calculated. By integrating the results on each unit, the whole thermal performance of the laser is achieved. Various commercial softwares, such as ESC TMG, Flothermal, ABAQUS, COSMOS, Ansys, ADINA, and ICEPAK, have been developed to simulate the thermal properties of a device. Ansys is one of the popular softwares to be used in this field. The program flow of thermal simulation for a semiconductor laser using FEM is shown in Fig. 3.9 [4, 17].

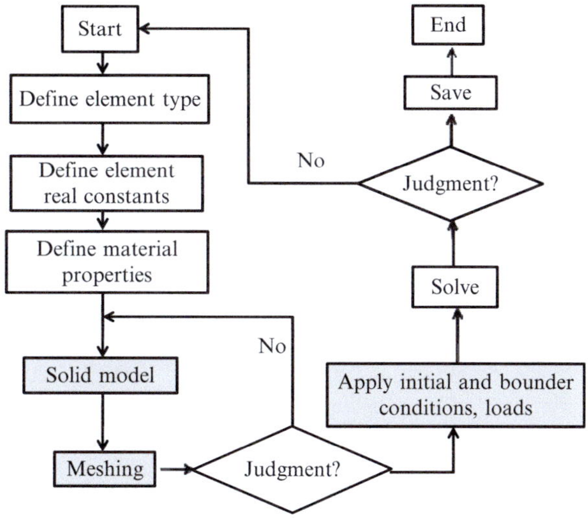

Fig. 3.9 The process flow of the FEM thermal design [4, 17]

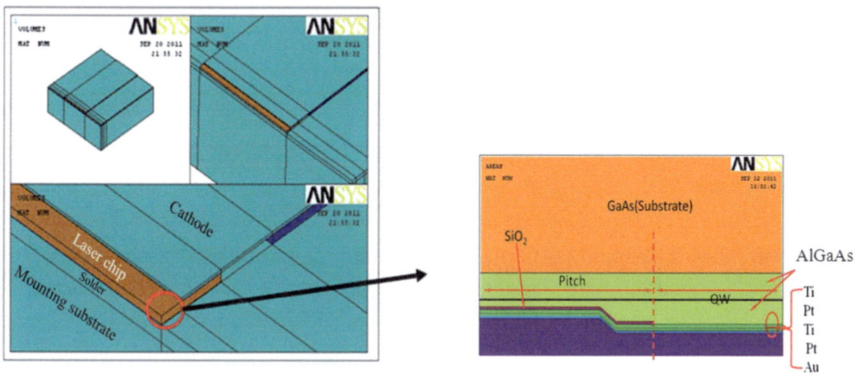

Fig. 3.10 An example of a solid model of a CS-packaged diode laser [4]

Among the steps in conducting FEM analysis for thermal simulation, the three steps highlighted in yellow shown in Fig. 3.9 are especially important. These steps are described below in detail.

1. Solid model
   In the step of solid model, the accurate structure of a semiconductor laser device needs to be built, in which every layer of a semiconductor laser structure is constructed. The size, material, and order of each layer must be in accordance with the actual laser device. It is worthwhile to address that the chip layer structure model should be consistent with the actual structure of the laser chip as much as possible.

   Figure 3.10 shows an example of a 3D solid model of a CS-packaged diode laser built in Ansys software [4]. The structure of CS device is divided into

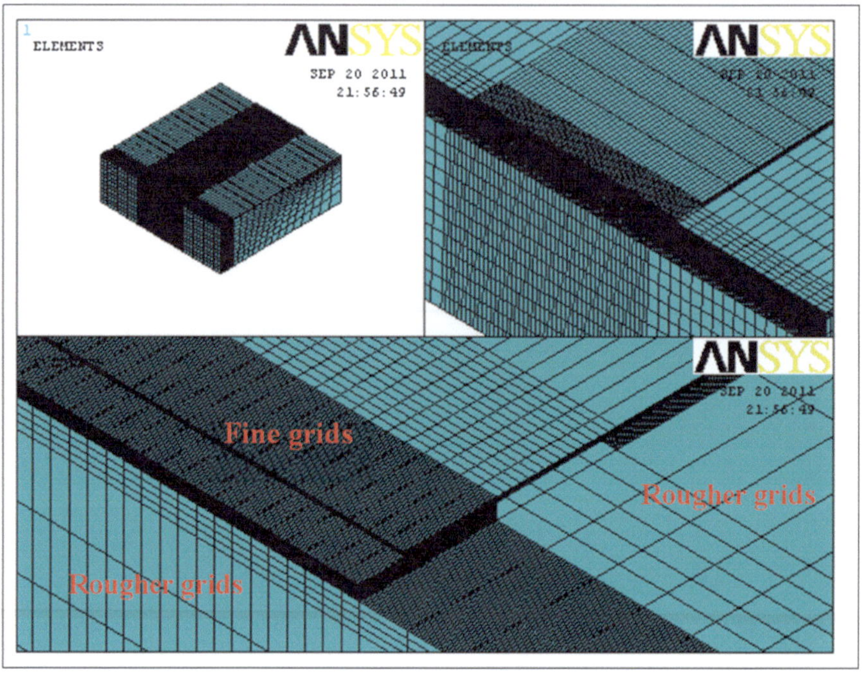

Fig. 3.11 An example of a meshed solid model of a CS-packaged diode laser [4]

several parts. The main parts include a cathode, a laser chip, solder layers, and mounting substrate. The whole structure and zoom-in structure are shown in Fig. 3.10. The most critical part is the laser chip, and the precise thermal stimulation of laser device is depended on the fine solid model of the laser chip which is shown on the right of Fig. 3.10.

2. Meshing
   Once the solid model of the semiconductor laser is built, the next step is to divide the solid model into many small grids. This step is called meshing. In the meshing process, finer grids must be constructed to accurately describe the fine structure at the critical parts, such as the laser bar and the solder layer, while rougher grids can be applied at the less critical locations, such as the mounting substrate and electrode.

   The solid model shown in Fig. 3.10 is meshed and presented in Fig. 3.11 [4]. As shown in the figure, the laser bar including the quantum well layer, cladding layers and metallization layers, and the solder layer are meshed with finer grids, while the other parts of the solid model are meshed with rougher grids.

3. Boundary conditions
   The boundary conditions include the heat generation of the diode laser and the environment temperature of the laser device. During the FEM analysis, the boundary conditions are very important to achieve good simulation which is consistent with the actual thermal distribution of the device. The boundary conditions include the initial temperature distribution and the heat flow density of the stimulated object. The boundary of temperature distribution is the temperature distribution on the surface of the object at the initial state and it can be expressed as

$$\begin{aligned} T|_{\Gamma} &= T_w \\ T|_{\Gamma} &= f(x, y, t), \end{aligned} \tag{3.17}$$

   where Γ is the surface of the object, T_w is the temperature constant of the surface, and f(x,y,t) is temperature function.

   The boundary of heat flow density condition is the heat flow density on the surface of the object and it can be calculated by [18]

$$-k\frac{\partial T}{\partial n}\bigg|_{\Gamma} = q, \tag{3.18}$$

   or

$$-k\frac{\partial T}{\partial n}\bigg|_{\Gamma} = g(x, y, t), \tag{3.19}$$

   where Γ is the surface of the object, q is the heat flow density of the surface, g(x, y, t) is the heat density function.

3.3.2 Thermal Design and Analysis of Single Emitter Semiconductor Lasers

1. Thermal modeling and design
   As introduced in Chap. 2, C-mount package is one of the typical single emitter laser packages available commercially. The laser chip is epi-down mounted onto a Cu or CuW substrate which is the p-side connection. A fly-lead is attached to the substrate as the n-side connection. The n-side of the chip is wire-bonded to the fly-lead. When the laser is operated, the heat is primarily conducted to the heat sink vertically, and then horizontally conducted to the cooler by heat sink. This detoured thermal path is not effective in dissipating heat generated from the chip and can lead to higher junction temperature rise of the device, and further influence the device performance.

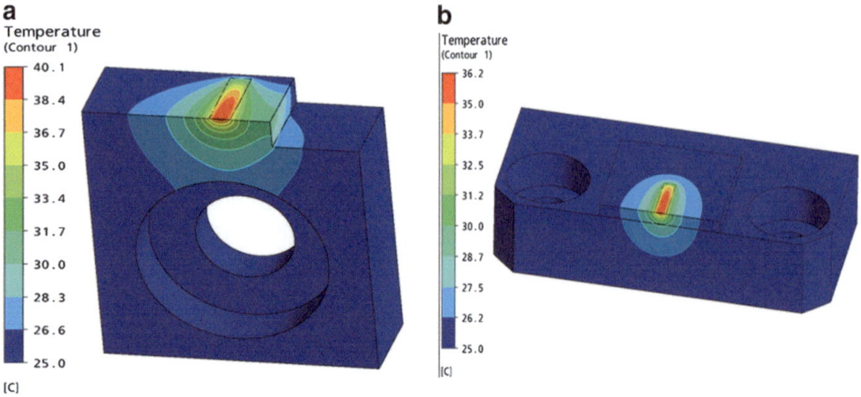

Fig. 3.12 Simulation of temperature distributions of C-mount and F-mount lasers in operation: (**a**) C-mount laser, (**b**) F-mount laser [20]

F-mount is another package style for single emitter semiconductor lasers [19, 20]. The flip-chip is mounted onto ceramic mounting substrate (e.g., AlN). The n-side is wire-bonded to the substrate placed on top of the cooler. When the laser is operated, the heat generated from chip can be easily dissipated downward to the heat sink through the mounting substrate. The straight-down thermal path of the F-mount laser package can improve the thermal dissipation efficiency significantly.

The thermal behavior is studied using FEM under the condition of 5 W output power and 50 % conversion efficiency at 25°C. Through the thermal analysis, the temperature distribution is obtained, as shown in Fig. 3.12 [20]. The analysis indicates the maximum temperature of F-mount structure is 36.2°C, while that of the C-mount structure is 40.1°C. According to Eq. (3.2), the thermal resistances of F-mount and C-mount laser are 2.24 K/W and 3.02 K/W, respectively, which suggest F-mount has better thermal management than C-mount [20].

2. Thermal analysis

The effects of thermal behavior on the performance of F-mount and C-mount semiconductor lasers are discussed in this portion.

LIV characteristics: Fig. 3.13a, b shows two typical LIV curves of C-mount and F-mount [20]. By calculation, the average slope efficiency of F-mount is about 1.22 W/A, compared to 1.19 W/A of C-mount. The average conversion efficiency of F-mount is 53.8 %, while that of the C-mount is 52.1 %.

Generally, the reliable output power P_{rel} is $P_{rel} = (0.3–0.5)\ P_{max}$. The maximum output power P_{max} is an indication of the diode laser design and its performance. Figure 3.14a shows the thermal rollover of the F-mount and C-mount lasers as a function of CW/QCW current. The lasers are 200 μm in stripe width and 2 mm in cavity length [20]. Under the CW condition the maximum power of F-mount reaches 12.6 W at 808 nm, while the output power only reaches 10.9 W for C-mount.

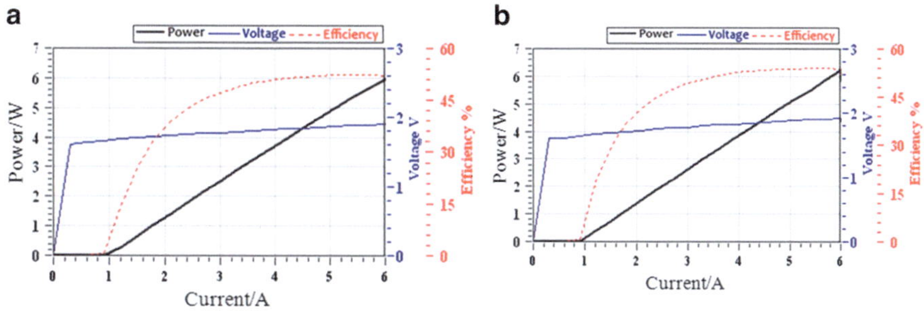

Fig. 3.13 LIV curves of C-mount and F-mount-packaged single emitter semiconductor lasers: (**a**) C-mount laser, (**b**) F-mount laser [20]

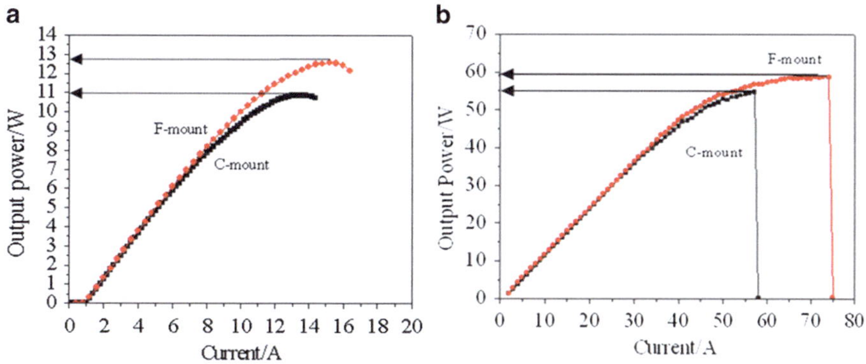

Fig. 3.14 Output power of 808 nm semiconductor lasers versus the operating current; (**a**) CW mode, (**b**) QCW mode [20]

As shown in Fig. 3.14a, b, both the maximum output power of F-mount in CW and QCW operation are much higher than that of C-mount [20]. The experimental results demonstrate that the maximum output power level is significantly affected by the package structure.

Thermal resistance: Using the relationship in Eq. (3.2), the thermal resistance can be calculated by

$$R_{th} = \left(\frac{\Delta\lambda}{\Delta T}\right)^{-1}\left(\frac{\Delta\lambda}{\Delta Q}\right) \tag{3.20}$$

where Q is generated heat, $\Delta\lambda/\Delta T$ is the wavelength temperature coefficient, and $\Delta\lambda/\Delta Q$ is the wavelength heat coefficient. Typically, for GaAs-based 808 nm laser devices, the wavelength shifts longer wavelength with 0.28 nm/K as the junction temperature increases. The wavelength heat coefficient can be obtained by measurement data. As shown in Fig. 3.15, the wavelength heat coefficient for F-mount laser is 0.832 nm/W and is 1.19 nm/W for C-mount laser [14, 20].

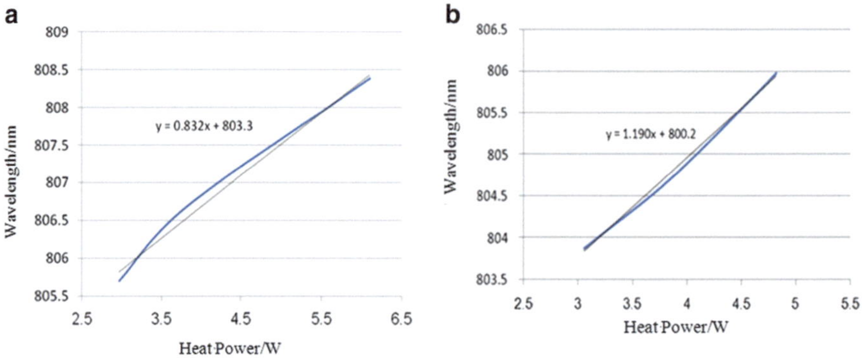

Fig. 3.15 Wavelength shift as a function of heat power; (**a**) F-mount laser, (**b**) C-mount laser [20]

Fig. 3.16 The wavelength difference between F-mount and C-mount in CW and QCW modes [20]

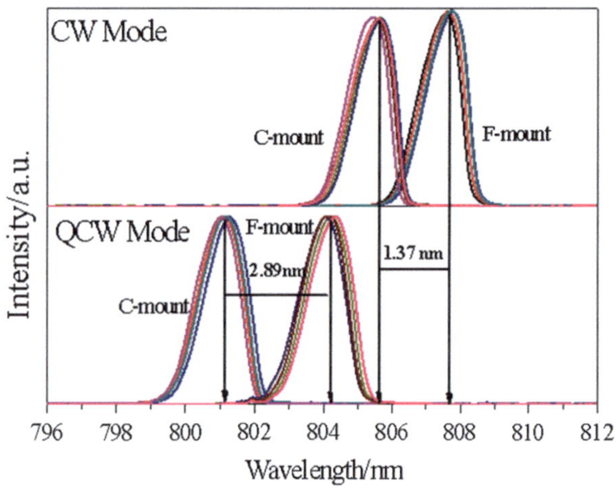

From Eq. (3.20), one can calculate that the thermal resistance of F-mount laser is only 2.97 K/W, while that of the C-mount is 4.25 K/W. These results further confirm that F-mount has better thermal management than C-mount. It should be noted that the measurement results of thermal resistance are larger than that of the simulated results. This is mainly because the simulation is conducted assuming perfect solder layer without any voids. But in practice, it is hard to completely avoid voids in the solder layer.

Wavelength: In order to compare the influence of different package structures on the wavelength of the laser, the spectral performance of 5 F-mount and 5 C-mount lasers using the neighbor chips from a same wafer have been tested under the same condition and the results are shown in Fig. 3.16 [20].

As shown in Fig. 3.16, the wavelength of C-mount is shorter than that of F-mount in both CW and QCW modes. There is a wavelength difference of ~1.37 nm between the two different package structures under CW mode and a

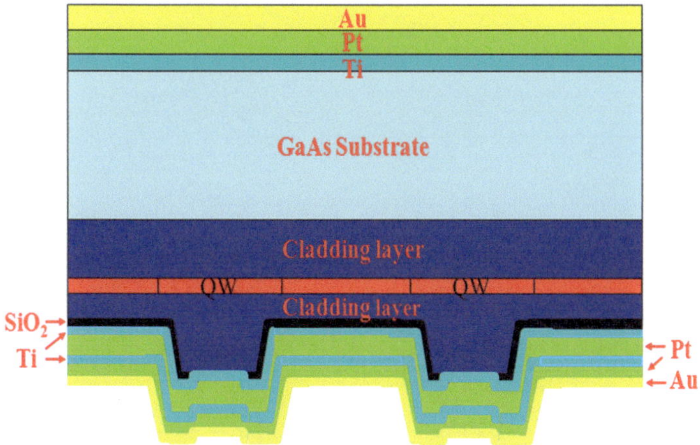

Fig. 3.17 Schematic diagram of a semiconductor laser array [21]

difference of 2.89 nm under QCW mode. It is well known that the wavelength depends directly on the bandgap of the laser's active region, which is influenced by temperature and thermal stress (due to the CTE mismatch between the laser chip and mounting substrate). When lasers are operated in QCW mode with low duty cycle and short pulse width there is nearly no heat generated on laser chip during operation, so the wavelength difference is not influenced by the heat accumulation. On the other hand, when lasers are operated in CW mode there is a great deal of heat generated, the wavelengths of both types of lasers will red shift due to the thermal effect, because the heat dissipation efficiency of C-mount is lower than that of F-mount, there is more red shift in C-mount.

3.3.3 Thermal Design and Analysis of Conduction-Cooled Semiconductor Laser Bars

Figure 3.17 shows the typical detailed schematic structure of the laser bar, which takes a laser array containing two emitters as an example [21]. The quantum well is the active region with 19 emitters in it. In this section, the modeling and analysis is based on this laser bar structure.

Thermal Modeling and Behavior of CS Package

1. Steady-state thermal behavior
   The basic structure of conduction-cooled semiconductor laser bar has been shown in Chap. 2. The steady thermal behavior of a CS-packaged semiconductor laser in CW mode was simulated using FEM. The 19 emitters in the quantum well are heat producers of the device. The output power of the device operated in

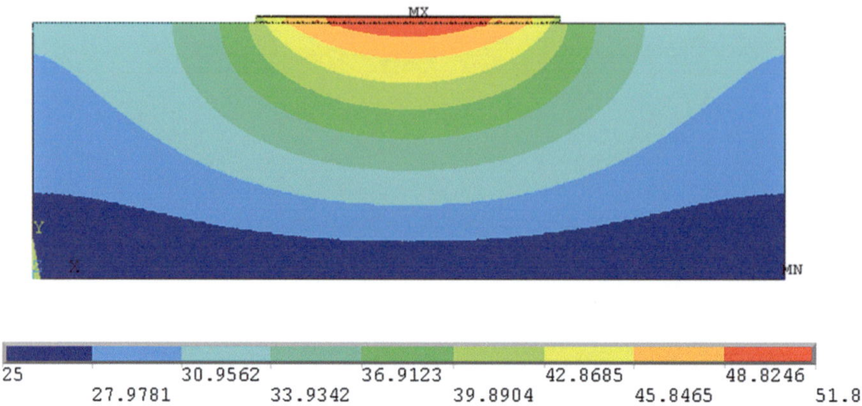

Fig. 3.18 The simulated steady-state temperature contour of a 60 W CS diode laser bar [4]

Fig. 3.19 Transverse temperature profile of quantum wells at steady state [22]

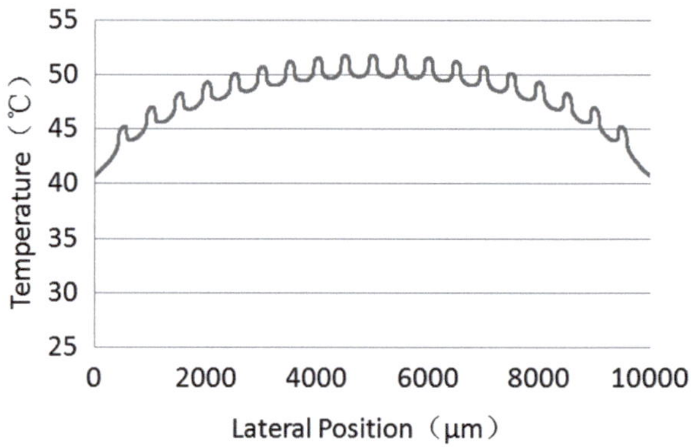

continuous wave mode is 60 W. The bottom side of the device is kept at 25°C. The result of the steady-state thermal simulation is shown in Fig. 3.18 [4]. From Fig. 3.18, one can see that the heat is primarily generated inside the active region. When the output power of the device operated in continuous wave mode is 60 W, the peak temperature of the active region is 51.8°C.

As shown in Fig. 3.19, the transverse temperature profile of quantum wells in steady state is basically parabola-shaped distributed [22]. At emitter regions, the heat flow is constrained in a small area and thus the temperature rises per unit length in these areas are higher than in the others. The local temperature rise in center emitters is greater than that in the side ones due to their longer heat transfer paths [22].

Based on the simulation result, the temperature rise in the device, ΔT, is 26.8°C (51.8°C minus 25°C). For a typical 808 nm semiconductor laser, the conversion efficiency is about 50 %. In the simulation, 50 % conversion efficiency is assumed. Therefore, the heat generation power of a 60 W (the optical power) 808 nm semiconductor laser, ΔQ, is 60 W. According to Eq. (3.2), the thermal resistance in the device is about 0.472 K/W [22].

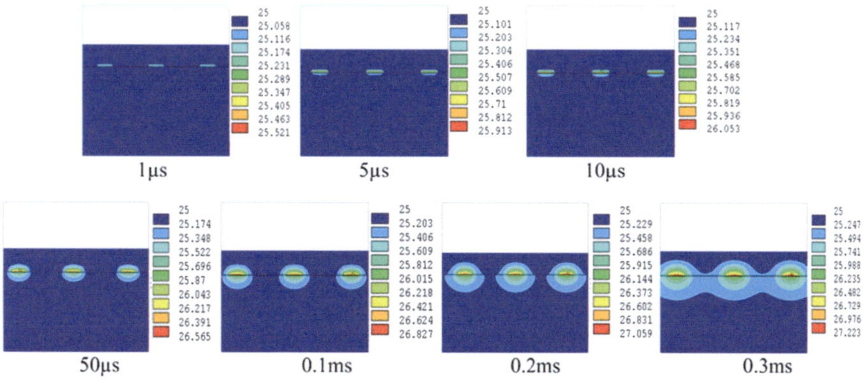

Fig. 3.20 FEA results illustrating of heat spreading during transient heating from 1 to 300 μs for a CS-packaged diode laser bar [4]

2. Transient thermal behavior
   The transient change of the temperature distribution of three emitters in the laser bar is shown in Fig. 3.20 [4]. As shown in the figure, seven snapshots of temperature distribution in time range from 1 to 300 μs were taken. The heat spread from active region to heat sink with time evolution. The boundary condition is the same as that in the steady-state thermal analysis. At the time between 200 and 300 μs, the heat spread cross to the adjacent emitters and thermal crosstalk between emitters happens. The temperature distribution of each emitter is affected by two parts when thermal crosstalk happens. One part is the heat generated by the emitter itself and the other part is the overlap of crossing heat transferred from adjacent emitters.

   The transient thermal behavior of the CS-packaged diode laser is shown in Fig. 3.21 [22]. As can be seen, it takes about 280 ns for the heat generated inside the active region to be transmitted to the upper side of the solder layer. Furthermore, it takes about 900 ns to transmit to the lower side of the solder layer. The simulation results suggest that the solder voids would not have any effect on the temperature distribution in the quantum well when the pulse width is shorter than 280 ns. For pulse longer than 280 ns, solder voids would gradually have some effect on the temperature distribution in the quantum well. For pulses longer than 900 ns, solder voids certainly would have obvious effect on the thermal performance of the laser device and the temperature distribution in the junction.

Thermal Analysis and Optimization of CS Package

1. Selection of operation mode
   It was reported that for a typical laser, its reliability is only good for controlled condition, for example, on short pulse QCW operation or CW operation [23]. When it works on long pulse QCW operation, the reliability reduces rapidly.

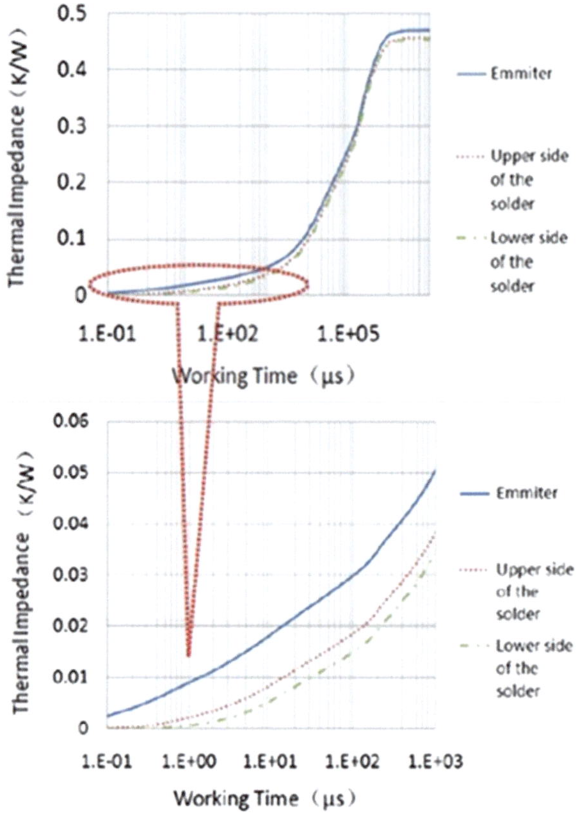

Fig. 3.21 Thermal impedance as a function of time of a CS-packaged semiconductor laser without voids in solder layer [22]

The thermal simulation results can explain this observation. When a semiconductor laser works in pretty short pulse QCW mode (pulse width $\tau < 280$ ns), there is not enough time for the heat to transmit to the solder layer and there is no temperature variation impact to the solder layer. That is, the solder layer does not "feel" the pulses and the temperature variation the pulses lead to. Therefore, the solder layer does not experience temperature cycling. When the device works in longer pulse width between 280 and 900 ns QCW mode, the heat is transmitted past the upper side of the solder layer and inside the solder layer, but not to the bottom side of the solder layer yet. This, accordingly, would cause large temperature gradient in the solder layer and could lead to serious electromigration, especially under high driving current [24]. However, for this range of pulse width, the solder layer barely "feels" the pulses and the temperature variation the pulses lead to. Temperature cycling on the solder layer is not severe and thus solder thermal fatigue is not of great concern. When the device works in even longer pulse QCW mode (pulse width $\tau > 900$ ns), temperature distribution in the solder layer has already been steady within a single pulse. The temperature gradient in the solder layer is stabilized. When the driving current is high, it could generate large temperature gradient in the solder layer. For pulse width longer

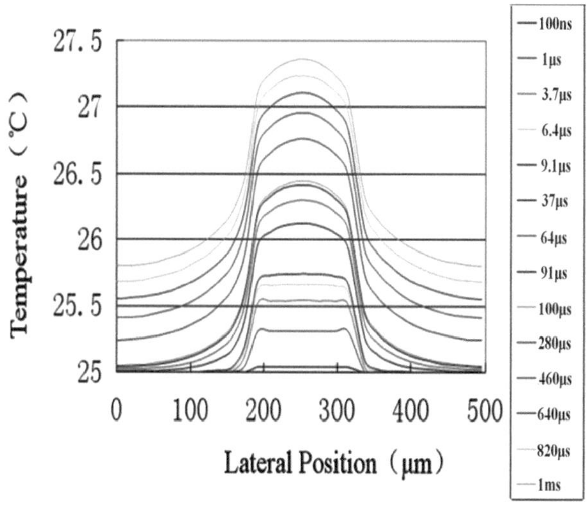

Fig. 3.22 Lateral temperature profiles of a single emitter (150 μm in width) [22]

than 900 ns, the solder layer completely "feels" the pulses and the temperature variation the pulses lead to. Temperature cycling on the solder layer is obvious. At this condition, both thermal fatigue and electromigration are of concern, especially under high driving current. It can be concluded that under the same driving current, the solder joint induced failure of a typical semiconductor laser is lower when the laser works in short pulse QCW mode and CW mode, while it is higher when the laser works in long pulse QCW mode [22].

Figure 3.22 gives the typical lateral temperature profiles of one of the 19 emitters [22]. As can be seen, the profile of lateral temperature in the quantum well changes with increasing heating time before it is stabilized at the heating time of 100–280 μs. It means that the device has entered into the stabilized status of transient heat transmission [25]. At this time, the temperature difference between the center of the emitter and the center of the pitch no longer changes although the temperature continuously increases with longer heating time. As discussed previously in Fig. 3.20, when the pulse is shorter than 200 μs, the heat of an emitter would not influence the neighboring ones. Therefore, even if one of the emitters does not work, or the heat of which could not be cooled, the heat of the emitter would not impact the neighbors around it. In this way, this device would be much more reliable [22].

Based on the transient numerical simulation, the single bar CS-packaged 60 W 808 nm semiconductor laser was tested in different pulses, as shown in Fig. 3.23 [22]. The tested thermal impedance is larger than the simulation results shown in Fig. 3.21. This is mainly because the simulation was conducted with perfect solder layer. In reality, there are always some solder voids which are inevitable in actual manufacturing processes. Additionally, defects of the materials and contact thermal impedance are also neglected in simulation. However, the trend of the impedance curves obtained from experiment and simulation is similar [22].

2. Effect of voiding on temperature rise

Some voids may appear in the solder layer in the packaging process. The heat generated in the quantum well tends to accumulate in the area near the voids, and

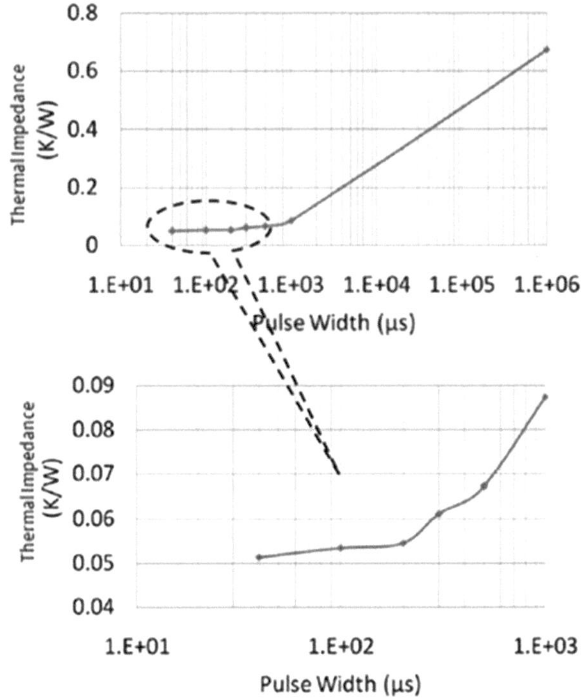

Fig. 3.23 Thermal impedance versus pulse width of a CS-packaged semiconductor laser [22]

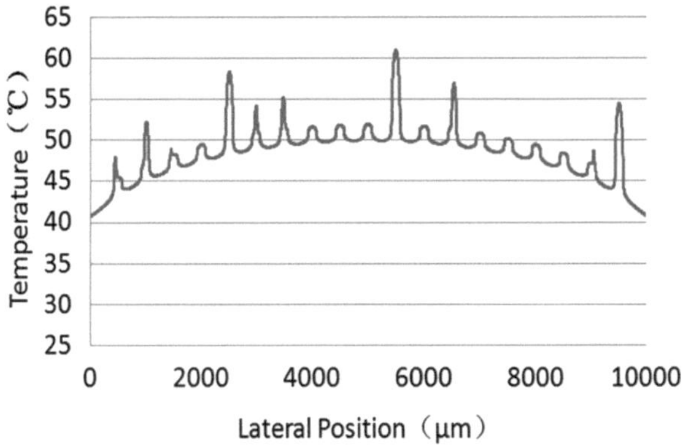

Fig. 3.24 Lateral temperature profile of quantum wells with different size voids in solder layer underneath emitters [22]

leads to local temperature rise. The temperature distribution in the quantum well may be changed significantly. As a result, the reliability and lifetime of semiconductor lasers can be reduced greatly. Additionally, the spectrum may be broadened seriously [22].

Figure 3.24 shows the lateral temperature profile of quantum wells with different size voids in solder layer underneath emitters for a diode laser bar [22]. The bar has 19 emitters in total. As can be seen, the peak temperature of the device could reach up to 61°C in the 11th emitter from left to right. The local

Table 3.1 Local temperature rise for different void [22]

Void no.	Void diameter (μm)	Local temperature rise (°C)
1	50	2.265
2	80	4.500
3	20	0.420
5	130	7.048
6	50	2.911
7	60	3.413
11	150	7.953
13	90	4.884
18	30	1.380
19	150	7.914

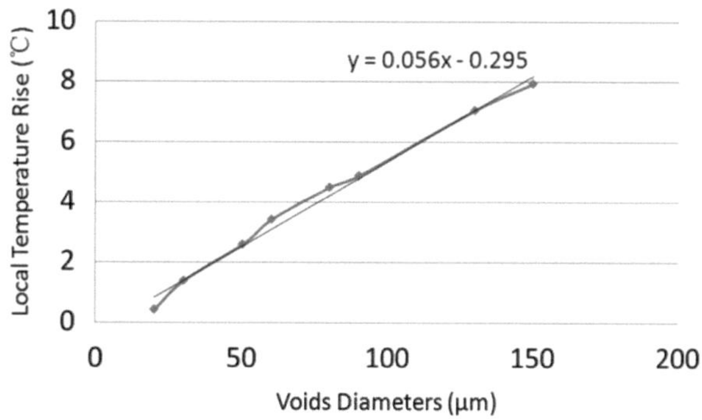

Fig. 3.25 Relationship of local temperature rise versus the diameter of solder voids [22]

temperature rise for the emitter with void of different sizes as shown in Fig. 3.24 is listed in Table 3.1 [22]. As can be clearly seen in the table, both void dimensions and location can affect local temperature rise. The larger the void is, the higher the local temperature rises. On the other hand, when the solder voids are in the same size, if the void is closer to the center of the bar, the local temperature rise will be higher [22].

The influence of void size on local temperature is quantitatively analyzed. Figure 3.25 gives the relationship of local temperature rise versus the diameter of solder voids [22]. The local temperature rise increases linearly with void size, and the local temperature rise rate versus the diameter of the void is 0.056°C/μm [22].

Thermal Modeling and Behavior of HCS Package

A typical sample of a high power hard soldered conduction-cooled semiconductor laser bar and its configuration is shown in Fig. 3.26 [21]. This packaged laser bar consists of four parts: the cathode, the laser bar, the mounting substrate, and the mounting heat sink. The bar is bonded to a CuW submount with AuSn solder first,

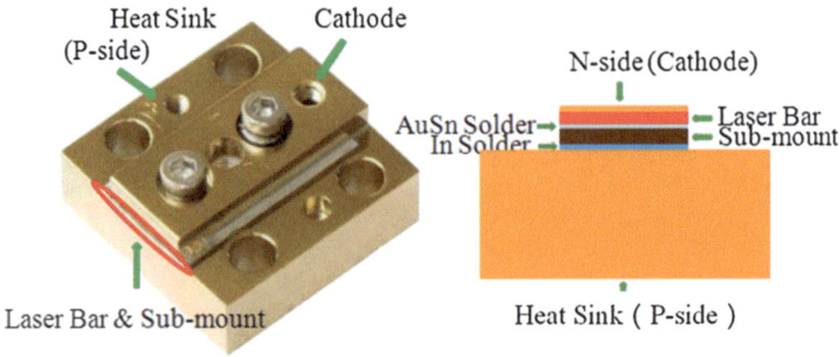

Fig. 3.26 A hard soldered conduction-cooled (HCS) 808 nm semiconductor laser package and the schematic diagram of the front view of device [21]

Fig. 3.27 Temperature contour of HCS package [4]

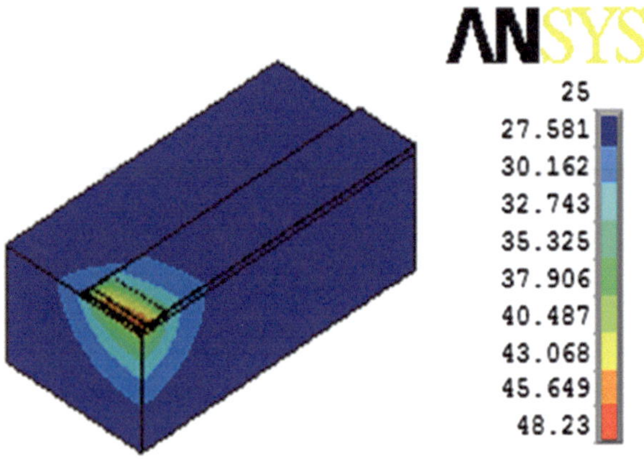

and then the CuW mounting substrate is attached to the copper heat sink using indium solder or other solder such as SnAgCu.

The steady state thermal characteristics are based on a hard soldered conduction cooled 40 W 808 nm semiconductor laser package with 30 % filling factor in CW mode. The electrical-optical conversion efficiency is assumed to be 50 %. In other words, if the output power of a device containing 19 emitters is 40 W, the heat generation of each emitter is 2.1 W. The bottom of mounting heat sink with a fixed temperature of 25°C is assumed. In addition, in order to reduce the computational time and to improve the analysis efficiency, half of package structure is used in the modeling due to the symmetrical design of the device. The results of the steady-state thermal simulation are shown in Fig. 3.27 [4]. From Fig. 3.27, one can see that the heat is primarily generated inside the active region. The peak temperature of the active region is 48.23°C, and the thermal resistance of the device is about 0.581 K/W by calculation.

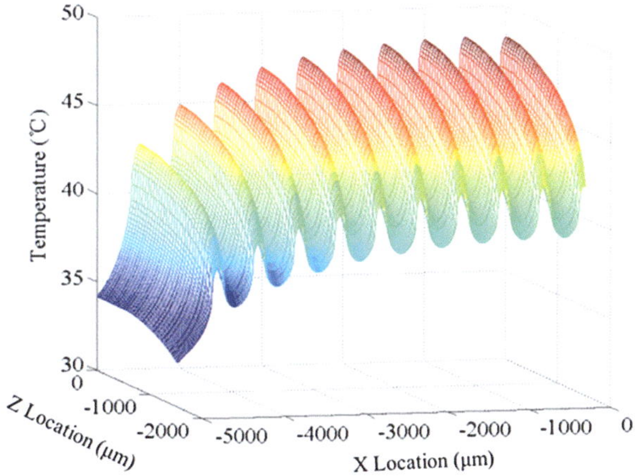

Fig. 3.28 Temperature profile of the active region of HCS package [21]

Figure 3.28 depicts the steady-state temperature distribution of quantum well in the laser bar [21]. For edge-emitting semiconductor laser array, the peak temperature of the active region is located at the central emitter of the semiconductor laser array and close to the front facet. The temperature of front facet of each emitter is higher than that of rear facet of each emitter. The lateral cycle variations of the temperature in the active region are consistent with the lateral pitch of the active region. Corresponding to each emitter in the quantum well, there is a sudden change in the temperature.

By extracting the x and y components of thermal flux of an emitter and its adjacent pitches in the quantum wells from the simulation results, the heat flow situation can be obtained. Figure 3.29 shows the x and y components of thermal flux as a function of lateral position [21]. The sign of value in the figure represents the direction of thermal flux in Fig. 3.29. In the emitter area, the x-component (parallel to the PN junction) of thermal flux approximately follows a tangent function law, and the y-component (perpendicular to the PN junction) of thermal flux roughly follows a quadratic function law. At the center of the emitter, the x-component of thermal flux is zero and the y-component absolute values of thermal flux reach the maximum. At the intersection of emitter and pitches, the x-component reaches the maximum value, while the y-component drops drastically and its absolute value almost reaches the minimum. Furthermore, it is also noticed that, in the pitch regions, the x-component follows a cotangent function law approximately and the y-component absolute values of thermal flux are low compared to that at the emitter regions. It is about one seventh of that in the emitter regions.

The vertical temperature profile of the device at steady state is shown in Fig. 3.30 [21]. The copper heat sink, the indium solder layer, and the CuW mounting substrate each contributes about 62.7 %, 1.72 %, and 31.51 %, respectively to the total effective thermal resistance of the device. This fact highlights that the CuW mounting substrate also plays a quite important role in the total effective thermal resistance of the device, besides the copper heat sink.

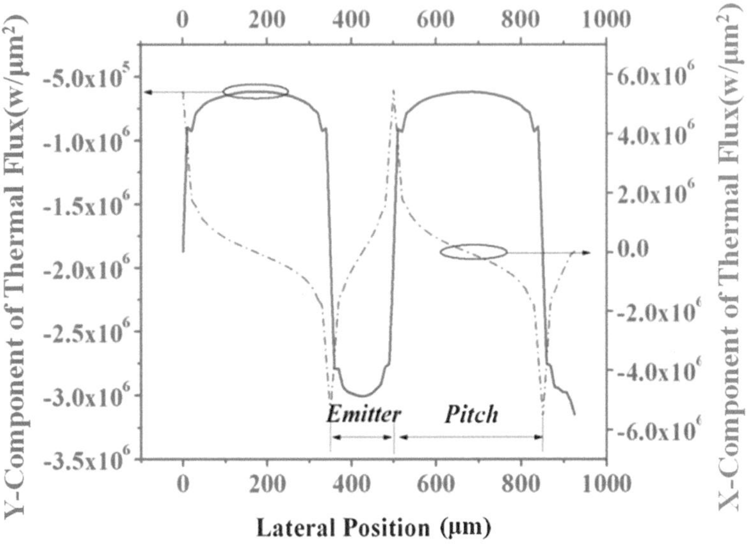

Fig. 3.29 x and y components of thermal flux in the quantum wells at emitter and adjacent region [21]

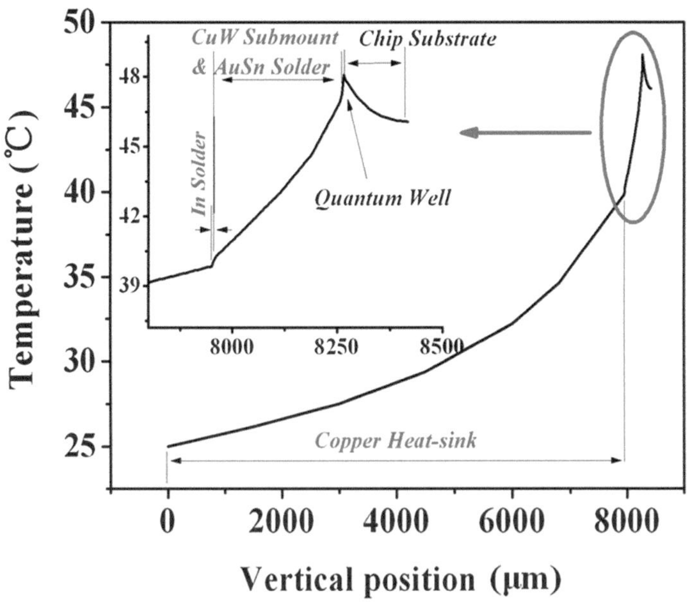

Fig. 3.30 The vertical temperature profile of a hard soldered conduction-cooled 40 W 808 nm semiconductor laser package [21]

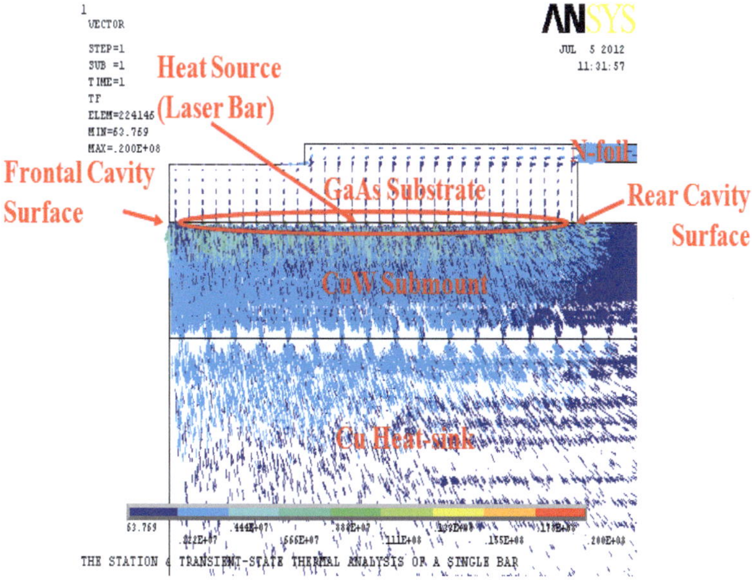

Fig. 3.31 The side-view of vector plot of thermal flux distribution of a hard soldered conduction-cooled 40 W 808 nm semiconductor laser package [21]

The side-view of vector plot of thermal flux distribution of the device is shown in Fig. 3.31 [21]. The result indicates that a large amount of heat generated in the laser bar is mainly dissipated through p-side, flowing through the AuSn solder layer, the CuW submount, the indium solder layer, and then into copper heat sink. According to the direction of thermal flux vector, it is obvious that the heat dissipation capability at the area near the front facet of resonant cavity is worse compared to that of the area near the rear facet of resonant cavity. This is mainly because the bar is attached to the edge of the heat sink and there is no heat flow path at the front side.

Thermal Analysis and Optimization of HCS Package

1. Selection of mounting substrate material
   Since the thermal resistance of the mounting substrate contributes significantly to the total effective thermal resistance, as elaborated above, the natural approach of improving thermal performance of HCS package is through selection of the mounting substrate material. Analysis as shown in Fig. 3.32 demonstrates that the steady-state temperature of device is obviously lower when the material of mounting substrate is copper-diamond composite material, instead of CuW [21]. By calculations, the thermal resistance could be reduced by about 30 % with the use of Cu-diamond mounting substrate.

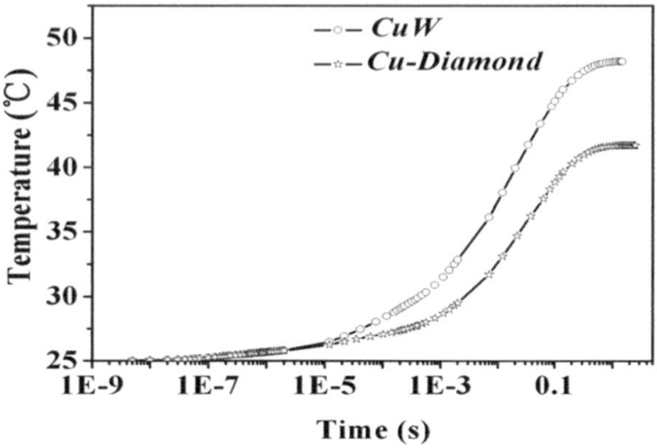

Fig. 3.32 Peak temperature of active region as a function of heating time of HCS package with different mounting substrate materials [21]

2. Optimization of packaging structure
   Different materials have different angles of thermal divergence due to the difference in the thermal conductivities of the materials. The relation of the divergent angles θ_1 of thermal flux to the thermal conductivities of materials is given by [26]

$$\theta_1 = 90\tanh\left\{0.355(\pi k/180)^{0.6}\right\} \tag{3.21}$$

Figure 3.33 shows the optimization of the structure design of the heat sink [21]. Instead of using a straight cut Cu heat sink, a tapered Cu heat sink is used. In this way, the diode laser bar submount is aligned to the intersection line of the tapered surface and the bonding surface and heat can be dissipated not only downwards but also to the front side. The diode laser bar is offset a distance d from the front edge of the heat sink. The taper angle θ depends on the fast axis divergence angle.

The relation of the divergent angles of thermal flux θ_1 to the taper angle θ is given by

$$\theta_1 + \theta = \frac{\pi}{2}, \tag{3.22}$$

Figure 3.34 shows the relationship between the offset distance d and the thermal resistance, by keeping the heat sink taper angle at 50° [21]. The thermal resistance of the device decreases gradually as the offset distance d increases. At d = 1,000 μm, the thermal resistance of the device comes down to 0.506 K/W from 0.58 K/W of the traditional aligned packaging structure design, which is a 12.9 % reduction in the thermal resistance.

On the other hand, by keeping d = 1,000 μm, as the thermal resistance as a function of the taper angle is shown in Fig. 3.35 [21]. The thermal resistance decreases slightly with the taper angle changing from 50 to 25°. When the taper angle is less than 25°, the thermal resistance is not sensitive to the taper angle.

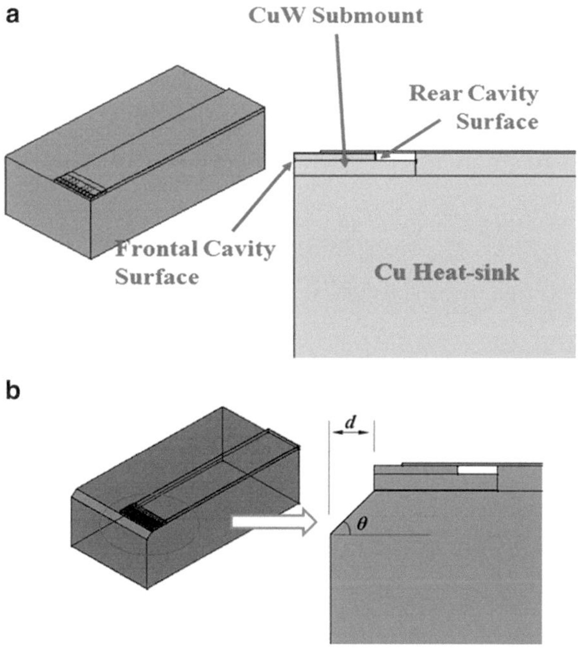

Fig. 3.33 Optimization of the HCS package structure [21]. (**a**) The initial packaging structure. (**b**) The optimized packaging structure

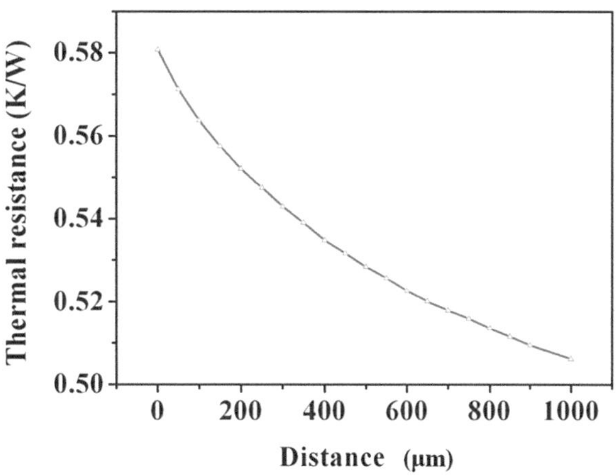

Fig. 3.34 Relation of the backward distance to the thermal resistance [21]

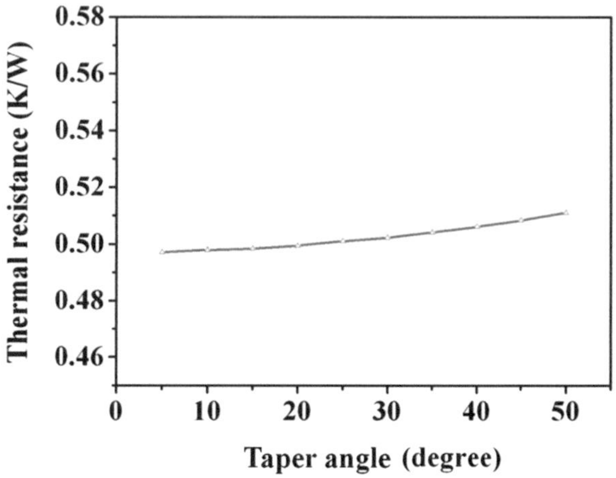

Fig. 3.35 Relation of the taper angle to the thermal resistance [21]

3.4 Thermal Management Techniques

3.4.1 Double-Sided Cooling

The cooling from the bottom side is the most common structure for a high power semiconductor laser. In order to improve the heat dissipation efficiency, double-sided cooling has been proposed. Figure 3.36 shows the traditional single-sided package structure and the double-sided packaging structure [27]. In the single-sided structure as shown in Fig. 3.36a, the heat generated from the laser bar is one-sided conducted to the heat sink with a low cooling efficiency, while in the double-sided structure in Fig. 3.36b, the heat can be conducted through both the anode and cathode. Therefore, the thermal dissipation efficiency is improved significantly [27].

The thermal behavior of double-sided cooling packaging has been studied using finite element analysis, as shown in Fig. 3.37 [27]. The simulation results indicate that the heat is not only conducted to the heat sink from bottom directly but also conducted by the cathode, which improves the cooling efficiency considerably. For the double-sided cooling packaging design, the heat dissipation efficiency from the cathode can be up to 20 %.

Packaged with double-side cooler, the output power of a single semiconductor laser bar can reach 150 W when the drive current is 144 A. Figure 3.38 shows the LIV curve of the single bar under the condition of CW current measured at room temperature [27]. The Full Width at Half Maximum (FWHM) of the single bar shown in Fig. 3.39 is only 2.79 nm and 90 % energy width is 4.07 nm [27].

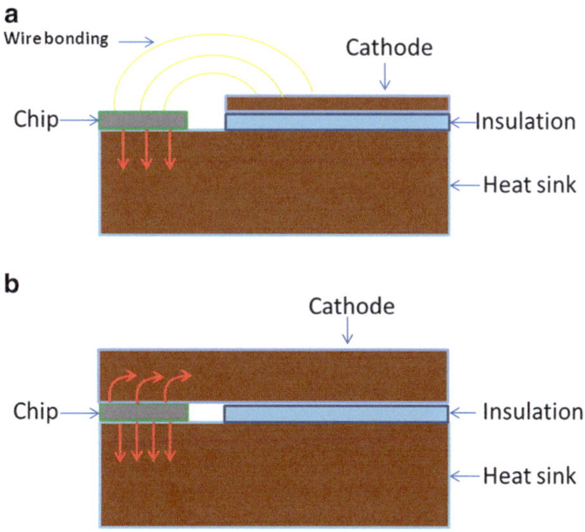

Fig. 3.36 The traditional (**a**) and double-sided (**b**) cooling package structure [27]

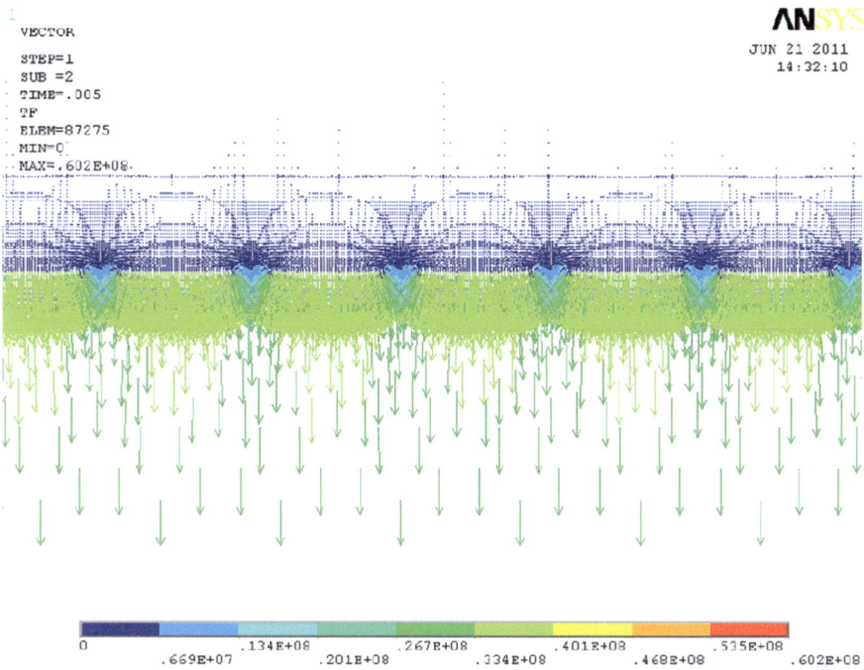

Fig. 3.37 Thermal flux vector graph of a double-sided-cooled high power semiconductor laser [27]

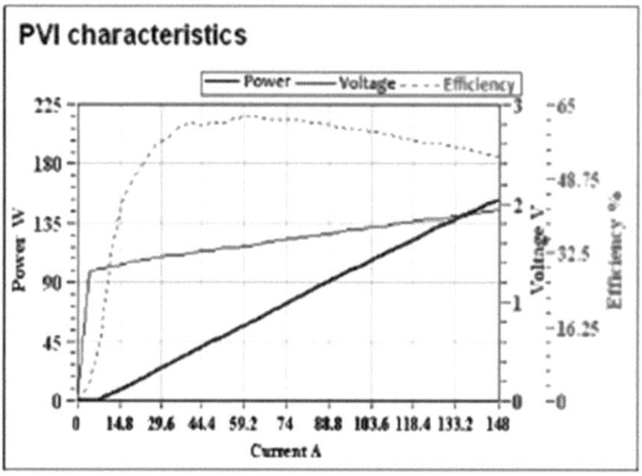

Pop(W)	150.00
Iop(A)	144.22
Ith(A)	8.20
Slope Eff.(W/A)	1.13
Eff.@Iop(%)	54.14
Max Eff.(%)	62.25
Rs(mOhms)	3.82
V @Iop(V)	1.92
TC	25.0

Fig. 3.38 LIV curve of single bar 976 nm CW Micro-channel water-cooled semiconductor laser [27]

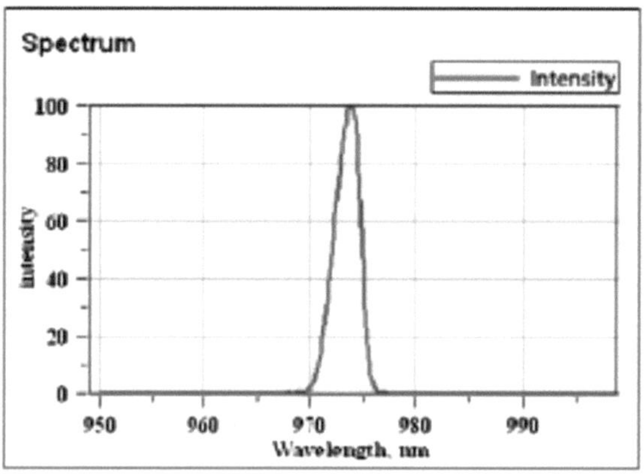

Peak wavelength(nm)	973.98
Centroid wavelength(nm)	973.93
FWHM(nm)	2.79
FW90% Energy(nm)	4.07

Fig. 3.39 Spectrum character of single bar 976 nm CW Micro-channel water-cooled semiconductor laser [27]

3.4.2 *Macro-channel Cooling*

A macro-channel-cooled (MaCC) hard soldered high power semiconductor laser package has been developed, as shown in Fig. 3.16 of Chap. 2. Compared with conventional MCC, the cost of MaCC is lower and the damage from electrochemical erosion is less severe [28, 29]. Moreover, the maintenance is much easier.

The transient thermal behavior of high power semiconductor laser arrays packaged using MaCC under QCW operation mode has been studied using finite element analysis. The structure of the single bar MaCC-packaged 250 W QCW semiconductor laser array is shown in Fig. 3.40 [30]. The transient thermal behavior of the package with output power of 250 W is presented in Fig. 3.41 [31].

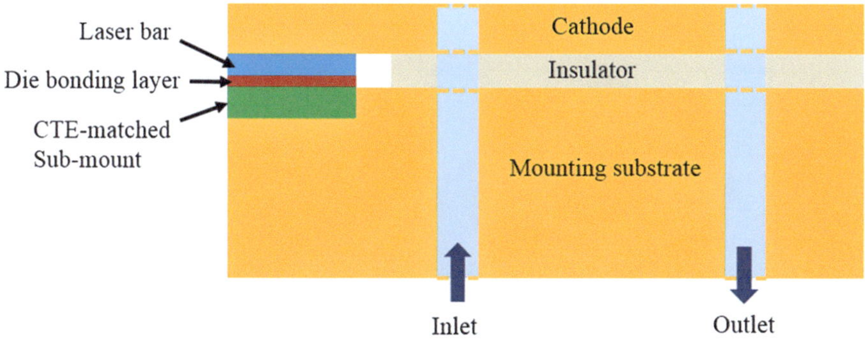

Fig. 3.40 Structure of a MaCC package of high power semiconductor laser bar [30]

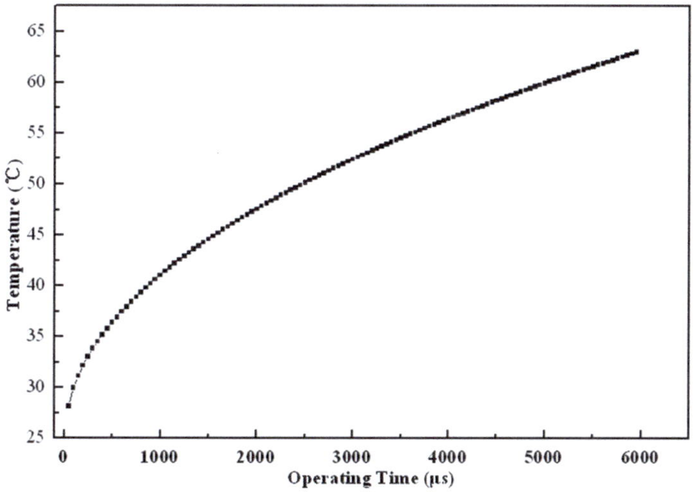

Fig. 3.41 Transient thermal behavior of 250 W MaCC-packaged high power diode laser [31]

The junction temperature increases rapidly initially. With the increase of operating time, the junction temperature rise slows down. The overall trend of junction temperature curve vs time is parabolic. Figure 3.42 shows the temperature contour of the package after temperature is stabilized when the device is operated at 250 W at 8 % duty cycle (200 μs, 400 Hz) [31]. The peak temperature is nearly 50°C in active region.

The LI curve of the QCW MaCC single bar is shown in Fig. 3.43 [31]. A power of 284 W and slope efficiency of 1.28 W/A at 250 A are obtained at 8 % duty cycle. The FWHM and full width at 90 % energy (FW90%E) of spectrum are 2.98 nm and 5.34 nm, respectively. These tested results are very close to that of copper micro-channel-cooled package [32].

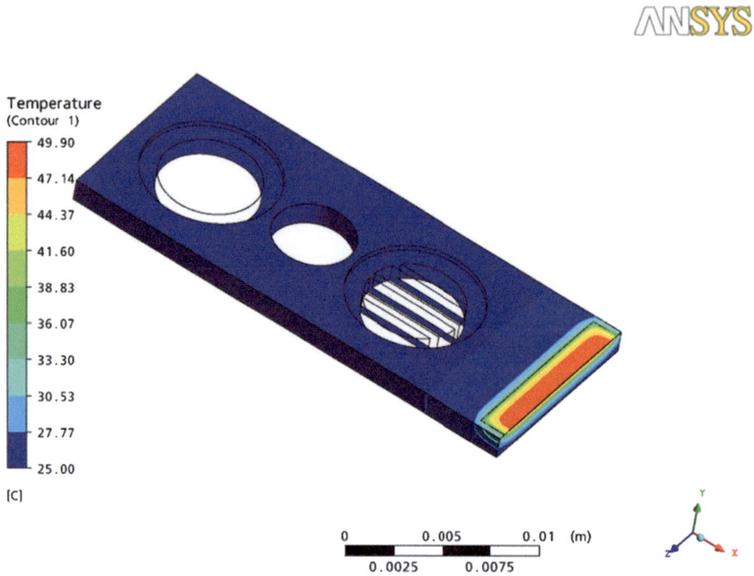

Fig. 3.42 Stabilized temperature contour of a single bar MaCC-packaged high power diode laser operated at 250 W QCW and 8 % duty cycle [31]

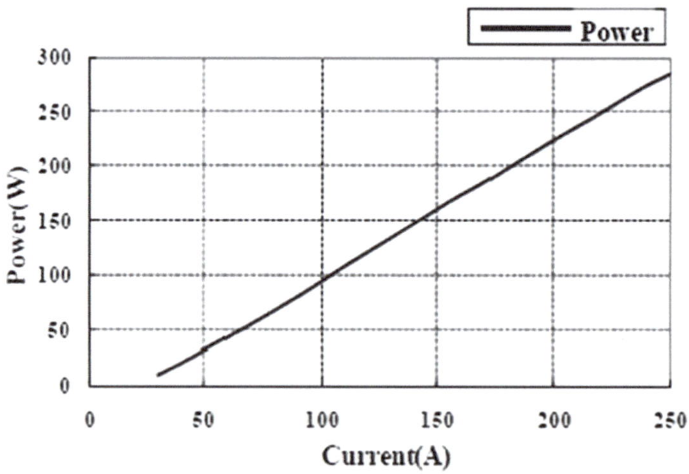

Fig. 3.43 LI curves of a QCW MaCC single bar [31]

3.4.3 Advanced Packaging Materials

To obtain good heat dissipation for semiconductor lasers, high thermal conductivity materials should be used as the material of heat sink. Due to the high thermal conductivity of ~400 W/(m∗K), copper has been widely used as heat sink material for commercial products for many years. In order to further improve the heat dissipation, heat sink materials with higher thermal conductivity have been studied. Owing to its high heat conductivity of 2,000 W/(m∗K), diamond has been the promising material for making new heat sink materials. The copper and diamond compound (CuD) has been proposed and studied in recent years as a new heat sink material [33], and many kinds of diamond and copper compound materials have been produced. The thermal conductivity of CuD has been improved greatly from 420 to 700 W/(m∗K), which is much higher than that of copper heat sink.

As discussed previously, packaging structure is critical for the thermal management of high power lasers. Many factors like the Epi-side Au metallization thicknesses, the width of the laser chip, and different metallization diffusion barriers structures all influence the thermal properties of the semiconductor laser [34].

References

1. R. Diehl, *High-Power Diode Laser* (Springer, Berlin, 2000)
2. http://ssl.xmu.edu.cn/download%5CStandards%5CJEDSD%5Cjesd51-1.pdf
3. G. Bacchin, A. Fily, B. Qiu, D. Fraser, S. Robertson, V. Loyo-Maldonado, S.D. McDougall, B. Schmidt, High temperature and high peak power 808 nm QCW bars and stacks. Proc. SPIE **7583**, 75830P(1–11) (2010)
4. J.W. Wang, X.S. Liu, *Thermal Simulation of High Power Semiconductor Lasers*. (Internal Talk from Focuslight Technologies Co., Ltd., 2010), pp. 15–27
5. http://course.ee.ust.hk/elec342/notes/Lecture%2013_laser%20diodes(2).pdf
6. L.A. Coldren, S.W. Corzine, *Diode Lasers and Photonic Integrated Circuits* (Wiley, New York, 1995)
7. Y.P. Varshni, Temperature dependence of the energy gap in semiconductors. Physica **34**(1), 149–154 (1967)
8. L. Fan, C.S. Cao, G. Thaler, D. Nonnemacher, F. Lapinski, I. Ai, B. Caliva, S. Das, R. Walker, L.F. Zeng, M. McElhinney, P. Thiagarajan, Reliable high-power long-pulse 8XX-nm diode laser bars and arrays operating at high temperature. Proc. SPIE **7918**, 791805(1–7) (2011)
9. M.H. Hu, X.S. Liu, C.E. Zah, Transient and static thermal behavior of high power single-mode semiconductor lasers. Proc. SPIE **4905**, 32–36 (2002)
10. J.S. Huang, Reliability-extrapolation methodology of semiconductor laser diodes: is a quick life test feasible. IEEE Trans. Device Mater. Reliab. **6**(1), 46–51 (2006)
11. P. Pobert, Sarzala, Thermal properties of buried-heterostructure diode lasers. Int. J. Optoelectron. **8**(5/6), 705–725 (1993)
12. B. Mrozewicz, M. Bugajski, W. Nakwaski, *Physics of Semiconductor Lasers* (Elsevier, Amsterdam, 1994)
13. T. Kobayashi, Y. Furukawa, Temperature distributions in the GaAs-AlGaAs double-heterostructure laser below and above the threshold current. Jpn. J. Appl. Phys. **14**, 1981–1986 (1975)
14. W. Nakwaski, E. Kvantovaya, Spontaneous radiation transfer in heterojunction laser diodes. Sov. J. Quant. Electron **9**(12), 1544–1545 (1979)
15. B. Mrozewicz, M. Bugajski, W. Nakwaski, J. Krauze, *Physics of Semiconductor Lasers* (North-Holland, Amsterdam, 1991)
16. W. Nakwaski, Dynamical thermal properties of stripe-geometry laser diodes. Proc. IEEE Inst. Electr. Eng. **131**, 94–102 (1984)
17. http://mostreal.sk/html/guide_55/g-bas/GBASToc.htm

18. E.L. Wilson, R.E. Nickell, Application of the finite element method to heat conduction analysis. Nucl. Eng. Design **4**, 276–286 (1966)
19. Y.X. Zhang, J.W. Wang, C.H. Peng, X.N. Li, L.L. Xiong, X.S. Liu, *A New Package Structure for High Power Single Emitter Semiconductor Lasers*. 2010 11th International Conference on Electronic Packaging Technology & High Density Packaging (2011), pp. 1346–1349
20. X.N. Li, Y.X. Zhang, J.W. Wang, L.L. Xiong, P. Zhang, Z.Q. Nie, Z.F. Wang, H. Liu, X.S. Liu, Influence of package structure on the performance of the single emitter diode laser. IEEE Trans. Compon. Packag. Manuf. Technol. **2**(10), 1592–1599 (2012)
21. Z.Y. Zhang, P. Zhang, X.N. Li, L.L. Xiong, H. Liu, Z.Q. Nie, Z.F. Wang, X.S. Liu, *Thermal Modeling and Analysis of High Power Semiconductor laser Arrays*. 2012 International Conference on Electronic Packaging Technology & High Density Packaging (2012), pp. 560–566
22. Z.B. Yuan, J.W. Wang, D. Wu, X. Chen, X.S. Liu, *Study of Steady and Transient Thermal Behavior of High Power Semiconductor Lasers*. 2009 I.E. Electronic Components and Technology Conference (2009), pp. 831–836
23. D. Schleuning, M. Griffin, P. James, J. McNulty, D. Mendoza, J. Morales, D. Nabors, M. Peters, H.L. Zhou, M. Reed, Robust hard-solder packaging of conduction cooled laser diode bars. Proc. SPIE **6456**, 645604(1–11) (2007)
24. X.S. Liu, R.W. Davis, L.C. Hughes, M.H. Rasmussen, C.E. Zah, A study on the reliability of indium solder die bonding of high power semiconductor lasers. J. Appl. Phys. **100**(1), 13104 (1–11) (2006)
25. S.M. Yang, W.Q. Tao, *Heat Transfer*, 3rd edn. (Higher Education Press, Beijing, 1998), pp. 63–100
26. E. Suhir, J.W. Wang, Z.B. Yuan, X. Chen, X.S. Liu, *Modeling of Thermal Phenomena in a High Power Diode Laser Package*. Proceedings of 10th on Electronic Packaging Technology & High Density Packaging (ICEPT-HDP) (2009), pp. 837–842
27. X.N. Li, C.H. Peng, Y.X. Zhang, J.W. Wang, L.L. Xiong, et al., *A New Continuous Wave 2500 W Semiconductor Laser Vertical Stack*. 2010 11th International Conference on Electronic Packaging Technology & High Density Packaging (2010), pp. 1350–1354
28. G. Truesch et al.., Reliability of water cooled high power diode laser modules. Proc. SPIE **5711**, 132–141 (2005)
29. R. Feeler, J. Junghans, G. Kemner, E. Stephens, Next generation micro-channel coolers. Proc. SPIE **6876**, 687608(1–8) (2008)
30. H.J. Zhong, J.W. Wang, D. Wu, Y.X. Zhang, Packaging of high power semiconductor laser arrays. Inner Report **1**, 34–42 (2010)
31. J.W. Wang, Z.B. Yuan, L. Guo, L.L. Xiong, Y.X. Zhang, C.H. Peng, X.N. Li, X.S. Liu, *Packaging of High Power Semiconductor Laser Arrays Using a Novel Macro-Channel Cooler*. 2010 11th International Conference on Electronic Packaging Technology & High Density Packaging (2010), pp. 92–97
32. http://www.jenoptik.com/cms/jenoptik.nsf/res/DL_Data_Sheets_Open_Heatsink_Packages.pdf/$file/DL_Data_Sheets_Open_Heatsink_Packages.pdf
33. C. Zweben, New, low-CTE, ultrahigh-thermal-conductivity materials for lidar laser diode packaging. Proc. SPIE 58870D-(1–10) (2005)
34. X.S. Liu, K. Song, R.W. Davis, M.H. Hu, C.E. Zah, *Design and Implementation of Metallization Structures for Epi-Down Bonded High Power Semiconductor Lasers*. 2004 Electronic Components and Technology Conference (2004), pp. 798–806

Chapter 4
Thermal Stress in High Power Semiconductor Lasers

Thermal stress occurring during the packaging and operating processes influences the performance and reliability of high power semiconductor lasers. The stress is mainly caused by the coefficients of the thermal expansion (CTE) mismatch between the mounting substrate and laser chip. Ideally, packaging materials with high thermal conductivities and CTEs matching those of the semiconductor materials such as GaAs, InP, and GaN are desired in high power diode laser packaging. Thermal stress is one of the most critical problems in packaging of high-power diode lasers. In this chapter, effects of the thermal stress on the optical, electrical, and mechanical performances of the semiconductor lasers are investigated, such as wavelength, polarization, smile, and cracking. Formation of thermal stress in high power semiconductor laser is discussed, and approaches to reduce the thermal stress are proposed.

4.1 Effects of Thermal Stress on Performances of High Power Semiconductor Lasers

The thermal stress causes the strain in the semiconductor and changes the band structure of the semiconductor, thus affecting the performance of the semiconductor laser. The thermal stress can have significant effects on threshold, wavelength, polarization, and "smile" of the device. In addition, induced thermal stress in the laser device may damage the laser chips/bars and decrease the lifetime of the device.

The strain induced by thermal stress splits the degeneration at the valence band edge maximum, and separates the heavy-hole E_{HH} and light-hole E_{LH} sub-bands. Figure 4.1 shows the strain influence on the bandgap of a zinc blende type semiconductor [1, 2]. Under compressive strain, the heavy-hole valence band receives less electrons because it has a higher energy level than the light-hole valence band (Fig. 4.1, *left side*). Therefore, the transition feasibility of the heavy-hole band is higher. If the crystal is strained in tension, the electrons from the conductive band are more likely to move to the light-hole valence band than to

X. Liu et al., *Packaging of High Power Semiconductor Lasers*, Micro- and Opto-Electronic Materials, Structures, and Systems, DOI 10.1007/978-1-4614-9263-4_4,

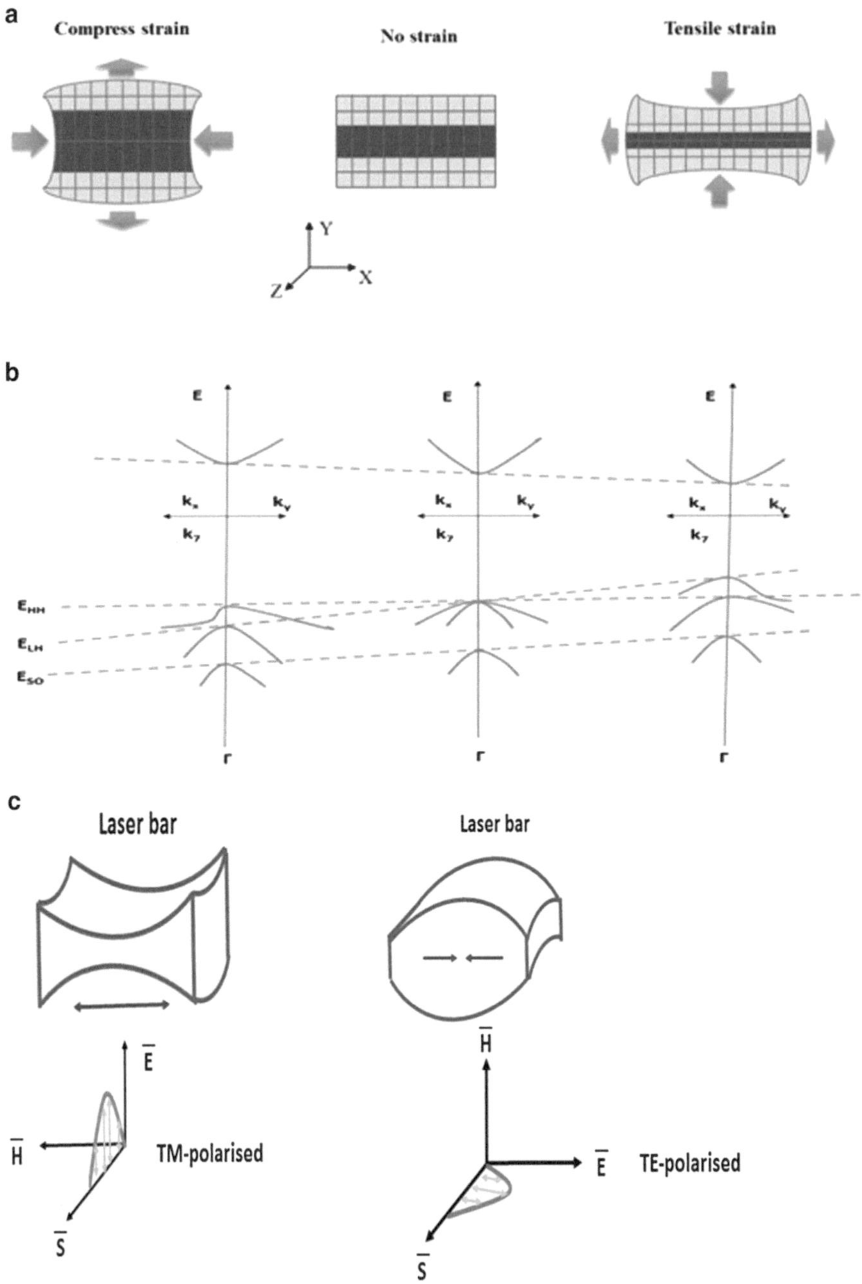

Fig. 4.1 (**a**) Schematic view of the lattice mismatch. (**b**) Schematic representation of the change in band structure caused by tensile and compressive strains. (**c**) Strain-induced polarization [1, 2]

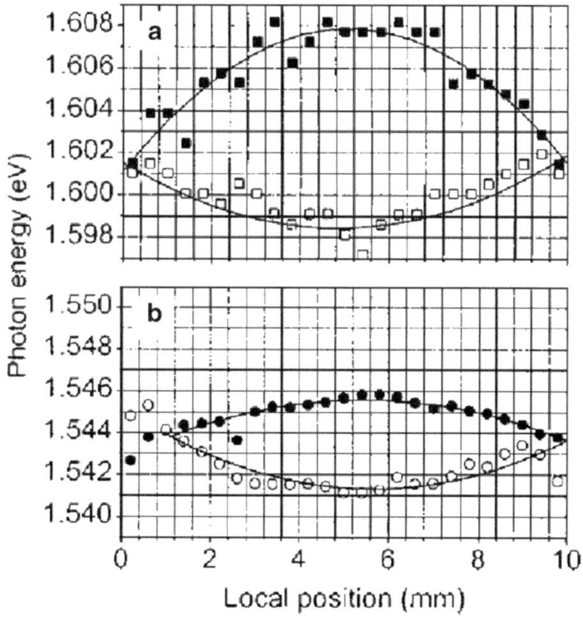

Fig. 4.2 The spectral positions of the (**a**) lh1-e1 and (**b**) hh1-e1 transitions as a function of position along two CBs (conductive band) from the same batch. One CB is soldered onto a copper heat sink (*filled squares or circles*), and the other onto a diamond heat spreader (*open squares or circles*) [3]

the heavy-hole valence band. The stress in the semiconductor laser will change the bandgap which can be seen through a change of the lasing wavelength of a semiconductor laser [1].

4.1.1 Wavelength

The emitting wavelength from individual emitters can be affected by packaging-related thermal and thermal stress effects. As the laser arrays are built with multiple individual emitters, the spectral broadening of laser arrays can happen owning to the wavelength nonuniformity of the individual emitters. Typically, the normal thermal stress of the emitter located in the center of a laser bar is higher than that at the edge. As a result, the wavelength of the emitters located in the center of the bar shows larger change than that located at the edge of the bar. Figure 4.2 shows the energy of the lh1-e1 and hh1-e1 interband transition as a function of position along two conduction bands (CB) from the same batch [3]. One of the CBs is soldered onto a copper heat sink (filled squares and filled circles), while the other is on a diamond heat spreader (open squares and open circles). The CB mounted on copper is expected to experience a compressive strain in the plane of quantum well (QW), while the CB on diamond experiences a tensile strain in the plane of the QW. The shift in the transition energies is toward lower values for the CB on the diamond heat spreader, and goes to the higher values for the CB on copper heat sink. These observations indicate that the redshift and blueshift of the output wavelength are closely related to the tensile and compressive strains, respectively.

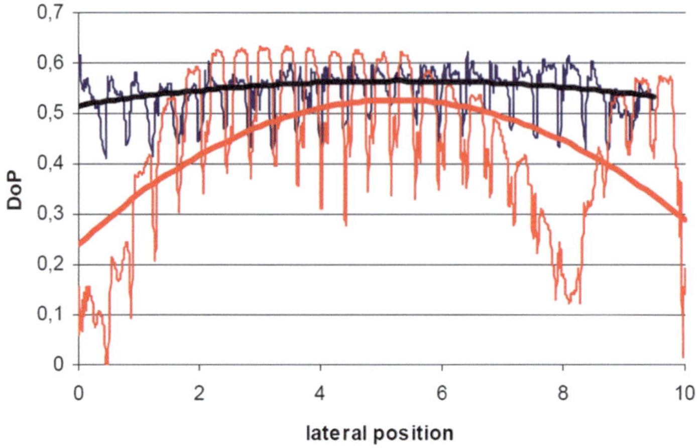

Fig. 4.3 Comparison of DoP results of the newly designed heat sink (*blue color*) and a standard copper heat sink (*red color*) with an indium mounted laser [5]

4.1.2 Polarization

The polarization of the emitted light depends on which valence band is involved in the transition. A compressively strained quantum well takes part in the transition between electrons and heavy holes, which leads to a polarization perpendicular to the quantum well plane (TE mode). Quantum wells under tensile strain undergo a transition between electrons and light holes, which results in a polarization in the quantum well plane (TM mode) [2]. The different transitions based on the Bloch theory and the polarizations states are described by Singh [4].

The degree of polarization (DoP) depends on the stress inside the quantum well mounted on different heat sinks [5]. An experiment has been designed to compare the effects of CTE mismatch between the laser device and the heat sink on the DoP of the emitters in the bar. In this experiment, the laser device has been attached to either the standard pure copper heat sink or a newly designed heat sink and tested at a current just below the threshold current. In the newly designed heat sink, a copper layer is added on both top and bottom sides of the silicon carbide diamond. This new heat sink has a better match of CTE (7–8 ppm/K) with the laser device and equivalent or even better thermal conductivity (~500 W/(m∗K)), compared to the standard pure copper heat sinks (CTE 16.8 ppm/K, thermal conductivity 400 W/(m∗K)). As shown in Fig. 4.3, the change of the DoP measured in device, which is mounted on the newly designed heat sink and has lower thermal stress, is much lower than that mounted on a standard copper heat sink which results in the higher thermal stress in the package [5].

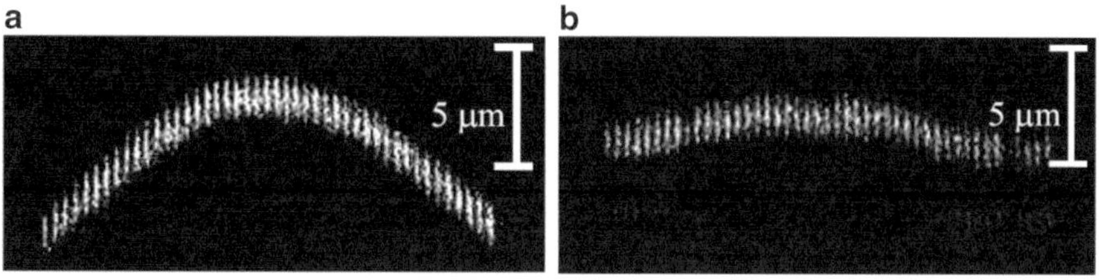

Fig. 4.4 Smile measurements of both bars taken just above threshold for (**a**) AuSn/CuW and (**b**) In mounted [7]

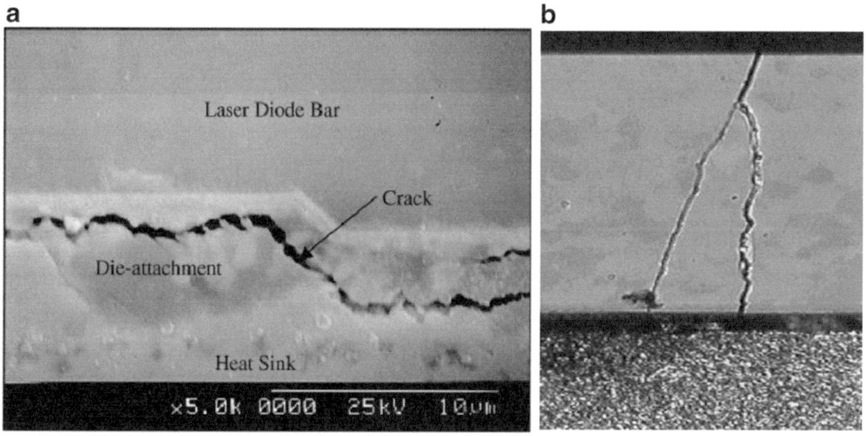

Fig. 4.5 Cracks running along (**a**) the die attachment and (**b**) bar [8, 9]

4.1.3 Smile

The smile refers to the bending of the line of emitters in transverse direction due to built-in and mounting-induced stress [6]. As shown in Fig. 4.4, under certain thermal stress the deformation (smile) can reach up to 5 μm, while with better controlled thermal stress the deformation can be reduced to the order of 1 μm [7]. Larger smile can induce a lower beam quality which causes difficulty on the collimation of laser beam.

4.1.4 Cracking

CTE mismatches between the metallization, die attachment, and copper heat sink are responsible for delamination and cracking that run along the interface, as shown in Fig. 4.5a [8]. In addition, the stress in the materials arising from the CTE mismatch between the diode and its mounting substrate can induce bar cracking, when the applied stress is greater than the yield strength of the laser material [9].

4.2 Analysis of Thermal Stress in High Power Semiconductor Lasers

The stress is a physical quantity that expresses the internal forces that neighboring particles of a continuous material exert on each other. For the semiconductor laser, the thermal stress is an inherent problem with the use of heat sink (for example: copper) since it has much larger CTE than the laser array which is essentially made of GaAs material. With the CTE difference of ~11 × 10^{-6}/°C between copper and GaAs material and the temperature difference of ~131°C between the indium solder freezing temperature (stress free point), and room temperature, there is a ~14 μm of contraction difference between the Cu heat sink and laser array along the length of the laser array for a standard 10 mm long laser array. The CTE mismatch is the root cause of thermal stress of laser diode array. Due to CTE mismatch, thermal stress is created during cooling process in the bar bonding procedure [10].

Figure 4.6 shows a simple model illustrating the formation of the thermal stress during the cooling from indium solder freezing temperature to room temperature [10]. Assuming CTE (a) > CTE (b) > CTE (c), when the temperature in the chamber of reflow oven cools down during the reflow process, materials a, b, and c are all contracts, but with different contraction magnitude. The material (a) has largest contraction, the material (b) is the second, and the material (c) has the smallest contraction. In this case, material (a) will impose a compress force on material (b), and material (c) will impose a tensile force on the material (b), and thus, the upper and lower surfaces of the material (b) are subjected to an asymmetric force. The asymmetric force produces a bending moment on the material (b), as illustrated with arrows in Fig. 4.6 [10]. This bending moment leads to material (b) to bend and residual thermal stress is formed.

The stress in the region of the joint between two isotropic materials one and two is a function of CTE and temperature of the two materials, as described by [10]

$$\sigma = \frac{E_1 E_2}{E_1 + E_2}(\alpha_1 - \alpha_2)(T_f - T_s), \tag{4.1}$$

where E is the modulus of elasticity of the materials one and two, α is the coefficient of thermal expansion of materials one and two, T_f is the freezing point of the solder, and T_s is the temperature at which the stress is measured (operating temperature). For GaAs-based laser chip and copper heat sink, the stress is described as follows [10]:

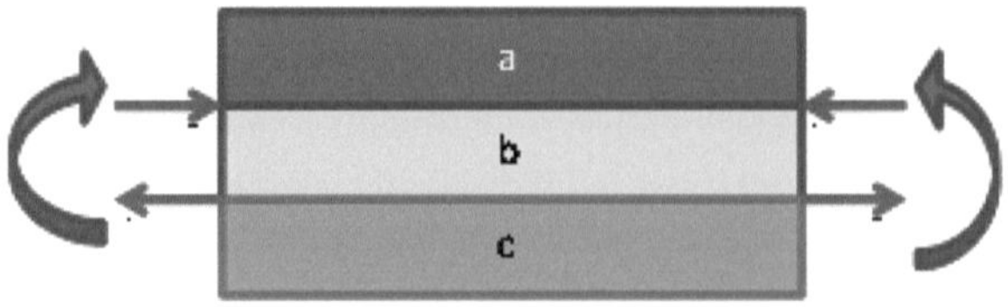

Fig. 4.6 A schematic diagram of thermal stress formation [10]

Fig. 4.7 Schematic of the residual packaging stress in a simplified model [11]

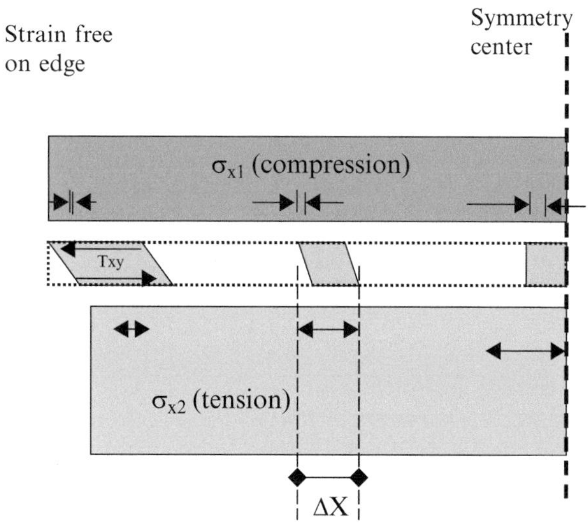

$$\sigma = \frac{E_{Cu}E_{GaAs}}{E_{Cu} + E_{GaAs}}(a_{Cu} - a_{GaAs})(T_f - T_s) \tag{4.2}$$

The copper-tungsten is alike [10]:

$$\sigma = \frac{E_{CuW}E_{GaAs}}{E_{CuW} + E_{GaAs}}(a_{CuW} - a_{GaAs})(T_f - T_s) \tag{4.3}$$

Figure 4.7 schematically shows the compressive normal strain that is induced in a semiconductor chip when it is mounted to a heat sink; the submount has a larger CTE which imparts compressive strain to the semiconductor chip [11]. In the stress/strain formation process, the solder acts to transfer the normal stresses between the chip and heat sink. For a 10 mm × 1.5 mm chip as shown in Fig. 4.7, the largest stresses are imparted along the longest dimension (i.e., σ_x).

Based on a simple model by Ohring [12], the uniaxial normal stress in the semiconductor can be quantified by

$$\sigma_1(x) = \frac{\tau_0}{\beta d_1}\left(1 - \frac{\cosh(\beta x)}{\cosh(\beta L)}\right) \tag{4.4}$$

$$\beta = \sqrt{\left(\frac{G}{d_0 d_1 E_1}\right)\left(1 + \frac{E_1 d_1}{E_2 d_2}\right)} \tag{4.5}$$

$$\tau_0 = \Delta\alpha\Delta T\left(\sqrt{\frac{G d_1 E_1}{d_0}}\right)\left(\sqrt{\frac{1}{1 + \frac{E_1 d_1}{E_2 d_2}}}\right) \tag{4.6}$$

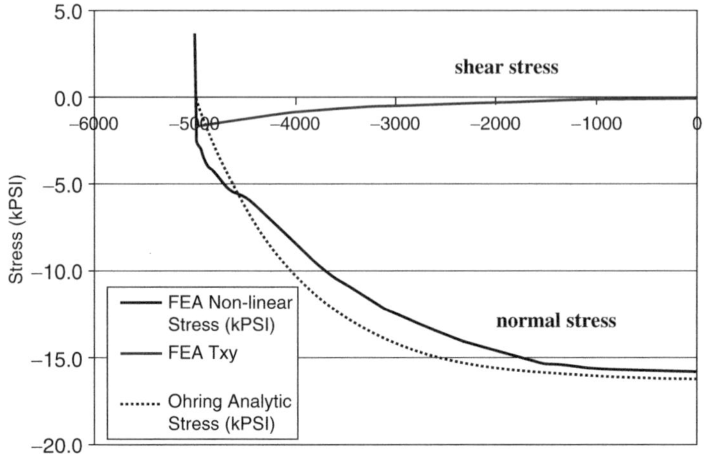

Fig. 4.8 Theoretical calculation for stress across a semiconductor laser bar mounted with indium solder on a copper heat sink (Analytic and FEA profiles of the normal stress in the semiconductor chip and shear stress in the solder layer) [11]

where E_x corresponds to Young's modulus, G is the shear modulus of the solder, d_x are the vertical dimensions, $\Delta\alpha$ is the CTE difference, and ΔT is the temperature difference. For a large CTE mismatch (i.e., GaAs to Cu) one must use a compliant solder such as indium that can relieve the thermal stress by plastic deformation. Figure 4.8 shows the resulting stress using the analytic model and the finite element analysis (FEA) model [11]. The curve in red gives the FEA results of the shear stress (x-direction) in the solder layer, showing that the shear stress increases from the center to the edge in the semiconductor chip. The plot in black shows that the normal stress in the diode laser chip is low close to the edge and increases toward center.

It would be preferred if the CTE of the heat sink in the laser package could perfectly match the CTE of the laser chip. However, there is still a significant challenge to have such a heat sink available in the industry. The CTE mismatch always exists between the laser chip and the heat sink in a laser package to different extents depending on the heat sink materials and the package design. In order to understand the effect of CTE mismatch on the residual stresses in the laser bar, simulations are conducted on a laser bar packaged on a heat sink, by varying the expansion coefficient of the heat sink around GaAs CTE value (6.7×10^{-6} K^{-1}). The dependence of the residual stress in the bar on the expansion coefficient mismatch is shown in Fig. 4.9 [13]. The resulting compressive stress is about 2 MPa if the CTE of the heat sink is 1×10^{-6} K^{-1} greater than that of the laser chip.

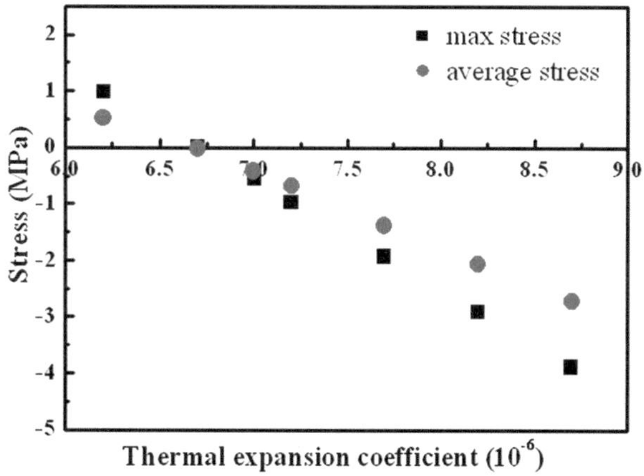

Fig. 4.9 Maximum stress and the average stress in dependence of the expansion coefficient mismatch between the heat sink and the laser bar [13]

4.3 Thermal Stress Minimization

Standard packaging of diode laser bars involves soldering the laser bars p-side down onto a heat sink. This process generally induces stress in the laser bar. Assuming the yield strength of the hard solder material is comparable to that of the heat sink, the upper limit of packaging-induced stress can be given by [14]

$$\sigma_{\max} = E \cdot \Delta\alpha \cdot \Delta T \tag{4.7}$$

where E is the Young's modulus of the heat sink, $\Delta\alpha = \alpha_1 - \alpha_2$ is the difference of the CTE α_1 of the laser bar and the CTE α_2 of the heat sink, and $\Delta T = T_s - T_a$ is the temperature difference between solder solidification temperature Ts and ambient temperature T_a. Eq. (4.7) suggests that selecting appropriate substrate and solder with high thermal conductivity is the key for reducing the thermal stress in the laser device.

4.3.1 Bonding Material

Table 4.1 lists the physical properties of a series of alloys used as bonding materials. In the packaging of semiconductor laser, two kinds of solders are commonly used: Indium solder which is a soft solder and AuSn solder which is a hard solder [14].

Indium has much lower yield strength and can incur plastic deformation more easily under stresses, compared with AuSn. With the Indium, the stress can be partly relieved by the plastic elongation of the solder during a thermal cycle. By increasing the Indium junction thickness, the level of maximum stress on the laser device is decreased. From the experimental results as shown in Fig. 4.10,

Table 4.1 Overview of the physical properties of different solder alloy [14]

Type of solder	Thermal conductivity (W/(m*K))	Max. allowed elongation (%)	Liquidus temperature (°C)
In	71–87	41	156
PbSn36Ag2	59	25–30	178
PbIn50	22	14–18	178–210
SnPb40	60–70	27	183
SnAg3.5	57–78	20–30	221
Sn	63	28	232
AuSn	46	1	283
PbSn5Ag2.5	44	20–30	280
PbIn5Ag2.5	42	28–34	307
Pb	37–42	17	327

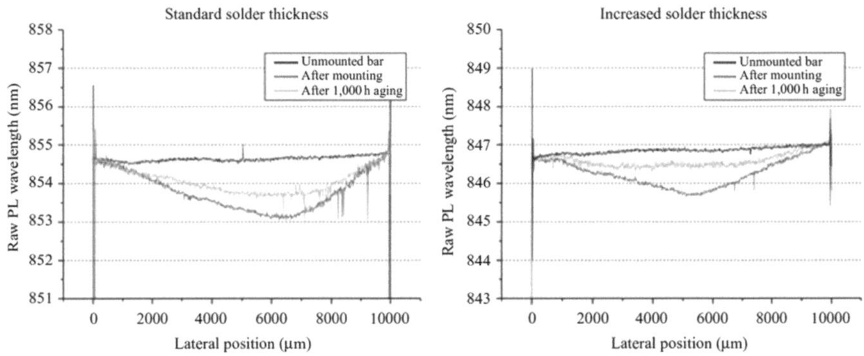

Fig. 4.10 Change of mounting induced stress with original solder junction thickness (left) and with increased solder junction thickness (right), indicated by the photoluminescence (PL) wavelength. In both cases a stress relaxation after aging can be seen [14]

the maximum stress level could be reduced by 35 % by increasing the Indium solder interface thickness by a factor of 1.5 [14].

Although the Indium can be used to reduce the thermal stress in high power semiconductor laser, the reliability of the semiconductor laser bonded with indium solder is largely decreased due to thermal fatigue, electro-migration, and thermo-electro-migration of the indium solder in the laser package. AuSn, which is much stiffer than Indium, can overcome these issues. However, the AuSn used as bonding material allows little stress and strain compensation (Fig. 4.11) [14]. Therefore, a buffer layer (for example: CuW) should be introduced to reduce the thermal stress of the laser chip when AuSn solder is used.

Similar to the Indium, the thickness of AuSn can also affect the interfacial stress in the laser package. Figure 4.12 shows the FEA analysis on the dependence of the maximum shear stress on the AuSn solder thickness in a GaAs chip on a silicon submount, when the maximum temperature changes between the room temperature

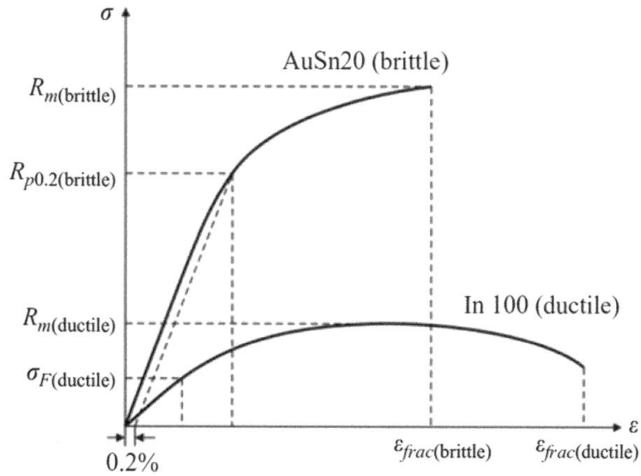

Fig. 4.11 Schematic illustration of stress strain characteristics of indium and AuSn hard solder [14]

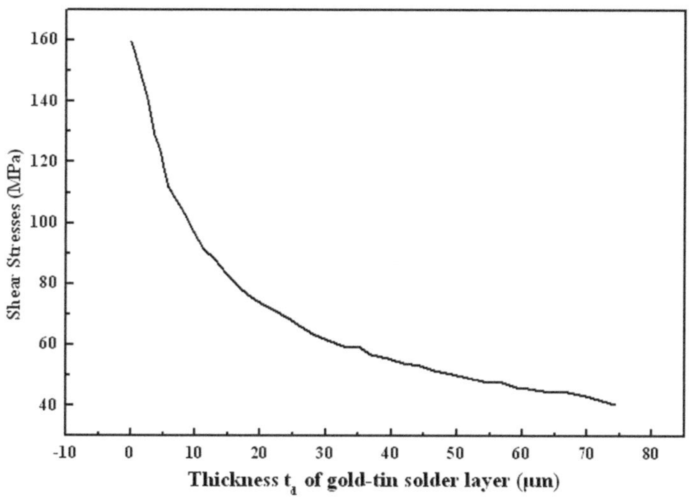

Fig. 4.12 The CTE mismatch induced shear stress as a function of the AuSn solder thickness [15]

and the soldering temperature 300°C [15]. The shear stresses at the substrate-solder and at the chip-solder interfaces are reduced by approximately 40 %, when the solder layer thickness is increased from 20 to 60 μm.

4.3.2 Mounting Substrate

Table 4.2 summarizes the properties of the chip and mounting substrate materials that are used in the laser package [15]. Typically, CuW, AlN, AlSiC, and Al_2O_3 are used as mounting substrate in packaging laser diodes. The CuW is a metal

Table 4.2 The mounting substrate material properties [15]

Mounting substrate	CTE (ppm/K)	Thermal conductivity (W/m*K)
Diamond	2	2,000
Cu	17.8	400
AlSiC	8	200
CuW	7	180
AlN	4.5	170
Si	4.2	150
Al_2O_3	6.7	21
Kovar	5.8	15

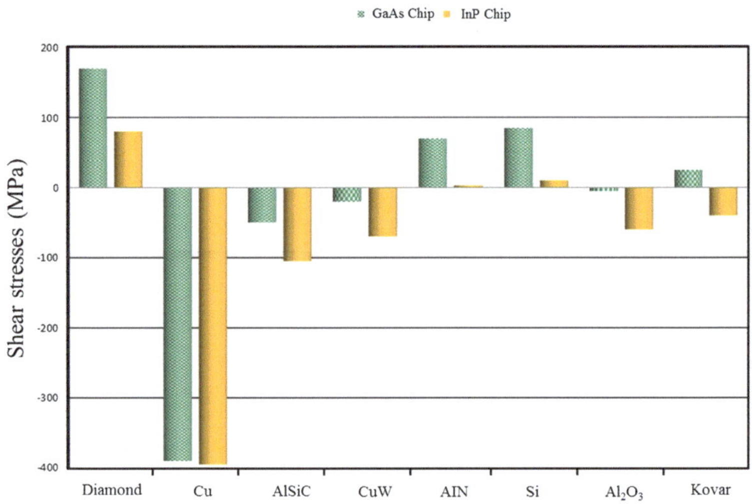

Fig. 4.13 Comparison of the CTE mismatch induced interfacial shear stress when a 2.4 mm GaAs or InP chip is soldered to different substrate materials [15]

composite, while the AlN, AlSiC, and Al_2O_3 are ceramic; they serve different purposes. The CuW composite is typically used for heat spreading purposes, while AlN and BeO are dielectric materials which are used for electrical insulation purposes.

The thermal stress in the laser device is closely related to the mounting substrate. The influence of the substrate on the maximum interfacial shear stress and normal stress has been numerically analyzed with the assumption that the maximum temperature variation is between room and soldering temperature of 300°C. Interfacial shear stress, induced by the CTE mismatch between the chip and the mounting substrate, is analyzed as shown in Fig. 4.13 in which AuSn is selected as the bonding material [15].

The estimated interfacial shear stresses are 387 and 168 MPa when the GaAs chip is bonded onto a copper and diamond substrate, respectively. As the estimated

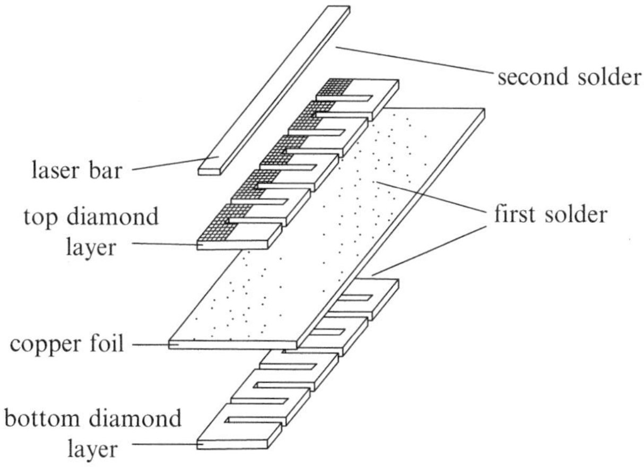

Fig. 4.14 Design of a CTE-matched substrate with high diamond content [16]

interfacial shear stress exceeds the yield stress of the gold-tin solder (275 MPa), a plastic deformation is induced in the solder during the cooling phase of the soldering process [15]. This plastic deformation redistributes the stresses in the solder layer and finally reduces the normal stresses in the GaAs chip.

Figure 4.13 gives the interfacial shear stress for laser chips soldered to different substrate materials [15]. In GaAs laser chips, Al_2O_3 substrate presents the lowest thermally induced shear stress than other substrates, while Cu shows significantly higher stress among all the substrates studied.

4.3.3 Advanced Composite Materials

In the current application, a tungsten copper composite material, which has a better CTE match to the laser chip than pure Cu, is used in gold-tin (AuSn) bonded diode laser packages. However, the thermal conductivity of CuW (195 W/(m∗K)) is lower than that of the Cu (394 W/(m∗K)). The low thermal conductivity of the CuW could lead to inefficient removal of waste heat, which can cause thermal stress in the laser package. In order for the CuW to be used as the mounting substrate, an additional copper plate needs to be attached to the CuW substrate away from the chip contact side to get the heat out faster.

In order to improve the thermal conductivity and achieve better CTE match with the chip, a novel sandwich design for the mounting substrate has been made. In this sandwich design, a substrate material with good thermal conductivity and better CTE match than copper is placed on the top and bottom, with a layer of copper placed in the middle of the sandwich mounting substrate. The desired CTE of the sandwich substrate can be achieved by choosing the right copper layer thickness [2, 16, 17].

Diamond-copper-diamond structure: Figure 4.14 shows a novel mounting substrate with diamond-copper-diamond structure [16]. With this sandwich structure, the mounting substrate demonstrates reasonably good thermal and CTE properties.

Table 4.3 CTE measurement results of 300 μm-diamond-copper–300 μm-diamond substrates [16]

Sample name	DCD25	DCD50	DCD100	DCD150	DCD200
Copper thickness (μm)	25	50	100	150	200
Average lat. CTE (ppm/K)	4.1	6.4	10.6	12.2	13.1

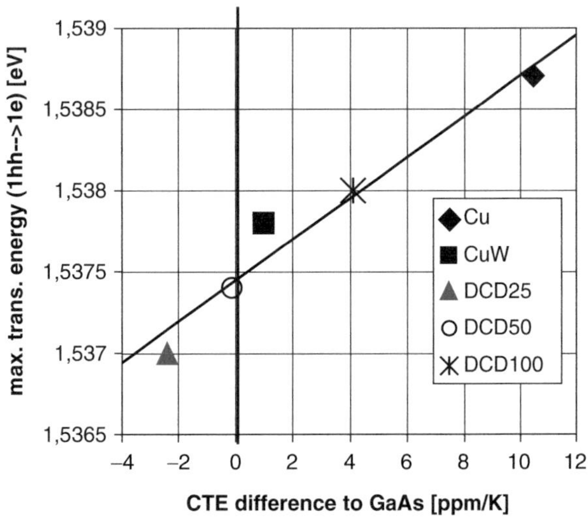

Fig. 4.15 Maximum transition energy near center of laser bars with different mounting substrates [16]

Table 4.3 shows CTE measurement results of 300 μm diamond-copper 300 μm-diamond substrates [16]. As the copper thickness varies from 25 to 100 μm, the CTE of the sandwich changes greatly. As shown in Table 4.3, the layer system (sample DCD50) of 300 μm diamond, 50 μm copper, and 300 μm diamond has the closest CTE match to that of GaAs, and the calculated effective thermal conductivity of this sandwich substrate nearly approaches 1,500 W/(m∗K).

The maximum energy transition values in laser bars depend on the thermal stress induced by the CTE mismatch between the mounting substrate and the laser chip. Figure 4.15 gives a comparison of maximum energy transition values with laser bars traditionally mounted on bare copper or CuW and on the designed mounting DCD substrates. One can see that the 50 μm-DCD substrates which has the best CTE match with the laser chip exhibits the minimum shift of the maximum energy transition value, while laser bars mounted on the bare copper mounting substrate which has the largest CTE mismatch give the biggest shift of the maximum energy transition values [16].

Silicon carbide diamond-copper-silicon carbide diamond structure: Silicon Carbide Diamond (ScD) is fabricated in an effort to achieve reduction of both CTE and thermal resistance [17]. On top and bottom sides of the ScD body, the copper is brazed. The manufacturing cost of ScD is much less than that for producing high-pressure diamond composites. In the ScD making process,

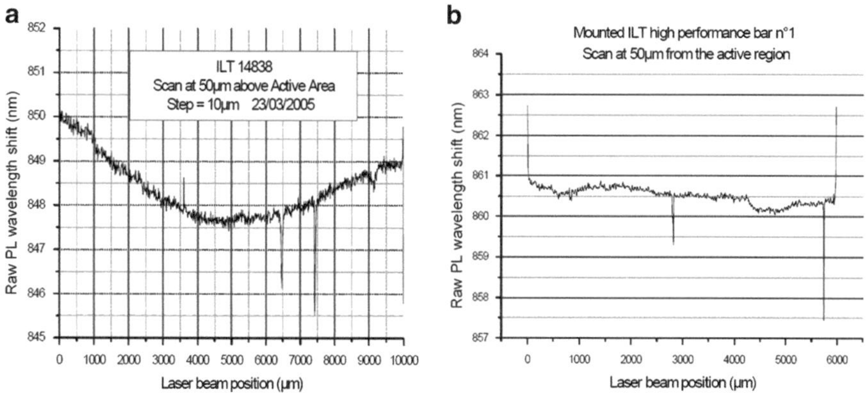

Fig. 4.16 Results of a µPL measurement of a copper (**a**) and ScD (**b**) heat sink with an indium mounted laser [17]

Powders are mixed and agglomerated in a clean room environment. First heat sink samples are designed and fabricated based on ScD material. By adding top and bottom copper layers of the appropriate thickness, a CTE of 7–8 ppm/K and a thermal resistance of 0.7 K/W have been achieved. This heat sink provides an expansion-matched and thermally equivalent or even better alternative to standard pure copper heat sinks.

The thermal stress in the laser package can be characterized by the microphotoluminescence spectroscopy (µ-PL). The precise wavelength of the PL signal depends on the lattice strain/stress in the laser package. Figure 4.16 gives µ-PL line scan on a laser bar mounted on a copper heat sink with indium solder. The higher wavelength indicates an increase of stress at the edges of the laser [17]. The wavelength shift is around 2 nm at the edges. In contrast, a laser bar mounted on the ScD heat sink shows a relative flat line. The edges of the µ-PL scan show no significant changes in the wavelength, and the wavelength shift is around 0.5 nm only. From these data, one can see the stress induced in the laser bar can be greatly reduced with ScD mounting substrate [17].

Molybdenum-copper-molybdenum structure: Figure 4.17 shows the copper heat sinks with molybdenum layers added on the top and bottom of the copper block. On the surface of the molybdenum, a thin layer of copper is electroplated [2, 17]. The thin copper layer acts as a head spreader, so that the thermal resistance is low. Due to the very small thickness, the CTE effect of the thin copper layer is negligible, compared with the molybdenum layer [2]. The CTE of the bulk molybdenum is approximately 5 ppm/K which is a little lower than that of GaAs (CTE $\approx$ 6.7 ppm/K). The sandwich design with symmetric structure is made to avoid bending of the heat sink during temperature changes. Compared to the standard copper heat sink, the molybdenum-copper-molybdenum-structured heat sink has a CTE smaller than 9 ppm/K and a thermal resistance lower than 1 K/W.

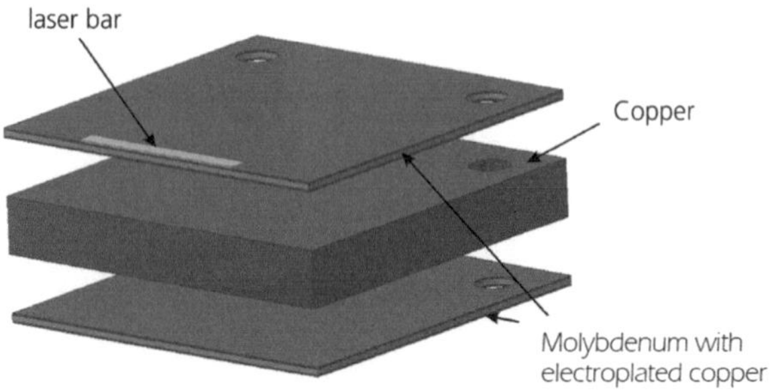

Fig. 4.17 Sandwich structure of Cu/Mo heat sink [17]

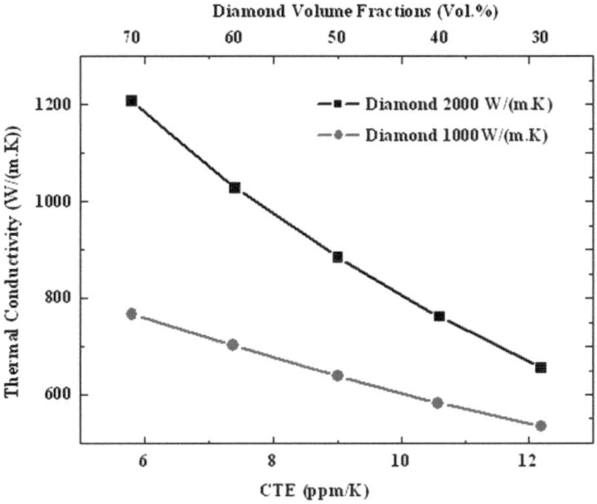

Fig. 4.18 Calculated thermal conductivity of a copper-diamond composite for different diamond volume fractions and two different conducting diamond qualities (assuming a perfect thermal interface) [18]

Copper-diamond fillers: Composite materials with copper filled with diamond fillers are used for mounting substrates to achieve a combination of good CTE match with the chip and high thermal conductivity. Copper-diamond filler (CD) composites have been manufactured by a powder metallurgical process. The thermal conductivity and CTE of the composite can be changed by varying the volume content percentage of diamond fillers. Figure 4.18 shows the changes of thermal conductivity and CTE with the diamond volume content percentage for two different diamonds [18]. The thermal conductivity goes up with the increase of the diamond content, while the CTE goes down as the diamond content gets higher. For the composite filled with the two different diamonds, the high purity diamond with the bulk thermal conductivity of 2,000 W/(m∗K) shows better thermal behavior in the final composite materials than the diamond with the bulk thermal conductivity of 1,000 W/(m∗K), when the CTE of the copper-diamond composites is the same. In the case that the mounting substrate matches the CTE of the chip, the high purity diamond composite mounting substrate exhibits around 50 % higher in thermal conductivity than that with the diamond of 1,000 W/(m∗K).

References

1. R. Diehl, *High-Power Diode Lasers: Fundamentals, Technology, Applications* (Springer, Berlin, 2000)
2. C. Scholz, Thermal and mechanical optimization of diode laser bar packaging, PhD thesis, RWTH Aachen University of Technology, 2007
3. M.L. Biermann, S. Duran, K. Peterson, A. Gerhardt, J.W. Tommaet, A. Bercha, W. Trzeciakowski, Spectroscopic method of strain analysis in semiconductor quantum-well devices. J. Appl. Phys. **96**(8), 4056–4065 (2004)
4. J. Singh, *Semiconductor Optoelectronics, Physics and Technology* (McGraw-Hill, New York, 1995)
5. M. Leers, K. Boucke, C. Scholz, T. Westphalen, Next generation of cooling approaches for diode laser bars. Proc. SPIE **6456**, 64561A(1–10) (2007)
6. http://www.dilasinc.com/gdresources/downloads/whitepapers/DILAS_DPAL_8241-25.pdf
7. J.L. Hostetler, C.L. Jiang, V. Negoita, T. Vethake, R. Roff, A. Shroff, T. Li, C. Miester, U. Bonna, G. Charache, H. Schlüter, F. Dorsch, Thermal and strain characteristics of high-power 940 nm laser arrays mounted with AuSn and In solders. Proc. SPIE **6456**, 645502(1–12) (2007)
8. A.R. Dhamdhere, A.P. Malshe, W.F. Schmidt, W.D. Brown, Investigation of reliability issues in high power laser diode bar packages. Microelectron. Reliab. **43**(2), 287–295 (2003)
9. A. Hodges, J. Wang, M. DeFranza, X.S. Liu, B. Vivian, C. Johnson, P. Crump, P. Leisher, M. DeVito, R. Martinsen, J. Bell, A CTE matched, hard solder, passively cooled laser diode package combined with nXLT™ facet passivation enables high power, high reliability operation. Proc. SPIE **6552**, 65521E(1–9) (2007)
10. J.W. Wang, Z.B. Yuan, L.J. Kang, K. Yang, Y.X. Zhang, X.S. Liu, *Study of the Mechanism of "Smile" in High Power Diode Laser Arrays and Strategies in Improving Near-Field Linearity.* IEEE 2009 Electronic Components and Technology Conference (2009), pp. 837–842
11. D. Schleuning, K. Scholz, M. Griffin, B. Guo, C. Luong, R. Pathak, C. Scholz, J. Watson, H. Winhold, T. Hasenberg, Material survey for packaging semiconductor diode lasers. Proc. SPIE **7198**, 71981K(1–9) (2009)
12. M. Ohring, *Reliability and Failure of Electronic Materials and Devices* (Academic Press, Boston, 1998)
13. C. Scholz, K. Boucke, R. Poprawe, Mechanical stress reducing heat sinks for high-power diode lasers. Proc. SPIE **5336**, 176–187 (2004)
14. F. Bachmann, P. Loosen, R. Poprawe, *High Power Diode Lasers Technology and Applications.* Springer series in optical sciences (2007), pp. 75–120
15. A.C. Pliska, J. Mottin, N. Matuschek, C. Bosshard, *Bonding Semiconductor Laser Chip: Substrate Material Figure of Merit and Die Attach Layer Influence*, Therminic, 2005, pp. 28–30
16. D. Lorenzen, P. Hennig, Highly thermally conductive substrates with adjustable CTE for diode laser bar packaging. Proc. SPIE **4945**, 174–185 (2003)
17. M. Leers, C. Scholz, K. Boucke, M. Oudart, *Next Generation Heat Sinks for High-Power Diode Laser Bars.* 23rd IEEE semi-therm symposium (2007), p. 105
18. E. Neubauer, P. Angerer, *Advanced Composite Materials With Tailored Thermal Properties for Heat Sink Applications.* 2007 European Conference: Power Electronics and Applications (2007), pp. 1–8

Chapter 5
Optical Design and Beam Shaping in High Power Semiconductor Lasers

With improving of the performances, semiconductor lasers are becoming attractive as light sources in many fields, such as direct material processing, high power solid-state laser and fiber laser pumping, medical and display applications. However, the poor beam quality is still the main bottleneck limiting their further applications. In order to expand the applications of semiconductor lasers, beam shaping and optical design are essential. Generally speaking, the poor beam quality is due to the waveguide properties of the active region of semiconductor lasers. In this chapter, we will discuss the beam performances of semiconductor lasers in Sect. 5.1. The beam shaping and fiber coupling principles and design methods of a single emitter, a bar, and a stack will be discussed in Sect. 5.2. Section 5.3 presents beam combining and high brightness techniques.

5.1 Optical Characteristics of Semiconductor Lasers

It is well known that a semiconductor laser has large divergence angles and asymmetrical far field. High power semiconductor lasers can be divided into a single emitter semiconductor laser and semiconductor laser arrays or bars. The more emitters a semiconductor laser has, the more complex the beam property will be.

For a single emitter semiconductor laser, the mathematic models to describe the source field have been studied, and there are several typical models such as Gauss model, exponent model, and Lorentz model. In practice, Gauss model has been widely used to describe the source field and it is denoted by $E(x_0, y_0)$ and expressed as [1–3]

$$\mathrm{E}(x_0, y_0) = A_0 \exp\left[-\left(\frac{x_0^2}{w_{0x}^2} + \frac{y_0^2}{w_{0y}^2}\right)\right] \tag{5.1}$$

where A_0 is an amplitude of the source field; $w_{0\mathrm{x}}$ and $w_{0\mathrm{y}}$ are the beam waist radius in fast and slow axes, respectively; and x_0 and y_0 denote fast and slow axes at source

X. Liu et al., *Packaging of High Power Semiconductor Lasers*, Micro- and Opto-Electronic Materials, Structures, and Systems, DOI 10.1007/978-1-4614-9263-4_5,

facet, respectively. The source field of a semiconductor laser bar or laser stack is the incoherence superposition of source fields of each emitters.

The far field can be calculated by scalar Rayleigh–Sommerfeld function [3, 4]:

$$E(x,y,z) = -\frac{iz}{\lambda r}\frac{\exp(ikr)}{r}\int_{-\infty}^{\infty}\int_{-\infty}^{\infty}E(x_0,y_0) \times \exp\left(ik\frac{x_0^2+y_0^2-2xx_0-2yy_0}{2r}\right)dx_0dy_0, \tag{5.2}$$

where z is the propagation axis; x and y are fast and slow axes along z axis, and $r=\sqrt{x^2+y^2+z^2}$.

Propagating in free space, the beam waist of semiconductor lasers satisfies the hyperbola propagation law and it can be calculated as

$$w^2(z) = {w_0}^2 + z^2\tan^2\theta_0, \tag{5.3}$$

where w_0 is the waist radius in 90 % energy of a semiconductor laser and w(z) is the beam width in free space.

The beam width of semiconductor lasers can be also defined by second-order moments of the intensity distribution as

$$w_x^2 = \frac{4}{p}\int_{-\infty}^{+\infty} x^2E(x,y,z)E^*(x,y,z)dxdy, \tag{5.4}$$

$$w_y^2 = \frac{4}{p}\int_{-\infty}^{+\infty} y^2E(x,y,z)E^*(x,y,z)dxdy, \tag{5.5}$$

where, p is presented as follows

$$p = \int_{-\infty}^{+\infty}\int_{-\infty}^{+\infty} I(x,y,z)dxdy. \tag{5.6}$$

The far-field divergence angle θ can be calculated as follows:

$$\theta = \lim_{z\to\infty}\frac{w(z)}{z}, \tag{5.7}$$

According to the ISO-standard this property can be characterized by the beam-parameter product (BPP) [5, 6]. BPP is the product of the waist radius w_0 and the far-field divergence θ of the beam:

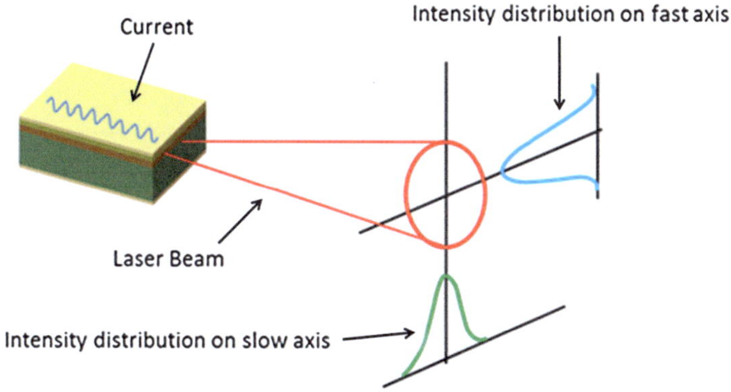

Fig. 5.1 The far field of a single emitter semiconductor laser [2]

$$\mathrm{BPP} = w_0 \times \theta, \tag{5.8}$$

In the best case, i.e., without any aberrations, this beam-parameter product remains constant when the beam is transformed by passive optical components such as lenses or mirrors. The minimum value for the beam-parameter product is the so-called diffraction limit. Besides BPP, the beam quality parameter, M^2 and optical power density denoted as B are important parameters to describe the beam quality and brightness of a beam. The beam quality M^2 has been widely used to characterize the beam quality of a Gaussian beam such as solid state and CO_2 lasers, and in some cases it is also used to describe the beam quality of semiconductor lasers. It can be calculated by the following equation:

$$M^2 = \frac{\pi}{\lambda} w_0 \theta, \tag{5.9}$$

For power density, namely, B is given by

$$B = \frac{P}{\pi^2 Q^2}, \tag{5.10}$$

where P is the total power of the beams, and $Q = \frac{w_0}{2\theta}$.

5.1.1 *Single Emitters*

Figure 5.1 shows a single emitter semiconductor laser with the far-field intensity [2, 3]. The far field of a single emitter is asymmetrical on fast and slow axes directions. Typically, the divergence angles FWHM (full width at half maximum)

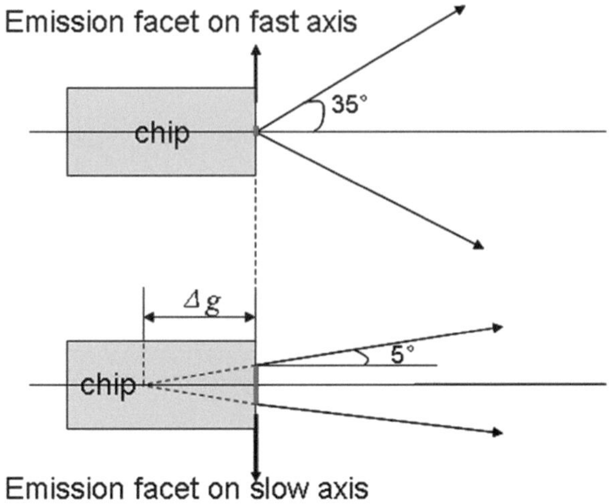

Fig. 5.2 The astigmatism of semiconductor laser [2]

are ~45° along the fast axis and ~8° along the slow axis, respectively [7]. For the special waveguide structure, e.g., photonic bandgap waveguide, the fast axis divergence is decreased down to ~8° [8]. In optical design of beam shaping for semiconductor lasers, the divergence angles of 90 % energy also are considered, and the typical divergence angles for the commercial laser products are 70° and 10° in the fast and slow axes directions, respectively.

For the asymmetry sizes of active region on fast and slow axes, semiconductor laser beam is astigmatism. Figure 5.2 shows an astigmatism example of semiconductor laser, and the astigmatism is denoted by Δg [2, 3]. Due to small beam size along fast axis, the beam along this axis is approximated to emit from active region. The sizes of an emitter beam are 1 μm and 100–200 μm in the fast and slow axes directions, respectively. The BPP of the fast and slow axes are denoted as BPP_f and BPP_s, respectively. w_{0x}, θ_{0x}, w_{0y} and θ_{0y} denote the waist radius and the divergences of a semiconductor laser in the fast and slow axes directions, respectively. Replacing w_0 and θ_0 in Eq. (5.5) with the exactly values $w_{0x} = 1$ μm, and $\theta_{0x} = 70°$ for the fast axe, and $w_{0y} = 100$ μm and $\theta_{0y} = 10°$ for the slow axis, we can get $BPP_f = 0.3$ mm*mrad and $BPP_s = 4.4$ mm*mrad. The value of BPP_s is 14.3 times larger than BPP_f. The beam size of a single emitter along the fast axis at $w_{0x} = 1$ μm is so small that it can be considered as a light source point, but for the slow axis the beam size is quite large.

5.1.2 *Diode Laser Bars*

A semiconductor laser bar is composed by multiple emitters arranged in a row as shown in Fig. 5.3 [9]. The emitter numbers of a semiconductor laser bar range from 19 to 69 with different fill factors. The radiation and the near-field pattern of an 808 nm semiconductor laser bar with 62 emitters are shown in Fig. 5.4 [10, 11].

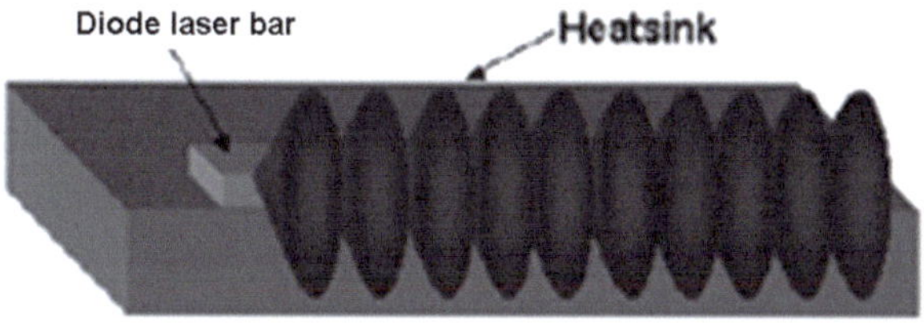

Fig. 5.3 A semiconductor laser bar [9]

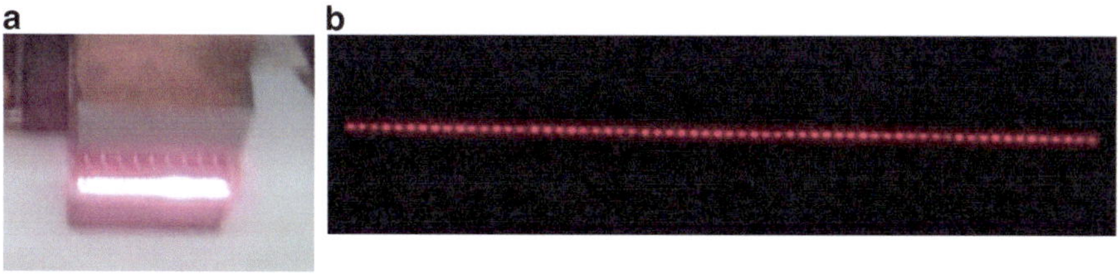

Fig. 5.4 The radiation and the near-field pattern of an 808 nm semiconductor laser bar [11]. (**a**) The radiation of a semiconductor laser bar. (**b**) The near-field pattern of a semiconductor laser bar

The beam size along the fast axis is 1 μm and 10 mm along the slow axis. The semiconductor laser bar has the same divergence angles as a single emitter. According to Eq. (5.1), we can obtain $BPP_f = 0.3$ mm*mrad and $BPP_s = 436.1$ mm*mrad for a semiconductor laser bar. The beam sizes of a semiconductor laser bar along the fast and slow axes are greatly asymmetric, since BPP_s is 1,400 times larger than BPP_f. The beam along the fast axis can also be considered as a point light source because of the small beam size, and therefore a semiconductor laser bar is considered as a line light source as a whole. To obtain high brightness, the asymmetry BPP of a semiconductor laser bar on the two axes should be reduced and the condition of $BPP_f = BPP_s$ need to be satisfied.

5.1.3 *Diode Laser Stacks*

A semiconductor laser stack is composed of multiple semiconductor laser bars arranged vertically, as shown in Fig. 5.5 [9]. For most of the applications, the bars in the vertical stack need to be collimated in the fast axis first. The far-field pattern of a semiconductor laser stack with fast axis collimation is a rectangle beam as shown in Fig. 5.6 [11].

In Fig. 5.6 the semiconductor laser bars are separated by heat sinks [11]. With the existence of heat sinks, the pitch between two bars is generally from 1 to 2 mm. The large pitch causes discontinuation of the intensity of a semiconductor laser stack, and the far-field pattern is light and dark stripes which are shown in Fig. 5.6. Hence, with the increasing number of laser bars the beam quality become worse.

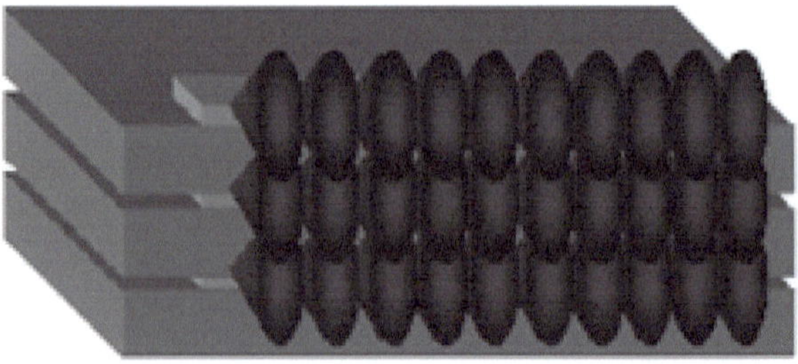

Fig. 5.5 A semiconductor laser stack [9]

Fig. 5.6 The far-field pattern of a semiconductor laser stack with fast axis collimation [11]

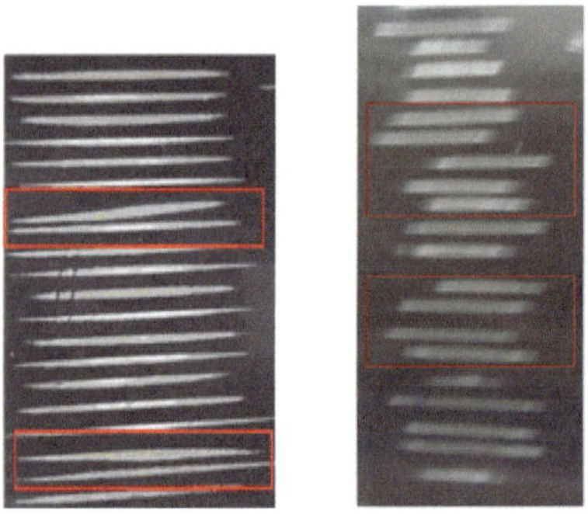

Fig. 5.7 Test result of beam point error along vertical axis [12]

Collimated beam pointing error (*CBPE*): A laser stack is composed of collimated laser bars with fast axis collimators (FACs). The laser stack has the beam problem which is called the CBPE [12]. The typical CBPEs are shown in Fig. 5.7 and they are marked with red rectangle [12]. Figure 5.8a shows the ideal far-field beam intensity of the laser stack, and the typical CBPEs are shown in Fig. 5.8b [12]. As shown in Fig. 5.8b, ΔD denotes the CBPE along fast axis, and δ denotes the CBPE rotated along slow axis, and b denotes the CBPE along slow axis.

The CBPE of the laser stack is caused by three factors. The first is the large smile of the laser bars. If the laser bars used in the laser stack have low smile less

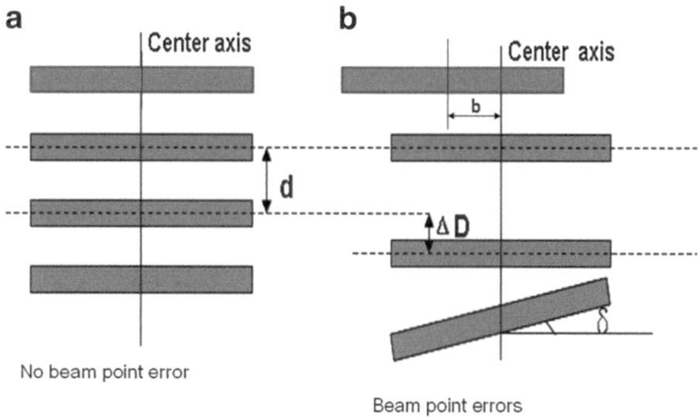

Fig. 5.8 The schematic of collimated beam point error [12]. (**a**) The ideal beam. (**b**) The typical collimated beam pointing errors

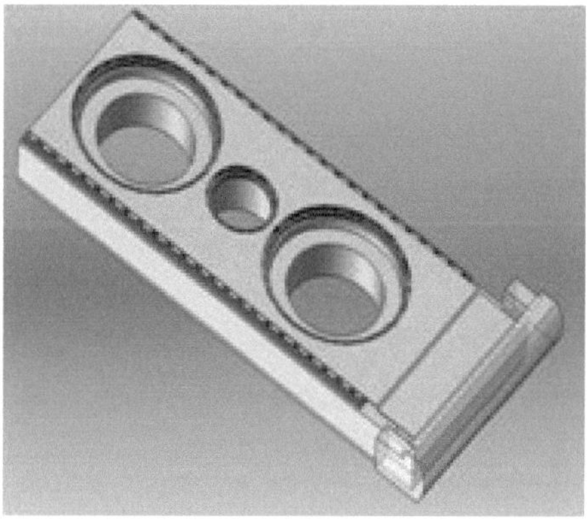

Fig. 5.9 The laser bar with a FAC [12]

than 1 μm, it does not reduce the collimated beam quality of the laser bars significantly [11]. The second factor is the installation error of the FAC. Figure 5.9 shows a laser bar with a FAC [12]. The third factor is the installation error of the stack during the laser stack assembling process.

The FAC is the micro-aspheric lens and the far-field intensity distribution of the laser bar is very sensitive to the position of the lens. The tiny installation errors of the FAC can reduce the collimated beam quality obviously. Several typical installation errors of the FACs are discussed below. The laser stack with seven laser bars is used as an example, and the FACs on the laser bars are arranged as "1, 2, 3, 4, 5, 6, 7" as shown in Fig. 5.10 [12]. The ideal far-field intensity distribution without any installation error of the FAC is shown in Fig. 5.10a. The six typical installation errors of the FACs are shown in the left of Fig. 5.10b and are denoted by $-\delta'$, δ', $-\Delta z$, Δz, $\Delta D'$, and $-\Delta D'$, respectively. Firstly, the first and last lenses are rotated as z axis with the angles of δ' and $-\delta'$, respectively, and the values of them are $-\delta' = -0.1°$ and $\delta' = 0.1°$; and secondly, the second and sixth lenses are ab-axial along fast axis with the distances of $\Delta D'$ and $-\Delta D'$, respectively, and then $\Delta D' = -0.01$ mm and $-\Delta D' = 0.01$ mm; and at last the positions of the third and fifth lenses are off the focus spot of the FACs with the distances of Δz and $-\Delta z$,

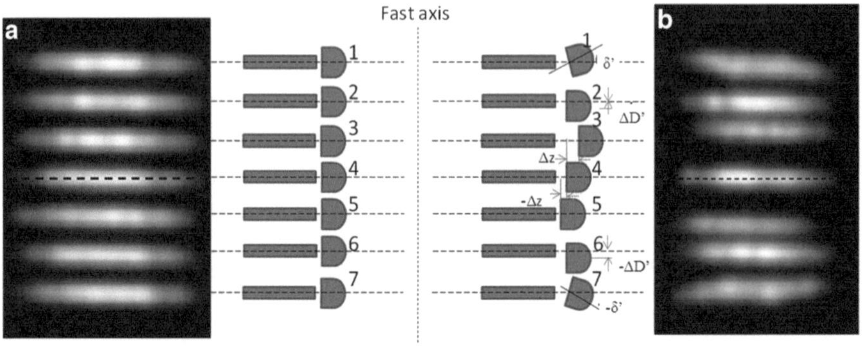

Fig. 5.10 The collimated beam error of the stack due to the installation error of FAC [12]. (**a**) The ideal beam with no installation error. (**b**) Typical installation and collimated beam errors

Fig. 5.11 The process of pointing direction control [12]

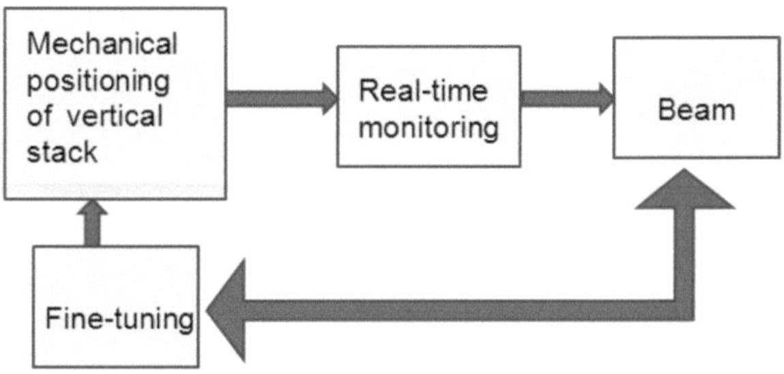

respective, and then $-\Delta z = -0.2$ mm and $\Delta z = 0.2$ mm. At right side of Fig. 5.10b, the stimulated far-field intensity of the laser bars with installation errors is showed, and the CBPEs which are caused by the installation errors of the FACs are observed obviously. So reducing the installation error of each laser bar with the FAC lens is an important step to control the collimated beam point error of the laser stack.

During the laser stack assembling process, the CBPE can also be caused by the installation error of the stack. It means that during the packaging there are position deviations between each laser bar of the laser stack, such as the tilting or displacement of laser bars along fast or slow axis. So reducing the installation error during the stack packaging is another important measure. To reduce the CBPE, the beam control method has been proposed, and the main steps are shown in Fig. 5.11 [11, 12]. First, fast axis collimation components are added to keep each bar pointing direction consistent. And then using advanced real-time monitoring equipment, the position of each bar is fine-tuned from vertical and horizontal orientation to ensure accurate positioning.

Based on this approach, the beam pointing error in a laser stack can be successfully corrected. As shown in Fig. 5.12, the value of parameter δ is below 0.1°; ΔD and b can be ignored mostly in applications [12].

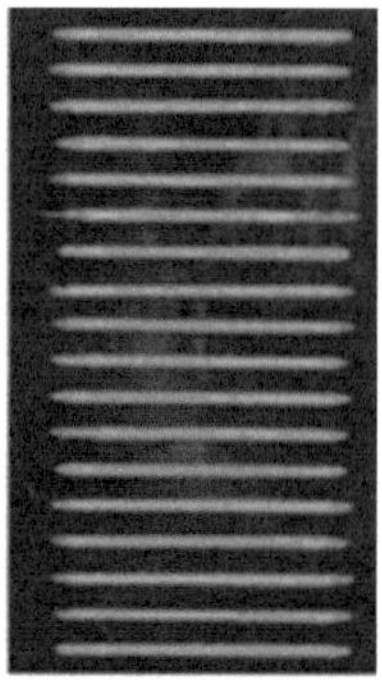

Fig. 5.12 The corrected result [12]

5.2 Beam Shaping and Fiber Coupling

According to the discussion in Sect. 5.1, the poor beam quality of semiconductor lasers should be improved before application. For some applications, the laser beam is required to be delivered by fiber. Generally, there are three steps to realize the improvement of beam quality of semiconductor lasers and fiber coupling. They are beam collimation, beam shaping, and fiber coupling. In this section, the optical principle and design for semiconductor lasers are discussed and analyzed, and the optical designs of beam shaping and fiber coupling of a single emitter, a single bar, and a single vertical stack are studied and presented.

5.2.1 Single Emitter Semiconductor Lasers

1. Direct fiber coupling
   The active region of a single emitter laser has small size which is 1 μm and 100–200 μm along fast and slow axes, respectively, and the FWHM divergence angle in slow axis is $\sim 8^\circ$. Hence after beam collimation in fast axis, single emitter laser can be coupled into a fiber with 100–200 μm diameter directly. Figure 5.13 shows wedge-shaped fiber coupling [2, 13]. The input end of the fiber is wedge shaped, and the divergence angle along fast axis is collimated by this end. The wedge-shaped tip of the fiber serves as a collimator. After collimation by the wedge, the beam is coupled into the fiber. However, in order to obtain high fiber coupling efficiency, the requirement of the wedge shape is very high and it is difficult to be fabricated precisely and with high yield. This significantly limits the application of this technology.
2. Free space coupling
   On the other hand, the beam can be collimated by a collimator before fiber coupling, and the requirements on component and fabrication can be decreased during processing. Due to small beam size, a cylinder lens is always used to collimate the beam along fast axis. The beam coupling scheme is shown in

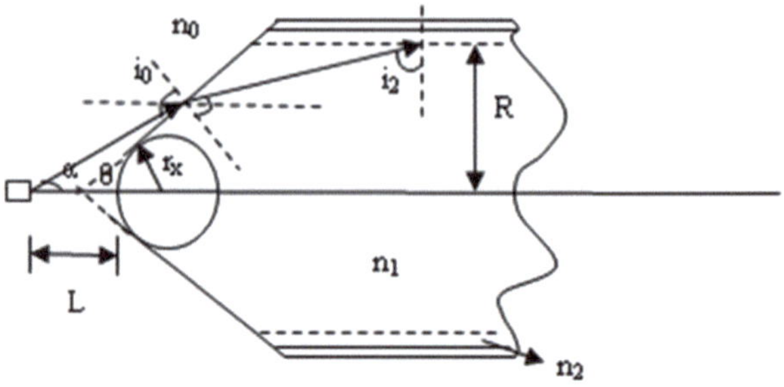

Fig. 5.13 Direct beam coupling scheme based on wedge-shaped fiber [2]

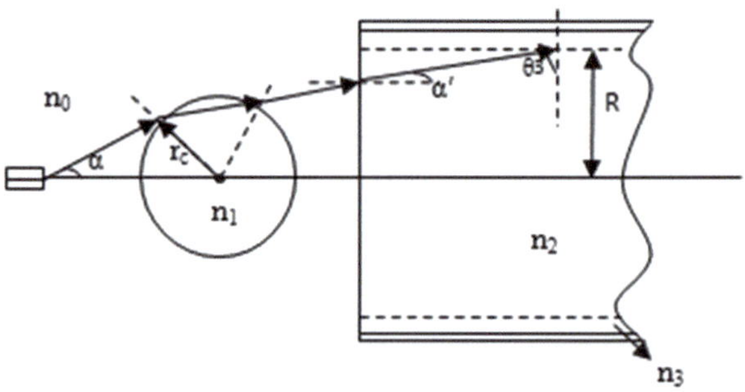

Fig. 5.14 Free space beam coupling scheme of a single emitter [2]

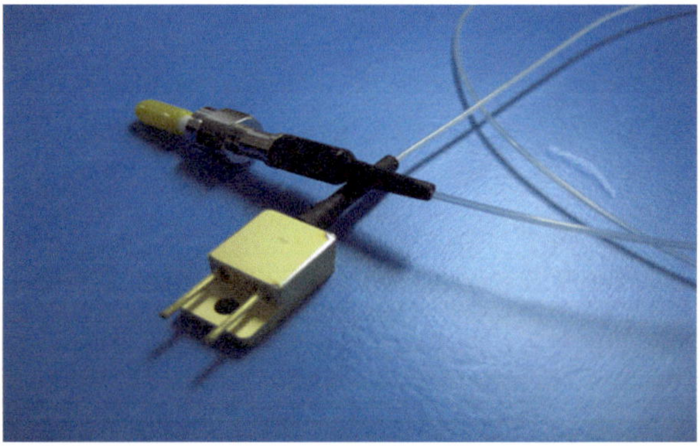

Fig. 5.15 A typical fiber-coupled single emitter semiconductor laser [18]

Fig. 5.14 [2, 14]. The cylinder lens is used to collimate fast beam, and n_0, n_1, and n_2 are refraction index of air, cylinder lens, and fiber core, respectively. R is diameter of the inner fiber core, and generally it is equal to the active region size along slow axis. Besides the cylinder lens, other optical lenses, such as spherical and aspherical lenses, are also used to collimate or focus the beam [15–17].

Based on this method, a 10 W semiconductor laser with active region size of 1 μm × 100 μm has successfully been coupled into a fiber core of diameter 100 μm as shown in Fig. 5.15 [18]. The coupling efficiency can be higher than 90 % and the output power is around 9 W.

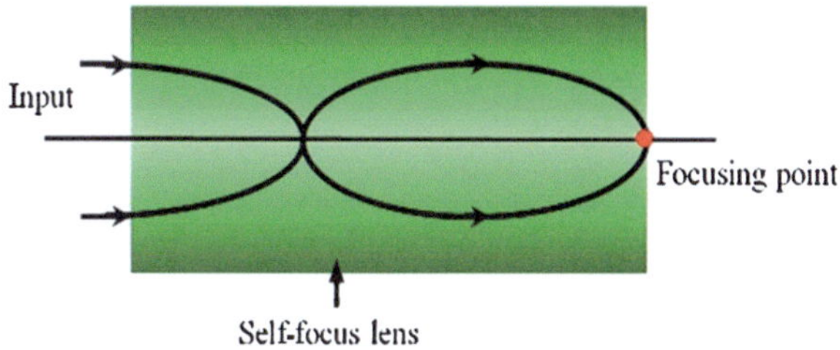

Fig. 5.16 The beam focusing principle of self-focusing lens [2]

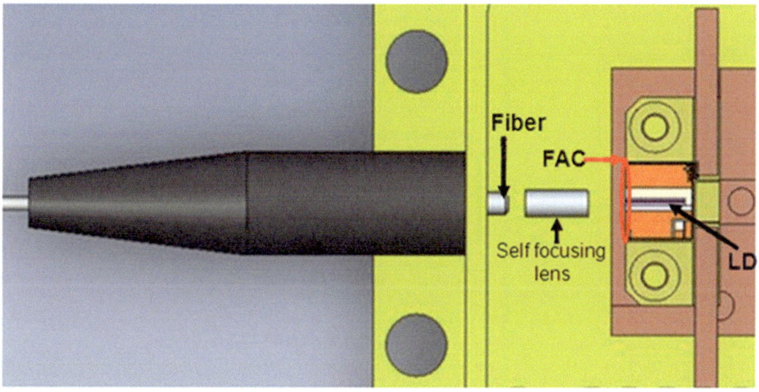

Fig. 5.17 The design of optical coupling system [20]

3. Graded-index lens

   The self-focusing lens is Gradient index lens, of which the inner index gradually reduces along the radial direction. The Graded-index lens is a common optical element to focus light beams. It has been used to couple semiconductor laser beams in recent years. The beam focusing principle of Graded-index lens is shown in Fig. 5.16, and a parallel beam translates as a stationary wave in the lens and focuses on the output end [2, 19].

   The Graded-index lens is applied in fiber coupling scheme of a single semiconductor laser, and the design of optical coupling structure is shown in Fig. 5.17 [20]. In this scheme, there are three important optical elements: FAC, self-focusing lens, and fiber. The FAC is a very small cylinder lens with hundreds micrometers bonded in front of the semiconductor laser, and the self-focusing lens is used after the FAC to further focus the beam.

5.2.2 *Single Bar Semiconductor Lasers*

The fiber coupling for single emitter laser is simple due to small beam size. However, the size of beam of semiconductor laser bar is 10 mm along slow axis, and beam quality of a laser bar is much worse than that of a single emitter laser.

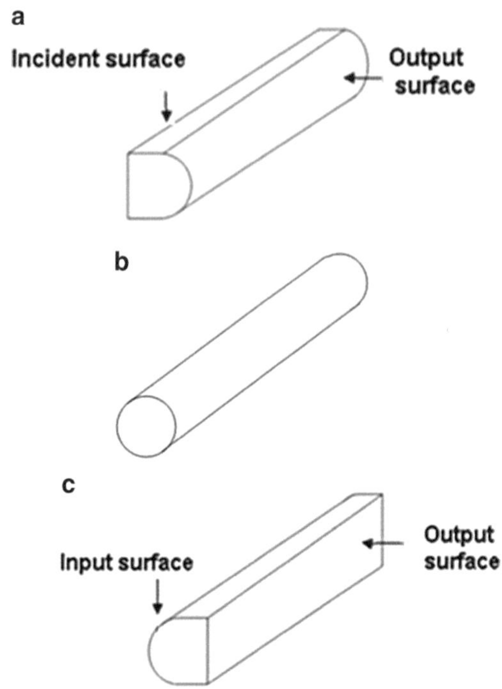

Fig. 5.18 Three collimation lenses for the fast axis [20]. (**a**) "D" type. (**b**) "O" type. (**c**) Inverse "D" type

Besides beam collimation along fast axis, the beam shaping of slow axis is also needed in many cases. The fiber coupling system for single diode laser bars includes three parts which are beam collimation, beam shaping, and fiber coupling, and they will be discussed in following section.

1. Collimation principle and design

   Fast axis: In general, the collimation of a semiconductor laser is divided into two steps. The beam of a semiconductor laser along the fast axis is collimated at first, and then along the slow axis. There are several kinds of collimation lenses for the fast axis, such as the cylindrical lens of "D" type, the cylindrical lens of "O" type. and the lens of the inverse "D" type, as shown in Fig. 5.18, where incident surfaces and output surfaces of the collimation lenses are marked, respectively [20]. The input surface of the "D" type lens is a plane and the output surface on the fast axis is curved. The inverse "D" type lens is the exact opposite. Both surfaces of the "O" type lens are curved on fast axis. However, the surfaces of all three lenses are plane in the slow axis. The shape and the curvature radius of the lens on the fast axis should be designed and optimized to achieve the collimation according to different applications.

   The size of a semiconductor laser source on the fast axis is so small that it can be considered as a point source. According to geometrical theory it can be

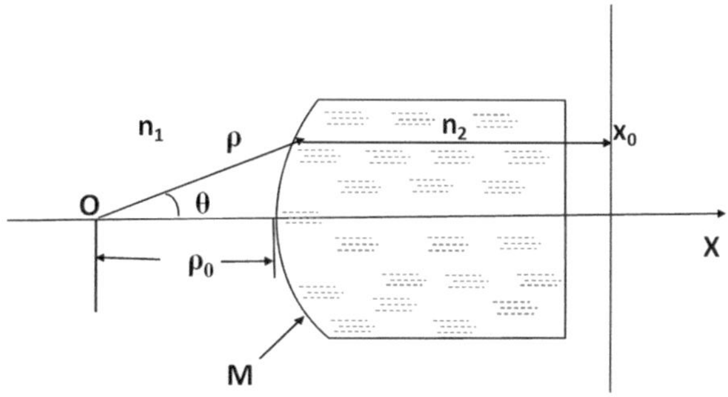

Fig. 5.19 The principle for collimation lens [2]

strictly collimated if the correct collimation lens is designed. In Fig. 5.19 surface M is denoted the interface of two optical media, and the refractive indexes of these two media are n_1 and n_2, respectively [1, 2]. The surface M is rotationally symmetrical. A semiconductor laser is put in the medium with refractive index of n_1 and the location is O. The light rays are refracted by surface M and propagate into the medium with n_2 refractive index. It is supposed that the beam is converted into a collimated beam.

According to Aplanatic principle, we can obtain [1]:

$$\rho(\theta) + n(x_0 - \rho(\theta)\cos\theta) = C, \tag{5.11}$$

where n denotes the value of n_2/n_1, $\rho(\theta)$ denotes the polar radius from the source point to the interface M, and C is a constant. Let $\rho(0) = \rho_0, \theta = 0$ and $x_0 - \rho(\theta)$ $\cos\,\theta > 0$, then we can obtain the value of C in Eq. (5.11).

$$C = \rho_0(1 - n) + nx_0, \tag{5.12}$$

Replaced C by Eq. (5.12), Eq. (5.11) can be rewritten as

$$\rho(\theta) = \frac{(1 - n)\rho_0}{1 - n\cos\theta}, \tag{5.13}$$

Equation (5.13) is a conic equation in polar coordinates. It expresses a ellipse equation for $n < 1$. When $n > 1$, it expresses an hyperbola equation. Generally, n_2 is always greater than n_1, and a hyperbola lens is used to collimate a point light source so the ideal collimation beam can be obtained. However, the semiconductor laser beam on the fast axis is not an ideal point source, so the collimated beam still has a divergence angle. In theory, the divergence angle of a semiconductor laser beam can be below 0.1 mrad on the fast axis after collimating. The profile of the collimation lens is shown in Fig. 5.20 [20].

Due to complex processing and high cost, the fast axis collimation of a semiconductor laser also uses cylindrical lens or spherical lens instead of hyperbola lens. Figure 5.21 shows a cylindrical lens of the "D" type to collimate a semiconductor laser bar, and the divergence angle is reduced from 35 to 2 [21].

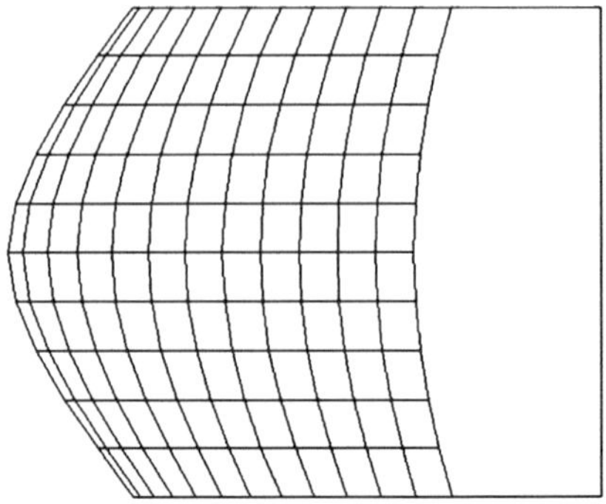

Fig. 5.20 The profile of the collimation lens [20]

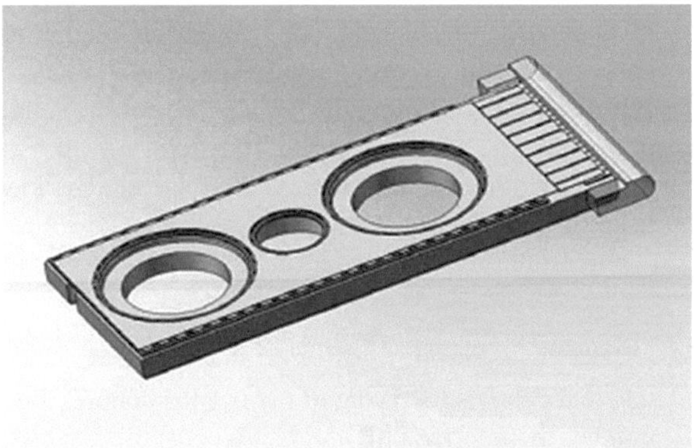

Fig. 5.21 A laser bar with "D" type collimated lens [21]

Slow axis: The beam on the slow axis is collimated by a cylindrical lens, which is composed of a number of micro-cylinders. Each micro-cylinder faces to one emitter of the semiconductor laser bar, so the number of the micro-lenses N equals to the number of the emitters of a semiconductor laser bar. A semiconductor laser bar has 19 emitters, and the collimation lens has 19 micro-cylinders as shown in Fig. 5.22 [21, 22]. The divergence angle of the slow axis of a laser bar with an emitter size of 100–200 μm can be reduced from 10 to 3° after collimation, which is much larger than the divergence angle of the collimated beam of the fast axis.

Integrated collimation lens: The fast and slow axes collimation lenses can be integrated together. Figure 5.23 shows the integrated lens system, where the fast axis collimation lens (arrow 1) is in the front and the slow axis collimation lens (arrow 2) is in the back [21]. A picture of a commercial collimation system is shown in Fig. 5.24 [23].

An optimized integrated lens can collimate the beam of a semiconductor laser in fast and slow axes simultaneously, which could replace the fast and slow axes collimation lenses, and therefore the collimation system is simplified. The design principle of the optimized lens is shown in Fig. 5.25, where "1" and "2" denote the input and output surfaces of the lens [20]. The input surface of the

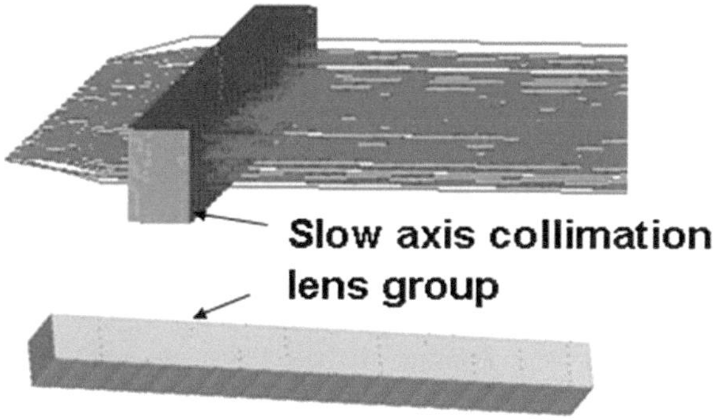

Fig. 5.22 The slow axis collimator [21]

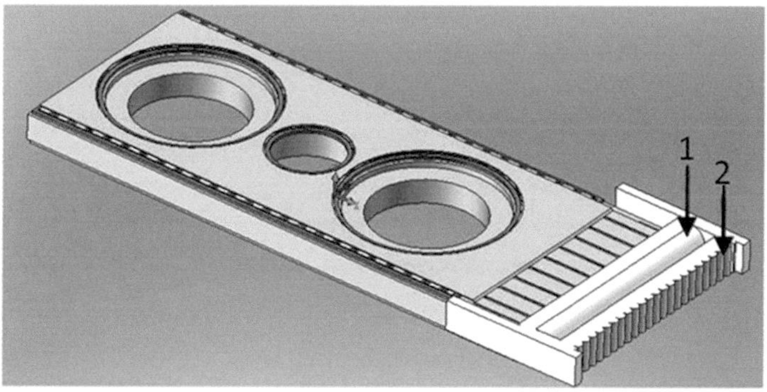

Fig. 5.23 A laser bar with fast and slow axes collimators [21]

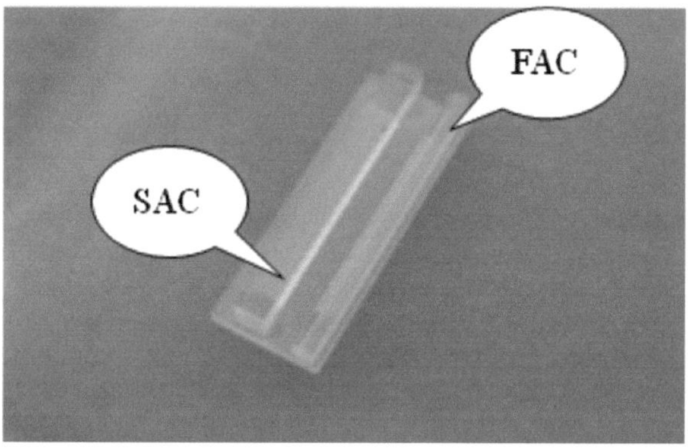

Fig. 5.24 A picture of an integrated collimation lens [23]

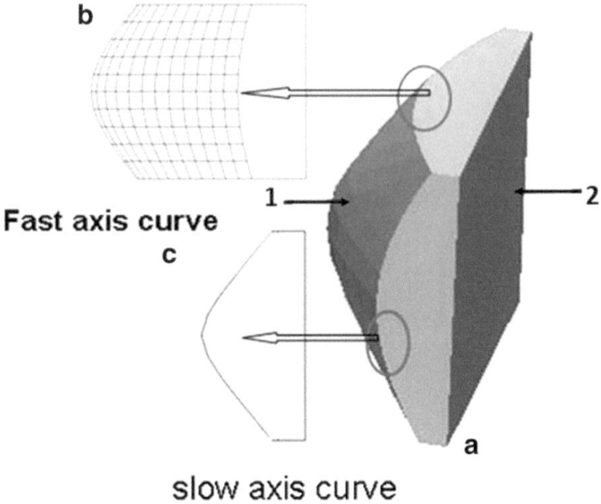

Fig. 5.25 Optimization principle for integrated collimation lens [20]

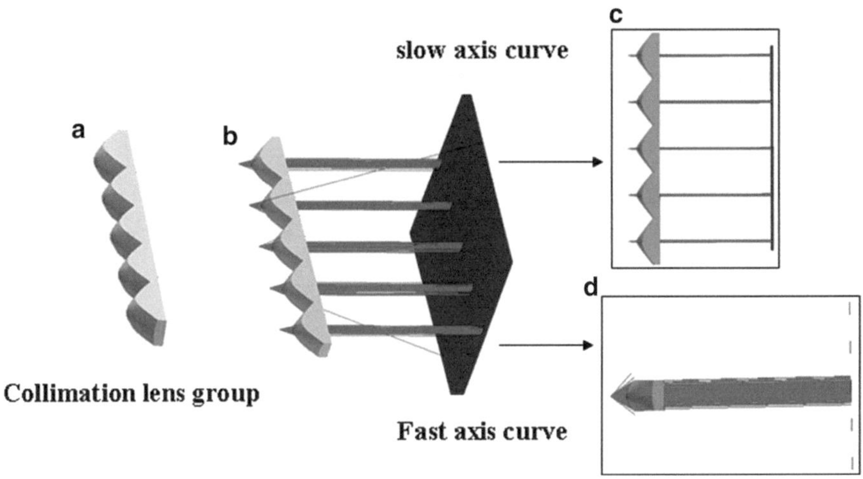

Fig. 5.26 The optimized collimation lenses array [21]

lens is a hyperboloid in the fast axis and a cylinder in the slow axis, and the output surface is a plane in both axes. Figure 5.25b, c shows the profiles of the lens on the fast and slow axes, respectively.

The collimation lenses are arranged in a row as shown in Fig. 5.26a [21]. The numbers of lenses are identical with the number of emitter in a semiconductor laser array. Here, five lenses are taken as an example to collimate a semiconductor array with five emitters. Each emitter is collimated by one lens as shown in Fig. 5.26b and the profiles of the collimation beam along fast and slow axes are shown in Fig. 5.26c, d, respectively.

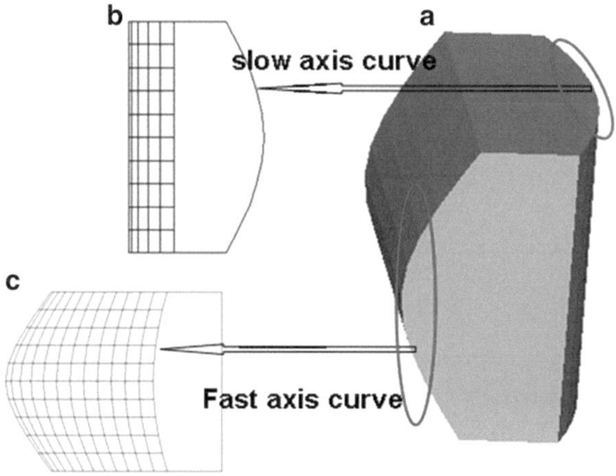

Fig. 5.27 Another efficient design for lens optimization [20]

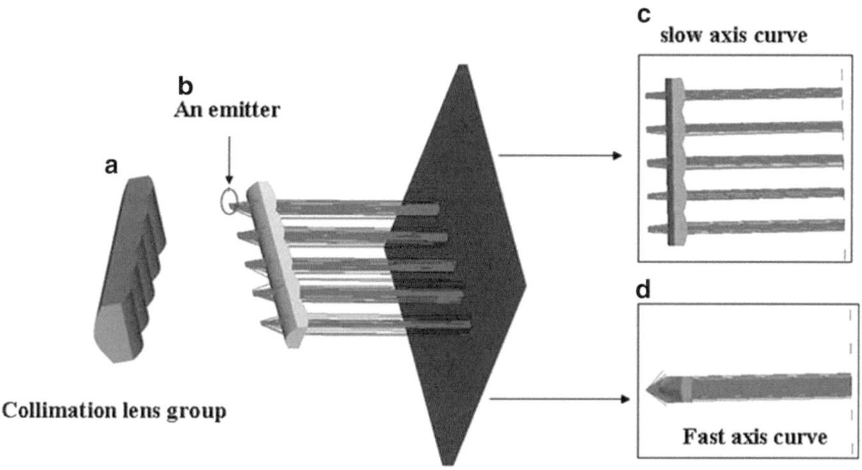

Fig. 5.28 The collimation lens array [21]

Another efficient design of the optimization lens is shown in Fig. 5.27a. On the incidence surface of the lens there are a hyperboloid on the fast axis and a plane on the slow axis, and the output surface is a plane on the fast axis and a cylinder on the slow axis. Figure 5.27b, c shows the profiles of the lens in the fast and slow axes directions, respectively [20].

Similarly, the lenses are arranged in a row as shown in Fig. 5.28a, and five lenses are taken an example to collimate a semiconductor array with five emitters shown in Fig. 5.28b [21]. The number of lenses is identical with the number of emitters for a semiconductor laser array. The profiles of the collimation beam in fast and slow axes are shown in Fig. 5.28c, d, respectively.

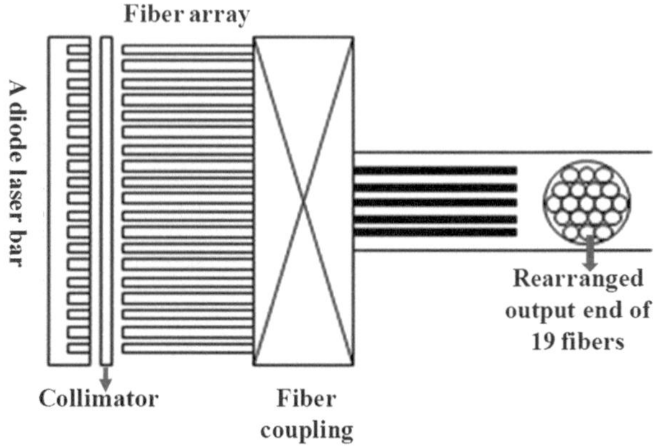

Fig. 5.29 The principle of fiber coupling by fiber array bundle [24]

Fig. 5.30 Fiber coupling module by fiber array bundle [18]

2. Fiber coupling by fiber array bundle
   After collimation, the beam of semiconductor laser bar can be coupled by fiber array bundle. Taking a laser bar with 19 emitters as an example, the principle is shown in Fig. 5.29 [24]. The fiber array shown in the figure is composed by 19 fibers, and 19 are equal to the number of emitters in laser bar, and each emitter is coupled into one fiber. After coupling the fibers are combined and rearranged as a circle at output end by a fiber combiner. And then the rearranged end is shown in Fig. 5.29. Generally, the diameter of the fibers is 100 μm and based on this method after rearranged the diameter of the end is 500 μm. Finally, a 600 μm fiber is used to couple the beam from the rearranged end.

   A 915 nm 50 W semiconductor laser bar has successfully been coupled into a fiber core of diameter 600 μm as shown in Fig. 5.30 [18]. The coupling efficiency and output power are 95 % and 48 W. However, the brightness cannot be improved more by the fiber array bundle approach. In order to improve brightness, the beam of a laser bar should be shaped before fiber coupling.

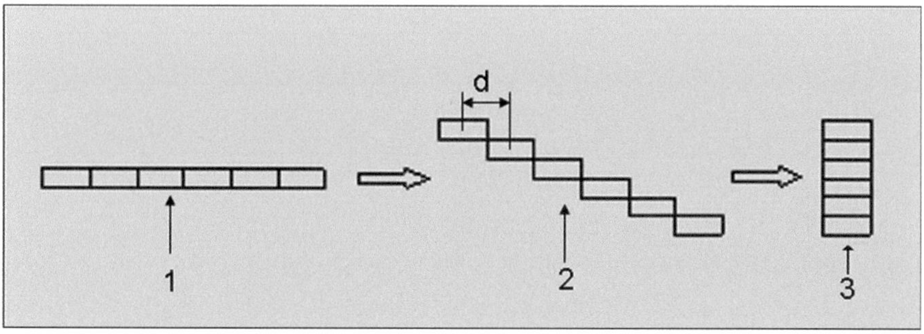

Fig. 5.31 The beam shaping processes of a semiconductor laser bar [23]

3. Optical design and beam shaping
   The poor beam quality of a semiconductor laser bar is on account of the large difference between the BPP on fast and slow axes directions, and the value of BPP_s is nearly 1400 times larger than BPP_f. To improve the beam quality, the principle of the shaping method is to eliminate the difference of BPP between two axes, so that BPP_f is similar to BPP_s. The process of beam shaping is that the beam of a semiconductor laser bar should be collimated on fast and slow axes at first, and then the collimated beam is cut into N sub-beams and each sub-beams is separated with a distance d along propagation axis as shown in Fig. 5.31 [23]. And then the sub-beams are rotated and rearranged to make the BPP of output beam symmetric in both fast and slow axes directions.

   The parameter N is defined as the cutting number and is decided by the following equation:

$$N^{*}BPP_{f} = BPP_{s}/N, \tag{5.14}$$

$$N = \sqrt{\frac{BPP_s}{BPP_f}}. \tag{5.15}$$

   After beam shaping, the BPP of a semiconductor laser bar in the fast axis direction is increased N times, while the BPP in the slow axis direction is decreased N times, then after rearrangement the value of the BPP in the fast axis direction is equal to that in the slow axis direction. The sketch of the shaping principle is shown in Fig. 5.31, where "1" expresses the collimation beam of a semiconductor laser bar, "2" expresses the cut and rotated beam, and "3" denotes the rearrangement beam. After shaping, the beam of a semiconductor laser bar can be focused into a small spot or coupled into a small diameter fiber core, such as 100 μm.

   Based on the reflection and refraction principles of geometrical optics, many kinds of shaping optical elements for a semiconductor laser bar are designed, such as micro-step mirrors, two high reflective parallel mirrors, and the micro-prism stack. The shaping principles and processes of these elements are

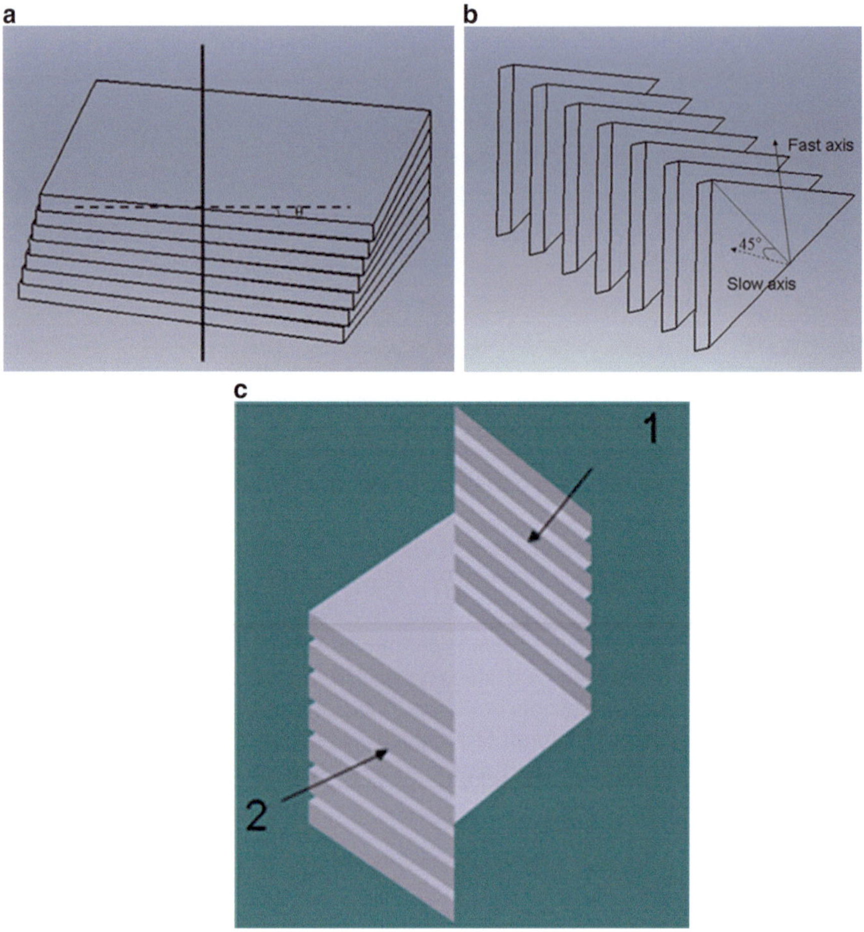

Fig. 5.32 The beam shapers of prism stacks. (**a**) Prism stack of parallelogram lens slices [24]. (**b**) Prism stack of isosceles right-angle triangle lens slices [26]. (**c**) Two prism stacks [20]

introduced as follows. The collimation of beam along fast and slow axes is the basic step, and the following discussions about the beam shaping of a semiconductor laser bar are based on the collimated beam. Taking an 808 nm semiconductor laser bar as an example, it has 19 emitters and the size of the source is 1 μm and 10 mm in the fast and slow axes directions, respectively. After collimating, the values of BPP in the fast and slow axes directions are 28.57 mm*mrad and 1,400 mm*mrad, respectively. According to Eqs. (5.14) and (5.15), the cutting number N is 7 and therefore the collimated beam should be cut into seven sub-beams, then the BPP of the semiconductor laser bar has the same value for both axes.

Prism lens stacks: Based on the shaping principle, prism stacks are proposed [20, 24–26]. In Fig. 5.32, three kinds of prism stacks shapers are depicted. Figure 5.32a shows the prism stack which consists of parallelogram lens slices,

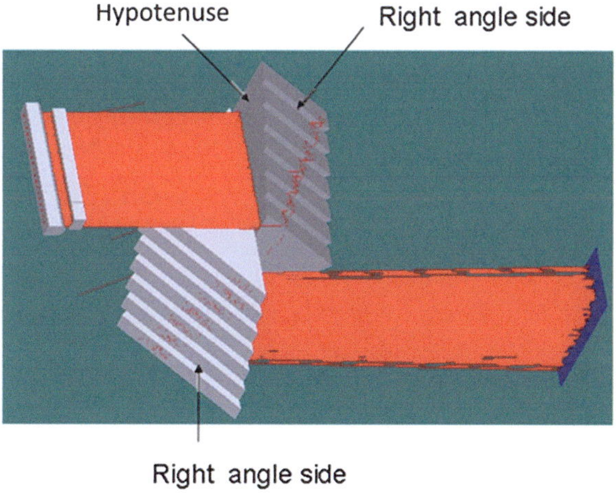

Fig. 5.33 The shaping processes of the prism stack pair [20]

and each slice is rotated with an angle along the slow axis [24]. Figure 5.32b shows the prism stack which is composed of isosceles right-angle triangle lens slices and each slice is titled 45° along the slow axis [26]. Two prism stacks are shown in Fig. 5.32c, they are perpendicular with each other and each stack is composed by the isosceles right-angle triangle lens slices [20].

The prism stack shapers are based on the reflection and refraction theory, the shaping process of the prism stack shapers in Fig. 5.32c is more complex than the former two shapers. As an example, the shaping principle of the prism stacks system is discussed in detail as follows. As shown in Fig. 5.32c, the prism stacks system is composed by two prism stacks, each stack consists of 7 isosceles right-angle triangle prisms. The planes of the prism stacks are composed of the hypotenuse of the isosceles right-angle triangle lenses, which are marked as "1" and "2" and they are perpendicular with each other and partly overlapped. The light propagation in the lens stacks is simulated in Fig. 5.33 [20]. The collimated beam incidents on the surface "1" of the right side stack and it is cut into seven sub-beams, meanwhile the sub-beams are rotated by an angle of θ along the slow axis. Then the rotated sub-beams are reflected by the right-angle side of the right side stack and incident into the left lens stack after passing through the overlapping interface of "1" and "2". Finally, the sub-beams are reflected by the right-angle side of the left lens stack and refracted by the plane "2", and then the beams are rearranged. After rearrangement, the BPP of the semiconductor laser bar on the two axes has the same value. The input and output intensity patterns of the semiconductor laser beam are shown in Fig. 5.34 [20]. Figure 5.35 shows the two prism lens stacks made by quartz glass [21].

Two plane parallel mirrors: Based on reflection principle, two highly reflective plane mirrors, aligned approximately parallel and separated by a small distance d, are shown in Fig. 5.36; "A" and "B" denote the two parallel mirrors, respectively [27]. The mirrors are transversely offset from each other in both the x′ and y′ directions defined in Fig. 5.36, so that small sections of each mirror

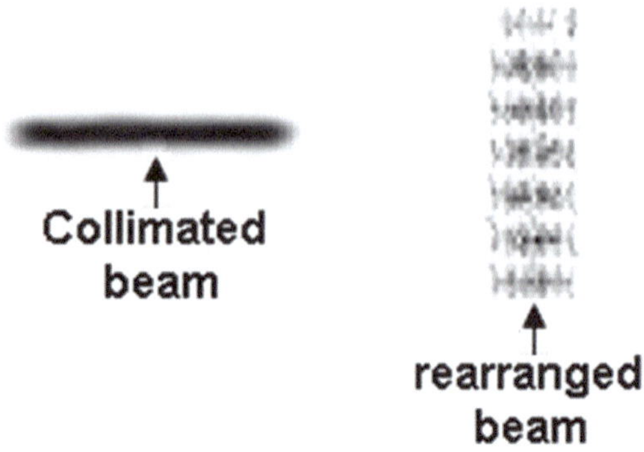

Fig. 5.34 The input and output intensity patterns [20]

Fig. 5.35 The two prism lens beam shaper [21]

are not obscured by the other. These unobscured sections from the input and output form apertures of the beam shaper.

Figure 5.36a, b, respectively, show plane and side views of the beam shaper. In each case the mirror surfaces are orthogonal to the plane of the figure. Light from a semiconductor laser bar with BPP_f and BPP_s in orthogonal planes of propagation x–z and y–z is incident obliquely at angles θ_f and θ_s in the orthogonal planes, x'–z' and y'–z', on the unobscured section of mirror B. The incident beam can be considered as composing of a number of adjacent beams. For the purpose of illustration, the incident beam has been arbitrarily chosen to be consist of N, N=5, parallel beams, (1)–(5), as shown in Fig. 5.36. Beam (1) is not incident upon either mirror A or B. Instead, it passes above mirror A and passes by the side of mirror B. Therefore, it does not change original direction. Beam (2) passes above mirror A but is incident upon mirror B and is reflected so

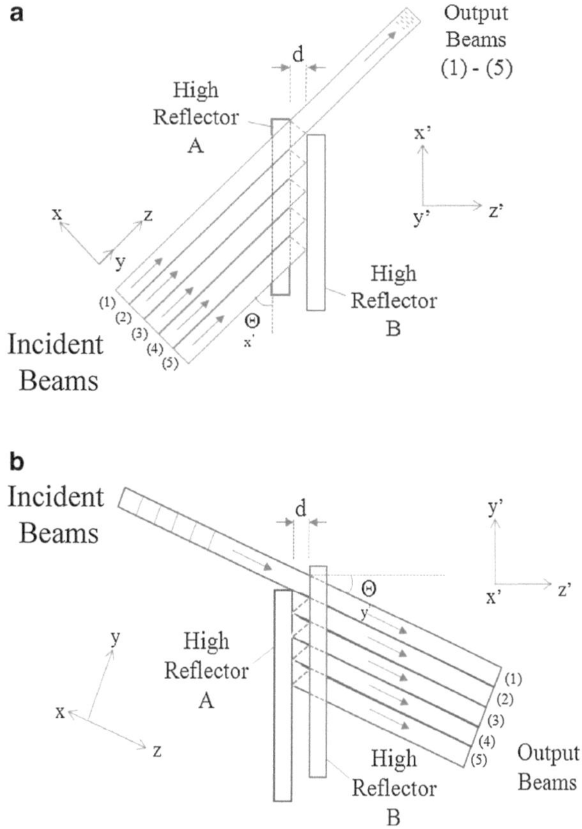

Fig. 5.36 Beam shaping principle of two plane parallel mirrors [27]

that it strikes mirror A immediately below beam (1). Beam (2) is then reflected at mirror A and emerges from the beam shaper in the direction of beam (1) but displaced beneath beam (1). Beam (3) after incidence upon mirror B, is reflected so that it strikes mirror A underneath beam (2) and is then reflected back to mirror B, where it is reflected onto mirror A, subsequently emerging parallel to beams (1) and (2) but displaced underneath beam (2). Beams (4) and (5) undergo similar multiple reflections at mirror A and B, and an arbitrary beam (n) finally emerges propagating beneath beam (n−1), as shown in Fig. 5.36. By this kind of beam shaper, if N=7, the beam of a semiconductor laser beam can be transformed as approximate square.

Step mirrors: Based on the reflection theory, step mirrors are also used as a beam shaper [28, 29]. The Cartesian coordinate system is built in Fig. 5.37, x and y axes denote the fast and slow axes, respectively, and z axis is the propagation axis. As described in Fig. 5.37, the step mirror system consists of two identical

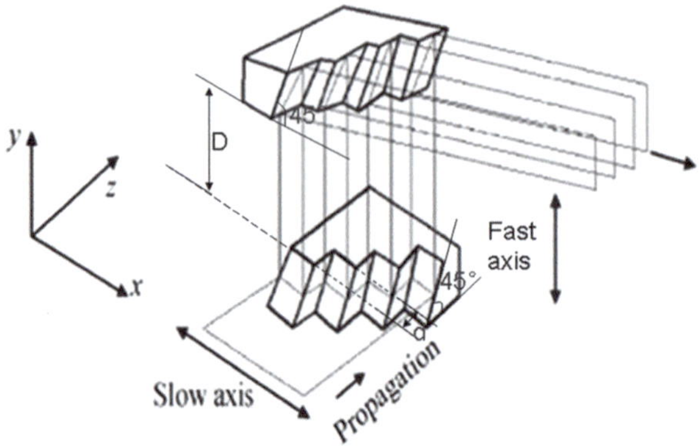

Fig. 5.37 The beam shaping principle of the Step mirrors shaper [28]

Fig. 5.38 The beam shaper of micro-step mirrors [20]

step mirrors, which are separated along the fast axis with a distance D [28]. Each mirror has N micro-mirrors which are arranged in a special manner as depicted in Fig. 5.37. The N micro-mirrors of the upper mirror are titled with 45° along the x axis and separated from the neighboring surfaces by a constant d along the x axis. This distance d corresponds to the width of a mirror surface. The N micro-mirrors of the under mirrors are titled along the z axis also with 45° and are separated from the neighboring surface by a distance of d along the z axis.

The beam shaping by the step mirrors are also shown in Fig. 5.37. The collimated beam is incident on the below step mirror, it is cut into N sub-beams along the slow axis. The sub-beams are reflected by each micro-mirror and the propagation direction is changed to parallel to the fast axis. Then the sub-beam is reflected by the above mirror and the propagation direction is changed again to parallel to the slow axis, so the beams are rearranged. After rearrangement, the beam quality is improved and the BPP on fast and slow axes are 135 mm*mrad and 140 mm*mrad, respectively. Figure 5.38 shows the beam shaper of micro-step mirrors [20].

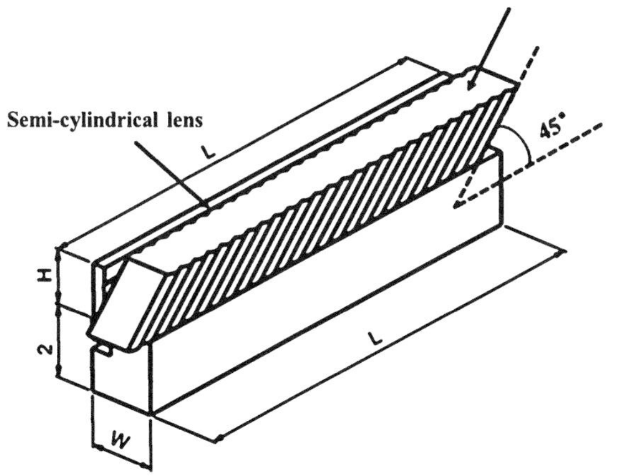

Fig. 5.39 The monolithic micro-optical system [30]

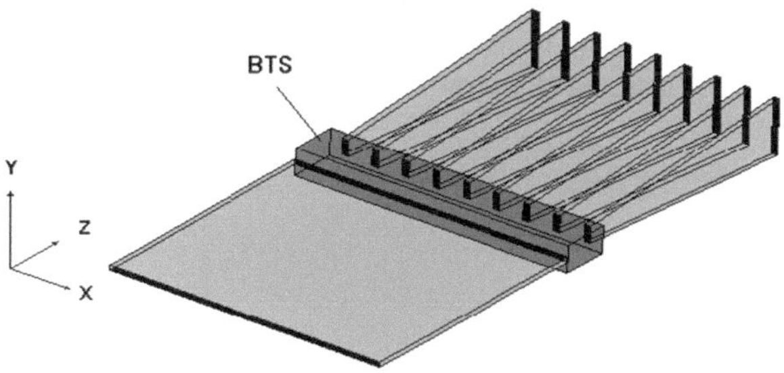

Fig. 5.40 The shaping process of the beam transformation system (BTS) [30]

The monolithic micro-optical system: The monolithic micro-cylinder system is another optical shaping element. It consists of N micro-cylinders and each cylinder is tilted along the slow axis as shown in Fig. 5.39 [30]. The system is usually made by quartz. Generally, the optical elements and the fast collimation lens are concentrated into an integrator, and the whole system is called as the beam transformation system (BTS), as shown in Fig. 5.39.

The shaping process of the beam transition in the BTS is simulated in Fig. 5.40 [30]. In the figure, the collimation beam is incident on the air–quartz interface, which is cut into N sub-beams along the slow axis and they are refracted in the cylinders. The sub-beams are oriented at an angle θ_1 along the slow axis. The rotated beams are incident on the quartz–air interface and then refracted into the air. The beams are rotated at an angle θ_2 along the slow axis, and the total value of $\theta_1 + \theta_2$ is 90°, then the beam are rearranged after refracting by the quartz–air interface.

4. Fiber coupling

   Based on the principle and technology of beam rearrangement of a semiconductor laser bar, the high brightness beam can be obtained by the focusing optical system.

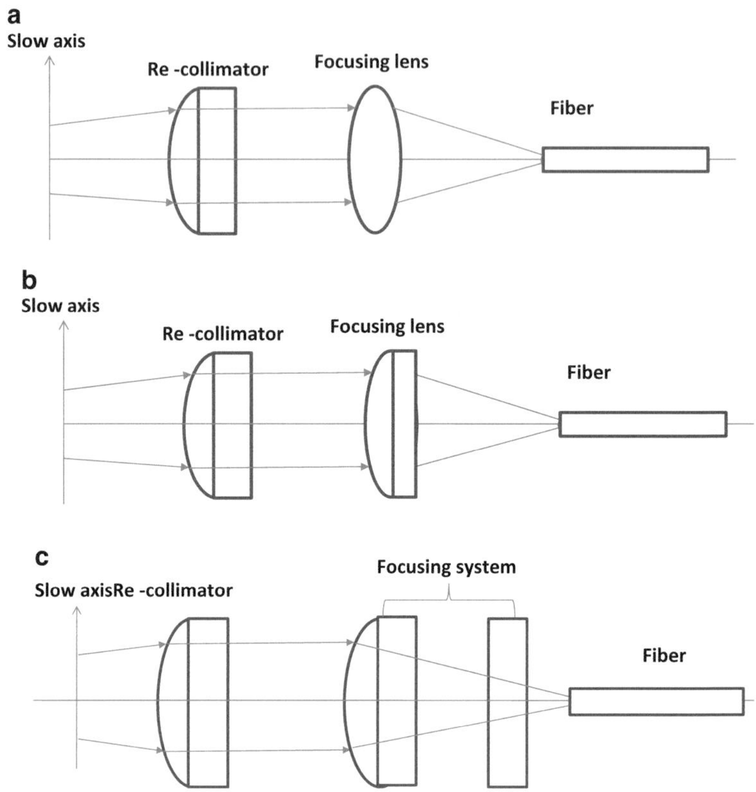

Fig. 5.41 The designs of focusing system [20]. (**a**) One focus lens with same focus length along two axes. (**b**) One focus lens with different focus length along two axes. (**c**) Two focus lens with different focus length along two axes

The focusing optical system should be designed according to the requirement of numerical aperture (NA) of the fiber. Generally, the beam along slow axis should be re-collimated after beam shaping, and cylinder lens always is used in this step. After that it is the focusing optical system, and Fig. 5.41 shows the designs of three typical focusing systems, respectively [20]. In Fig. 5.41a, the beam along fast and slow axes can be focused by one lens which has the same focal length along the two axes. In Fig. 5.41b, the beam along fast and slow axes can be focused by one lens, and however, the lens has different focal lengths along the two axes. And based on this design, the shaped beam can be focused into the fiber with smaller NA. In Fig. 5.41c, the beam along fast and slow axes can also be focused by two lenses which have different focus length.

Hence, the integrated optical system of fiber coupling module of a semiconductor laser bar is composed of the collimation lenses, the rearrangement element, and the focusing optical system. Taking step mirrors as an example,

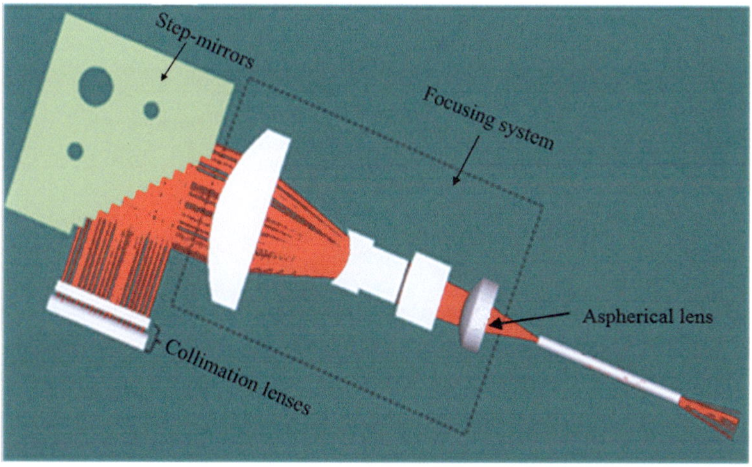

Fig. 5.42 The coupling system based on step mirrors [20]

the design of the high brightness system is depicted in Fig. 5.42 [20]. The collimation system, rearrangement element, and the focusing optical system are marked in the figure, respectively. The beam is successfully coupled into a 200 μm diameter fiber with NA 0.22 and the coupling efficiency is above 88 % in theory.

5.2.3 *Single Vertical Stack Semiconductor Lasers*

As discussed in Sect. 5.1, the semiconductor laser stack has large beam size, great beam divergence angles, and asymmetrical emission. Beam shaping and fiber coupling for vertical stacks are more sophisticated. In this section, we discuss some of the more practical approaches.

1. The telescope ocular principle
   The beam of a semiconductor laser stack is collimated along the fast and slow axes before shaping. After the collimating, the divergence angle on the fast axis is below 0.25°, so the beam on the fast axis can be treated as parallel light source approximately. The beam propagation principle through the telescope ocular is effectively used to shape the laser stack beam. Figure 5.43 shows the beam transition principle of a Galileo telescope ocular system [1, 2]. The input parallel light source transmits through the system and the beam is compressed into a narrow beam. It means that the beam size can be reduced after passing through a telescope ocular system.

   Taking an 808 nm semiconductor laser stack with ten semiconductor laser bars as an example, the beam shaping of the semiconductor laser stack is shown in Fig. 5.44 [31]. In the figure, the fast axis collimation lenses are marked

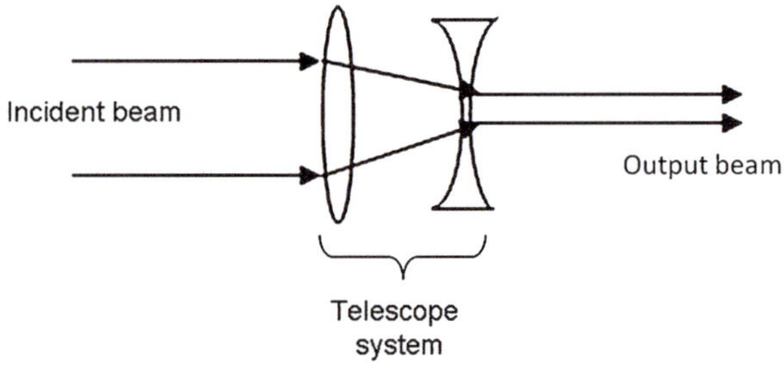

Fig. 5.43 The beam transition principle passing through a Galileo telescope ocular system [2]

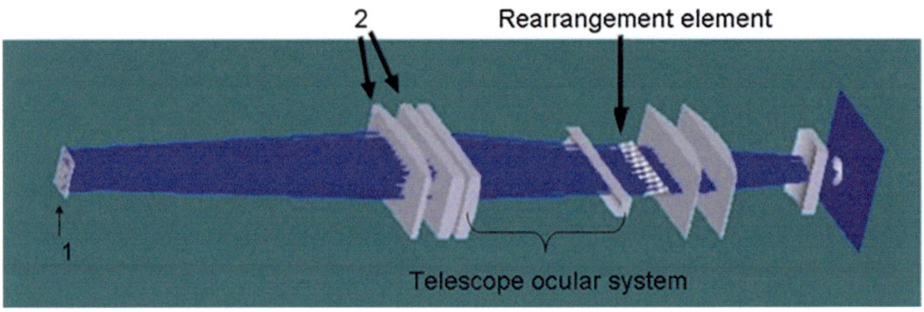

Fig. 5.44 The beam shaping of 10 bar semiconductor laser stack [31]

as 1, and the slow collimation lenses are composed with two cylinder lenses marked as 2, and the beam size of the collimated beam is 30 mm × 10 mm. A Galileo telescope ocular system is used to reduce the beam size at fast axis. After passing through the ocular system, the beam size on the fast axis is reduced from 30 to 1 mm, and the integer beam size of the stack is reduced to 1 mm × 10 mm. The beam rearrangement element is used and the rearrangement process has the same principle as the semiconductor laser bar. Then the beam of the laser stack is focused into a small spot and coupled into an optical fiber which has a diameter of 2 mm. The theoretical coupling efficiency can be as high as 90 %.

2. Beam rearrangement principle
   The second scheme is to reconfigure the collimated beam on the slow axis first, and then focus the reconfigured beam to a small spot.

   Optical rectangular cube. The design of the shaping method is shown in Fig. 5.45 [32]. The principle of beam reconfiguring of a semiconductor laser stack is similar to the beam rearrangement of a semiconductor laser bar. Two optical rectangular lenses are used for beam shaping as shown in Fig. 5.45. The beam is collimated at first, and the collimated beam of each high power semiconductor laser bar is separated into three parts with a vertical height shift of (bar)/3. For each part of the beam, it is separated by the two inclined optical rectangular cubes. The shift was precisely determined by the thickness and the inclined angel of the two optical rectangular cubes. The base holder of the two

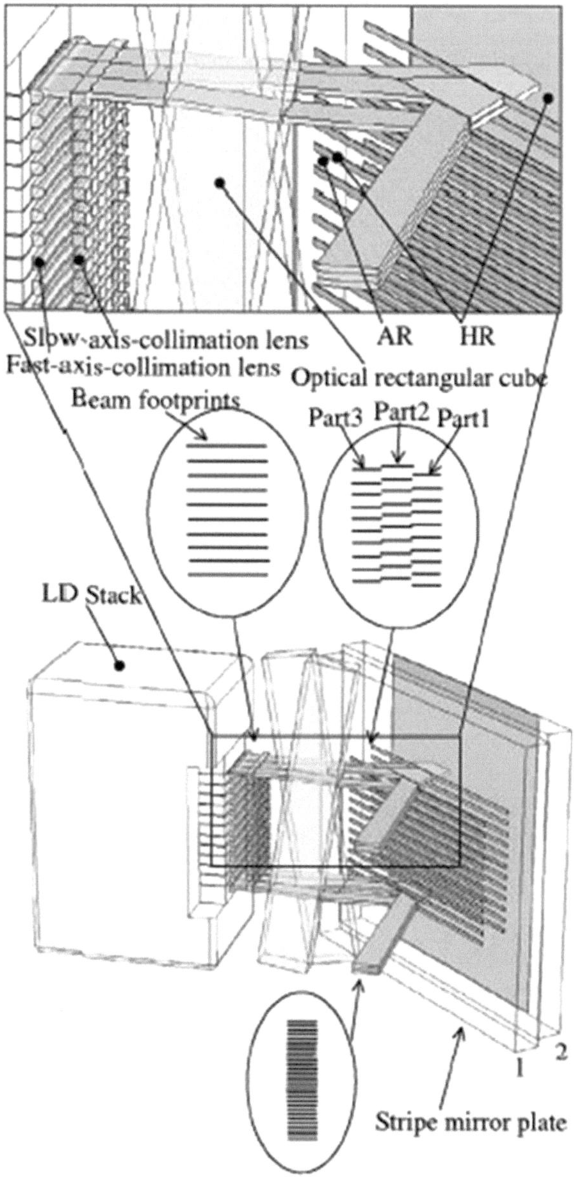

Fig. 5.45 The design of the shaping method of semiconductor laser stack [32]

optical rectangular cubes holds the cubes in place. As we can seen in the zoom of Fig. 5.45, part 1 is the lowest; part 2 is highest; and part 3, which does not pass through any rectangular cube, is in the middle of the vertical direction.

Based on this method, the fiber coupling system of a semiconductor laser stack with six bars is designed. In Fig. 5.46 the stack beam are collimated with FAC and SAC, and then the collimated beam is rearranged after propagating

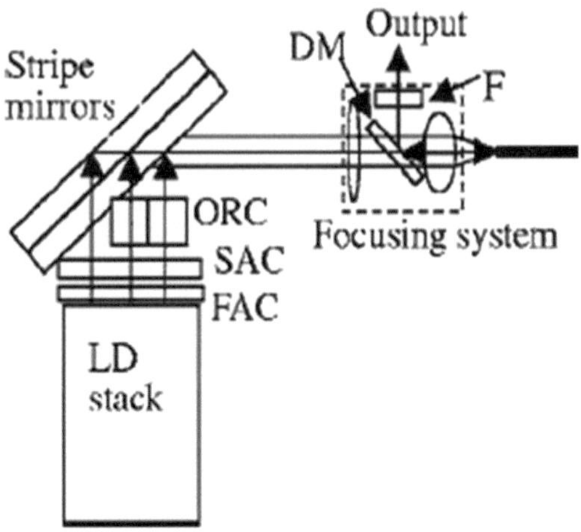

Fig. 5.46 The fiber coupling system of a semiconductor laser stack [32]

through the stripe mirrors [32]. Then rearranged beam is then focused into a small spot by the focusing optical system. As an example, a semiconductor laser stack with output power of 245 W can be coupled into a fiber with a core diameter of 400 μm and the launched power of the fiber module is 199 W which corresponds to a launch efficiency of 81 %.

Double-cutting beam shaping method. In order to evaluate the beam quality of a laser diode array, the most convenient way is to characterize the BPP, defined as $\theta \times W$, where W is the half beam width and θ is the beam divergence at half angle in the far filed. Due to the highly asymmetric output beam profile from the diode laser bar, efficient fiber coupling is only possible if the beam quality is adapted by cutting the beam into N_{fold} parts in slow axis and deflecting them to proper places in fast axis, and then meets the requirement given by Eqs. (5.14) and (5.15).

When N_{fold} is an even number, an example construction of new beam shaper with schematic diagram of transformation process is illustrated in Fig. 5.47 [33]. The device is organized with three main components. The first one is array1 of $N_{fold}/2$ pieces of prism, with an identical increment in length of adjacent prisms. While the incident beam passes through prism array1 vertically, it carries out the first cutting, tailoring the beam into $N_{fold}/2$ segments along slow axis and compressing the beam width in fast axis by refracting on different inclined planes. The second one is array2 consisting of $N_{fold}/2 + 1$ pieces of prism that has two kinds of length placed alternately and set at an appropriate angle to the X–Z plane.

According to the design parameters, the beam shaping elements were fabricated, as shown in Fig. 5.48 [33].

The simulated schematic of beam combination system is shown in Fig. 5.49 [33]. The laser source is a three-bar LD stack with the maximum 60 W (CW) per bar at a central wavelength of 976 nm. The output beams after transformation

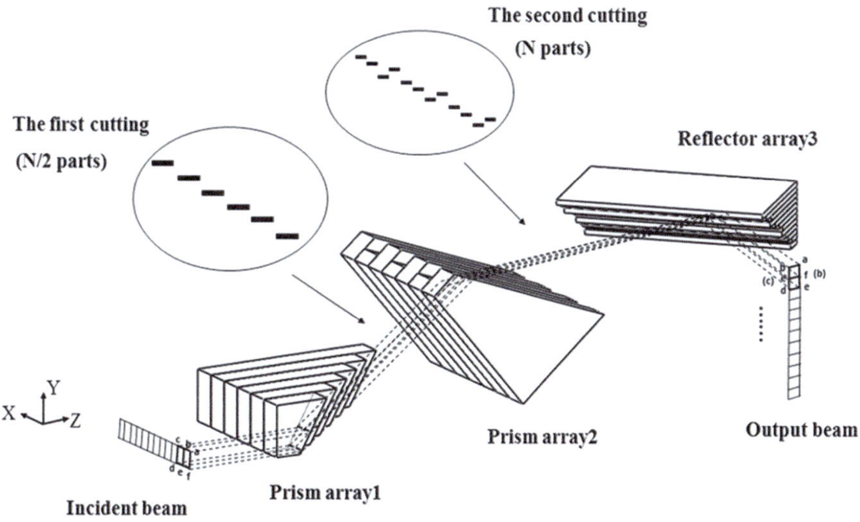

Fig. 5.47 Principle of double-cutting beam shaping method [33]

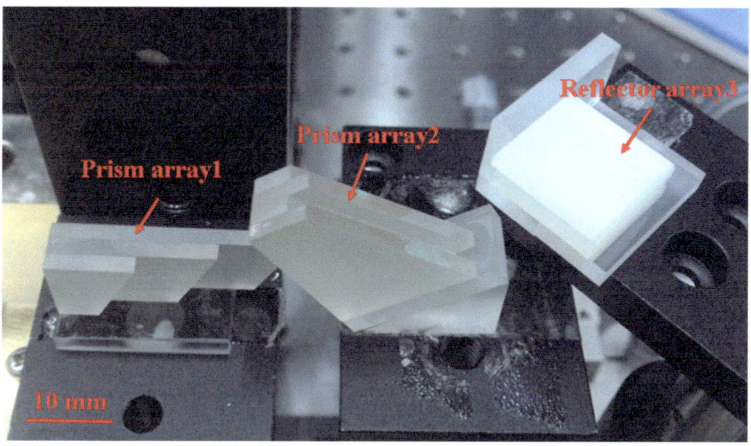

Fig. 5.48 Fabricated double-cutting beam transforming elements [33]

goes through a set of lenses to be focused into an optical fiber of standard 400 μm core diameter and NA 0.22.

In the same operating situation described above, the measured maximum power after collimating was 166.8 W while the fiber output was 129.8 W, the results of which are given in Fig. 5.50 [33]. The electrooptical conversion efficiency of 40.5 % (whole system) and a slope efficiency of 2.24 W/A (fiber output) testing from 7 A to 65 A were obtained. Considering all factors, the low

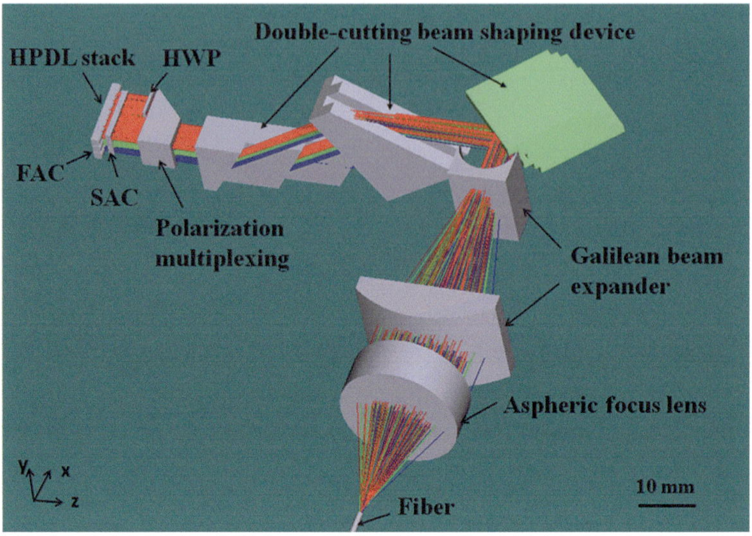

Fig. 5.49 Layout of beam combination in fiber-coupled module [33]

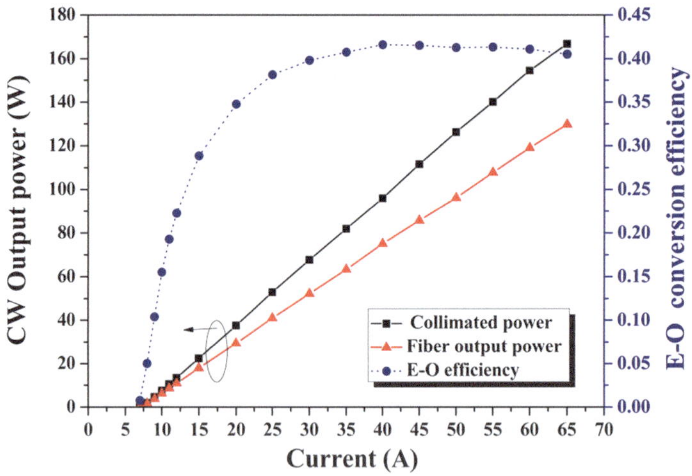

Fig. 5.50 Measured power and electrooptical conversion efficiency of fiber output in different current situations [33]

polarization degree (only about 93 %) of incident beam is one reason for the loss of the optical power in this system. Selecting the LD with higher polarization degree can further enhance the E-O efficiency. Besides, the directional error of LD stack also reduces the beam quality and leads to the energy loss in focusing the beam into fiber.

5.3 Beam Combining and Fiber Coupling Techniques

In many applications such as laser marking, laser cladding, and other material processing, high output power and brightness of semiconductor lasers are needed. The beam combination techniques, such as fiber array scaling, spatial, polarization, or wavelength superposition, are used to scale the output power or improve the beam brightness of semiconductor lasers.

5.3.1 Basic Beam Combining Principles

In order to increase output power, beam combination is needed. However, if the individual laser beams are simply set side by side in a one- or two-dimensional arrays, the size of the beam will be also increased. To achieve relatively high output power at high system efficiency, the basic physical principles used in all systems are spatial combining, polarization, and wavelength multiplexing superposition. The later two complementary techniques can increase the output power at constant beam size.

1. Free space beam combining
   In order to increase the output power, multiple semiconductor lasers can be combined by reasonable space arrangement or spatial combining, as shown in Fig. 5.51 [20]. The interval distance of b' between neighbor lasers should be precisely calculated. The parameter b' is denoted the center interval distance between two neighbor beams. In order to obtain high power and brightness, the value of b' should be as small as possible, meanwhile the beam should not be shielded by each other.
2. Polarization beam combining
   Laser beams generally have two polarization states. If the oscillation direction of optical field vector is perpendicular to the junction plane of the laser diode,

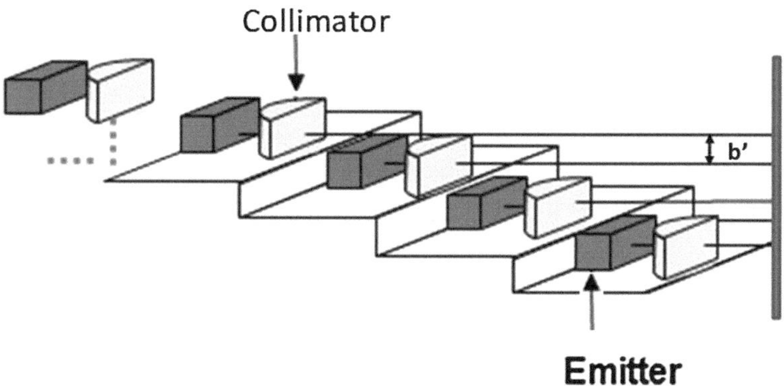

Fig. 5.51 The principle of free space combining [20]

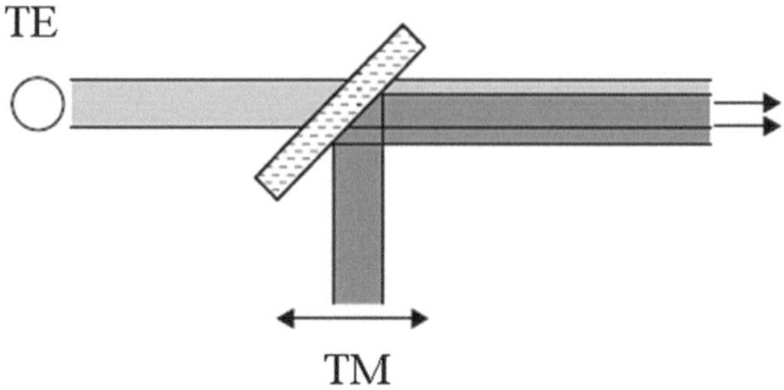

Fig. 5.52 The basic physical principle of beam polarization multiplexing superposition [7]

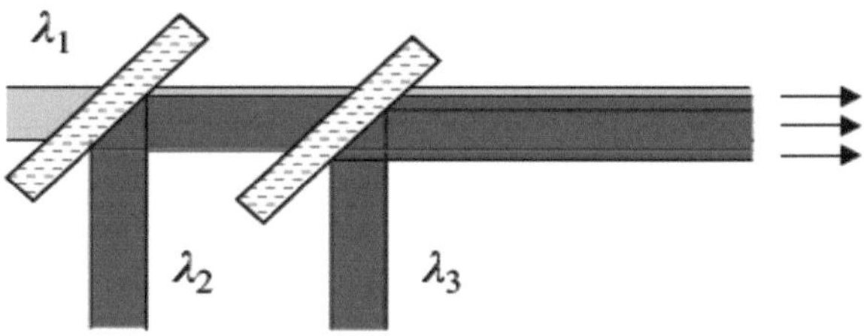

Fig. 5.53 The principle of the wavelength multiplexing superposition [7]

the light is in TE polarization state; On the other hand, if the oscillation direction of optical field vector is parallel to the junction plane of the laser diode, the light is in TM polarization state. Two polarization beams which are mutually perpendicularly polarized can be coupled via a polarization coupler, thus output power increases twice at constant beam size. The basic physical principle sketch of beam polarization multiplexing superposition is shown in Fig. 5.52 [7]. The TE beam passes through the polarization coupler directly and the TM beam is reflected by the coupler, two beams are combined after the combiner. Then the output power increases twice at constant beam size. The light sources can be two single emitters, two semiconductor laser bars or two semiconductor laser stacks.

If two light sources have the same linearly polarization state, a $\lambda/2$ retardation plate can be used to convert the polarization state. After the polarization state is converted for one light source, then the polarization multiplexing superposition can be used.

3. Wavelength beam combining

For the wavelength multiplexing superposition technique, multiple beams with different wavelengths can be coupled by the multi-wavelength coupler, thus the combined beams have the same beam size with the input beam and the output power is increased. Figure 5.53 shows the principle of the wavelength multiplexing superposition, and the output power can be increased by n times, where n is decided by the number of the different semiconductor laser wavelengths in the combining system [7].

To obtain even higher output power, any two of the above mentioned combining methods can be used in one system simultaneously. Figure 5.54 illustrates one example using the combination of polarization combining and wavelength

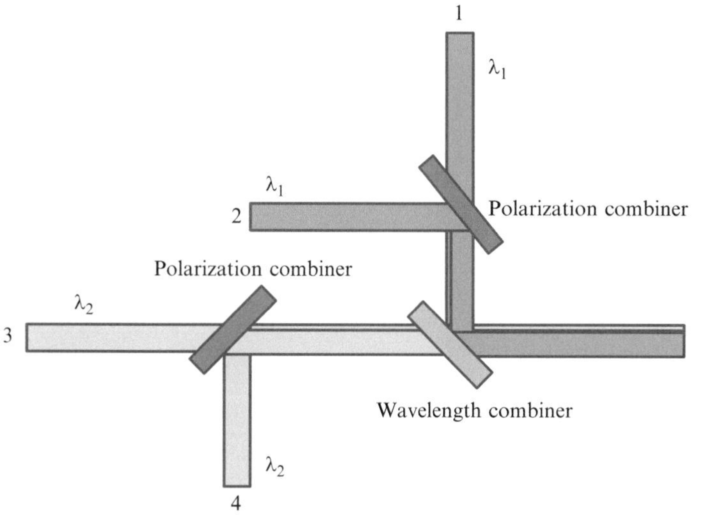

Fig. 5.54 The principle of beam combining method based on multiple waves and polarization multiplexing super-positions [2]

multiplexing [2, 7]. In the figure, the light sources "1" and "2" with same wavelength λ_1 are TE and TM polarization states, respectively. The light sources "3" and "4" with same wavelength λ_2 are TE and TM polarization states, respectively. The light sources "1" and "2" are combined by a polarization coupler to be a new beam with wavelength λ_1, and the light sources "3" and "4" are combined by another polarization coupler to be a new beam with wavelength λ_2. Then the two beams are combined by a multi-wavelength coupler, and the output power is increased by four times with the constant beam size.

5.3.2 *Single Emitter-Based Beam Combining and Fiber Coupling*

1. Fiber bundle
   There are N single emitter semiconductor lasers, and using the shaping and coupling technology discussed in Sect. 5.2 each laser is coupled into an optical fiber. Then the ends of the fibers are bundled together and the fiber is arranged as shown in Fig. 5.55 [20]. After bundling the output power is N times of the one coupling model, but the beam diameter of output beam is also increased. Taking 61 fiber modules with the wavelength of 975 nm as an example, the output power of each module is 6 W, the fiber diameter is 100 μm and the intensity density of the module is 0.76×10^5 W/cm^2. The fibers are bundled and rearranged into a pan as shown in Fig. 5.55. The fiber pan consists of nine groups and in the middle

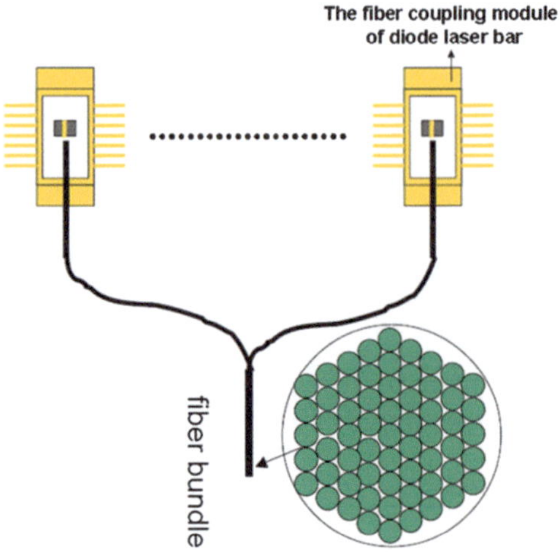

Fig. 5.55 The fiber bundle of 61 semiconductor laser bars [20]

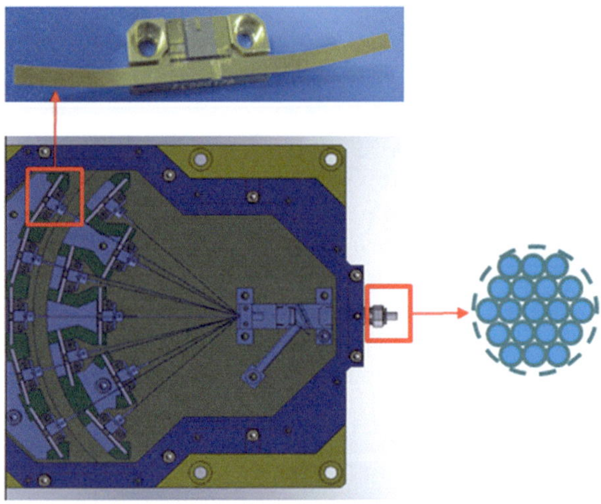

Fig. 5.56 Fiber coupling modules of single emitter semiconductor lasers using fiber bundle [20]

of the pan the number of the fiber is 9. The distribution of fibers is symmetrical and the fiber numbers of each group are 5, 6, 7, 8, 9, 8, 7, 6, and 5, respectively. In theory, the output power could be increased to 366 W which is 61 times of one fiber module, and at the mean time, the diameter of output beam is increased to 900 μm which is as large as nine times of one module. The intensity density of the beam keeps the same as one fiber module.

As shown in Fig. 5.56 19 fiber coupling modules of single emitter semiconductor lasers are bundled together [20]. The fiber core is 100 μm diameter and the output power is 8 W. In Fig. 5.56, there are 19 single emitter lasers of F-mount coupled into 19 fibers with 100 μm core diameter, and at output end these fibers are bundled and ranged as a pan. After fiber bundle, the total output power is 152 W and the beam size at the bundle end is 500 μm.

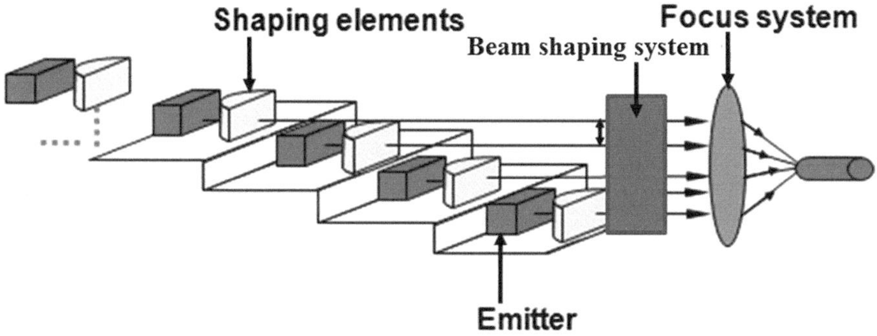

Fig. 5.57 Single emitter-based spatial beam combining and coupling [20]

2. Free spatial beam combination and fiber coupling
   As shown in Fig. 5.57, based on the free space combining principle, the semiconductor lasers are arranged in a row, and the beam shaping and focusing optical system should be designed [20]. Based on this method, the output power and brightness can be improved efficiently, and the output power of thousand watts from a fiber of 600 μm can be produced.
3. Polarization and wavelength beam combination and fiber coupling
   The polarization and wavelength beam combinations have been successfully used to achieve higher output power. The fiber coupling system designs of polarization beam combiners for multiple semiconductor lasers are shown in Fig. 5.58 [20]. In the figure, the semiconductor lasers are divided into two groups of G1 and G2, and they are TE and TM polarizations, respectively. If the two groups have two different wavelengths of λ_1 and λ_2, the wavelength beam combiner can be used to replace the polarization combiner.

5.3.3 Bar-Based Beam Combining and Fiber Coupling

It is known that the power of fiber-coupled module of a single laser bar is usually less than 100 W. In order to improve the output power and brightness, two shaping methods and techniques for multiple semiconductor laser bars are proposed, and the principle of these two methods will be discussed in the following paragraphs.

1. Fiber bundle of multi semiconductor laser bars
   The fiber bundle also can be used in semiconductor laser bars based on the same principle, and the output power can be reach thousand watts. The high power 1,900 W semiconductor laser system was fabricated with 19 fiber modules with output power of 100 W [34]. The 19 fiber modules are small fiber core with diameter of 100 μm and numerical aperture (NA) of 0.12. After fiber bundling, the output power increased 19 time more than that in one fiber modules and the diameter of the output beam is 500 μm which is more than five times.

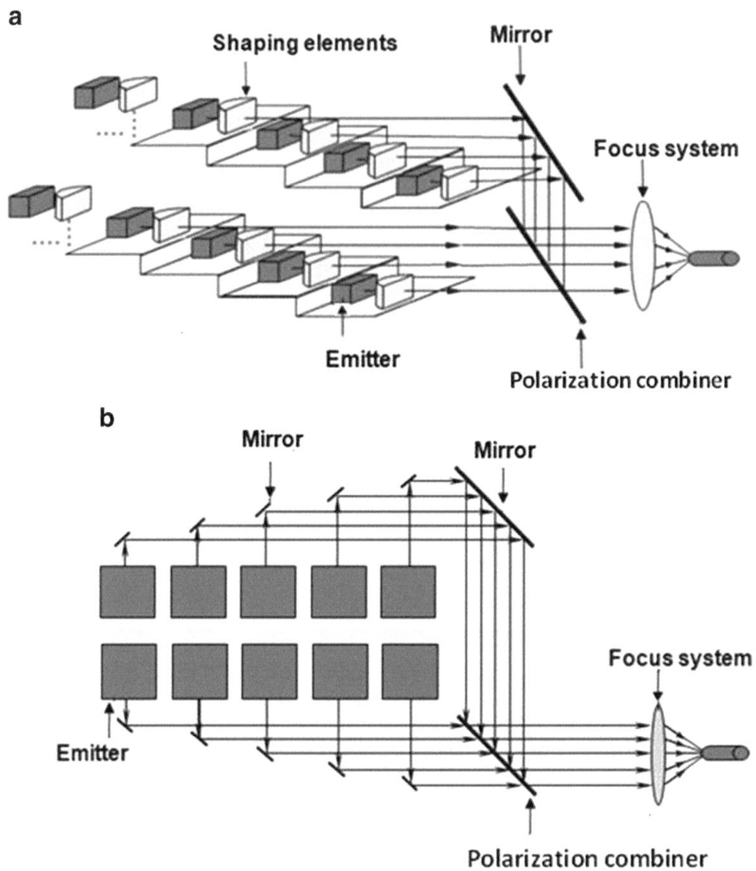

Fig. 5.58 Single emitter-based optical beam shaping and coupling system using spatial and polarization superposition [20]. (**a**) Step spatial beam combining design. (**b**) Reflecting mirror beam combining design

This method is simple, but with the increasing number of the bundle fibers, the volume of fiber system is increased rapidly. Due to the bundle fibers, the high power source system has large volume and the brightness is identical to that of a single fiber.

2. Spatial coupling for multiple semiconductor laser bars

   The spatial coupling technique of semiconductor lasers is an important method to improve both the output power and beam brightness. The main steps of this method are collimating beams in the fast and slow axes, rearranging the collimated beam of each semiconductor laser bar, and beam focusing. To obtain high brightness and high coupling efficiency, semiconductor laser bars should be placed as close as possible, and the beam should be concentrated into the focusing system as much as possible. In the third step, the arrangement

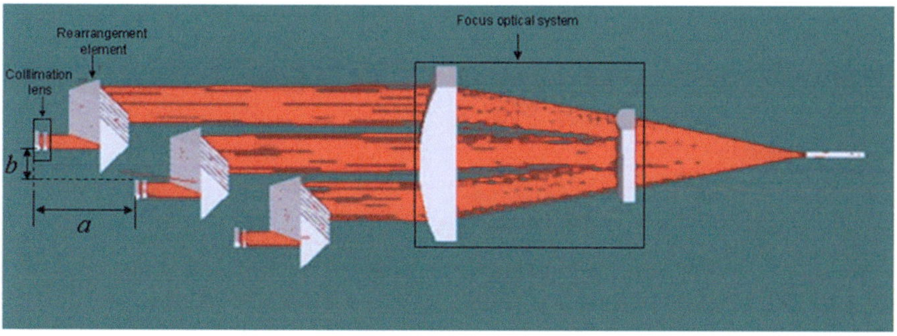

Fig. 5.59 The fiber coupling optical system of three 808 nm semiconductor laser bars [21]

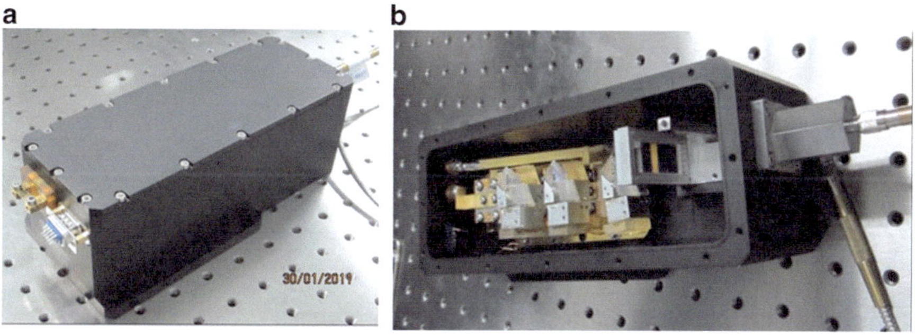

Fig. 5.60 Spatial coupling of three diode laser bars [21]. (**a**) The fiber-coupled module using three laser bars. (**b**) The inside of the module suing spatial coupling

parameters of semiconductor laser bars such as the relative positions and distances between each bar should be exactly determined. The fiber coupling optical system of three 808 nm semiconductor laser bars is shown in Fig. 5.59, where a and b denote the relative position and distance between each laser bars, respectively, and the collimating lenses, the rearrangement elements for the semiconductor laser bars, as well as the focusing optical system are marked [21].

In theory, the three semiconductor laser bars can be coupled into a single fiber with the core diameter of 200 μm. Based on this design, a coupling system is made and shown in Fig. 5.60 [21]. The beams are successfully coupled into a single fiber with 600 μm core diameter and 0.22 NA, and the output power is 150 W.

Based on the same principle, a 400 W coupling system of 808 nm semiconductor laser bars is also designed and fabricated. The diameter of the fiber is 400 μm and NA is 0.22. The sketch of that system is presented in Fig. 5.61 [21], and the actual optical system is shown in Fig. 5.62 [21].

3. Polarization and wavelength beam combination and fiber coupling
Polarization combination and wavelength multiplexing superposition are also used to improve the output power of the coupling system, where Fig. 5.63 shows

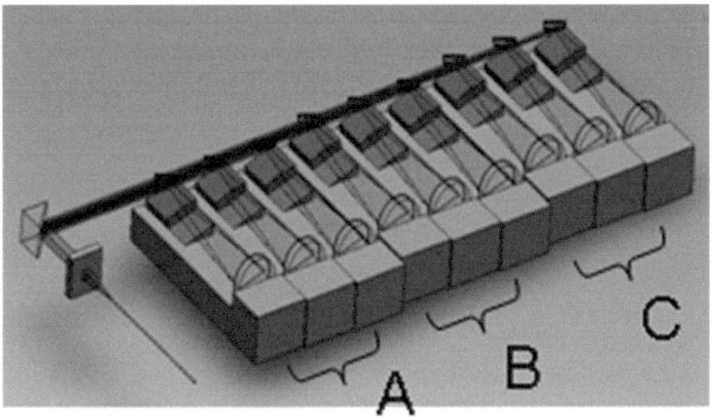

Fig. 5.61 The shaping and coupling design of nine semiconductor laser bars [21]

Fig. 5.62 The actual optical system [21]

an example of polarization combination [21]. The combination principles of these two methods are similar. Taking the polarization combination as an example, according to the polarization state, the semiconductor laser bars are divided into two groups, the TE and TM polarization beams, which are marked with 1 and 2 in Fig. 5.63. Beams from each semiconductor laser bar in each group are collimated and rearranged at first. Then the TE beams from the group 1 are incident on the polarization coupler and reflected by the coupler, while the TM beams from the group 2 are reflected by a mirror and then transmit through the coupler directly. The TE and TM beams are combined after propagating through the coupler. Finally, the focusing optical system is used to focus the combined beams into a small spot.

The wavelength and polarization multiplexing coupling technologies are used together frequently [7, 23]. As shown in Fig. 5.64, beams from six linearly polarized semiconductor laser bars with three different wavelengths were superimposed by wavelength and polarization multiplexing coupling [7].

The six semiconductor laser bars are divided into two groups according to the polarizations: the first and the second groups are the TM and TE polarizations, respectively. Each group has three semiconductor laser bars with the wavelengths of 808, 910, and 980 nm. Three beams from semiconductor laser bars

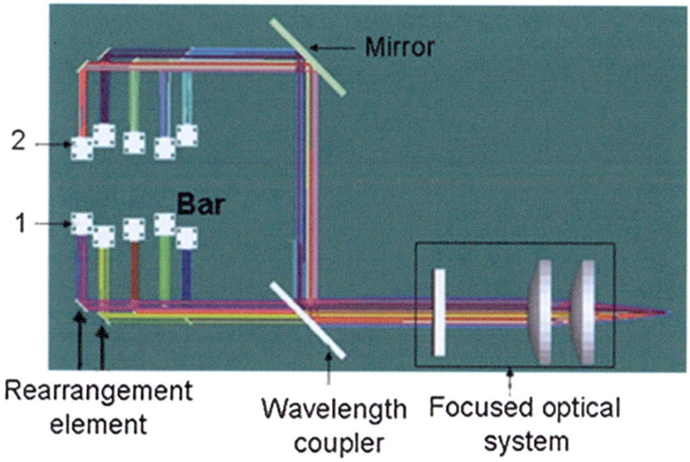

Fig. 5.63 Polarization multiplexing superposition in the coupling system [21]

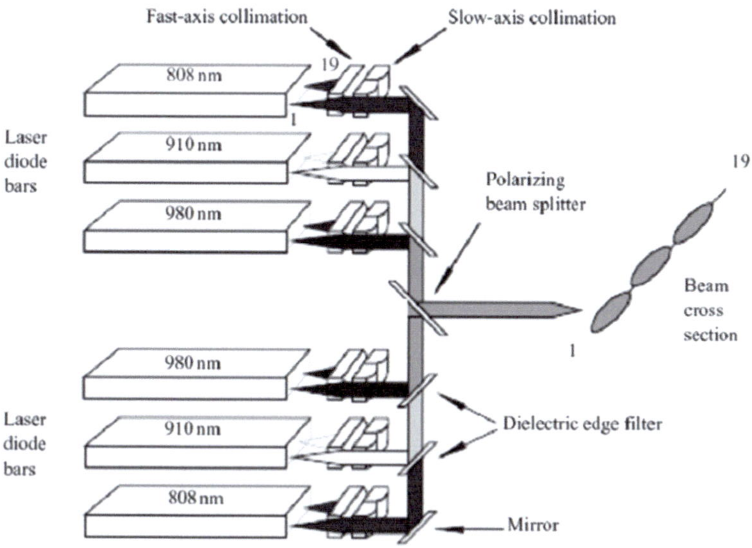

Fig. 5.64 The wavelength and polarization multiplexing coupling technologies [7]

in each group are combined by wavelength combiner and form two new optical beams, and after that the two beams are combined by the polarization beam splitter. Finally, that beam is focused into a small spot by a telescopic system, which is shown in Fig. 5.65 [7].

It was reported that the output power of six semiconductor laser bars combined using the scheme shown in Figs. 5.64 and 5.65, can be 140 W at 50 A.

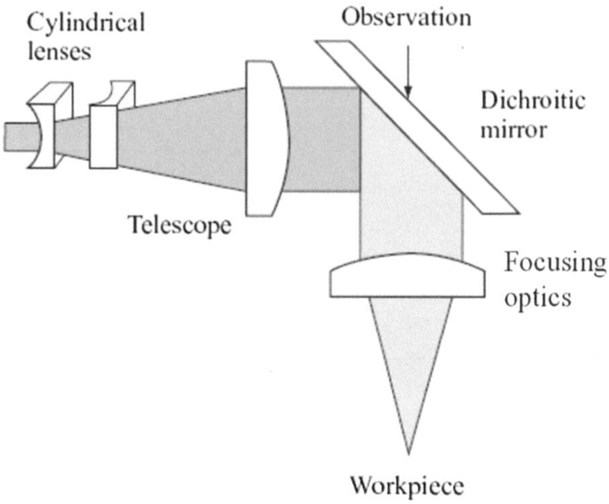

Fig. 5.65 That beam focus system by telescopic system [7]

The spot size is 0.2 mm × 0.25 mm, and the beam quality is 20 mm × 22 mrad. Cutting application tests was successfully performed with this lasers system, and mild steel up to 1 mm thickness was cut with a power of 125 W with the assistance of oxygen [7].

5.3.4 Stack-Based Beam Combining and Fiber Coupling

The beam combining technologies, such as spatial, polarization, and wavelength beam combining, have been used for improving of the output power and beam brightness of semiconductor laser stacks [7, 35–37]. Furthermore, any two or even three of the beam combining methods can be combined to scale the output power and improve the brightness [7, 37].

1. Spatial beam combining
   Due to the large vertical distance between adjacent laser bars, the vertical fill factor of the radiation is generally below 50 %. This allows the beams from two or more stacks to be spatially interleaved. Figure 5.66 shows the principle of superposition of the beams of two vertical stacks. The beams of the two laser stacks are redirected by optical devices, and are combined after a beam combiner. After combining, the beam of one laser stack is inserted into the beam space of the other laser stack. At the output, the beam size of the two stacks is reduced to 1/2 of original beams [21]. There are different optical approaches to achieve spatial beam combining. Figure 5.67 shows one example which uses a pile of glass plates with reflective sides to reflect the beams from the two stacks and then combine them. Since this incoherent coupling will neither increase the divergence nor the near-field extension [35], combining the two stack elements only slightly increases the beam propagation factor compared to a single

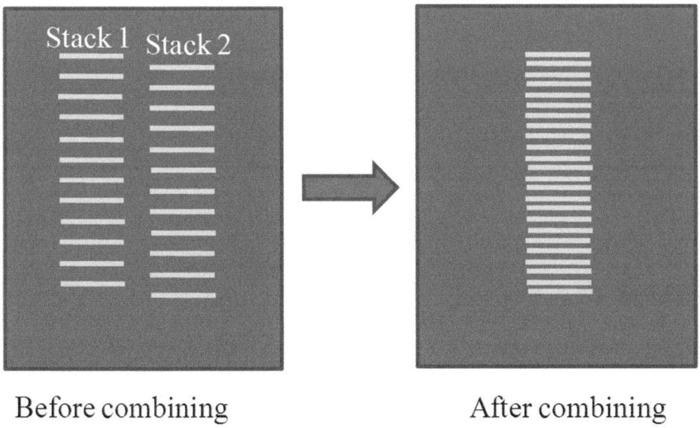

Fig. 5.66 Illustration of the spatial beam combining principle [21]

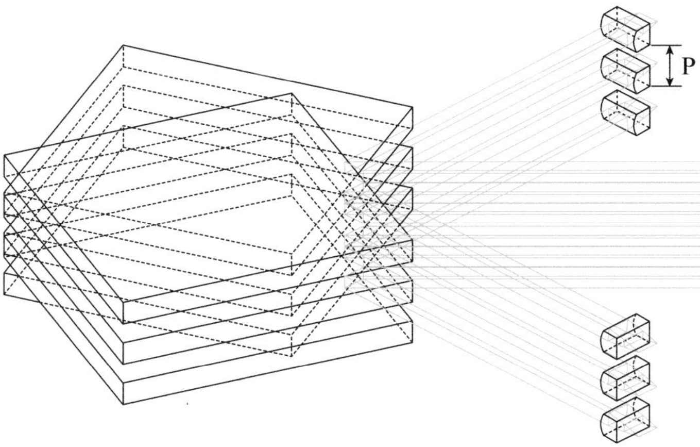

Fig. 5.67 Example of a spatial beam combining of two stacks [35]

stack but double the output power. Based on this beam combining approach, fiber-coupled power of 2,000 W from 1.2 mm fiber with 0.22NA was demonstrated [35].

2. Polarization beam combining
   The polarization beam combining technology has already been used in beam shaping systems of vertical stack semiconductor lasers. The design of a 3,000 W laser cladding system is taken as an example in Fig. 5.68 [12]. Two 976 nm semiconductor laser stacks with 20 semiconductor laser bars are used in this system, and the total power of those two stacks is 4,000 W with 100 W per bar. In Fig. 5.68, the two laser stacks with TE polarization are arranged in a row along slow axis, and the beam from the semiconductor laser stacks are collimated with FAC on the fast axis. The polarization of one TE laser stack is

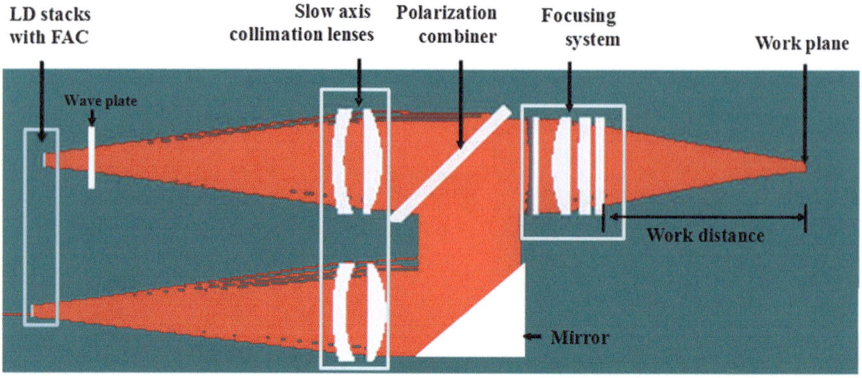

Fig. 5.68 Stacked-based beam combining using polarization coupling [12]

Fig. 5.69 The designed beam size of the system based on polarization beam combining [12]

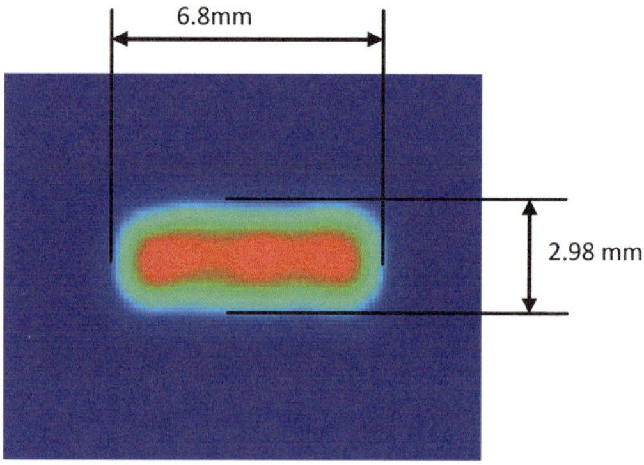

changed to be TM polarization beam by the wave plate shown in the figure. The beams of each laser stack on the slow axis are collimated by two lenses which are a flat-concave cylinder lens and a double convex cylinder lens. After collimation, The TM beam of the semiconductor laser stack transmits through the polarization coupler directly, while the TE beam is reflected by a mirror and then reflected again by the coupler. After that, those two beams are combined. The combined beam is then focused by a focusing system which is composed of four cylindrical lenses.

According to the above design, the beam size is 3 mm × 7 mm and the intensity distribution is a symmetrical rectangle at the work surface, the simulated result is shown in Fig. 5.69 [12]. The construction of the system is shown in Fig. 5.70 [31].

The experimental setup is shown in Fig. 5.71 and the performances of this laser system were tested at 25 °C [12]. When the current were driven to 84 A, the total output power was 3,738 W with a optical transmission efficiency of 93.7 %. In Fig. 5.72, the beam shape at work plane is shown [12]. Comparing to the

Fig. 5.70 The construction of the system based on vertical stack polarization beam combining [31]

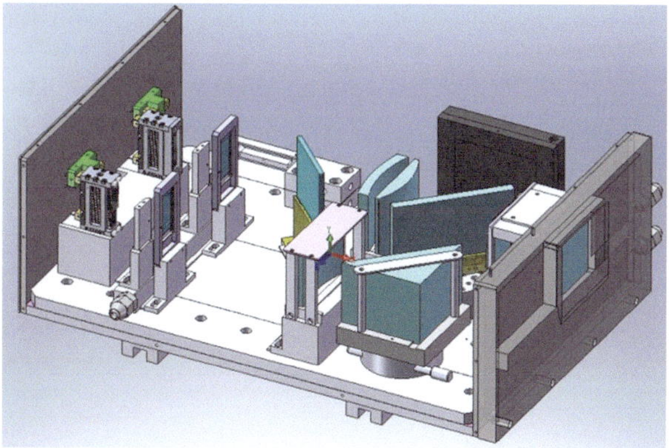

Fig. 5.71 The experimental setup of polarization beam combining of two stacks [12]

Fig. 5.72 The beam size at work plane [12]

simulation result shown in Fig. 5.69, the experiment result agrees well with the theoretical value.

3. Wavelength beam combining
   The beams of diode laser stacks with different wavelengths can be superimposed by various methods such as dielectric edge-filters, diffraction, prisms, or other

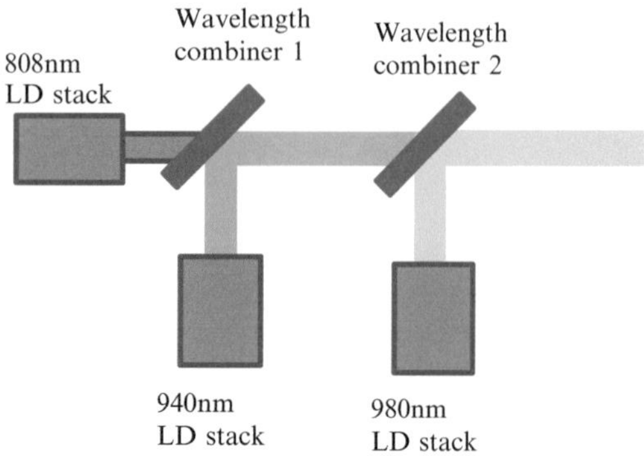

Fig. 5.73 The radiation of three collimated laser stacks combined by wavelength multiplexing [31]

optical components. With regard to efficiency, adjustment and compact dimensions dielectric edge-filters are generally used for wavelength multiplexing. Figure 5.73 shows the beam combining by wavelength combiners [31]. As shown in Fig. 5.73, the radiation of three collimated stacks with three wavelengths, which are 808, 940, and 976 nm, respectively, are arranged and superimposed by two wavelength combiners with low loss. The 808 nm beam transmits through the wavelength combiner 1 and 940 nm beam is reflected by it, and they are combined after the combiner 1. The combined beam also transmits through the wavelength combiner 2, and 980 nm beam is reflected by combiner 2. After combiner 2, the three beams are combined.

References

1. M. Born, E. Wolf, *Principles of Optics* (Pregamon Press, Oxford, 1964)
2. M. Mu, K. Yang, K. Luan, Beam shaping design of high power semiconductor lasers. Focuslight Inner Reports **3**, 10–17 (2010)
3. K. Duna, B. Lu, Propagation properties of vector elliptical gaussian am beyond the paraxial approximation. Opt. Laser Tech. **36**, 489–496 (2004)
4. X. Kang, B. Lu, Vectorial nonparaxial flattened gaussian beams and their beam quality in terms of the power in the bucket. Opt. Commun. **262**, 1–7 (2006)
5. W. Sokolowski, D. Wolff, P. Hennig, *Beam Shaping and Fiber Coupling of High Power Diode Lasers*. Proceedings of the Symposium on Photonics Technologies for 7th Framework Program Wroclaw, vol. 129 (2006), pp. 12–14
6. F.Z. Fang, Z. Xiong, X.T. Hu, Ultra-precision machining of reflector array for laser diode beam shaping. Optoelectron. Lett. **3**(2), 141–143 (2007)
7. R. Diehl, *High Power Diode Lasers: Fundamentals, Technology, Applications* (Springer, Berlin, 2000)
8. M.V. Maximov, Y.M. Shernyakov, I.I. Novikov, S.M. Kuznetsov, L.Ya. Karachinsky, N. Yu. Gordeev, I.P. Soshnikov, Yu.G. Musikhin, N.V. Kryzhanovskaya, A. Sharon, U. Ben-Amic, V.P. Kalosha, N.D. Zakharov, P. Werner, T. Kettler, K. Posilovic, A. Shchukin, N.N. Ledentsov, D. Bimberg, Longitudinal photonic bandgap crystal laser diodes with ultra-narrow vertical beam divergence. Proc. of SPIE **6115**, 611513(1–13) (2006)
9. W.S. Cai, K. Yang, M.G. Mu, Optical shaping for semiconductor laser. Internal Report **6**, 40–45 (2011)

10. J.W. Wang, Z.B. Yuan, Y.X. Zhang, E.T. Zhang, D. Wu, X.S. Liu, *250 W QCW Conduction Cooled High Power Semiconductor Laser*. IEEE, 2009 International Conference on Electronic Packaging Technology & High Density Packaging (ICEPT-HDP) (2010), pp. 451–455
11. X.N. Li, C.H. Peng, Y.X. Zhang, J.W. Wang, L.L. Xiong, P. Zhang, X.S. Liu, *A New Continuous Wave 2500 W Semiconductor Laser Vertical Stack*. Proceedings of 11th on Electronic Packaging Technology and high Density Packaging (ICEPT-HDP) (2010), pp. 1350–1354
12. L.L. Xiong, M. Wang, X.B. Wang, Y.F. Zheng, D. Wu, P. Zhang, X.N. Li, Z.F. Wang, X.S. Liu, 3000 W CW diode laser cladding system [C]. Proc. SPIE **8241**, 824106(1–5) (2012)
13. S.M. Yeh, S.Y. Huang, W.H. Cheng, A new scheme of conical-wedge-shaped fiber end face for coupling between high-power laser diodes and single-mode fibers. J. Light Wave Technol. **23**(4), 1781–1786 (2005)
14. Y.L. Tang, Beam collimation of high-power laser diode array with graded-index fiber lens array. Opt. Eng. **47**(5), 054202(1–4) (2008)
15. K. Kawano, Coupling characteristics of lens systems for laser diode modules using single-mode fiber. Appl. Opt. **25**(15), 2600–2605 (1986)
16. O. Mitomi, K. Kawano, Coupling characteristics of laser diode to multimode fiber using separate lens methods. Appl. Opt. **25**(1), 136–141 (1986)
17. T. Horimatsu, M. Sasaki, K. Aoyama, Stabilization of diode laser output by beveled-end fiber coupling. Appl. Opt. **19**(12), 1984–1986 (1980)
18. http://www.focuslight.com.cn/products.asp
19. J.C. Palais, Fiber coupling using graded-index rod lenses. Appl. Opt. **19**(12), 2011–2018 (1980)
20. H. Liu, J.W. Wang, Z.B. Yuan, Y.X. Zhang, Packaging technology of single emitter semiconductor lasers. Inner Report **6**, 21–28 (2011)
21. J.W. Wang, Z.B. Yuan, Y.X. Zhang, Packaging of semiconductor laser arrays. Inner Report **12**, 12–17 (2011)
22. J.W. Wang, Z.B. Yuan, *Study of the Mechanism of "Smile" in High Power Diode Laser Arrays and Strategies in Improving Near-Field Linearity*. IEEE Proceedings of 59th Electronic Components and Technology Conference (ECTC) (2009), pp. 837–842
23. Y.Q. Liu, Y.H. Cao, C.X. Xu, W.B. Qin, Z.Y. Wang, *1000 W Compound Coupling High Beam Quality Diode Laser*. International Symposium on Photo-electronic Detection and Imaging, vol. 7382 (2009), pp. 738231(1–8)
24. X.W. Wang, J.W. Xiao, X.Y. Ma, Z.M. Wang, G.Z. Fang et al., Fiber coupling of laser diode bar to multimode fiber array. Chin. J. Semicond. **23**(5), 4–7 (2002)
25. P. Shi, X.L. Li, G.F. Zhang, M.X. Guo, Y.T. Lu, Micro-prism stack beam shaper for high power laser diode array. Acta Opt. Sin. **20**(11), 1544–1547 (2000)
26. D. Brown, Laser diode bar integrator/reimager, United states patent application publication, US 2005/0264893 Al, vol. 12 (2005)
27. W.A. Clarkson, D.C. Hanna, Two-mirror beam-shaping technique for high-power diode bars. Opt. Lett. **21**(6), 375 (1996)
28. B. Faircloth, High-brightness high-power fiber coupled diode laser system for material processing and laser pumping. Proc. SPIE **4973**, 34–41 (2003)
29. K.M. Du, M. Baumann, B. Ehlers, H.G. Treusch, P. Loosen, Fiber-coupling technique with micro step-mirrors for high-power diode lasers bars. OSA TOPS **10**, 390 (1997)
30. X.H. Ma, L. Xu, G.J. Liu, Q.L. Shi, D.S. Xin, J.J. Zhang, Beam shaping and fiber coupling of high power laser diode arrays. Proc. SPIE **5644**, 545 (2005)
31. J.W. Wang, Z.B. Yuan, Y.X. Zhang, Packaging and fiber coupling of semiconductor laser stacks. Inner Report **7**, 32–37 (2012)
32. X. Gao, H. Ohashi, H. Okamoto, M. Takasaka, K. Shinoda, Beam-shaping technique for improving the beam quality of a high-power laser-diode stack. Opt. Lett. **31**(11), 1654–1656 (2006)

33. Z.H. Huang, L.L. Xiong, H. Liu, Z.F. Wang, P. Zhang, Z.Q. Nie, D.H. Wu, X.S. Liu, Double-cutting beam shaping technique for high-power diode laser area light source. Opt. Eng. **52**(10), 106108(1–7) (2013)
34. D. Havrilla, M. Holzer, S. Strohmaier, High-power diode lasers-low-numerical-aperture direct diode laser maintains high power—A passively cooled, fiber-coupled diode module has a previously unachieved combination of high power and high beam quality. Laser Focus World **47**(3), 46–48 (2011)
35. C. Fiebig, B. Eppich, W. Pittroff, G. Erbert, Stable and compact mounting scheme for >1 kW QCW diode laser stacks at 940 nm. Proc. SPIE **6876**, 68760J(1–10) (2008)
36. B. Kőhler, J. Biesenbach, T. Brand, M. Haag, S. Huke, A. Noeske, G. Seibold, M. Behringer, J. Luft, *High-Brightness High-Power kW-System With Tapered Diode Laser Bars, Lasers and Applications in Science and Engineering*. International Society for Optics and Photonics (2005), pp. 73–84
37. F. Dorsch, F.X. Daiminger, P. Hennig, V. Blümel, 2kW CW fiber-coupled diode laser system. Proc. SPIE **3889**, 45–53 (2000)

Chapter 6
Materials in High Power Semiconductor Laser Packaging

The performance of semiconductor lasers is greatly affected by the properties of packaging materials, which mainly consist of diverse bonding solders, mounting substrates [1–4]. The selection of packaging materials is multidisciplinary and involves achieving a balance among device performance, reliability, manufacturability, and cost-effectiveness. In this chapter, the properties of solder materials, as well as mounting substrates employed in the packaging of high power semiconductor lasers are presented and the effects of material properties on the performance of semiconductor lasers are analyzed in details.

The typical electrooptical conversion efficiency is about 50 % for a 808 nm high power semiconductor laser. For example, with the 50 % conversion efficiency, 100 W waste heat can be generated from the active region of a 808 nm single bar high power semiconductor laser with a optical output power of 100 W. Large amount of heat should be dissipated in time from the semiconductor laser chip to reduce the junction temperature. The lower the junction temperature is, the higher the reliability of the semiconductor laser will be. Solder materials and mounting substrates should efficiently transfer the heat from the semiconductor junction to the ambient environment. The properties, such as thermal conductivity and thermal expansion, should be considered significantly in design of the devices.

6.1 Solder Materials

The selection of solder material is based on the consideration of a number of properties, including melting temperature, thermal conductivity (TC), wettability, coefficient of thermal expansion (CTE), yield strength, Young's modulus, thermal fatigue, and creep properties. Table 6.1 shows the physical properties of some common solder materials used in packaging process [5].

X. Liu et al., *Packaging of High Power Semiconductor Lasers*, Micro- and Opto-Electronic Materials, Structures, and Systems, DOI 10.1007/978-1-4614-9263-4_6,

Table 6.1 Overview of the physical properties of different solder materials [5]

Solder materials	Thermal conductivity (W/(m∗K))	Density (kg/m^3)	Specific heat (J/(kg∗K))	CTE (ppm/K)	Melting point (°C)
PbSn	50.6	1,080	128	24.7	183
In	83.7	7,290	233	33	156.61
Sn	63	7,300	226	22	232
Ag	420	10,500	269.5	19.5	961.78
$Au_{80}Sn_{20}$	57.3	19,720	129	15.9	283
SnAgCu	33	7,400	212	19.1	217–220
InSn	34	7,294	–	20	118

In general, the solder materials ideally must satisfy the following requirements:

1. Have the desired processing temperature to establish an effective assembly process temperature hierarchy and meet field application temperature and environment requirements
2. Provide sufficient wetting characteristics to form metallurgical bond between diode laser and mounting substrate
3. Provide an efficient heat dissipation channel from the diode laser die to the mounting substrate
4. Relieve thermal-induced stresses arising from the mismatch of thermal expansion between the diode laser and mounting substrate
5. No or low deformation during the long-term operation of semiconductor laser

Among the solder materials used for the high power semiconductor lasers, indium (In) and gold-tin (AuSn) are most commonly used in die bonding [6–8]. SnAgCu (SAC) and InSn alloys are more commonly used in the next level packaging, e.g., bonding the mounting substrate to heat sink or attaching a submount to a package house.

Indium solder is ductile and has plasticity property, which can reduce the thermal stress on the laser diode caused by the CTE mismatch between the laser diode and the mounting substrate. The conventional method of laser diode packaging is to bond a diode laser chip onto a mounting substrate made of copper due to its high thermal conductivity. Due to the large CTE mismatch between Cu and the GaAs, indium solder is the preferred die bonding materials. However, due to its plasticity property, indium is prone to thermal fatigue under temperature cycling conditions. Also indium solder can have electromigration and electrothermal migration under high driving current density.

Compared to the conventionally used indium solder, AuSn solder is a material with high yield strength and low ductility. This results in high creep resistance and can prevent thermal fatigue problem. In addition, $Au_{80}Sn_{20}$ solder possesses properties of high anticorrosion, high melting temperature, and high stability. However, due to its low ductility, AuSn solder only can be used when the diode laser chip is bonded to a closely CTE-matched mounting substrate.

Table 6.2 The physical properties of indium [9]

Material	Melting point (°C)	CTE (ppm/K)	Thermal conductivity (W/(m*K))	Elastic modulus (GPa) at 20 °C	Density (kg/m^3)
Indium	156.6	33	83.7	12.74	7,290

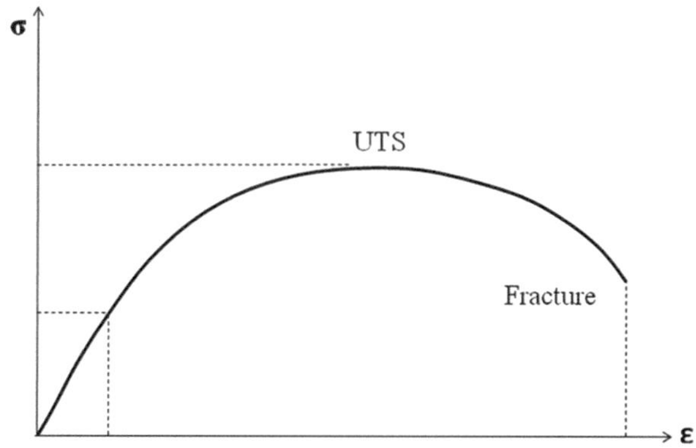

Fig. 6.1 Schematic diagram of stress–strain curve for indium [5]

Compared to indium, InSn reveals the property of good oxidation resistance, but it has low melting temperature and is susceptible to corrosion in the humid environment. InSn is also a very soft material and has the tendency to creep.

SAC solder is the most promising replacements for traditional eutectic SnPb solder alloy due to its high wettability, good strength, and creep resistance. It can be used as the bonding material between heat sink and submount in high power semiconductor laser packaging.

In this section, we mainly introduce the properties of In, AuSn, InSn, and SAC solders.

6.1.1 Indium Solder

Indium is most widely used in high power semiconductor laser die bonding. Indium is a soft, silvery-white metal with a bright luster. Some properties of Indium are shown in Table 6.2 [9]. Indium has good metal malleability and plasticity property, and its melting point is 156.6°C.

Figure 6.1 shows the schematic diagram of stress–strain curve for indium [5]. It can be seen that the rate of strain hardening diminishes up to the point of UTS (ultimate tensile strength) and beyond the point, indium appears to strain soften. Indium has low strength and high ductility. The readiness to creep of indium material means that solder joints are unlikely to fail under thermal stress. Indium solders are suitable for making joints between dissimilar materials that will be subject to thermal stress due to CTE mismatch. Creep in indium solder joints can usually take place sufficiently fast to ensure that the stress is always close to zero, with roughly 80 % stress relaxation occurring within seconds of a step change in strain [10]. If joints are optimized to take advantage of these beneficial thermomechanical characteristics, indium solders can provide superior life. Hence historically indium solder was dominant in bonding high-power diode lasers. Due to its high thermal conductivity and manufacturability, Cu has been widely used as mounting substrate for high power diode lasers. Indium has been the preferred choice because it can reduce the stress caused by the CTE mismatch between laser chip and Cu mounting substrate by deforming the solder layer itself [7]. From Fig. 6.1 it can be concluded that the thermal stress experienced by the laser die cannot exceed UTS of indium. In fact, during laser die bonding process or during temperature or power cycling, indium solder transmit very little stress to the laser die as far as the bonding indium solder layer remains to be pure indium material. Therefore, indium can be used to join much larger die such as laser bars which are typically 10 mm by 2 mm or minibars which are 3.5 mm by 4 mm.

The stress–strain curve in Fig. 6.1 indicates mechanical failure of indium solder joints tends to be caused by stress overload or unidirectional creep. Stress overload failure actually rarely happens in high power semiconductor lasers unless external mechanical force is applied to the indium solder joint. When indium solder joints fail, the mechanism is predominantly by classical creep rupture, which has its origins in the nucleation and coalescence of cavities that arise from the material redistribution associated with stress relaxation [10]. Indium solder thermal fatigue failure under thermal cycling condition is commonly observed in high power semiconductor lasers. Fatigue is the weakening of indium solder caused by repeatedly cyclic applied thermal stress during thermal cycling. It is a progressive and localized structural damage which can be caused by a stress much less than the strength of ultimate tensile stress limit. Thermal cycling refers to situations where the environment surrounding the diode laser package undergoes cyclic thermal excursions. The cycle consists of a dwell at a low temperature, a temperature ramp to the high temperature, a dwell at a high temperature, and a temperature ramp to the low temperature. In high power diode lasers, the thermal cycling could be ambient environment temperature or more commonly power cycling when the diode laser is operated at on-off mode. The thermal fatigue condition and life are determined by the following factors: (1) thickness of indium solder layer, (2) the temperature cycling range ΔT, (3) the two temperature extremes, and (4) the temperature ramp rate, and dwell time. The above factors are related to the magnitude of cyclic stress/strain due to CTE mismatch between diode laser and the mounting substrate. Indium-bonded semiconductor laser usually fails more likely in long pulse operation mode, such as one second on/one second off cycling, than short pulse width and high frequency conditions such as 200 μs and 400 Hz. This is because large temperature cycling range is created and thermal gradient is formed in the indium solder layer under the conditions of long pulse operation [8].

Indium solders for high power semiconductor lasers can also fail by a process known as phase segregation [5]. This develops when the electrical current density is high and it accelerates when there is large thermal gradient across the indium solder joint. The result is migration of indium toward mounting substrate side and

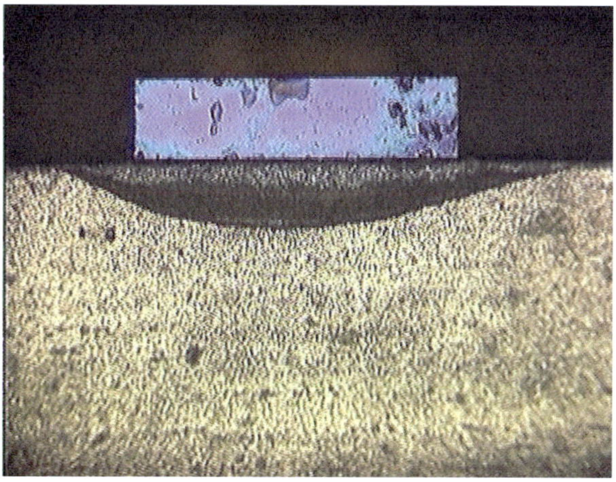

Fig. 6.2 An optical microscope image of a single emitter diode laser after 300 h testing total at 7 A and 40 °C [7]

the accumulation of voids at the diode laser side due to the direction of the current flow. Failure analysis has demonstrated that the diode lasers packaged with indium solder suffer electromigration and thermo-electromigration process at the current density of ~10^3 A/cm^2 or higher [7]. Three major factors, which are current crowding, localized high temperature, and large temperature gradient, could contribute to electromigration in a high power laser assembly. Taking the major driving forces for atomic electromigration into account, the flux divergence of atoms is expressed as [7]

$$\nabla \cdot J = \frac{ND_0}{kT(r)} Z^* q\rho_0 [1 + \alpha \Delta T(r)][j(r) - j_c] \Delta Y e^{-Q_0/kT(r)} \tag{6.1}$$

where N is atom concentration, D_0 is intrinsic diffusivity, k is Boltzmann's constant, $T(r)$ is position-dependent temperature, Z^*q is the effective charge, ΔY is the structure factor, j(r) is the position-dependent current density, j_c is the threshold current density referring to the stress-induced mass backflow, Q_0 is activation energy, α is the temperature coefficient of the resistivity, and ρ_0 is the resistivity. $\nabla \cdot J > 0$ and $\nabla \cdot J < 0$ represent mass depletion and accumulation, which lead to the formation of void and hillock, respectively [7]. Figure 6.2 shows the optical image for typical diode laser after 300 h testing total at 7 A and 40 °C. It can be seen from the optical image that there was a pile of indium material on the front surface of the diode laser below the laser chip due to electromigration of the indium solder [7].

Indium intermetallics are formed during the die bonding soldering process and can also gradually grow during the device life span, especially when the device experiences high temperatures. Indium can form intermetallics with Au, Cu, and other metals which are used in high power diode laser packaging. Intermetallics often have higher melting temperatures than pure indium. In the intermetallics, strong bonding between unlike atoms is formed which results in high creep resistance. Thus, intermetallics at the bonding interfaces can strengthen the solder joint.

However, intermetallics can also become brittle and excessive, especially at the bonding interfaces, which can result in bond fracture failure. As discussed above, indium solder is the preferred choice when a diode laser chip is mounted on a Cu mounting substrate. Indium/copper intermetallics and indium oxide are easily formed in the indium soldering process, which will affect the reliability of high power semiconductor lasers [11].

Indium/copper intermetallics: Studies have been done to track the diffusion rate of indium into copper. At room temperature, over a period of 5 years, indium and copper will completely diffuse into one another [11]. The rate of diffusion will increase with temperature rise. Inter-diffusion between indium and copper atoms can form a brittle intermetallic layer. This intermetallic layer is sensitive to thermal shock or thermal cycling, and could cause laser failure in applications. An effective way of preventing this diffusion is to deposit a diffusion barrier layer such as a Ni layer or Pt layer onto the copper. The diffusion barrier layer acts as a barrier to prevent the solid state diffusion from occurring in the indium solder interface.

Indium oxidation: Indium oxide is extremely tenacious and thermodynamically stable. Indium oxidation on the surface of indium needs to be effectively removed in order to form a uniform and defect-free bonding interface at both the laser die side and mounting substrate side. The presence of voids and defects in the solder joint will potentially lead to an increase in local thermal resistance and local stress concentration which consequently results in the degradation of the performance of the device.

The formation of metal oxidations on indium is self-passivating. A thickness of 80–100 Å of oxide will form on the surface. Prior to using indium solder in packaging process of high power semiconductor lasers, such as indium film deposition or perform soldering, it is strongly recommended that this oxide layer should be removed. Hydrochloric acid solution is commonly used to remove the indium oxide layer on the surface of the bulk indium material.

6.1.2 AuSn Solder

AuSn solder has been successfully used for highly reliable die attach and fluxless soldering in the packaging process of high power semiconductor lasers. AuSn solder has good thermal and electrical conductivity, high corrosion resistance, no thermal fatigue and the possibility of fluxless soldering. In general, a composition of AuSn at or close to the eutectic point with 80 wt% Au and 20 wt% Sn is commonly recommended. The main physical properties of 80 wt%Au20 wt%Sn are shown in Table 6.3 [12].

Figure 6.3 shows the schematic diagram of stress–strain curve for AuSn alloy [5]. AuSn solder has a higher Young's modulus than indium and the strain is lower than that of indium at the same stress level. However, AuSn solder does not have a UTS point as indium does, which makes it unable to release the stress due to the

Table 6.3 The physical properties of Au_{80}/Sn_{20} solder at 20 °C [12]

Properties	Value
Density (g/cm^3)	14.7
Coefficient of thermal expansion (ppm /K)	16
Thermal conductivity (W × m^{-1} × K^{-1})	57
Tensile strength (MPa)	275
Young's modulus (GPa)	68
Shear modulus (GPa)	25
Poisson's ratio	0.405
Electrical resistivity (10^{-8} Ω × m)	16.4
Elongation (%)	2

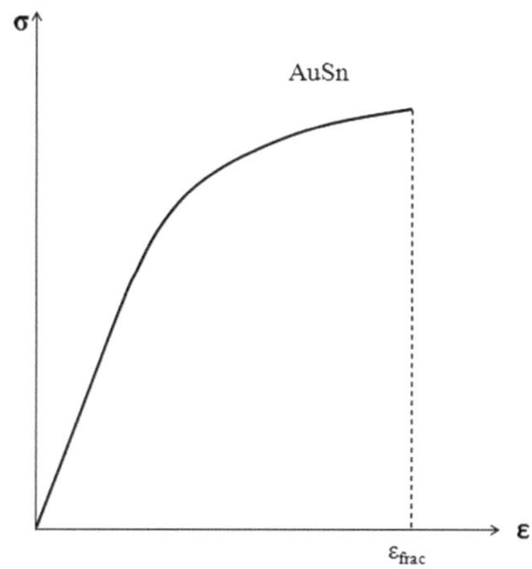

Fig. 6.3 Schematic diagram of stress–strain curve for AuSn alloy [5]

CTE mismatch between laser diode chip and heat sink. AuSn solder have high strength and does not undergo stress relaxation at ordinary temperatures. As a consequence, AuSn solder layer may transmit more stress to the laser chip and may even provoke cracks in laser chips. The high temperature required for AuSn die attachment would also aggravate stress when assemblies are cooled to low temperatures. Therefore, when AuSn solder is used in packaging of high power semiconductor lasers, there must have a CTE-matched buffer layer (the so-called submount, which CTE is close to laser diode) between the heat sink which is typically Cu due to its high thermal conductivity and laser diode chip.

As can be seen from Fig. 6.3, AuSn solder does not incur plastic deformation. Hence, AuSn solder does not quickly degrade from fatigue damage and generally has no fatigue problem even under thermal cycling. This makes AuSn solder the preferred choice when the diode laser is operated under harsh environmental conditions or pulsed mode.

Figure 6.4 shows the AuSn equilibrium phase diagram [13]. It contains four different stable intermetallic compounds (IMCs) between Au and Sn at room temperature: Au_5Sn, AuSn, $AuSn_2$, and $AuSn_4$. Among them, eutectic 80 wt%Au and 20 wt%Sn composition with a melting point of 280°C is known to be the most

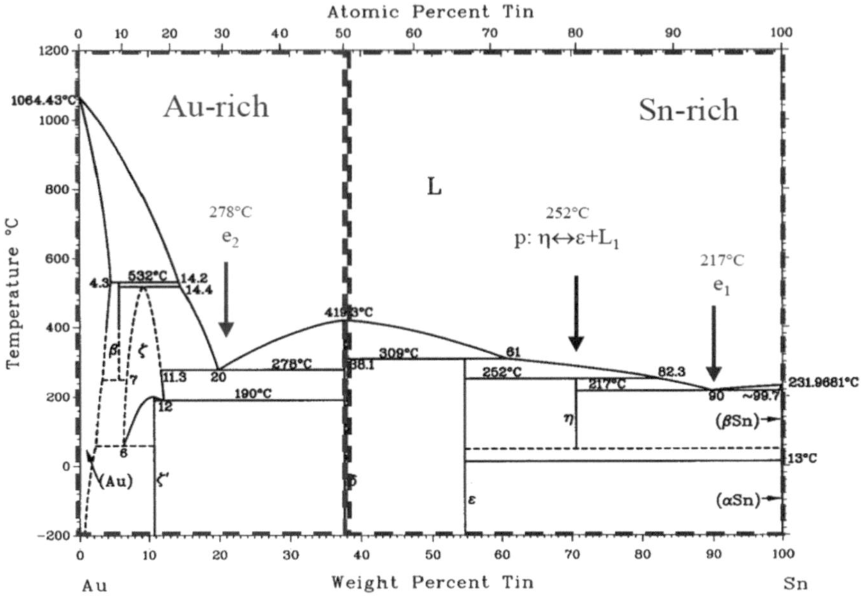

Fig. 6.4 Phase diagram of binary AuSn system [13]

appropriate composition for the high bonding quality because of its high melting temperature, good creep behavior, and good corrosion resistance [14].

According to the phase diagram as shown in Fig. 6.4, the $Au_{80}Sn_{20}$ has a eutectic ξ-Au_5Sn + δ-AuSn microstructure at room temperature. Experimentally, the microstructure of the as-cast AuSn eutectic alloy is comprised of ξ and ξ′-Au_5Sn dendrites surrounded by the matrix of the AuSn eutectic structure (ξ + δ). ξ′-Au_5Sn is the lower temperature phase while the ξ phase is stable at higher temperatures. Meanwhile, the ξ phase has a wider composition range of 10–18.5 at.% Sn. During the AuSn bonding process, when the temperature is cooled down below the eutectic point of 278°C, ξ phases will solidify simultaneously in the solder joint [15].

From AuSn binary phase diagram in Fig. 6.4, there is a tin-rich eutectic phase of 90 wt% tin and 10 wt% gold, which has an eutectic transformation at 217 °C [16]. There exists a transformation between tin-rich phase and gold-rich phase during the reflow process, as shown in Fig. 6.5 [17]. The tin-rich eutectic (η + Sn → L_1) forms first at 217 °C, the peritectic reaction (η + L_1 → ε) takes place at 251 °C and the gold-rich eutectic transformation (δ + ξ → L_2) at 281 °C. Obviously, an exothermic reaction occurs just before the second eutectic transformation and is contributed to the mixing of two liquids of different composition: a tin-rich L_1 and a gold-rich L_2 [17].

The enthalpy of formation of four intermetallic phases in the AuSn binary system with Sn concentration greater than 1.5 at.% were measured by Debski et al. [18]. As shown in Fig. 6.6 [19], the enthalpy of formation of δ-AuSn is the lowest and independent of temperature, when compared to other IMCs that were

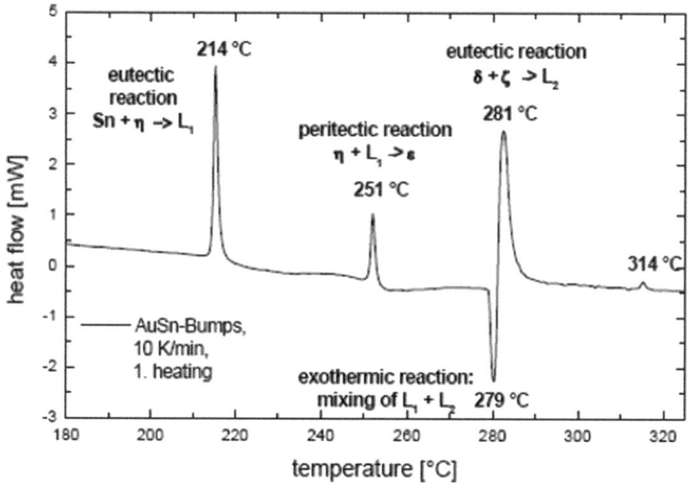

Fig. 6.5 Transformation of AuSn during reflow [17]

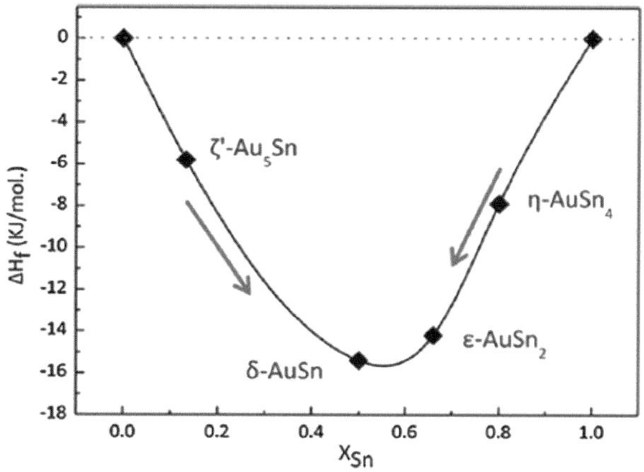

Fig. 6.6 The formation enthalpy of intermetallic phases from the AuSn system at 27 °C [19]

investigated in the system. From Fig. 6.6, the δ-AuSn is the most stable phase in AuSn system at room temperature [19].

Ti/Pt/Au metallization and AuSn solder systems have been widely used for the packaging of semiconductor laser, as shown in Fig. 6.7, a major concern with the Ti/Pt/Au/AuSn die bonding scheme is the reactive nature of the system [20]. The metallurgical interaction of AuSn solder with Au layer is to form the Au_5Sn, $AuSn_4$, and other IMCs.

When the thin Au layer is dissolved, the interaction between the Pt and Sn creates an excessive buildup of IMCs, such as PtSn and $PtSn_4$. As a result, the AuSn ratio in the solder changes to a more Au-rich composition, which significantly increases the liquidus temperature for the AuSn alloy [20, 21].

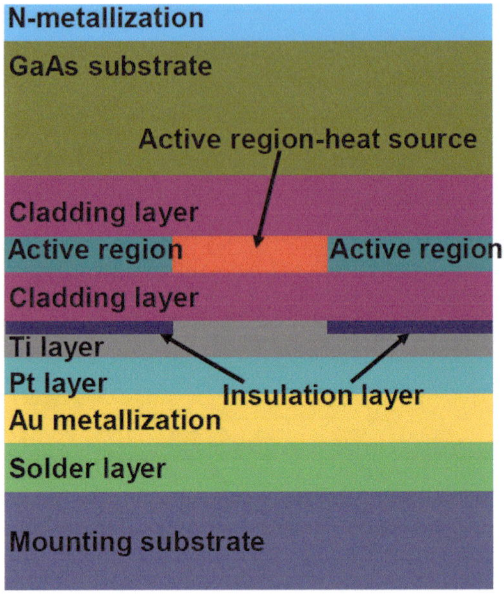

Fig. 6.7 Schematic structure of an epi-down bonded semiconductor laser [20]

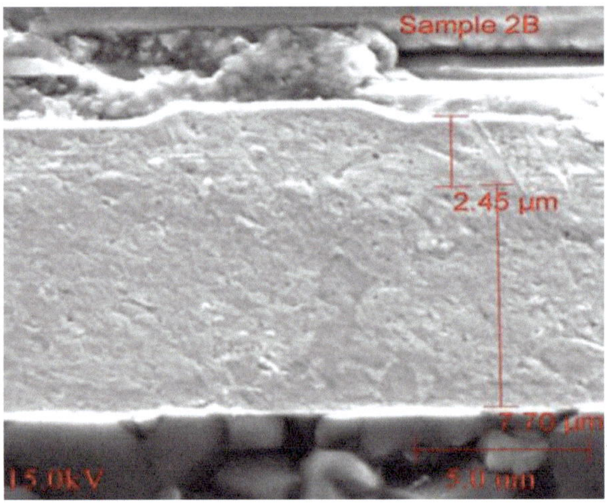

Fig. 6.8 Typical cross-sectional SEM images of the die bonding solder interface of a laser chip having the traditional Ti/Pt/Au (thick Au of 2–3 μm) metallization system [20]

Figure 6.8 shows the typical cross-sectional microstructure of the as-reflowed samples having the traditional Ti/Pt/Au system, but with thick Au of 2–3 μm [20]. EDX analysis showed that Sn is present inside the Au layer and ζ-phase (Au_5Sn) is formed, as shown in the EDX map of Sn in Fig. 6.9 and EDX spectrum of the material inside the Au layer in Fig. 6.10 [20]. This indicates that some of the Au, if not most, is consumed by the reaction of Au layer with the Sn in the AuSn solder during the reflow process. From the AuSn phase diagram, it is clear that when the AuSn solder is molten at temperatures higher than 283 °C, the epi-side Au

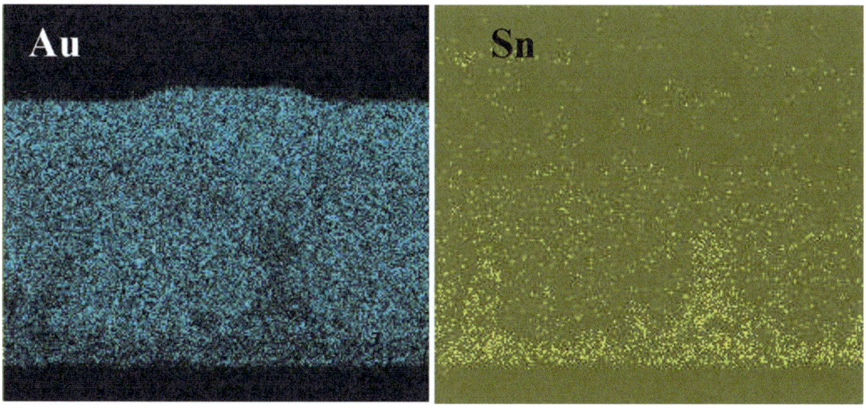

Fig. 6.9 EDX maps of Sn and Au from the cross-section of the die bonding solder interface of an as-reflowed sample shown in Fig. 6.8 [20]

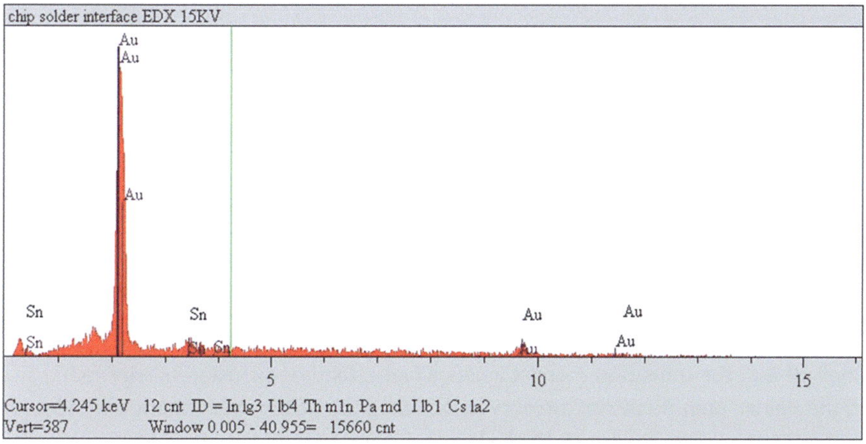

Fig. 6.10 EDX spectra of the material close to the chip and solder interface (inside the epi-side Au layer) [20]

layer dissolves and increases the Au content of the liquid AuSn solder besides the diffusion of the Sn into the Au layer, leading to the Au-rich ζ-phase during solidification. During aging, more Sn gradually diffuses into the Au layer and forms intermetallic phases [20].

As a result, an undesirable joint structure is formed, and this structure may degrade the performance of the bonded diode laser. Different diffusion barrier metals were studied [20, 21]. It was found that, without a diffusion barrier, the thick Au layer in the epi-side metallization would be mostly consumed and form intermetallics with the Sn from the AuSn solder during soldering and thermal aging. An effective diffusion barrier is needed to preserve the thick Au layer from forming

Table 6.4 The physical properties of InSn solder [22]

Material	Melting point (°C)	Tensile strength (MPa)	Thermal conductivity (W/(m∗K))	Electrical resistivity (μΩ cm)
$In_{52}Sn_{48}$	118	11.9	34	15

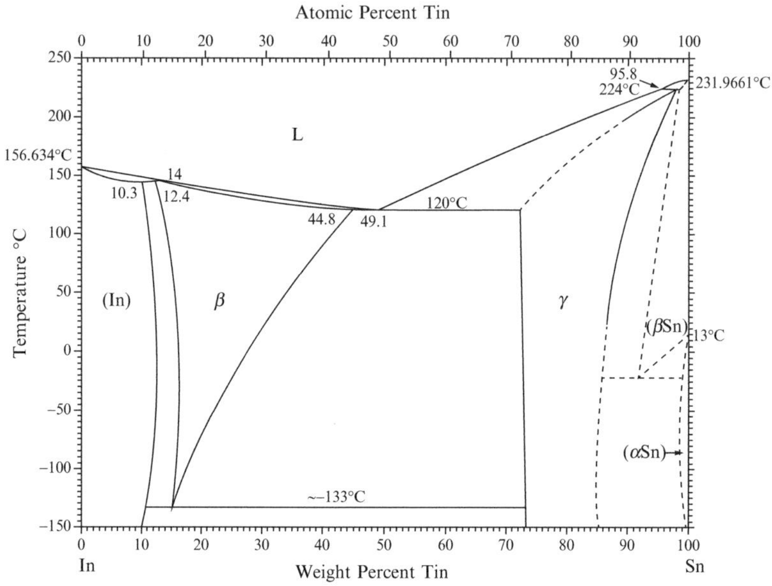

Fig. 6.11 Indium-Tin phase diagram [23]

intermetallics in order to achieve the desired thermal performance. The study showed that the transition metal Cr acts as an effective diffusion barrier metal for AuSn solder and, therefore, preserves the integrity of the metallization system. It was concluded that the Ti–Pt–thick Au–Ti–Cr–Au metallization scheme is a promising candidate for the epi-side metal of a junction-down bonded high power semiconductor laser [20, 21].

6.1.3 InSn Solder

The melting point of InSn alloy makes it suitable for low temperature applications such as second level packaging. The properties of InSn solder are shown in Table 6.4 [22].

According to the InSn phase diagram shown in Fig. 6.11, the eutectic point occurs at 120°C and corresponds to the $In_{52}Sn_{48}$ alloy [23]. The two phases that form are intermetallic phases Ð an In-rich, pseudo-body-centered tetragonal phase, β, which has 44.8 wt% Sn, and a hexagonal Sn-rich phase, γ, with 77.6 wt% Sn [24].

Table 6.5 The typical properties of SnAgCu solder alloys [22]

Material	Melting point (°C)	CTE (ppm/K)	Elastic modulus (GPa) at 20 °C	Thermal conductivity (W/(m*K))	Density (kg/m^3)
SnAgCu	218	21.7	51	61.5	8,930

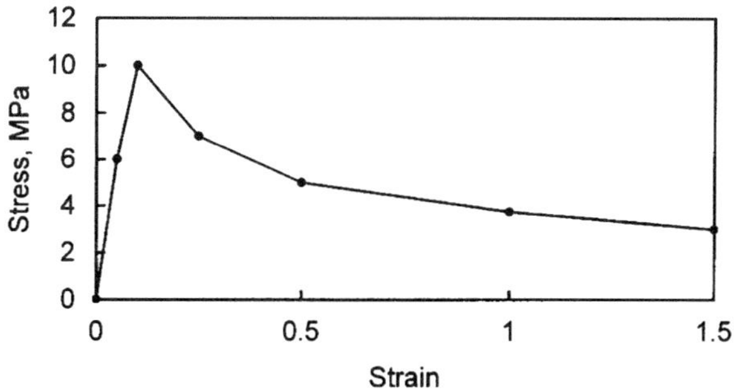

Fig. 6.12 Shear stress strain curve for a 500 μm thick indium-tin (52In–48Sn) joint held at 40°C ambient and strained at a rate of 5 × 10^{-4}/s [29]

Mei and Morris described the microstructure of $In_{52}Sn_{48}$ solder on a Cu substrate as having lamellar features [25]. The Sn-rich phase is composed of equiaxed grains. The In-rich phase contains Sn precipitates. A similar structure with less irregularity was observed by Freer and Morris on a Ni substrate [26], and significant microstructural coarsening was observed by Seyyedi [27], after prolonged aging of the solder joints made on a Cu substrate (Table 6.5) [22].

The melting temperature of this alloy makes it suitable to second level packaging process at lower temperature. Compared to indium, it displays good oxidation resistance, but is susceptible to corrosion in a humid environment. It is also a very soft metal and has a tendency to cold weld. In addition, the $In_{52}Sn_{48}$ alloy displays rather poor high temperature fatigue behavior, due to its low melting point. The high indium content limits the widespread use of this alloy due to cost and availability constraints [28].

The stress–strain curve of a thick indium-tin soldered joint and the continuum between stress–strain and creep data are given in Figs. 6.12 and 6.13, which show that recovery and recrystallization occur as fast as work hardening is induced, and mechanical failure of joints made using indium-base solders tends to be stress overload or unidirectional creep [29].

Figure 6.14 shows the schematic structure of a three-bar horizontal array package which uses InSn solder to attach the anode to the ceramic base [30]. As the melting temperature of 52In-48Sn is lower than that of In, it is suitable for second level packaging to prevent the In from re-melting after first level soldering. Also, the 52In-48Sn alloy has excellent low temperature malleability, which allows the intermediate layer to absorb any stress generated upon bonding of dissimilar substrates due to the difference in the CTE. The use of InSn alloy requires accurate control over stoichiometry of the deposited solders.

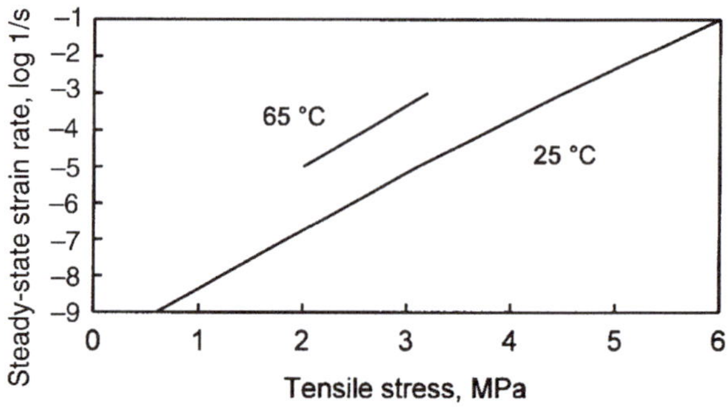

Fig. 6.13 Continuum between stress–strain and creep data for indium-tin eutectic solder (52In–48Sn) at room and elevated temperature [29]

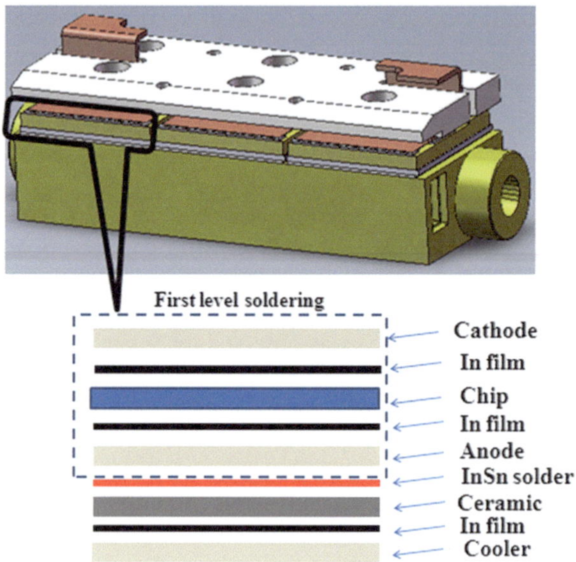

Fig. 6.14 Schematic structure of a three-bar horizontal array package with InSn solder as the second level bonding material [30]

6.1.4 SAC Solder

The most important Pb-free solders are binary SnCu eutectic, binary SnAg eutectic, and the ternary SAC eutectic. For SAC solder, the exact composition is likely to be in the range Sn (3.6–3.8) wt.%Ag (0.7–1) wt.%Cu [22]. The CTE value for SAC is 21.4–21.8 ppm/K. The thermal conductivity of SAC is 61.1–63.2 W/(m∗K).

As shown in Fig. 6.15 [31] of the SnAg binary system, there are two IMCs, ζ-Ag_4Sn and ε-Ag_3Sn, two peritectic reactions, and one eutectic reaction, liquid = Ag_3Sn + Sn [31]. The eutectic composition is at Sn3.5wt%Ag and its melting temperature is at 221°C. The SnCu system is a complicated binary system. As shown in Fig. 6.16, there are seven IMCs (β, γ, ζ, ε-Cu_3Sn, η-Cu_6Sn_5 and η′-Cu_6Sn_5), and 13 invariant reactions [32]. For soldering application purposes,

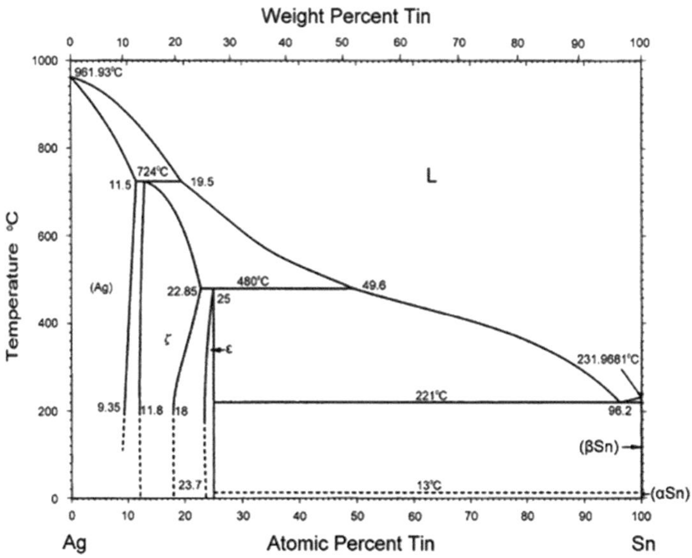

Fig. 6.15 The SnAg binary phase diagram [31]

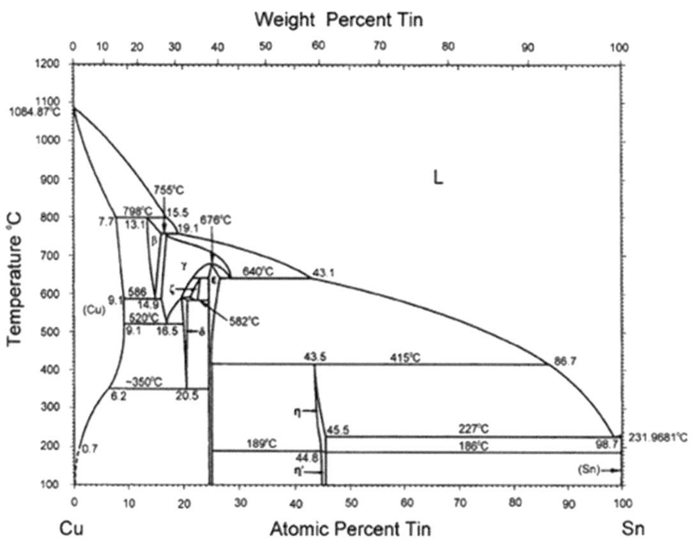

Fig. 6.16 The SnCu binary phase diagram [32]

phase equilibria at the Sn-rich corner are more important, and there are three IMCs, ε-Cu_3Sn, η-Cu_6Sn_5 and η′-Cu_6Sn_5, one eutectic, liquid = Sn + g–Cu_6Sn_5, and one possible eutectoid, η-Cu_6Sn_5 = Sn + η′–Cu_6Sn_5. The liquid composition and the temperature of the eutectic are at Sn0.7 wt%Cu and 227°C, respectively.

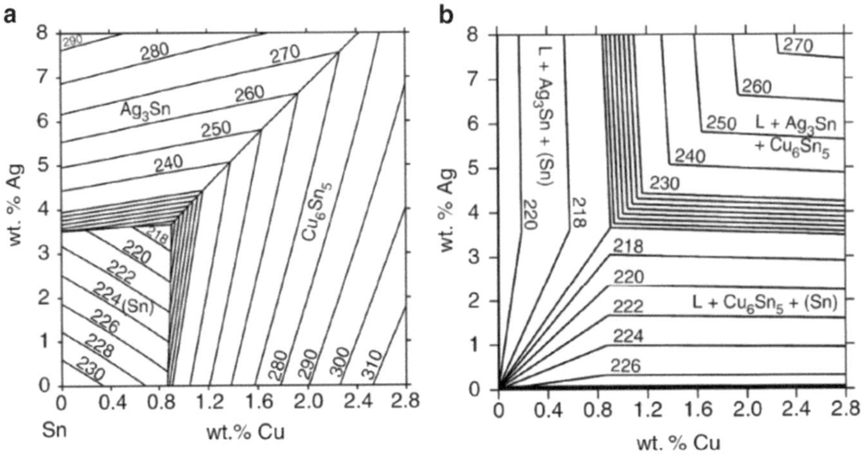

Fig. 6.17 (**a**) Calculated liquidus surface. (**b**) Calculated surface of secondary solidification. Note that the calculated ternary eutectic composition of 3.66 wt% Ag, 0.91 wt% Cu differs from the experimentally determined value of 3.5 wt% Ag, 0.9 wt% Cu [35]

Up to now, there are several reports on the composition of the liquid phase at the eutectic temperature. Miller et al. found that the SAC solder exhibited a ternary eutectic reaction and placed its composition at 4.7 wt% Ag, 1.7 wt% Cu [33]. Loomans et al. studied the phase diagram of SAC tenary alloy and confirmed that the eutectic composition was 95.6 wt% Sn3.5 wt%Ag0.9 wt%Cu [34]. The experimental work of Moon et al. confirmed the composition of the liquid phase as $x_{Ag} = 0.035$ and $x_{Cu} = 0.009$ and the temperature as 217.2 °C [35]. The calculated Sn-rich liquidus surface and surface of secondary crystallization by Moon et al. are shown in Fig. 6.17 [35]. Among the SAC alloys, there is no big difference in the mechanical and thermal properties [33–38].

As for wetting, SAC solders do not wet Cu as well as Sn/Pb using commercial fluxes. However, good wetting formation can be easily achieved by using special fluxes suitable for higher temperature use. Soldering in nitrogen atmosphere also improves wettability without fluxes. The copper dissolution experiment provides a relative measurement of the solder's tendency to dissolve Cu from the base metal and form the Cu_6Sn_5 IMC [28].

The stress–strain curve of SAC is presented in Fig. 6.18 [39]. Comparing to AuSn alloy, the stress–strain relationship of SAC is similar to that of AuSn alloy in elastic region. However, the SAC exhibits more ductile and could relieve the stress beyond the elastic region. Because of the low Ag and Cu solubility in solid Sn, most of the Ag and Cu in SAC solder react with Sn to form Ag_3Sn and Cu_6Sn_5 in the matrix of β-Sn. In the eutectic structure of $Ag_3Sn + Cu_6Sn_5 +$ β-Sn, these two phases are in the form of fine fibers and rods. The presence of the fine IMC crystals has both advantages and disadvantages for interconnection reliability. On one hand, IMC crystals increase the fatigue resistance of the bulk solder, which is desired for

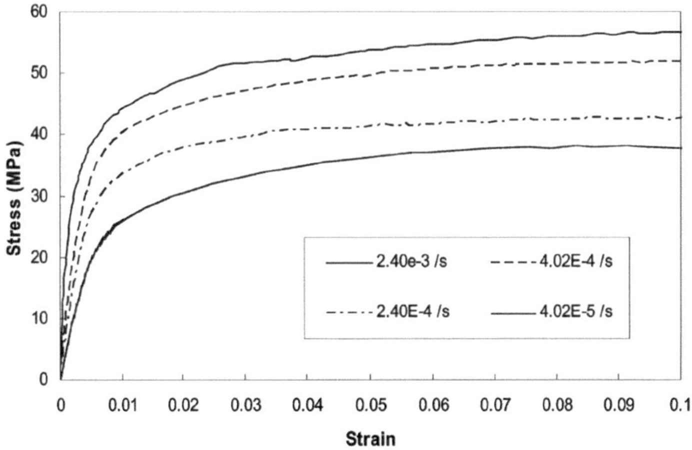

Fig. 6.18 Monotonic test results at constant strain rate at 25°C for Sn3.8Ag0.7Cu alloy [39]

Fig. 6.19 Creep rupture for Sn/40Pb and 95.5Sn/4Ag/0.5Cu [28]

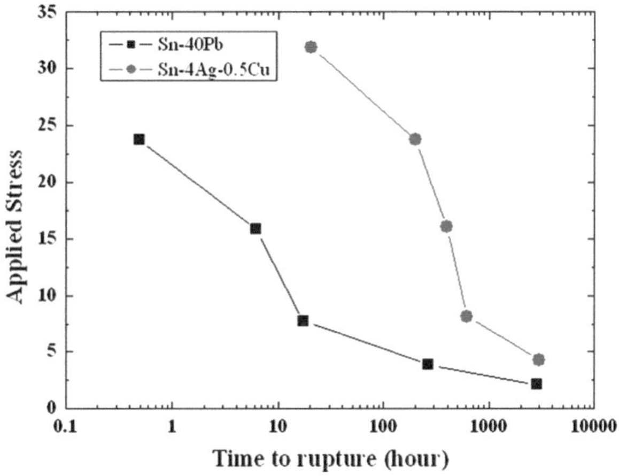

thermal cycling performance of packaging interconnections. Hence, SAC exhibits high reliability compared to some soft solders, As shown in Fig. 6.19, at an applied stress of 4 MPa, 60 wt%Sn40 wt%Pb failed after 265 h, whereas for the 95.5 wt% Sn4 wt%Ag0.5 wt%Cu alloy, it took 3,000 h for the failure to occur [28]. On the other hand, the SAC solder is much stiffer than SnPb solder because of their hardening effect. This makes it more difficult to deform in the solder joints. Thus, mechanical stress is not easily relieved, resulting in more interfacial failures in drop test.

Due to its lower melting point compared to AuSn solder and high reliability, SAC solder is suitable for second level packaging as an alternative to indium solder when AuSn solder is used in die bonding first level packaging. Figure 6.20 shows the schematic structure of a hard solder CS (HCS) semiconductor laser using SAC

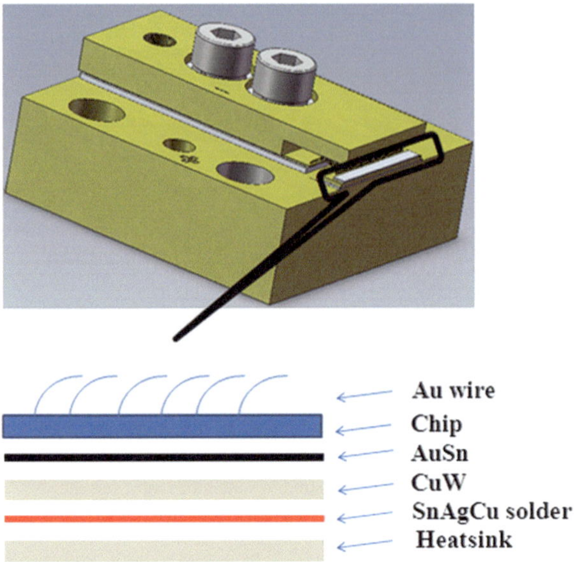

Fig. 6.20 Schematic structure of HCS package using SnAgCu solder in second level packaging [30]

solder to attach the CuW surmount to Cu heat sink [30]. This kind of semiconductor lasers with complete indium-free packaging exhibits high reliability under power cycling operation.

6.2 Mounting Substrates

The performances of mounting substrate are different for different materials, and each material has its advantages in certain aspects. The selection of the mounting substrate is driven both by the need of matching the CTE of diode laser chip and by providing high thermal conductivity in order to achieve the ability of highly effective heat dissipation and minimize the junction temperature. Table 6.6 lists a series of commonly used materials for making mounting substrates [40–42].

Copper is widely used as mounting substrate because of its high thermal conductivity. However, its CTE is large and poor performance of high power semiconductor lasers such as "smile" is usually generated as a result of the CTE mismatch between Cu and GaAs-based diode laser chip [43].

CuW, which is a metal composite, is also commonly used as mounting substrate due to its CTE closely matched with GaAs-based diode laser chip. The thermal conductivity of CuW is much lower than that of copper. For mounting high power laser bars, CuW is generally only served as a stress relief buffer layer and it is bonded with high thermal conductivity copper in order to increase the heat dissipation.

Table 6.6 Thermomechanical parameters of commonly used mounting substrate materials [40–42]

Material	Thermal conductivity (W/(m∗K))	Coefficiency of thermal expansion (ppm/K)
Cu	393	17
CuW	180–240	5.6–10.2
CuMo	190–250	7.8–10.7
Cu-diamond	470	6.7
Diamond	900–2,320	1–1.5
Si	83	7.5
AlN	170–200	4.3
Al_2O_3	30	7.3
BeO	250	7

Table 6.7 The properties of oxygen-free copper [9, 41]

Material	Melting point (°C)	CTE (ppm/K)	Thermal conductivity (W/(m∗K))	Elastic modulus (GPa) at 20 °C	Density (kg/m^3)
Copper	1,083	17	393	129.8	8,930

Cu-diamond offers not only higher thermal conductivity but also matched CTE with GaAs-based diode laser chip. However, complex and immature fabrication process and high cost prevent it from being widely used today in packaging of high power semiconductor lasers.

Aluminum Nitride (AlN) and BeO are ceramic materials. Their CTE are more closely matched to that of diode laser and the thermal conductivity of the materials is decent for low to medium power applications. These materials are used as a mounting substrate for semiconductor lasers especially when electrical insulation is needed [44].

In this section, discussions are mainly focused on the mounting substrate materials Cu, CuW, AlN, BeO, and Cu-diamond in the packaging of high power diode lasers.

6.2.1 Copper

The basic properties of copper are shown in Table 6.7 [9, 41]. As referred above, the high thermal conductivity of 393 W/(m∗K) and manufacturability make it to be the most widely used mounting substrate material in high power semiconductor lasers. Some of the application is limited by the high CTE of 17 ppm/K [45]. Copper is a metal material and thus the Cu mounting substrate is also used as the electrode for semiconductor lasers. Cu mounting substrate is dominantly used in CS and micro-channel-cooled high-power diode laser packages, especially when indium solder is used.

Due to the high thermal conductivity and electric conduction, copper has been widely used as mounting substrates in semiconductor laser bars and stacks. However, besides of the mismatch of CTE to GaAs, there are still some limitations that prevent copper from being applied as a mounting substrate material in high power semiconductor lasers [44, 45].

The CTE of Cu is much higher than that of GaAs-based diode laser. This CTE mismatch can cause large thermal stresses that lead to major compromises, such as use of "soft," indium-based solders, which however have thermal fatigue and metallurgical problems [46]. Hard solder cannot be directly used on copper mounting substrate to bond diode lasers.

Copper is easy to be oxidized and the oxide CuO is covered on the surface. The electrical and thermal performances of the copper mounting substrate are declined obviously. More importantly, the wettability and adhesion are compromised significantly due to oxidation as well.

In micro-channel coolers (MCCs), copper material serve as the anode and it is in direct contact with the cooling water. This requires the use of deionized water in order to ensure that no electricity flows through the cooling water. Corrosion is a big issue and challenge in copper-based MCCs. The reliability of copper MCC and thus the packaged device is greatly reduced due to susceptible corrosion of copper. Experimental work has been conducted to check the erosion properties of the coolers [45]. In the experiment, a vertical stack of six copper MCCs was subjected to a flow rate of 0.2 g per minute for each cooler, which was approximately four times greater than the flow rate recommended by the manufacturer of the copper coolers. No voltage or current was applied to the diode bars in the stack in order to eliminate the effects of galvanic corrosion [45]. Figure 6.21 shows the appearance of the coolers before and after testing [45]. The blue arrows mark the direction of water flow in the testing. The zigzag pattern in the cooler is designed to promote turbulent flow and therefore improves the cooling properties of the MCC.

The effect of the high flow rate is clearly seen on the right-hand side of Fig. 6.21b [45]. Most of the structure that existed in the front of the cooler has been completely eroded by the water. The small channels in the front of the cooler that transport the water to the cooling layer has greatly expanded in size, and a significant portion of the zigzag pattern has been eroded away. These phenomena are representative of what has been observed in other tests conducted at high flow rates [45].

There are various types of copper material, such as red copper, fine copper, and beryllium bronze. High purity of copper is needed in high power semiconductor laser packaging. However, it is impossible to remove impurities completely. The performances of copper are declined by impurities, and hence, some impurities, such as Fe and Ni, should be eliminated as much as possible due to the decline of thermal conductivity and electrocorrosion. The effect of Zr and Cr impurities on the thermal conductivity and electrocorrosion of copper is not significant. On the other hand, Zr and Cr impurities can increase the strength of copper. Due to high thermal and electric conductivity, oxygen-free copper with high purity above 99.95 % and low oxygen and impurities below 0.003 % and 0.05 %, respectively, has been used and is preferred as the packaging material for high power semiconductor lasers.

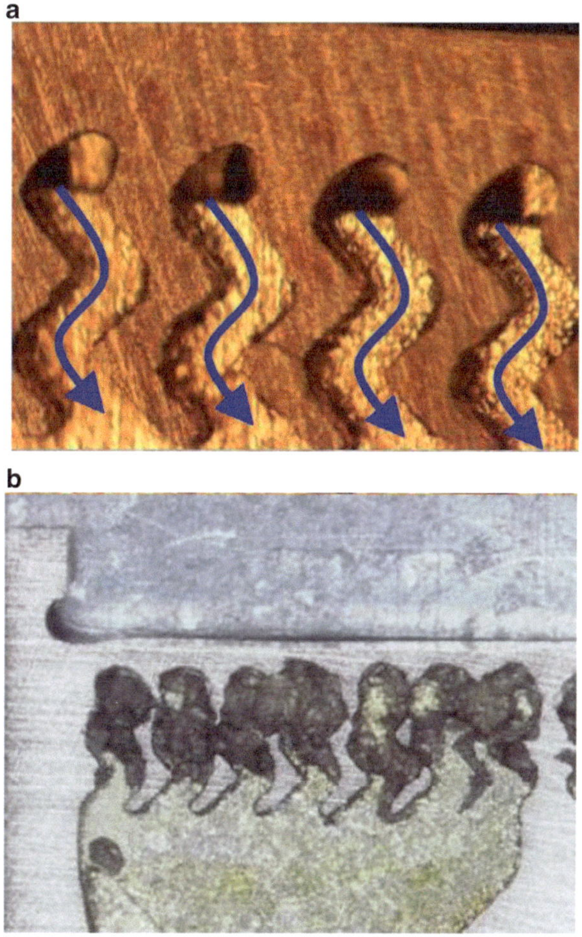

Fig. 6.21 Examples of the internal structure of a copper micro-channel coolers before (*top*) and after (*bottom*) a high-flow-rate test [45]. (**a**) Before test. (**b**) After test

Figure 6.22 shows a typical copper MCC high power diode laser package [30]. The MCC is made of oxygen-free copper. As discussed above, copper is prone to oxidation and corrosion during packaging process and operation. In order to prevent oxidation and corrosion of the MCC, the cooler is cleaned and metallized first during the packaging process. Diluted hydrochloric acid is used to remove the copper oxide. Typically both the inner and outside surfaces are coated with metal films of Ni and Au. Ceramic MCCs have been developed to overcome the drawbacks of copper MCCs. These coolers can provide excellent thermal performance and thermal-mechanical performance due to the lower CTE of ceramic material. Also these coolers do not require the use of deionised water since it is insulated by the ceramic material. However, the ceramic MCCs have not been widely used so far in packaging of high power diode lasers yet due to immature ceramic MCC technology, packaging process, and high cost [47].

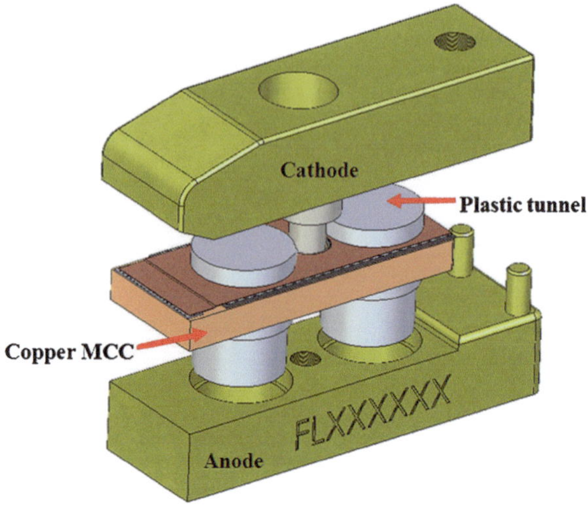

Fig. 6.22 A typical copper MCC package for high power semiconductor laser bars [30]

6.2.2 *Copper Tungsten*

Tungsten has a very high melting point, high density, and low CTE. Copper is a good electrical and thermal conductor. The combination of tungsten and copper (i.e., copper–tungsten composite) enjoys the low CTE of tungsten and high thermal conductivity of copper. Meanwhile, the CTE and electrical conductivity can be adjusted by varying the copper-to-tungsten ratio. The CTE of Tungsten Copper (CuW) can be designed to closely match the CTE of GaAs. CuW mounting substrate is widely used in G-stack and HCS high power diode laser packages. The typical properties of CuW are shown in Table 6.8 [44].

CuW provides a much lower thermal expansion coefficient compared to pure copper while maintaining a necessary thermal conductivity. For binary CuW alloys, copper generally accounts for about the ratio of 10–50 % in weight. The higher the tungsten content, the lower its thermal expansion rate is. The copper/tungsten mounting substrate with gold-tin solder is a widely used method that can provide good CTE match to the GaAs device used to make high power and high reliability semiconductor lasers [48]. Among the different compositions, W90Cu is the most commonly used composition for high power diode lasers. W90Cu has a CTE of 6.5 ppm/K and the maximum degree of tensile stress experienced by a GaAs diode laser array when it is mounted on a W-90 CuW mounting substrate is 6 MPa. In the case of CuW mounting substrate with W-80, the CTE is 8.3 ppm /K. In this case, the maximum tensile stress encountered by the GaAs diode laser array is ~43 MPa [44].

The conventional CuW submount provides a thermal conductivity of 180–240 W/(m∗K) with a CTE of 5.6–10.2 ppm/K that matches the chip of laser diodes [41]. Even though CuW has a good CTE match with GaAs, its relatively poor thermal performance hinders its application in high power lasers and high duty

Table 6.8 Typical properties of CuW [44]

	Composition	Properties		
Type	Tungsten content (wt%)	Density (g/cm^3)	CTE (ppm/K)	Thermal conductivity (W/(m∗K))
W90Cu	90	17	6.5	180–190
W85Cu	85	16.3	7	190–200
W80Cu	80	15.6	8.3	200–210
W75Cu	75	14.9	9	220–230

Table 6.9 Typical properties of Cu-diamond with a CTE matched to GaAs [41]

Material	CTE (ppm/K)	Thermal conductivity (W/(m∗K))	Density (g/cm^3)
Cu-Diamond	6.7	470	5.35

cycle operations [49]. However, with newly improved solutions such as the finite boundary value method and functionally graded materials, the thermal conductivity of copper tungsten can be pushed up to approximately 320 W/(m∗K). With the newly improved CuW materials, most of the thermal management solutions can be pursued and achieved using tungsten copper for high power diode lasers [44]. The trend of increasing diode laser chip/bar size and heat dissipation requirements has made CuW the material of choice for diode laser packaging.

6.2.3 Copper Diamond

Diamond has a very high thermal conductivity, high melting point, and low CTE while copper is a good electrical conductor and has large CTE. The combination of diamond and copper (i.e., copper-diamond composite and denoted by Cu-diamond) enjoys high thermal conductivity and a CTE closely matched to GaAs material. The CTE and electrical conductivity can be adjusted by varying the copper-to-diamond ratio [45, 48, 50, 51]. The CTE of Cu-diamond can be designed to match the CTE of diode lasers. Typical properties of Cu-diamond which has a CTE matched to that of diode lasers are shown in Table 6.9 [41].

Copper-diamond composites consisting of a high volume fraction of diamond provide a fitting solution to existing thermal management issues with their excellent thermal conductivity (>450 W/(m∗K)) and a tailored CTE matching that of semi-conductor materials. These composites are available with a machinable surface for further surface operations such as milling, lapping for realizing surfaces of optical quality. The effective thermal conductivity of Cu-diamond can be changed by different volume fraction of diamond and the diamond particle size. Figure 6.23 shows the relationship among thermal conductivity, volume fraction of diamond, and the diamond particle size [41]. The copper-diamond material is conventionally produced by high temperature, high pressure sintering of small diamond particles

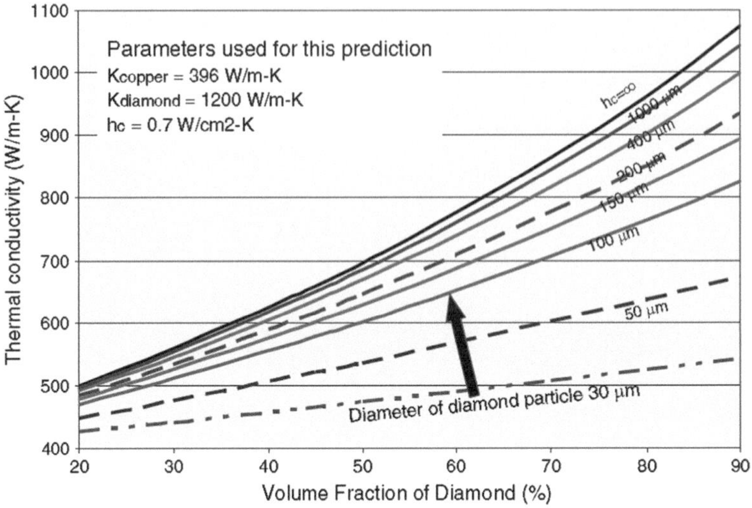

Fig. 6.23 Illustration of thermal conductivity variation with volume fraction and particle size of diamond for copper-diamond material [41]

to form a homogenous intergrown diamond lattice structure. The fine grain structure and uniform distribution of the binder makes this material ideal for heat spreader applications [41]. The most commonly used copper-diamond heat spreaders typically offer a thermal conductivity of 600 W/(m∗K) and an average CTE of 3.0 ppm/K in the range of 20–300°C. The surface of the copper-diamond heat spreader can be polished to a high degree of surface quality and flatness, and also excellent edge qualities can be achieved by high precision cutting. These attributes are required for bonding edge emitting diode lasers and arrays. Packaging technologies using copper-diamond as mounting substrate are being developed. However, complex and immature fabrication process and high cost prevent it from being widely used today in high power semiconductor laser products.

6.2.4 Aluminum Nitride

AlN is a stable compound and exists in crystal structure (wurtzite, hexagonal). Pure AlN has a density of 3.26 g/cm^3. AlN ceramic has attracted much attention in electronic industries in recent decades due to the exceptional properties such as high thermal conductivity, low dielectric constant, high electrical resistivity, and nontoxicity. Table 6.10 shows the major physical properties of AlN ceramic [41, 52]. The thermal conductivity of AlN is ~180 W/(m K) and the volume resistivity is higher than 10^{13} Ω cm. Although the CTE of AlN is lower than GaAs, it is more closely matched than most of other mounting substrates. When AlN is used as a mounting substrate to bond

Table 6.10 Physical properties of AlN ceramic [41, 52]

Aluminum nitride	AlN
Density (g/cm^3)	3.33
Young's modulus (GPa)	310
Thermal conductivity (W/(m*K))	170–200
CTE (ppm/K)	4.3
Melting point (°C)	2,500
Specific heat (J/kg/K)	738

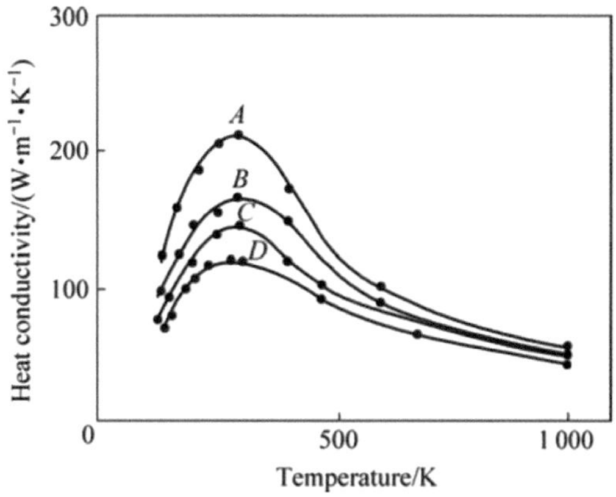

Fig. 6.24 Thermal conductivity of AlN ceramic as a function of temperature [53]

GaAs diode lasers, the diode laser experiences tensile stress as compared to compressive stress for most of the mounting substrates such as copper and CuW.

The thermal conductivity of AlN is affected by temperature. Figure 6.24 shows the temperature dependence of AlN thermal conductivity. The thermal conductivity increases rapidly at low temperature as the temperature increases. It reaches a maximum at around 300 K and then decreases thereafter with increasing temperature [53]. In order to reduce the thermal resistance of AlN in high power semiconductor laser application, yttrium compounds are commonly used to act as the additives to improve the thermal performance of AlN ceramic [54, 55], which makes it become a superior ceramic material for application in packaging of high power semiconductor lasers.

In addition, the thermal conductivity of AlN ceramic is greatly affected by the presence of metallic impurities [56] and oxygen impurity [57, 58]. Al_2O_3 is formed on the surface of the initial AlN powder due to spontaneous reaction with air and it also dissolves in the AlN lattice. The lattice distortion due to aluminum vacancies resulting from incorporation of oxygen into the lattice leads to increased phonon scattering, thus lowering the thermal conductivity of the material [59–63].

AlN is an attractive, nontoxic alternative to BeO for high-power electronic substrates, packages, and heat spreaders, where thermal management is a significant design issue [56, 64, 65]. AlN also has good oxidation resistance, due to the formation of a passivating alumina coating, even as high as 1,400°C [56]. AlN mounting substrate is mainly used in single emitter diode lasers, QCW diode laser

Table 6.11 Physical properties of BeO ceramic at 298 K [42]

Beryllium oxide	99.5 % (BeO)
Density (g/cm^3)	2.85
Young's modulus (GPa)	345
Thermal conductivity (W/(m K))	250
CTE (ppm/K)	7
Melting point (°C)	2,820

bars, or low power CW bars (generally 40 W or lower) where total thermal load is not high and insulation is needed. AlN is also widely used as a second level packaging substrate where thermal conduction and electrical insulation is needed in high power diode laser packages.

6.2.5 *Beryllium Oxide*

BeO ceramic is highly conductive thermally with thermal conductivity of 250 W/(m K) and CTE of 7 ppm/K as shown in Table 6.11 [42]. It has higher thermal conductivity than AlN ceramic and the CTE is also closely matched to GaAs. Due to its high thermal conductivity, it allows for quick heat transfer onto copper heat sink underneath it, while helping to maintain electrical insulation between the diode and the heat sink. BeO has an excellent thermal shock resistance, despite a high CTE than that of GaAs and only moderate strength. Different from AlN material, diode laser experiences compressive stress when BeO is used as a mounting substrate to bond GaAs diode lasers.

The thermal conductivity of BeO at room temperature is the highest among the ceramics, close to that of high conductivity metals, but decreases rapidly between room temperature and 200°C [56]. The CTE of BeO increases with increasing temperature. Figure 6.25 gives a comparison of several widely used ceramic materials [66]. As can be seen in Fig. 6.25, the BeO has the highest thermal conductivity and the largest CTE among the several ceramic materials (BeO, AlN, Al_2O_3). All the ceramics have similar trend in thermal conductivity and CTE dependency on temperature, but BeO changes faster than others. BeO is toxic and hazardous for human health and environments.

The performances of BeO, such as thermal conductivity and stability, are declined by the impurities in the material. Typical impurities present in BeO powders include Fe, Mg, Ca, and Al [56]. SO_3 and SO_4 impurities may also be present in sulfate-derived powders, such as Brush-Wellman grade UOX, remaining after the production of the BeO powder by decomposition of beryllium sulfate salt [56]. BeO typically also has a number of significant anionic impurities, including F, P, S, and Si, which is typically not stated by manufacturers in the total impurity content [67]. These anionic impurities contribute to the reduction in strength of BeO at moderate and high temperatures [67]. Same as AlN, BeO mounting substrate is mainly used in single emitter diode lasers, QCW diode laser bars, or low power CW bars where total thermal load is not high and insulation is needed. As BeO has higher thermal conductivity, BeO mounting substrate can be used at higher power level than AlN substrate. Due to its toxic nature, BeO is generally not used as a second level packaging substrate in high power diode laser packages.

References

1. C. Zweben, New, low-CTE, ultrahigh-thermal-conductivity materials for lidar laser diode packaging. Proc. SPIE 58870D(1–10) (2005)
2. V. Von, *Thermal and Mechanical Optimisation Diode Laser Bar Packaging*, PHD paper, Herstellung und Verlag: Books on Demand GmbH, Norderstedt (2007)
3. www.torreyhillstech.com Understanding of Laser, Laser diodes, Laser diode packaging and its relationship to Tungsten Copper; 6370 LUSK BLVD, SUITE F-11
4. A.C. Pliska, J. Mottin, N. Matuschek, C. Bosshard, Bonding semiconductor laser chips: substrate material figure of merit and die attach layer influence, Belgirate, Italy, (2005), pp. 28–30
5. F. Bachmann, P. Loosen, R. Poprawe, *High Power Diode Lasers Technology and Applications* (Springer Science + Business Media, LLC, New York, 2007)
6. D. Lorenzen, M. Schrer, J. Meusel, P. Hennig, H. Kig, M. Phillippens, J. Sebastian, R. Hülsewede, Comparative performance studies of indium and gold-tin packaged diode laser bars. Proc. SPIE **6104**, 610404 (2006)
7. X.S. Liu, R.W. Davis, L.C. Hughes, M.H. Rasmussen, R. Bhat, C.E. Zah, A study on the reliability of indium solder die bonding of high power semiconductor lasers. J. Appl. Phys. **100**, 013104(1–11) (2006)
8. J.L. Hostetler, C.L. Jiang, V. Negoita, T. Vethake, R. Roff, A. Shroff, C. Miester, U. Bonna, G. Charache, H. Schlüter, F. Dorsch, Thermal and strain characteristics of high-power 940 nm laser arrays mounted with AuSn and In solders. Proc. SPIE **6456**(645602) 645602(1–12) (2007)
9. M. Wakaki, K. Kudo, T. Shibuya, *Physical Properties and Data of Optical Materials* (CRC Press, Boca Raton, FL, 2009)
10. G. Humpston, D.M. Jacobson, Advanced materials and processes. Indium Solders **163**(4), 45–47 (2005)
11. www.indium.com
12. http://www.coininginc.com/files/admin/english_gold_tin_paper.pdf
13. H. Okamoto, T.B. Massalski, *The Au-Sn(Gold-Tin) System in Phase Diagram of Binary Gold Alloys* (ASM International, Metals Park, OH, 1987), pp. 278–289
14. Q. Wang, S.-H. Choa, W. Kim, J. Hwang, S.K. Ham, C. Moon, Application of Au-Sn eutectic bonding in hermetic radio-frequency microelectromechanical system wafer level packaging. J. Electron. Mater. **35**(3), 425–432 (2006)
15. H. Okamoto, T.B. Massalski, *Binary Alloy Phase Diagrams* (ASM International, Metals Park, OH, 1990)
16. S. Zama, D.F. Baldwin, T. Hikami, H. Murata, *Flip Chip Interconnect Systems Using Wire Stud Bumps and Lead Free Solder*. Proceedings of 50th Electronic Components and Technology Conference, Las Vegas, May 2000
17. H. Oppermann, *The Role of Au/Sn Solder in Packaging Materials for Information Technology* (Springer, London, 2005), pp. 377–390
18. A. Debski, W. Gasior, Z. Moser, R. Major, Enthalpy of formation of intermetallic phases from the Au–Sn system. J. Alloy Compd. **491**(1–2), 173–177 (2010)
19. G. Zeng, S. McDonald, K. Nogita, Development of high-temperature solders: review. Microelectron. Reliab. **52**(7), 1306–1322 (2012)

20. X.S. Liu, K. Song, R.W. Davis, M.H. Hu, C.E. Zah, *Design and Implementation of Metallization Structures for Epi-Down Bonded High Power Semiconductor Lasers*. 2004 Electronic Components and Technology Conference, vol. 1 (2004), pp. 798–806
21. X.S. Liu, K.C. Song, R.W. Davis, L.C. Hughes, M.H. Hu, C.E. Zah, A metallization scheme for junction-down bonding of high-power semiconductor lasers. IEEE Trans. Adv. Pack. **29**(3), 533–541 (2006)
22. http://www.indium.com/products/alloy_sorted_by_temperature.pdf
23. D.P. Seraphim, R. Lasky, C.Y. Li, *Principles of Electronic Packaging* (McGraw-Hill, New York, 1989)
24. J. Glazer, Metallurgy of low temperature Pb-free solders for electronic assembly. Int. Mater. Rev. **40**(2), 65–93 (1995)
25. Z. Mei, J.W. Morris Jr., Superplastic creep of low melting point solder joints. J. Electron. Mater. **21**(4), 401–407 (1992)
26. J.L. Freer, J.W. Morris Jr., Microstructure and creep of indium/tin on Cu and Ni substrates. J. Electron. Mater. **21**(6), 647–652 (1992)
27. J. Seyyedi, Thermal fatigue behavior of low melting point solder joints. Soldering Surf. Mount Technol. **5**(1), 26–32 (1993)
28. http://tersted.home.xs4all.nl/PDF_files/Heraeus/SMI98NoPb.pdf
29. G. Humpston, D.M. Jacobson, *Principles of Soldering* (ASM International, Metals Park, OH, 2004)
30. J.W. Wang, D. Hou, X.S. Liu, Introduction of packaging materials for high power semiconductor lasers. Internal Talk from Focuslight Technologies Co., Ltd. (2011), pp. 18–27
31. I. Karakaya, W.T. Thompson, The Ag-Sn (Silver-Tin) system. Bull. Alloy Phase Diagrams **8**(4), 340–347 (1987)
32. N. Saunders, A.P. Miodownik, The Cu-Sn (Copper-Tin) system. Bull. Alloy Phase Diagrams **11**(3), 278–287 (1990)
33. C.M. Miller, I.E. Anderson, J.F. Smith, A viable tin-lead solder substitute Sn-Ag-Cu. J. Electron. Mater. **23**, 595–601 (1994)
34. M.E. Loomans, M.E. Fine, Tin-silver-copper eutectic temperature and composition. Metall. Mater. Trans. **31**, 1155–1162 (2000)
35. K.W. Moon, W.J. Boettinger, U.R. Kattner, F.S. Biancaniello, C.A. Handwerker, Experimental and thermodynamic assessment of Sn-Ag-Cu solder alloys. J. Electron. Mater. **29**, 1122–1136 (2000)
36. http://cmst.be/projects/imecat/documents/08_2004_Eurosime_Vandevelde_paper.pdf
37. J. Bartelo, S.R. Cain, D. Caletka, K. Darbha, T. Gosselin, D.W. Henderson, D. King, K. Knadle, A. Sarkhel, G. Thiel, C. Woychik, *Thermomechanical Fatigue Behavior of Selected Lead Free Solders*. 2nd Electronics Assembly Process Conference (2001)
38. P. Chalco, E. Blackshear, *Reliability Issues of BGA Packages Attached With Lead-Free Solder*. Proceedings InterPack01, The Pacific Rim/ASME International Electronic Packaging Technical Conference (2001), pp. 8–13
39. D. Bhate, D. Chan, G. Subbarayan, T.C. Chiu, V. Gupta, D.R. Edwards, Constitutive behavior of Sn3.8Ag0.7Cu and Sn1.0Ag0.5Cu alloys at creep and low strain rate regimes. IEEE Trans. Compon. Pack. Technol. **31**(3), 622–633 (2008)
40. http://www.et-trends.com/files/Heat_Sinks_Question.pdf
41. X.C. Tong, *Advanced Materials for Thermal Management of Electronic Packaging* (Springer Science + Business Media, LLC, New York, 2011)
42. G.P. Akishin, S.K. Turnaev, V.Y. Vaispapir, M.A. Gorbunova, Y.N. Makurin, V.S. Kiiko, A.L. Ivanovskii, Thermal conductivity of beryllium oxide ceramic. Refract. Ind. Ceram. **50**(6), 465–468 (2009)
43. J.W. Wang, Z.B. Yuan, L.J. Kang, K. Yang, Y.X. Zhang, X.S. Liu, *Study of the Mechanism of "Smile" in High Power Diode Laser Arrays and Strategies in Improving Near-field Linearity*. 2009 Electronic Components and Technology Conference (2009), pp. 837–842
44. G.S. Jiang, L.Y. Diao, K. Kuang, *Advanced Thermal Management Materials* (Springer, Berlin, 2013)

45. R. Feeler, J. Junghans, G. Kemner, E. Stephens, Next-generation micro-channel coolers. Proc. SPIE **6876**, 687608(1–8) (2008)
46. C. Zweben, New, low-CTE, ultrahigh-thermal-conductivity materials for lidar laser diode packaging. Proc. SPIE 58870D (1–10) (2005)
47. http://132.228.182.183/products/ceo_micro_cooled_diodes/assets/Ceramic_coolers_paper.pdf
48. K.E. Goodson, K. Kurabayashi, R. Fabian, W. Pease, Improved heat sinking for laser-diode arrays using micro-channels in CVD diamond. IEEE Trans. Compon. Pack. B **20**(1), 104–109 (1997)
49. E.C. Yu, A.J. Przekwas, *Thermomechanical Design of a Microchannel Cooled Semiconductor Laser Diode Array Package*. Part of the SPIE Conference on Physics and Simulation of Optoelectronic Devices VII, vol. 3625 (1999), pp. 535–542
50. C. Zweben, New, low-CTE, ultra high-thermal-conductivity materials for lidar laser diode packaging. Proc. SPIE 58870D(1–10) (2005)
51. M. Leers, C. Scholz, K. Boucke, M. Oudart, *Next Generation Heat Sinks for High-Power Diode Laser Bars*. 23rd IEEE Semi-Thermal Symposium (2007), pp. 105–111
52. http://www.hfmorke.cn/alncer.html
53. K. Watari, K. Ishizaki, F. Tsuchiya, Phonon-scattering and thermal conduction mechanisms of sintered aluminum nitride ceramics. J. Materi. Sci. **28**(14), 3709–3714 (1993)
54. H. Nasery, M. Pugh, M. Medraj, Novel fabrication process of AlN ceramic matrix composites at low temperatures. Sci. Eng. Compos. Mater. **18**(3), 117–125 (2011)
55. A. Hafidi, M. Billy, J.P. Lecompte, Influence of microstructural parameters on thermal-diffusivity of aluminum nitride-based ceramics. J. Mater. Sci. **27**(12), 3405–3408 (1992)
56. D. de Faoite, D.J. Browne, F.R. Chang-Díaz, K.T. Stanton, A review of the processing, composition, and temperature-dependent mechanical and thermal properties of dielectric technical ceramics. J. Mater. Sci **47**(10), 4211–4235 (2012)
57. J.H. Harris, R.A. Youngman, R.G. Teller, On the nature of the oxygen-related defect in aluminum nitride. J. Mater. Res. **5**(8), 1763–1773 (1990)
58. H. Buhr, G. Muller, H. Wiggers, F. Aldinger, P. Foley, A. Roosen, Phase composition, oxygen content, and thermal conductivity of A1N(Y203) ceramics. J. Am. Ceram. Soc. **74**(4), 718–723 (1991)
59. T.B. Jackson, A.V. Virkar, K.L. More, R.B. Dinwiddie, R.A. Cutler, High thermal conductivity aluminum nitride ceramics: the effect of thermodynamic, kinetic, and microstructural factors. J. Am. Ceram. Soc. **80**(6), 1421–1435 (1997)
60. G.A. Slack, Nonmetallic crystals with high thermal conductivity. J. Phys. Chem. Solids **34**(2), 321–335 (1973)
61. G.A. Slack, L.J. Schowalter, D. Morelli, J.A. Freitas, Some effects of oxygen impurities on AlN and GaN. J. Cryst. Growth **246**(3–4), 287–298 (2002)
62. W. Koji, High thermal conductivity non-oxide ceramics. J. Ceram. Soc. Jpn. **109**(1), S7–S16 (2001)
63. J.P. Sachet, J.Y. Laval, F. Lepoutre, A.C. Boccara, Thermal behaviour of grain boundaries in aluminium nitride ceramics. J. Phys. Colloq. **51**(C1), 617–622 (1999)
64. K.J. Lodge, J.A. Sparrow, E.D. Perry, E.A. Logan, M.T. Goosey, D.J. Pedder, C. Montgomery, Prototype packages in aluminum nitride for high performance electronic systems. IEEE Trans. Compon. Hybr. **13**(4), 633–638 (1990)
65. L. La Spina, E. Iborra, H. Schellevis, M. Clement, J. Olivares, L.K. Nanver, Aluminum nitride for heat spreading in RF IC's. Solid-State Electron. **52**(9), 1359–1363 (2008)
66. http://www.ntktech.com/AlN/ALN%20for%20web.pdf
67. S.C. Carniglia, R.E. Johnson, A.C. Hott, G.G. Bentle, Hot pressing for nuclear applications of BeO; process, product, and properties. J. Nucl. Mater. **14**, 378–394 (1964)

Chapter 7
Packaging Process of High Power Semiconductor Lasers

Despite the many advances in manufacturing of high power semiconductor lasers, the basic packaging process has not been changed significantly. This chapter reviews the steps used to package and assemble high power semiconductor lasers in open packages and fiber-coupled modules. The fabrication procedure of open packages contains a sequence of processes, which involve incoming materials inspection, raw materials cleaning, metallization, solder deposition, die bonding, wire bonding, assembling, screening, before burn-in (BBI) test, burn in (BI), after burn-in (ABI) test, and final inspection. The details are presented in Sects. 7.1–7.10. Fiber-coupled modules typically use an open package, optics, and a fiber to couple the light into the fiber. The packaging process for fiber-coupled module is introduced in Sect. 7.11.

Figure 7.1 shows the flowchart of packaging process for high power semiconductor laser open package [1]. Among the packaging processes, incoming material inspection is the first step to sort out the unqualified materials for quality control and cost saving. All the raw materials are cleaned in different cleaning process to eliminate the contamination and oxidation before further steps. After cleaning, metallization and solder layers are deposited on the targeting part for soldering. Die bonding, also known as die attachment, is the process of bonding a laser chip to a submount or heat sink. For some structures, wire bonding is used to connect the laser chip with the cathode. The step of assembling is to assemble the laser device with a set of packaging components. After the assembling, screening is carried out to examine the laser defects, such as facet defects, solder overflow, soldering voids, and misalignment. Subsequently, the performances of the diode lasers are tested and characterized BBI. As an essential and useful approach to screen unqualified products in short time, burn in is applied in all of the products for quality control. ABI, the performances of the diode lasers are tested and characterized again. Final inspection is the last step to make sure no extra issue is involved for final product. The packaging process steps are discussed in detail in following sections.

X. Liu et al., *Packaging of High Power Semiconductor Lasers*, Micro- and Opto-Electronic Materials, Structures, and Systems, DOI 10.1007/978-1-4614-9263-4_7,

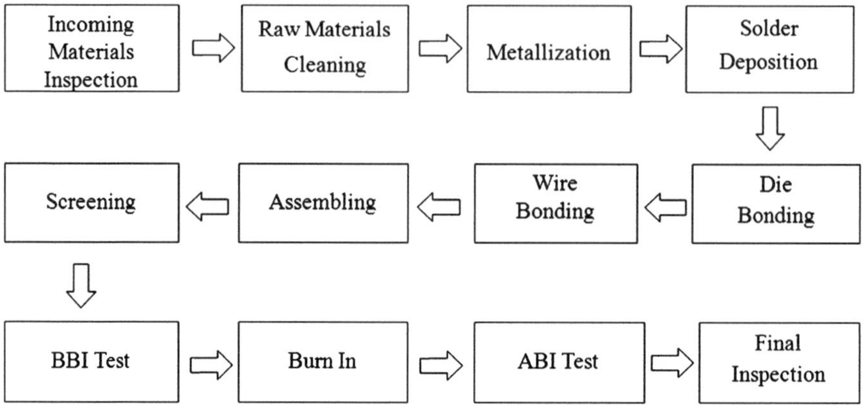

Fig. 7.1 The flowchart of packaging process for a high power semiconductor laser open package [1]

7.1 Incoming Material Inspection

Incoming material inspection (IMI) or receiving inspection is the most common and important step of quality and yield control in industrial manufacturing. As the first gate of product manufacturing, IMI is to ensure the incoming materials to meet the material specifications and manufacturing requirements. It helps to reduce the manufacturing costs and eliminate scrap losses. Moreover, it plays an important role in maintaining the consistency of incoming raw materials and improving quality and reliability of the products. IMI also collects the information regarding to the material property to extend the technical understanding of incoming materials.

7.1.1 Main Inspection Items

The main inspection items and corresponding inspection instruments used in the IMI process are summarized in Table 7.1 [1]. The IMI process is separated into two categories, the appearance inspection and the physical property measurement, as stated in Table 7.1.

7.1.2 Appearance Inspection

The received package is inspected visually to have an overview of the package condition and to make sure no damage to the package and the parts inside. Profile projectors are used to inspect and measure the dimensions of the components and substrates. Surface profiling systems are used to measure the surface roughness and

Table 7.1 IMI items [1]

Categories	Inspection items	Instruments
Appearance inspection	Package as received	Visual
	Surface state of parts as received	Visual/Microscope
	Dimension	Profile projector/Micrometer/Vernier caliper/Surface optical profiling system
	Critical Area	Microscope
Physical property measurement	Melting temperature	Differential scanning calorimetry (DSC)
	Metallization adhesion	Tape and Blade, High temperature test
	Thickness of thin film	X-Ray fluorescence (XRF) tester
	Solderability	Reflow, Die shear, and Visual
	Wettability	Reflow and visual

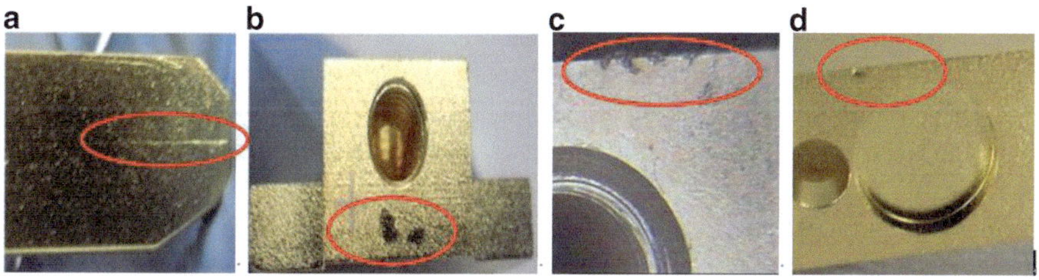

Fig. 7.2 The typical surface defects of incoming materials [1]. (**a**) Scratch. (**b**) Contaminations. (**c**) Metallization burr. (**d**) Chipping

flatness of the critical areas. The appearance of incoming materials, such as mounting substrates, coolers, laser chips, electrodes, and insulators, are inspected by visual and microscopy. All the surface defects, typically including scratch, contaminations, metallization burr, and chipping are identified in this process as shown in Fig. 7.2 [1].

Mounting substrate, whether it is a copper block, a copper-tungsten (CuW) substrate, a ceramic substrate or a micro-channel cooler (MCC), is used to bond the laser chip on it. The surface quality of the critical area where the laser chip is bonded is important to achieve a high quality die bonding. The inspection criterion of the critical area is stricter than the other area and needs to pay more attentions to it.

7.1.3 Physical Property Measurement

Melting temperature is an important parameter for solder materials. In actual manufacturing process, the melting temperature of solder materials is tested and verified for material composition, oxidation, and contaminations. The melting

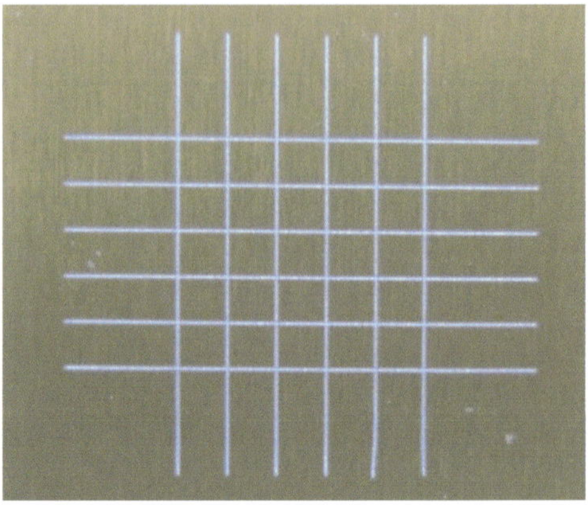

Fig. 7.3 Cross-cut and tape test on the surface of a metallized heat sink [1]

temperature can be estimated by simply observing the melting and freezing process in a reflow oven with certain preset temperature profile. More accurate melting temperature can be characterized by differential scanning calorimetry (DSC). DSC is a thermoanalytical technique in which the difference in the amount of heat required to increase the temperature of a sample and reference is measured as a function of temperature [2]. The basic principle of measuring melting point of solder materials is that when the solder material sample undergoes a phase transition from solid to liquid, more heat will be needed than the reference to maintain both at the same temperature due to the absorption of heat by the sample as it undergoes the endothermic phase transition from solid to liquid. The melting temperature is measured as onset temperature of the endothermic event.

The adhesion of the metallization to the body material, no matter it is the mounting substrate or the heat sink, is crucial to achieve high quality diode laser packaging. In high power semiconductor laser packaging, the laser chip is bonded to a mounting substrate using a solder material and the mounting substrate is then sometimes attached to a heat sink or cooler using metallurgical bonding as well. This requires the mounting substrate and heat sink material to be solderable. Therefore, the surface of the mounting substrates and heat sink materials are required to be metalized. The adhesion of the metallization significantly affects the quality and performance of high power diode lasers. Figure 7.3 illustrates the adhesion test on the surface of a heat sink with nickel and gold metallization using the cross-cut and tape test [1]. Metallization with good adhesion has no material peeling off or detached when sticky tape is placed and peeled at the cross-cut area.

The thickness of the metallization layers and the thickness of the solder thin-film layer for die bonding are required to be controlled within a certain range. X-ray fluorescence (XRF) tester can be applied to measure the thickness and component of the coating layers. XRF is the emission of characteristic "secondary" (or fluorescent) X-rays from a material that has been excited by bombarding with high-energy X-rays. Coating thickness is determined from the energy and intensity

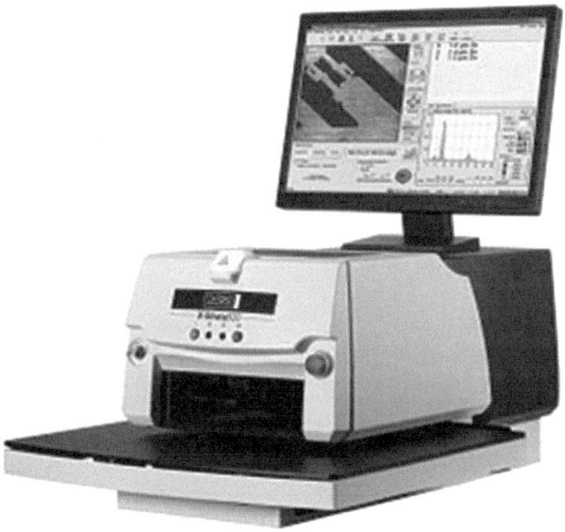

Fig. 7.4 X-ray fluorescence tester for measuring thin films [4]

of the respective X-ray emission. The thickness measured ranges from 10 Å to many micrometers [3]. Figure 7.4 shows one example of XRF tester for the measurement of thickness of thin films [4].

The wettability is usually estimated by the value of contact angle θ. The wettability of the same metal material may vary considerably, as it is strongly influenced by the surface state of the metal. Thin films of oxide, grease, or organic contaminants on the surface of the metal submount can severely affect its wettability. Contact heating method can be used to evaluate the wettability. In contact heating, a solder perform is placed on the metallized substrate or solder film is deposited on the metallized substrate and the substrate is heated up (typically in an inert nitrogen-filled environment) to the melting point to visually identify whether the solder spreads over the substrate or pull up from the substrate.

7.2 Cleaning

In order to eliminate the oxidation and contaminations on the surface of materials and improve solderability, incoming parts should be cleaned up according to the requirements of packaging process. It is extremely important to get rid of organics, dust particle, and other contaminations listed in Table 7.2 that cause degradation of device quality and affect production yield. The particle contamination mainly comes from fabrication process, airflow in a clean room, transportation, and storage. Organic contamination includes oil or grease from machining process, skin oil, polymers, or photo resist. Oxide could be caused by nature oxidation, for example, indium and copper could be easily oxidized in open air at room temperature. It could also be caused by chemical oxidation by chemicals.

Table 7.2 Types of contaminations for cleaning [1]

Type of contamination
Particle contamination
Organic contamination
Natural and chemical oxide films (e.g., moisture, oxygen)
Heavy metal
Acids (e.g., SO_x, HCl)
Alkali

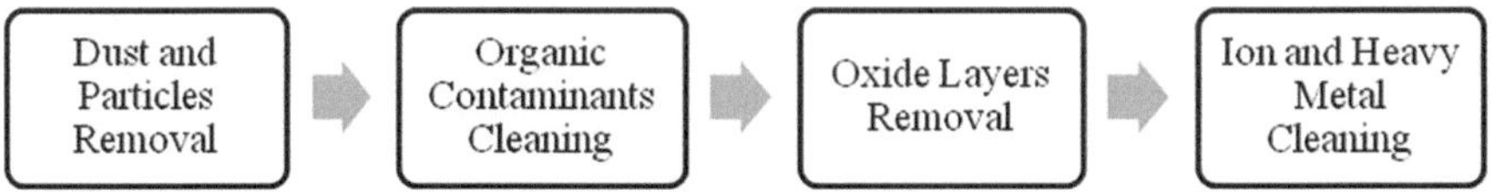

Fig. 7.5 The main procedure of cleaning process [1]

The main cleaning procedure for the laser components and packaging fixtures is shown in Fig. 7.5 [1]. At first dust and particles are removed by ultrasonic cleaning with ethyl alcohol. Then organic contaminants, such as particles, grease, or silica gel are chemically eliminated from the surface of mounting substrate. The oxide layer is removed by the plasma cleaning. Finally the ion and heavy metal contaminants are cleaned by chemical solution. Ultrasonic cleaning, chemical cleaning, and plasma cleaning are the three main cleaning methods used for semiconductor laser packaging process. The detailed descriptions of the three cleaning approaches are introduced in the following.

All the parts used for laser device packaging should be cleaned, such as mounting substrates, heat sinks, coolers, insulators, solder, electrodes, and packaging fixtures. As different materials and parts used in high power diode laser packaging have different structures, properties, and coating layers, the incoming parts should be cleaned up by different equipments and technical processes. Some parts may only need simple cleaning using acetone and rinse with isopropyl alcohol. For most of the materials and parts, a more thorough cleaning is needed.

7.2.1 Ultrasonic Cleaning

Ultrasonic cleaning uses cavitation bubbles induced by high frequency pressure (sound) waves to agitate a liquid. The agitation produces high forces on contaminants adhering to substrates like metals, plastics, glass, rubber, and ceramics. This action also penetrates blind holes, cracks, and recesses. The intention is to thoroughly remove all traces of contamination tightly adhering or embedded onto solid surfaces. Water or other solvents, such as acetone and anhydrous ethanol can be used, depending on the type of contamination and the work piece. Ultrasonic

Fig. 7.6 An example of an ultrasonic cleaner [5]

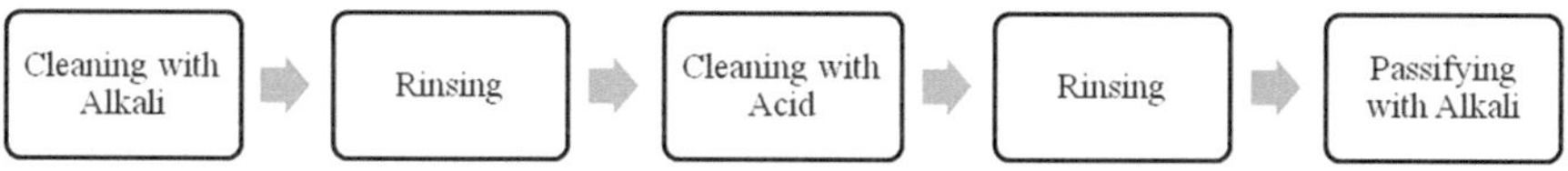

Fig. 7.7 Five steps in the chemical cleaning process [1]

cleaning utilizes ultrasound (usually from 20 to 400 kHz) and an appropriate cleaning solvent (sometimes ordinary tap water) to clean delicate items. Figure 7.6 shows an example of an ultrasound cleaner [5]. Cleaning normally lasts a few minutes, but can also up to 20 min, depending on the object to be cleaned. Ultrasonic cleaning can be used for a wide range of work piece shapes, sizes, and materials, and may not require the part to be disassembled prior to cleaning [6].

7.2.2 Chemical Cleaning

The parts used in high power laser packaging typically have organic, oxide, and metallic contaminants on them, coming from various lubricating oils, natural or chemical oxidation, corroding metal, machining process, and deposits from hard water. Normally five steps are followed in the chemical cleaning process [7]. Figure 7.7 shows the five steps in the chemical cleaning process [1].

The first step is cleaning with alkali. The purpose of this step is to remove insoluble organic contaminants, such as dust particles, grease, or silica gel. Different chemical solutions may be used. The mixture of 5:1:1 H_2O:H_2O_2:NH_4OH is one of solution commonly used. The second step is rinsing which loosens the deposits on the parts and washes away both the alkali and the substances by deionized water. The third step is cleaning with acid. The choice of acid depends on what material to

Fig. 7.8 The cleaned surfaces of anode copper heat sinks before (*left*) and after (*right*) chemical cleaning [1]

be cleaned and what kind of contaminants to be removed. For example, hydrochloric acid (HCl) solution is used for the removal of indium oxide layer, ionic, and heavy metal atomic contaminants, while hydrofluoric acid (HF) solution is more effective for the removal of a silicon dioxide layer. The fourth step is rinsing which is similar as the second step, but to wash away the acid solution. Also the rinsing liquid in this step can be deionized water, acetone, and anhydrous ethanol depending on the material and function of the parts to be cleaned. The cleaned metal is now in a highly reactive state, and so it must be coated with a protective layer to prevent corrosion. In the fifth step, the metal is reactivated with an oxidizing agent (usually an alkali) to form an oxide layer on the metal. This layer cannot be further oxidized and so protects the metal underneath. Figure 7.8 shows the top view of anode copper heat sink used for CS laser bar packages after chemical cleaning [1].

7.2.3 Plasma Cleaning

The application of plasma cleaning is to remove very thin films, especially hydrocarbonats and oxides, which remain after conventional cleaning [8, 9]. Ultra-fine plasma cleaning has the advantages: (1) plasma removes organic contamination from the surface, (2) plasma cleaning makes the surface more wettable, (3) typically twofold to a tenfold improvement in lap-shear strength and peel-strength, and (4) improved the adhesion on polymers, glass, ceramics, and metal.

Plasma cleaning involves the removal of impurities and contaminants from surfaces through the use of an energetic plasma or dielectric barrier discharge plasma created from gaseous species. Gases such as argon and oxygen, as well as mixtures such as air and hydrogen/nitrogen are used. The plasma is created by using high frequency voltages to ionize the low pressure gas (typically around 1/1,000 atmospheric pressure) [8].

It is important to choose the correct plasma gas since gases react and work in different ways to remove the contaminants. When oxygen gas is used, the plasma

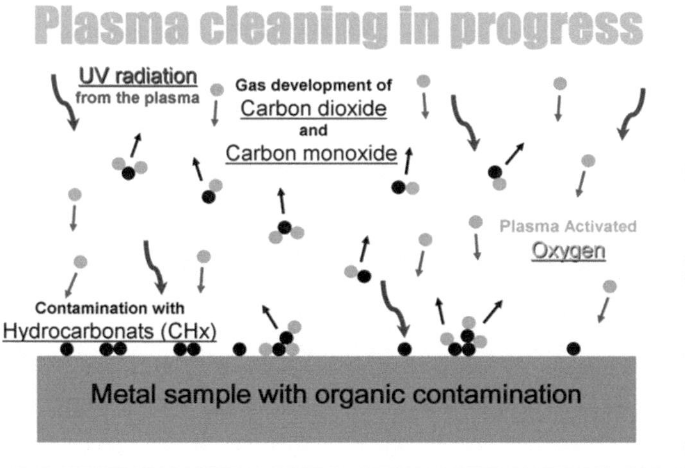

Fig. 7.9 Oxygen plasma removes the organic contamination on metal sample [9]

is an effective, economical, environmentally safe method for critical cleaning. The vacuum ultraviolet energy is very effective in the breaking of most organic bonds (i.e., C–H, C–C, C=C, C–O, and C–N) of surface contaminants. This helps to break apart high molecular weight contaminants. A second cleaning action is carried out by the oxygen species created in the plasma (O^{2+}, O^{2-}, O^{+}, O^{-}, ionized ozone, metastable excited oxygen, and free electrons). These species react with organic contaminants to form H_2O, CO, CO_2, and lower molecular weight hydrocarbons. These compounds have relatively high vapor pressures and are evacuated from the chamber during processing. The resulting surface is ultra-clean. Oxygen plasma is very effective in removing organic films and residues. Oxygen removes organic contaminants as shown in Fig. 7.9 [9]. If the part to be treated consists of easily oxidized materials such as silver or copper, inert gases such as argon or helium are used instead. The plasma-activated atoms and ions behave like a molecular sandblast and can break down organic contaminants. These contaminants are again vaporized and evacuated from the chamber during processing. The oxidation on indium layer can be removed by argon plasma cleaning before soldering.

Whether or not organic removal is complete can be assessed with contact angle measurements. When an organic contaminant is present, the contact angle of water with the surface will be large. After the removal of the contaminant, the contact angle will be reduced, as shown in Fig. 7.10 [10].

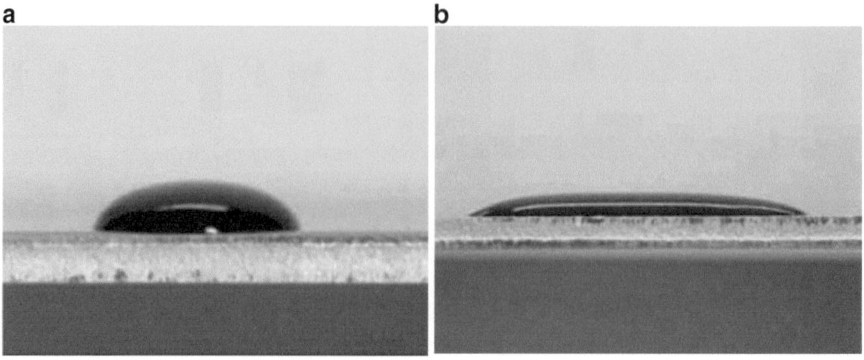

Fig. 7.10 Water beads on (a) an untreated surface but spreads out (has lower contact angle) on (b) a plasma-activated surface [10]

7.3 Metallization

Metallization is required on mounting substrates, heat sink, and other parts when soldering process is necessary. Metallization serves as a diffusion barrier and wetting layer between the solder material and the substrate material. It also prevents oxidation or moisture from the atmosphere, decreases the contact resistance, increases the soldering strength, and enhances the device reliability. The flatness and roughness of the target surface is required to be well controlled before metallization, as unqualified surface states may lead to pinholes, defects, or nonuniformity on the coating layer.

For a mounting substrate on which the diode laser chip is directly bonded, the metallization is generally more sophisticated. Metallic layer coated on the surface of mounting substrate consists of an adhesion layer, a diffusion barrier layer, and a wetting and oxidation prevention layer. The function of the adhesion layer is to provide good adhesion to the mounting substrate and to the barrier layer that can withstand high temperatures, low temperatures, and temperature cycling.

There are two common metallization structures utilized in manufacturing the high power diode lasers, as shown in Fig. 7.11. One is Ni/Au structure and the other is Ti/Pt/Au structure. Ni/Au metallization layer is deposited on the parts used in high power diode laser packaging, such as mounting substrate, cooler, heat sink, insulator, and electrode. For the latter metallization structure, Titanium (Ti) is widely used as adhesion layer as it has good adhesion to many kinds of metal materials, such as copper and platinum and ceramics. A diffusion barrier layer is deposited on the adhesion layer to prevent or slow down the diffusion between the solder material and the bonding substrate. Good barrier prevents intermetallic formation of the solder material and the mounting substrate materials. Pt, Ni, W, and Cr are the typical diffusion barrier layer in semiconductor laser packaging. The wetting and oxidation prevention layer is a sacrificial layer which provides good

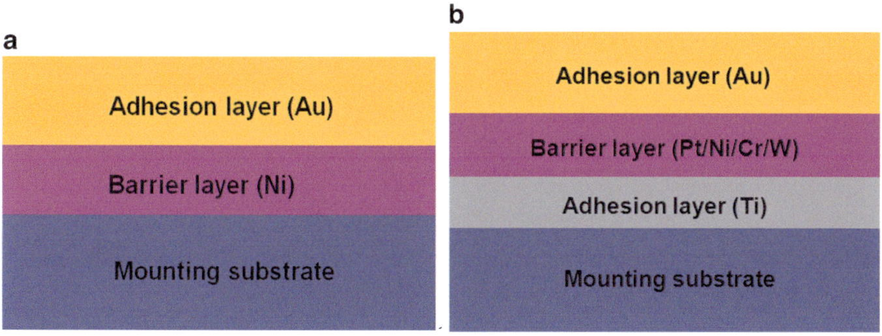

Fig. 7.11 Metallization structure of typical mounting substrate. (**a**) Ni/Au structure. (**b**) Ti/Pt/Au structure [1]

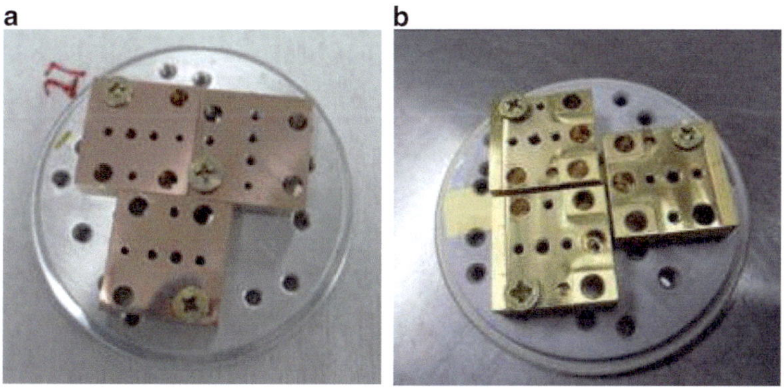

Fig. 7.12 A CS package anode copper heat sink before and after Ti/Pt/Au metallization [1]. (**a**) Before metallization. (**b**) After metallization

wetting to the solder material and prevent the barrier layer to be exposed to air and oxidized. The wetting and oxidation prevention layer is generally consumed or partially consumed during die bonding process. Au is the most common wetting and oxidation prevention layer in semiconductor laser packaging.

The metallization structures for Indium solder and AuSn solder could be different. The preferred metallization for indium soldering is Au with an underlying diffusion barrier of nickel or platinum as shown in Fig. 7.11a [1]. The metallization for AuSn soldering typically consists of an Au top layer and a Ni, Pt, Cr, or W diffusion barrier. The thickness of the Au layer can vary between 50 nm and more than 1 μm. The adhesion layer Ti is present between the barrier and the base material in Fig. 7.11b. Figure 7.12 shows a CS package copper heat sink before and after Ni/Au metallization [1].

The metallization process in high power semiconductor laser packaging generally is achieved by physical vapor deposition (PVD) or electroplating. PVD is a

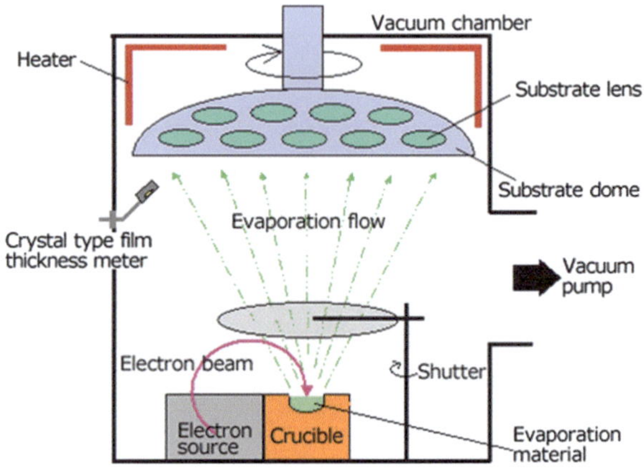

Fig. 7.13 The schematic illustration of electron beam evaporation [12]

vacuum deposition method used to deposit thin films by the condensation of a vaporized film material onto the critical surface of a work piece. This technology is widely used in metallization process, which includes electron beam evaporation, thermal evaporation, and sputtering deposition. Electroplating is an electro deposition process which utilizes electrolytic cells in which a thin layer of metal is deposited onto an electrically conductive surface.

7.3.1 Electron Beam Evaporation

Electron beam evaporation, frequently called "e-beam" evaporation, uses a focused beam of electrons to heat the evaporation material to achieve a controlled deposition of thin films, as shown in Fig. 7.13 [11, 12]. Both dielectrics and metals can be deposited using e-beam evaporation. The target material is kept in a water-cooled crucible and exposed to the electron beam, causing it to vaporize and condense on the substrate being deposited. By using a multiple crucible e-beam gun, several different materials can be deposited without breaking the vacuum. Planetary substrate rotation provides uniform deposition. Usually radiant heaters are provided for substrate heating. Ion source-assisted deposition is also an option. In an e-beam evaporation system, the deposition chamber must be evacuated to a pressure of at least 1×10^{-4} mbar to allow passage of electrons from the electron gun to the evaporation material [13]. Metallization layers such as Ti/Pt/Au used on the mounting substrates in high power semiconductor lasers can be deposited by e-beam evaporation. The materials are deposited consecutively without breaking the vacuum using multiple crucible e-beam guns. The thickness of the layers can be controlled precisely and the deposited layers are generally of high density and good adhesion to the substrates. Besides the metallization layers, solder materials used in diode lasers such as AuSn can be deposited by e-beam as well.

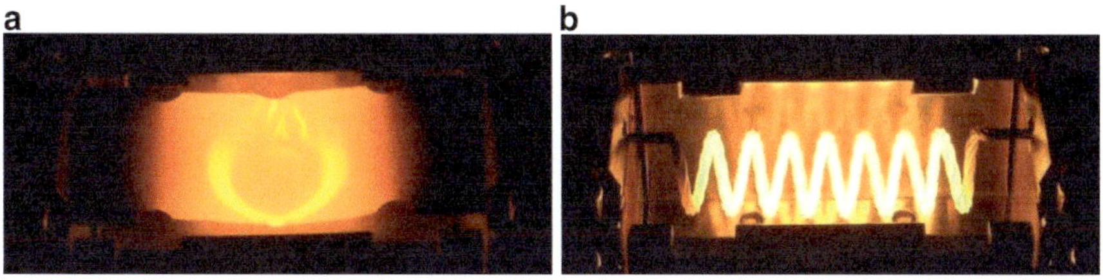

Fig. 7.14 Resistance heaters used for thermal evaporation [14]. (**a**) A molybdenum boat. (**b**) A filament [14]

7.3.2 Thermal Evaporation

Thermal evaporation or resistive evaporation uses an electric resistance heater to melt the target material and raise its vapor pressure to a useful range. There are many different physical configurations available for resistance heater, such as filaments and "boats," which are basically thin sheet metal pieces of appropriate high temperature metals like Tungsten or Molybdenum with formed indentations or troughs to place the material, as shown in Fig. 7.14 [14]. Again the process is done in a vacuum, both to allow the vapor to reach the substrate without reacting with or scattering against other gas-phase atoms in the chamber, and to reduce the incorporation of impurities from the residual gas in the vacuum chamber. Thermal evaporation is the simplest way of depositing material onto a substrate. Two major disadvantages of this technology are the high material loss in the process and the limitation of low melting temperature for materials to be deposited [15]. In semiconductor laser packaging, thermal evaporation is dominantly used for deposition of the indium solder.

7.3.3 Sputtering Deposition

Sputtering deposition is a widely used technique to deposit thin films on substrates. The technique is based upon ion bombardment of a source material, or the target [16]. The sputtering process is illustrated in Fig. 7.15 [17]. A target, or source of the material desired to be deposited, is bombarded with energetic ions, typically inert gas ions such as Argon (Ar^+). The forceful collision of these ions onto the target ejects target atoms into the space. These ejected atoms then travel some distances until they reach the targeting substrate and start to condense into a film. As more and more atoms coalesce on the substrate, they begin to bind to each other at the molecular level, forming a tightly bonded atomic layer. One or more layers of such atoms can be created depending on the sputtering time, allowing for production of precise layered thin-film structures. Sputtering deposition is often used in metallization of the mounting substrates in semiconductor laser packaging.

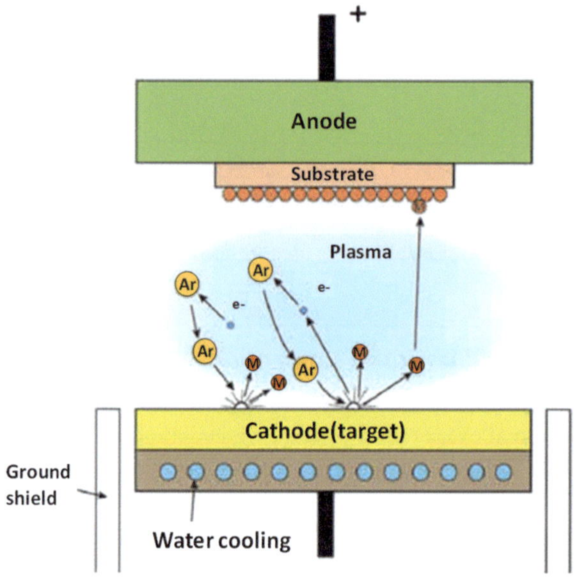

Fig. 7.15 Schematic illustration of the sputtering process [17]

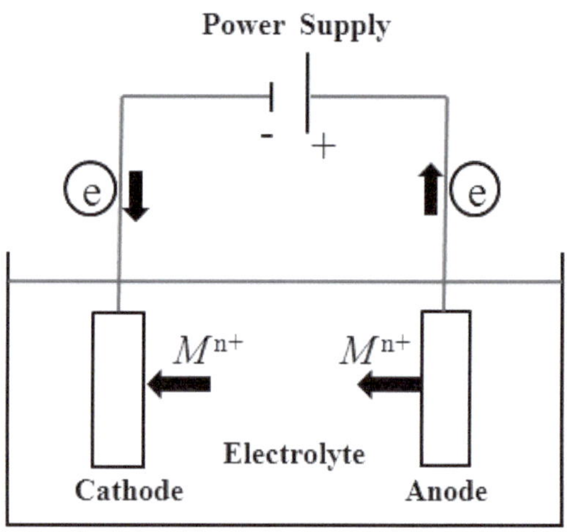

Fig. 7.16 Schematic of electroplating process [1]

7.3.4 Electroplating

In general, electroplating uses electrical current to reduce cations of a desired material from a solution and to coat a conductive object with a thin layer of the material, such as Ni and Au. Electroplating is primarily used for depositing a layer of material to obtain a desired property (e.g., abrasion and wear resistance, corrosion protection, lubricity, aesthetic qualities) to a surface that otherwise lacks that property. In other applications, electroplating is used to build up the thickness of undersized parts [18].

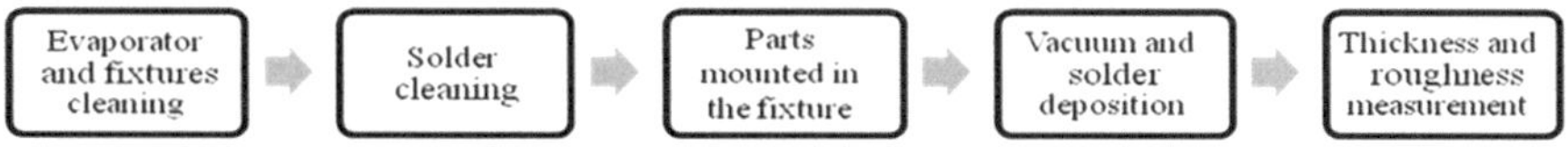

Fig. 7.17 The major deposition process [1]

Figure 7.16 shows the schematic of the electroplating [1]. In the electroplating process, the anode and cathode in the electroplating cell are both connected to an external power supply of direct current—a battery or, more commonly, a rectifier. The anode is connected to the positive terminal of the supply, and the cathode (the object to be plated) is connected to the negative terminal. When the external power supply is switched on, the metal (M) at the anode is oxidized from the zero valence state to form cations (M^{n+}) with a positive charge (n). These cations associate with the anions in the solution. The cations are reduced at the cathode to be deposited in the metallic, zero valence state. The reaction at the cathode during electroplating is shown in Eq. (7.1).

$$M^{n+} + ne^{-} = M \tag{7.1}$$

7.4 Solder Deposition

The main objective of solder deposition is to deposit the solder layer on the targeting substrate for the application of die bonding. The substrate is pre-metalized with certain structures as discussed in Sect. 7.3 to bond the solder layer and prevent the solder diffusion. The thickness of the solder layer should be well controlled. The deposited layer will be thick enough to bond the die with minimal void, but not too thick to yield solder overflow after melting. Indium and AuSn are commonly used solder materials for the bonding of laser diode bars to substrate. Due to the different material property, the deposition of Indium and AuSn applies quite different processes and equipment. This section will introduce the deposition processes of these two solder materials separately.

7.4.1 Indium Solder Deposition

Thermal evaporation and electron beam evaporation are two different techniques for indium solder deposition. Thermal evaporation is the most widely used process of indium deposition in laser diode packaging. The Indium layer is normally deposited on the substrates, such as the heat sink or cooler. Typical Indium solder thickness ranges from 5 to 15 μm [19].

The major thermal evaporation process of Indium solder is shown in Fig. 7.17 [1]. At first the evaporator chamber and the fixtures for product clamping should be well cleaned to eliminate the contaminants. The Indium raw material must be

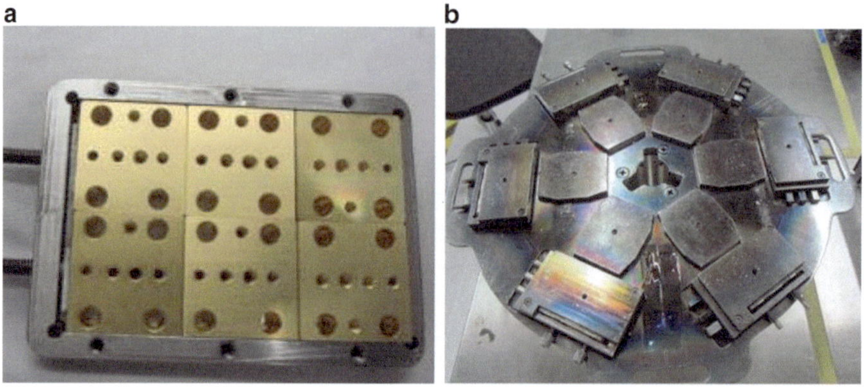

Fig. 7.18 Illustration of the fixtures and installation for solder deposition [1]. (**a**) The clamping fixture of the substrates to be deposited. (**b**) The installation of the clamping fixtures in the evaporator

Fig. 7.19 Example of Indium solder deposited on a CS package substrate/heatsink [1]

cleaned through cleaning process to remove the oxide layer. The substrates to be deposited is fixed in the specially designed fixture and mounted inside the evaporator as illustrated in Fig. 7.18 [1].

The non-deposition area should be covered by fixture mask to avoid contamination and electric short. The solder deposition starts when the vacuum of the thermal evaporator is less than 10^{-6} mbar, the deposition speed is typically in the range of 2–10 Å/s. The base vacuum pressure, evaporation pressure, deposition speed and sequence should be well controlled and optimized to ensure the quality of the solder layer. After solder deposition, as shown in Fig. 7.19, the thickness and roughness of the solder layer are measured by XRF tester and surface profiler for quality assurance purpose.

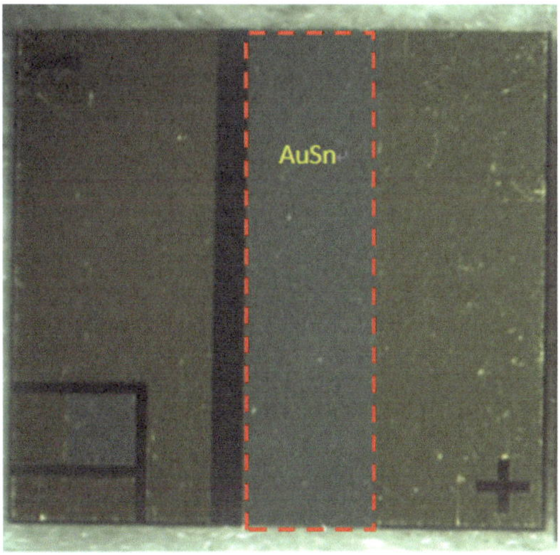

Fig. 7.20 AuSn solder deposited on a AIN mounting substrate by electron beam evaporation [1]

7.4.2 AuSn Solder Deposition

Electron beam (E-beam) evaporation and electroplating are the major techniques of AnSn deposition in laser diode industry. E-beam evaporation provides much better process control, reduced oxidation, and good uniformity of deposited layer compared to electroplating. Electroplating is possesses the advantage of low cost and less time consuming. The typical thickness of the solder layer for pre-deposited solder is in the range of 5–10 μm [19]. In general, a composition at or close to the eutectic point at 80 % Au and 20 % Sn (weight percent) is used, with a melting point of 283°C. Due to the steep increase of the liquidus temperature to both sides of the eutectic point, the composition of the solder has to be very precisely controlled. Pre-deposited solder can be covered with an Au top layer to prevent oxidation. As one kind of the "hard solder" materials, AuSn solder is not appropriate for the compensation of the thermal stress between the diode laser bar/chip and the mounting substrate. Therefore, AuSn should be deposited on a submount (that has a CTE closely matched to the CTE of the diode laser). Figure 7.20 shows AuSn solder deposited on a AIN mounting substrate by electron beam evaporation.

There are mainly three different methods using electron beam evaporation for AnSn deposition.

Method 1: Pure gold and pure tin are separately mounted in the evaporator as the target materials for the electron beam evaporation. During deposition process, Au and Sn are alternatively evaporated and deposited on the submount or heat sink, as shown in Fig. 7.21 [1]. The thickness of Au and Sn layer can be separately controlled according to the requirement of eutectic material composition. The AuSn solder layer deposited by this method has good uniformity of the thickness and materials composition. This method is more commonly utilized in packaging of high power diode lasers.

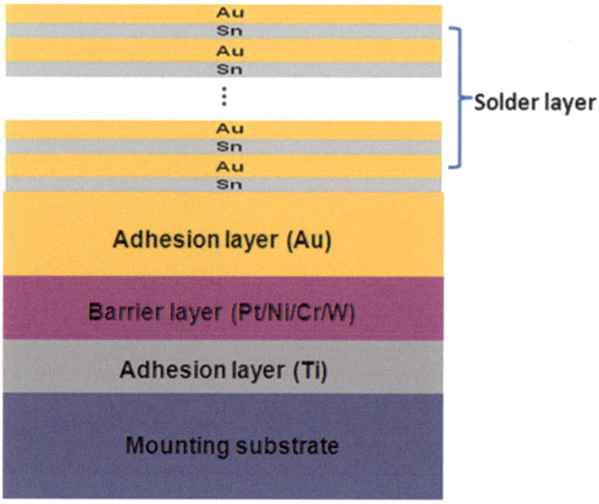

Fig. 7.21 The scheme of the AuSn deposition method [1]

Method 2: Pure gold and pure tin are separately mounted in the evaporator as the target materials, same as method (1). However, Au and Sn are evaporated and deposited on the mounting substrate simultaneously. The amount of the applied materials is pre-decided by the required thickness and composition of solder layer. The two materials should be completely evaporated when the deposition is completed in order to control the composition of the solder. This process is simple. In this method, the composition and uniformity of Au/Sn alloy is difficult to control and optimize.

Method 3: The AuSn solder preform with certain composition (e.g., Au80Sn20) can be directly evaporated and deposited as one source material on the substrate. Same as Method 2, the amount of the applied material is pre-decided by the required thickness of the solder layer. Again, the source material should be completely evaporated when the deposition is completed in order to control the composition of the solder. This process is even simpler than Method 2, but the composition and uniformity of Au/Sn alloy is more difficult to control and optimize. In this method, the initial deposition layers are more likely to be Sn-rich and the final layers are more likely to be Au-rich.

The Au and Sn can be deposited on the substrate sequentially using electroplating process. A pure Sn layer is deposited on top of a pure Au layer through two separated plating baths. The disadvantage of this technique is that a post-deposition anneal is required to homogenize the composition through inter diffusion [20].

7.5 Die Bonding

Die bonding is to bond the diode laser bar/chip on a mounting substrate or a submount. The quality of the die bonding critically influences major characteristics of the diode laser, such as thermal behavior, output power, wavelength, spectrum, lifetime, and even the polarization properties [21–35].

Fig. 7.22 Direct laser bar mounting to Cu heat sink using soft solder [21]

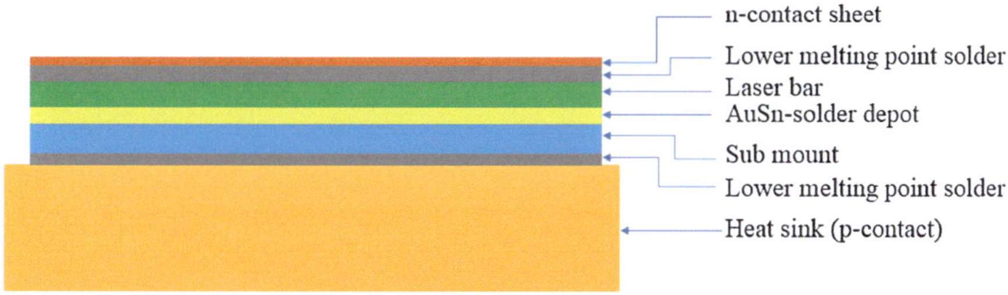

Fig. 7.23 Submount setup for laser bar mounting with hard solder [21]

A variety of die bonding technologies have been developed. Some are only suitable for a certain operational conditions, e.g., operation in quasi-continuous wave (QCW). Some can only be applied to a CTE-matched bonding substrate material. The most common techniques are listed below in order of frequency of application [21]:

- Soldering with ductile or soft solders, like indium, to a copper heat sink on the p-side (see Fig. 7.22 [21]).
- Soldering with harder solder, like AuSn alloys, to CTE-matched submounts on the p-side, made of CuW (see Fig. 7.23 [21]).
- Soldering with ductile solders, such as indium or a comparable solder, to a ceramic submount which is nearly expansion matched to GaAs with separate grooves for each bar.
- Clamping the p- and n-side between two very precisely lapped and possibly thick Au-plated, stiff copper heat sinks, by applying a very uniform pressure over the bar contact surface and so providing a homogeneous contact without any filler material.

After die bonding, the laser device is functional. Different packaging structures (See Chap. 2) require different die bonding processes. Generally, there are two typical die bonding processes used in the manufacturing of high power semiconductor lasers. The first one is the process which is carried out by using a die bonder system. It includes an individual pick up, place, and melting process (PPM process) with an integrated heating station with shielding protection gas. Therefore, the first die bonding process is called PPM process. The second is reflow process which is

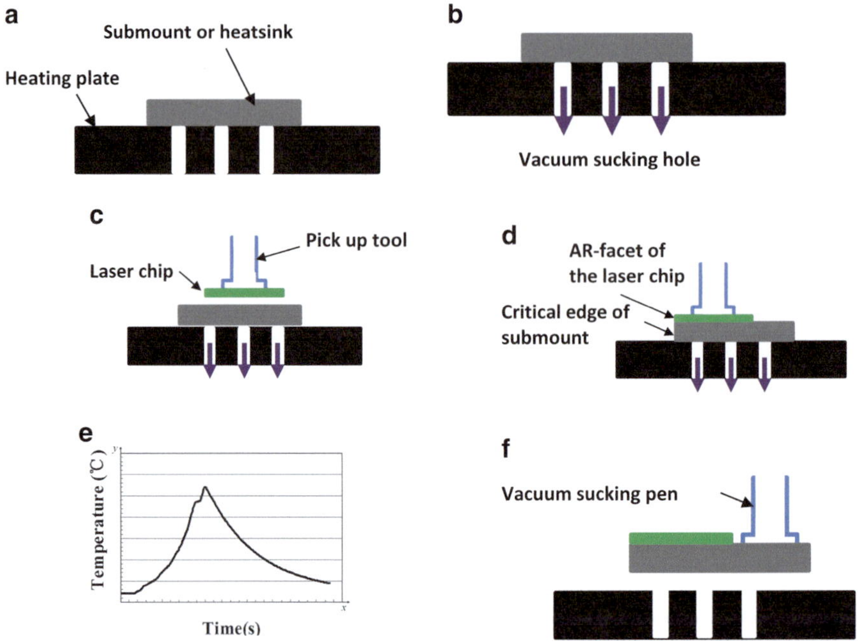

Fig. 7.24 Schematic of die bonding process [1]. (**a**) Place a submount onto the heating plate. (**b**) Suck the submount on the heating plate through vacuum. (**c**) Pick up a laser chip. (**d**) Align and place the laser chip at the critical edge of the submount (side view). (**e**) Soldering process under preset temperature profile. (**f**) Removal of the bonded die on submount

carried out by using a reflow oven. Diode laser bar/chip is bonded on a mounting substrate in the reflow oven controlled under a set of software which presets temperature profile, pressure profile, reactive vapor flow, nitrogen or other inert gas flow and other parameters [19].

7.5.1 PPM Process

For PPM process, it consists of six critical steps: (a) place a submount on the heating plate, (b) suck the submount on the heating plate through vacuum, (c) pick up a laser die and move the die onto the submount by pick-up tool, (d) align and place the laser die at the critical edge of the submount, (e) run the solder melting process under preset temperature profile, and (f) remove the bonded die on submount, as illustrated in Fig. 7.24 [1]. Among the above steps, step (d) and (e) are the key procedures. Step (d) directly influences the alignment of diode laser chip. In actual manufacturing, the semiconductor laser and the submount or heat sink are mechanically adjusted to a relative positioning precision of ±5 μm [19, 31]. Additionally, the soldering parameters of temperature profile and bond load on the laser chip are the key factors in reducing voids in soldering interface and the thermal stress applied on the laser chip in the step (e).

Fig. 7.25 An example of clamping tools for batch reflow [21]

7.5.2 *Reflow Process*

The other die bonding process is reflow process which is conducted by a reflow oven. Compared to PPM process, reflow process is a batch process run in a reflow oven. In reflow process, a specially designed clamping tool set is used to hold and accurately position the laser chip and the mounting substrate. The whole process is carried out in an environmental controlled chamber, and different packaging structures of semiconductor lasers may need different clamping tools. Figure 7.25 shows an example of clamping tools for the diode laser devices with MCC and conduction cooler [21].

The reflow process consists of several critical steps: (a) place a mounting substrate or heat sink on a fixture, (b) pick up a laser chip by vacuum sucking pen and place it on the heat sink, adjust and align the laser chip at the critical edge of the submount, (c) place an insulator on the heat sink, (d) pick up a cathode (e.g. copper foil) on the heat sink, (e) fix the heat sink, laser chip, insulator, and copper foil, (f) run the solder melting process under preset temperature profile, pressure profile and gas flow; (g) remove the bonded chip on submount or heat sink. The above described processes are illustrated in Fig. 7.26 [1].

Among the above steps, step (b) and (f) are the key procedures. Step (b) directly influences the alignment of diode laser chip. Similar to PPM process, in actual manufacturing, the semiconductor laser chip and the submount or heat sink are also mechanically adjusted to a relative positioning precision of ± 5 μm by the accurate clamping tools. Similarly, the step (f) is the most important step which should take into considerations of the reflowing parameters such as temperature profile, pressure on the laser chip, and vacuum conditions. These parameters directly influence the voids in soldering interface and stress applied on the diode laser. As reflow die bonding process is a batch process, it offers high productivity and high consistency. Figure 7.27 [1] shows an example of the reflow batch process run in a reflow oven.

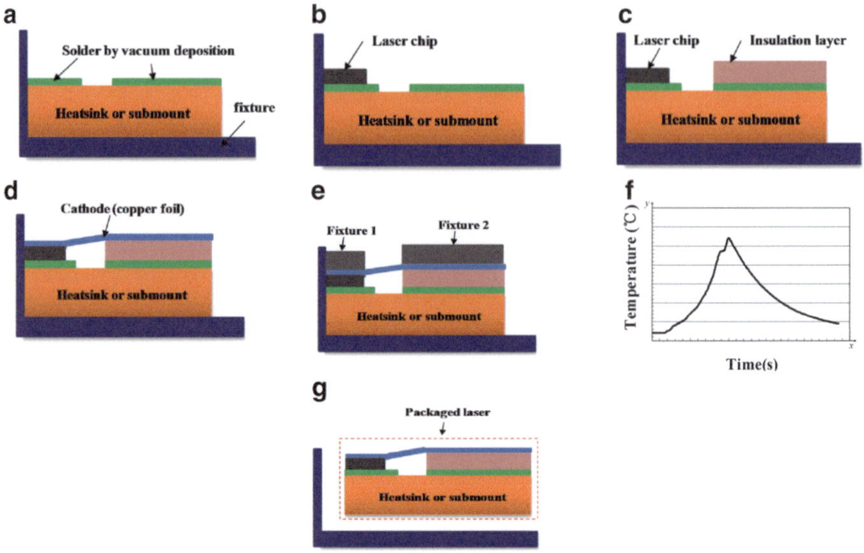

Fig. 7.26 Example of a reflow process for diode laser bar/chip bonding [1]. (**a**) Place a heat sink on a fixture. (**b**) Pick up a laser chip by vacuum sucking pen and place on the heat sink. (**c**) Place an insulator on the heat sink. (**d**) Pick up a cathode (copper foil) on the heat sink. (**e**) Fix the heat sink, laser chip, insulator, and copper foil. (**f**) Soldering the assembly under preset temperature profile. (**g**) Remove the packaged laser device

Fig. 7.27 An example of batch process using reflow die bonding process [1]

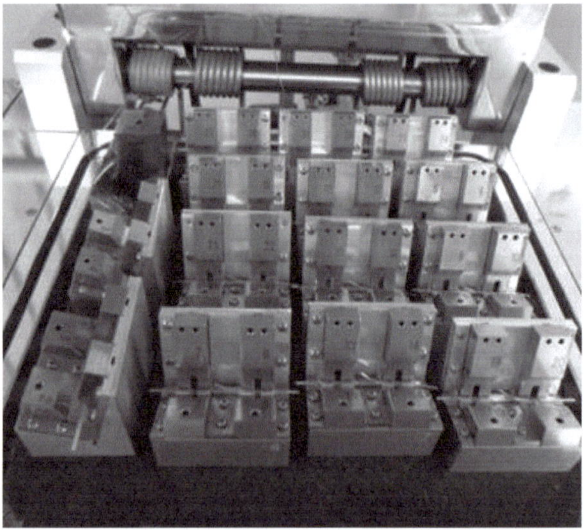

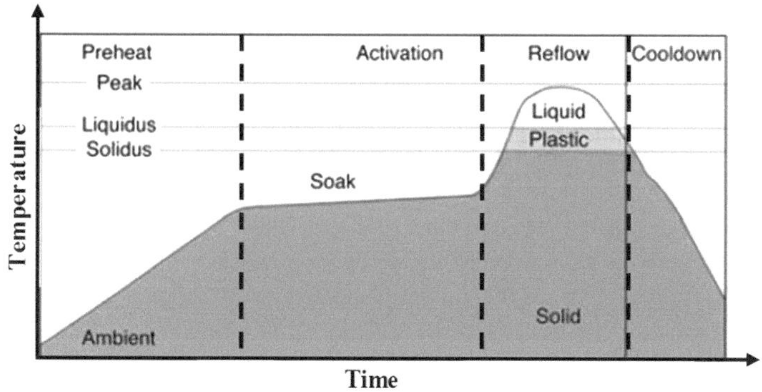

Fig. 7.28 The schematic diagram of temperature-time curve for a typical reflowing process [1]

Voids may be generated during the die bonding process, and voids are likely to grow in the span of field service due to mechanical stress, thermal stress, thermal fatigue, electro-migration and other factors. Voids should be eliminated or minimized to improve the performances of laser during die bonding process [22]. The principle of "void-free" bonding process is based on the following equation to eliminate void generation or minimize void size.

$$PV = nRT \tag{7.2}$$

where P, V, and T are pressure, volume, and temperature of the air bubble which eventually would be voids in solder layer, respectively, and n is the amount of substance of air bubble (measured in moles), and R is gas constant. For a given air bubble, when pressure is lower and temperature is higher, volume V is larger. When volume is large enough and in additional to the low pressure, the air bubble may burst. In this way, the potential void is eliminated. To minimize void size, that is to reduce volume V, higher pressure and lower temperature are desired. However, due to the phase change of the solder material and the bonding process, minimizing the void size by controlling the pressure and temperature is more sophisticated. For a given air bubble and at a given reflow temperature, the temperature T and amount of gas n remain unchanged, the volume is inversely proportional to the pressure and this is called Boyle's Law which is presented in Eq. (7.3).

$$P_1V_1 = P_2V_2 \tag{7.3}$$

On the other hand, as shown in Eq. (7.4), when P and n are constant, the volume is proportional to the temperature according to Eq. (7.2).

$$\frac{V_1}{V_2} = \frac{T_1}{T_2} \tag{7.4}$$

Figure 7.28 illustrates a typical the temperature curve during reflowing process. As shown in the figure, the whole reflowing process is divided into four phases:

preheat, activation, reflow, and cool down. In the first phase, the heat sink/ submount, laser chip, and clamping tools are preheated to a certain temperature below the melting temperature of the solder material. Meanwhile, the chamber is pumped and purged with inert gas. In the second phase, the assembly is soaked at the present temperature which is also below the melting temperature of the solder material. The whole assembly almost reach a homogenous temperature and is activated. During this phase, vacuum and activation gas may be applied. The temperature is rapidly increased to the peak temperature which is higher than the melting temperature of the solder material within a few seconds and consequently the solder is quickly melted when the reflow temperature reach the melting temperature in the third phase (reflow). In this phase, the pressure of the reflow chamber may be varied to control the voids. With the temperature fall below the freezing temperature of the solder material in the fourth phase, the solder material solidifies. If there are air bubbles in the solder material, voids is formed.

7.6 Wire Bonding

In high power semiconductor laser packaging, wire bonding provides electrical connection between the n-side metallization of a laser die and the cathode electrode of the packaged device. Wire bonding is a solid phase welding process, where the two metallic materials (wire and pad surface) are brought into intimate contact [23]. Once the surfaces are in intimate contact, electron sharing or inter diffusion of atoms takes place, resulting in the formation of wire bond. In wire bonding process, bonding force can lead to material deformation, breaking up contamination layer and smoothing out surface asperity, which can be enhanced by the application of ultrasonic energy. Heat can accelerate interatomic diffusion, thus the bond formation.

7.6.1 Process

Thermosonic gold ball bonding process is the most widely used wire bonding technique in semiconductor laser packaging. In this technique shown in Fig. 7.29, wire is passed through a hollow capillary, and an electronic-flame-off (EFO) system is used to melt a small portion of the wire extending beneath the capillary [24]. The surface tension of the molten metal from a spherical shape, or ball, as the wire material solidifies. The ball is pressed to the bonding pad on the die with sufficient force to cause plastic deformation and atomic interdiffusion of the wire and the underlying metallization, which ensure the intimate contact between the two metal surfaces and form the first bond (ball bond). The capillary is then raised and

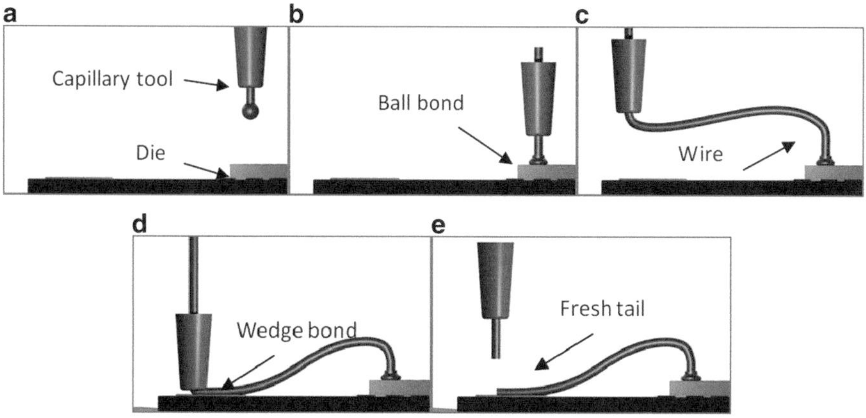

Fig. 7.29 Illustration of a typical wire bonding process for high power semiconductor lasers [24]. (**a**) Wire ball forming. (**b**) Ball bond contact. (**c**) Wire looping. (**d**) Wedge bond contact. (**e**) Wire breakage

Table 7.3 The parameters of thermosonic gold ball bonding [23]

Wire bonding	Pressure	Temperature	Ultrasonic energy	Wire	Pad
Thermosonic	Low	100–150°C	Yes	Au	Au

repositioned over the second bond site on the substrate; a precisely shaped wire connection called a wire loop is thus created as the wire goes. Deforming the wire against the bonding pad makes the second bond (wedge bond or stitch bond), having a crescent or fishtail shape made by the imprint of the capillary's outer geometry. Then the wire clamp is closed, and the capillary ascends once again, breaking the wire just above the wedge, an exact wire length is left for EFO to form a new ball to begin bonding the next wire [25].

This technique requires a high temperature raging from 100 to 150°C shown in Table 7.3. Heat is generated during the manufacturing process either by a heated capillary feeding the wire or by a heated pedestal on which the assembly is placed or by both depending on the bonding purpose and materials. Typically small gold wire with a diameter between 25 and 50 μm is mostly used in this technique because of its easy deformation under pressure at elevated temperature, its resistance to oxide formation, and its ball formability during a flame-off or electronic discharge cutting process. In general, the wire bonds are arranged in a dense pattern on the n-side metallization of the laser die in order to attach as many wire bonds as needed to achieve a sufficiently low ohmic resistance of the n-side current path. This increases the efficiency of the laser device and reduces resistive heating of the wire bonds. An example of a wire-bonded high power diode laser bar is shown in Fig. 7.30 [1].

Fig. 7.30 Example of a wire-bonded laser diode bar [1]

Table 7.4 Minimum bond strength requirement (MIL-STD-883) [36]

Wire diameter	Minimum bond strength (g)
0.0007″	2.0
0.0008″	2.4
0.001″	3.0
0.002″	7.0

7.6.2 Current-Carrying Capability

The Preece equation can be applied to estimate the current-carrying capability of the wire. The Preece equation relates the fusing current in amperes to the diameter of the wire in inches [26]:

$$i = kD^{3/2} \tag{7.5}$$

where i is the current, k is a constant corresponding to wire composition, and D is the diameter of the wire in inch. For gold and copper, k = 10,244. The amount of the wires can be calculated when the current and diameter of the wire are given.

7.6.3 Inspection

Quality inspection of the wire bonding process should be conducted on a dummy sample before running the process on products. The wire bonding quality is mainly determined by the bond strength from wire pull test. Table 7.4 shows the minimum bond strength requirement according to MIL-STD-883 [36].

Visual inspection under low magnification microscope is required after wire bonding process. Any sign of bond cracking, tearing at the wedge, misplacing wires, inconsistent wire placement etc. indicates failure in wire bonding.

7.7 Assembling

For high power diode lasers, the epitaxy side of laser die is normally bonded to the mounting substrate, since the epi-side generates more heat than n-side of the die. The mounting substrate could be a heatsink or a submount or carrier which is then attached to a heatsink or a package base. We use a hard solder CS package to illustrate the assembling process. Figure 7.31 shows the schematic structure of a hard solder CS package. The high power laser die is bonded to the submount using AuSn solder with a die bonding process discussed previously. The semi-product is called Chip-on-Submount (COS) or Chip-on-Carrier (COC). This process can also be called the first level packaging. Next the COS or COC is assembled on the heat sink. This process is called the second level packaging. A lower melting temperature solder such as indium, SnAgCu (SAC) or InSn is commonly used to solder the COS or COC to the heat sink in the assembling. Figure 7.32 demonstrates hard solder CS assembling process [1, 27].

Fig. 7.32 shows the hard solder CS diode laser assembling process. The exploded view of the structure is shown in Fig. 7.32a. In the assembling process, the COS or COC which a single bar bonded on a CTE matched submount by die bonding process and the insulator is attached to the heat sink. After the attachment, the n-metal of laser die is connected to the copper N-foil using wire bonding or copper foil direct contact method according to the design and process. The cathode block, non-conductive washer, and screws are installed onto the heat sink to complete the assembling process, as shown from Fig. 7.32b–f.

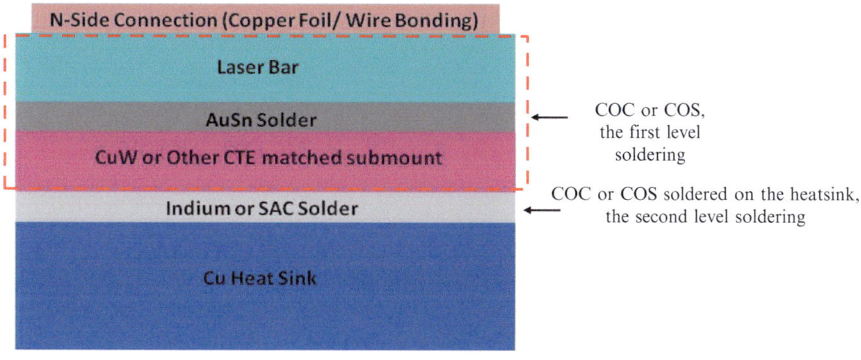

Fig. 7.31 The schematic structure of hard solder CS package [1]

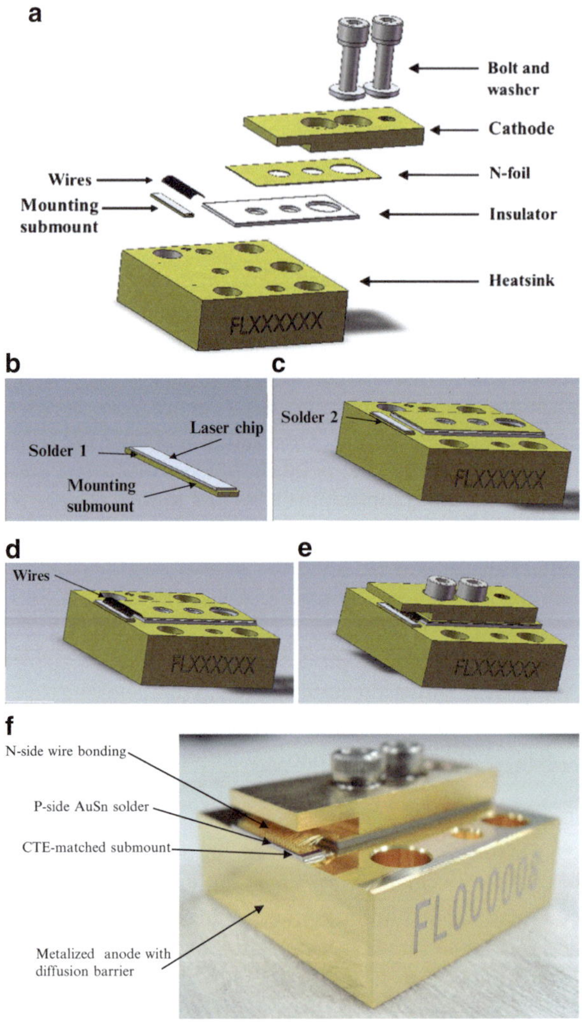

Fig. 7.32 Hard solder CS diode laser assembling process [1]. (**a**) Explosive view. (**b**) COS. (**c**) COS and insulator soldered on the heat sink. (**d**) Wire bonding. (**e**) Installing the cathode block. (**f**) A finished final single bar hard solder CS-package

7.8 Screening

Screening is an in-process step to verify whether the laser device has severe defects or damages and if the device should proceed to the next manufacturing steps. Different from incoming material inspection and final inspection and testing before shipment, screening is to identify obvious defective device and prevent those

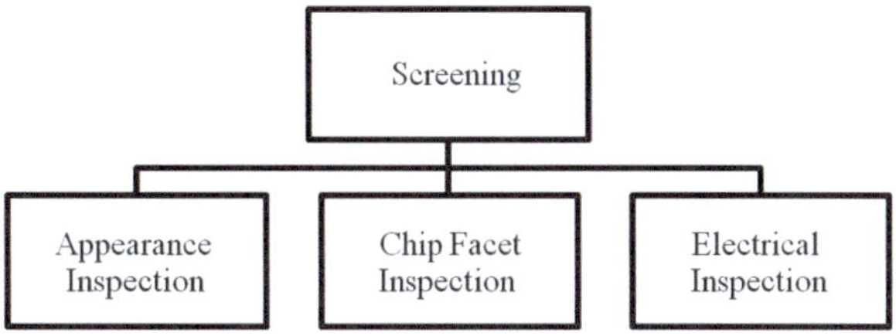

Fig. 7.33 Screening steps in high power semiconductor laser manufacturing [1]

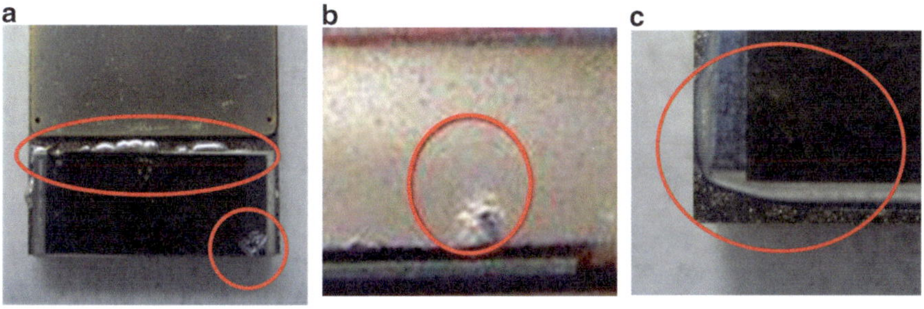

Fig. 7.34 Examples of defects in diode laser products that can be screened during production [1]. (**a**) Solder overflow and chip damage. (**b**) Bump on the cooler. (**c**) Submount and chip misalignment

devices flow into the next manufacturing procedures and reduce manufacturing cost. Therefore, screening is a rough inspection and testing procedure. As shown in Fig. 7.33, screening mainly includes three parts: device appearance inspection, diode laser facet inspection, and electrical inspection [1]. In the procedure, microscope is used to inspect the defect or damage of laser diode and substrate, while multimeter is applied to check whether the diode voltage is in normal state and makes sure no short circuit occurs between the anode and cathode.

7.8.1 Appearance Inspection

The appearance inspection is to inspect visible defects of the product, such as solder overflow, chip damages, bump or pits, and obvious misalignment, as shown in Fig. 7.34 [1]. The appearance inspection can be conducted visually by naked eyes or low magnification microscope (1.5–10×). Any defect that is observed during the inspection is commonly recorded electronically or in production traveler sheet associated with the particular laser diode device.

7.8.2 Chip Facet Inspection

Chip facet inspection is normally conducted by a metallurgical microscope. Devices with facet contamination, solder overflow, soldering void, and chip damage are

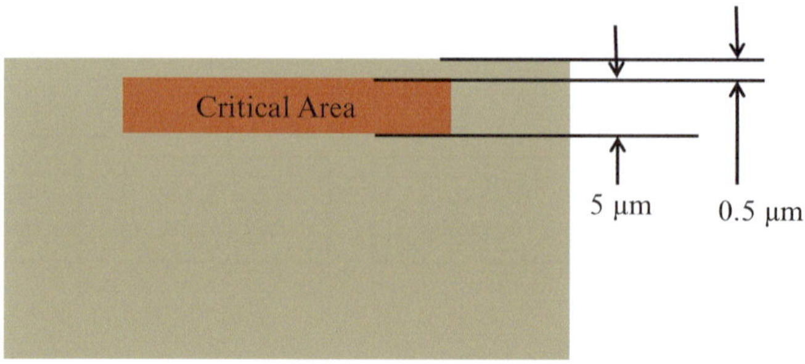

Fig. 7.35 The schematic drawing of a diode laser chip facet [1]

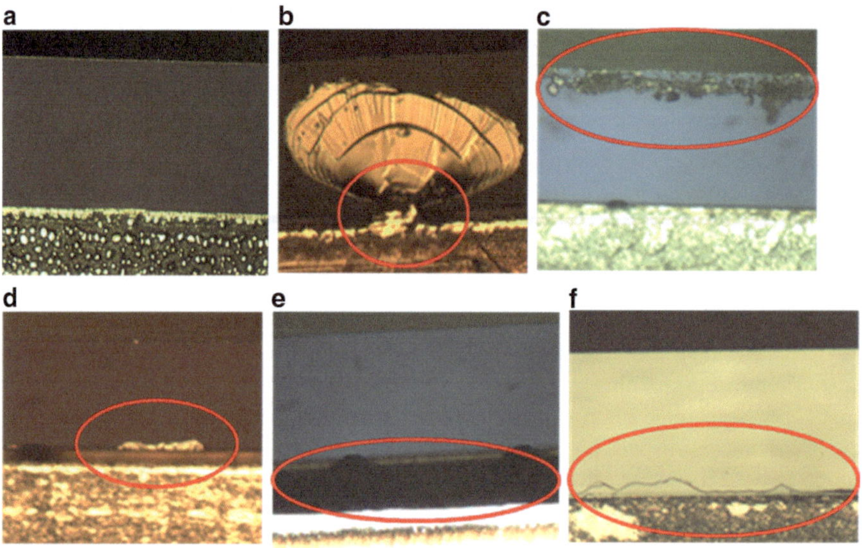

Fig. 7.36 Different types of defects on the chip facet [1]. (**a**) Normal facet after packaging. (**b**) Chipping on the facet. (**c**) Contamination on the facet. (**d**) Solder overflow on active area. (**e**) Solder voiding. (**f**) Horizontal cracking

screened out in this process. As shown in Fig. 7.35, the active area, which is also named as emitting area, is positioned 2–5 μm from the anode of the laser diode [1]. Severe contamination or solder overflow covered on the active area will lead to burns or catastrophic optical damage (COD) after functional testing or during field application. Contamination or solder overflow on non-active area on the chip facet is not a focus in the screening process. Severe solder voiding and obvious horizontal facet cracking are fatal to the laser diode, and they are also screened in the process. A normal chip facet after packaging is shown in Fig. 7.36a [1]. The major defects related to facet inspected in screening process are shown from Fig. 7.36b–f.

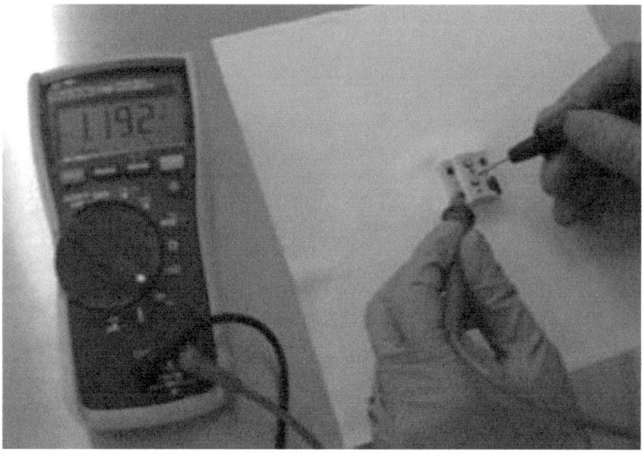

Fig. 7.37 Electrical inspection for shorting of the devices [1]

7.8.3 Electrical Inspection

The electrical inspection is mainly to test the voltage of the laser diode device by a multimeter to determine whether the anode and cathode are isolated from each other as shown in Fig. 7.37. If there are some solder particles, such as Indium solder, which is soft and very viscous, in the gap of structure layers, the anode and cathode may be connected which leads to a short circuit of the laser diode. Especially when the laser device is made of multiple bars such as a stack structure, this kind of electrical short is very hard to notice visually. Therefore the voltage check is particularly necessary for multi-bar devices.

7.9 BBI Test and ABI Test

The packaged laser device should be conducted before burn-in (BBI) and after burn-in (ABI) testing in the manufacturing process. The aim of BBI test and ABI test is to verify the device meet the specifications and also compare the performance differences of laser device before and after BI for quality control. Generally if a diode laser device degrades 5% in operation power or operation current or slope efficiency after 24 or 48 hours of burn-in, it is deemed unsafe for long term operation and should be screened out. The typical BBI and ABI test items include LIV, wavelength and spectrum tests in standard production process, which are listed in Table 7.5 [37]. Near field, far field and other parameters are tested for special cases.

The test flow of laser device consists of laser device installation, inspection of electric short circuit, device cooling and temperature stabilization, and laser device measurement as shown in Fig. 7.38 [1].

Table 7.5 Typical BBI and ABI test items for diode laser device in standard production process [37]

Items	Description
LIV test	Characterization of light output power (L) and voltage (V) of the laser vs. injection current (I)
Wavelength Test	Measurement of wavelength at the operation current of the laser diode
Spectrum Test	Measurement of the optical output spectrum at the operation current of the laser diode

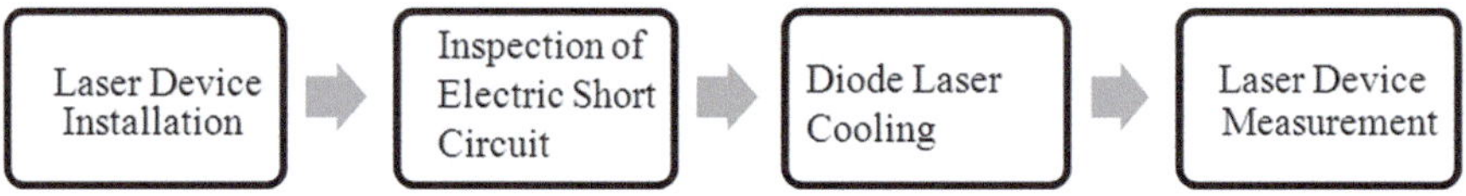

Fig. 7.38 BBI and ABI test flow of a diode laser [1]

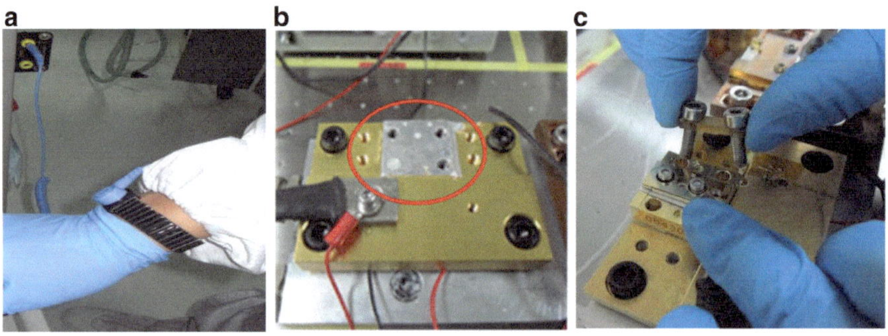

Fig. 7.39 Installation of laser device [1]. (**a**) Static discharge protection; (**b**) Placing Indium film on cooling plate. (**c**) Fixing device on cooling plate

7.9.1 Laser Device Installation

The operator should wear an anti-static wrist strap during installation to avoid device damaged by static discharge, as shown in Fig. 7.39a [1]. Normally an indium foil is pre-spread on the target cooling plate to improve the thermal contact of the laser device and cooling plate as shown in Fig. 7.39b. The facet of laser diode is forbidden to touch by finger or other fixing tools, which may introduce damage or contamination on the laser device. The laser device is fixed on the cooling plate by screwing or pressure in Fig. 7.39.

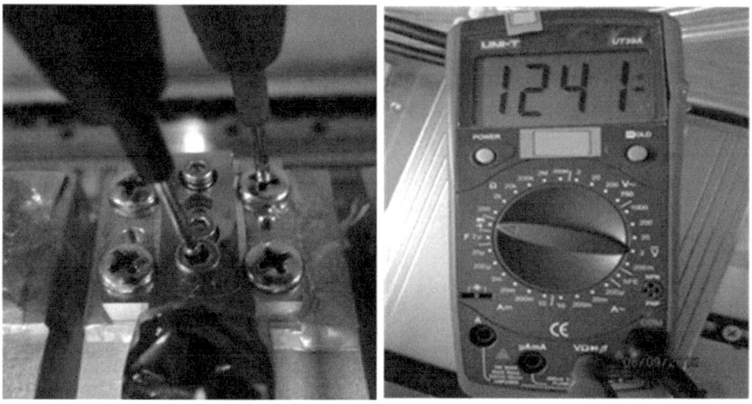

Fig. 7.40 Inspection of electric short circuit [1]

7.9.2 Inspection of Electric Short Circuit

One important step is to inspect the electric short circuit of anode and cathode of laser device after installation. The inspection of short circuit is normally conducted by the multimeter with the diode testing mode as shown in Fig. 7.40 [1].

7.9.3 Diode Laser Cooling

Cooling of measurement station is started before the measurement of the laser diode. For most of the cases, temperature stabilization is required to more accurately measure the wavelength and spectrum. Cooling and temperature stabilization can be achieved by thermoelectric cooler (TEC). TEC uses the Peltier effect to create a heat flux between the junction of two different types of materials or by direct liquid cooling which uses a chiller to control temperature. For more accurate wavelength measurement and at lower power levels, TEC cooling is preferred. For high power levels, direct liquid cooling is the choice. The measurement will not start until the temperature of cooling plate is stabilized.

7.9.4 Measurement of Laser Device

The power of laser device is measured with a laser power meter as illustrated in Fig. 7.40a or a integrating sphere with a detector as shown in Fig. 7.40b. The power meter is closely placed in front of the diode laser for maximum light catching. The spectrum can be measured by a spectrometer connected to the integrating sphere as shown in Fig. 7.40b [1]. A typical testing data of LIV and spectrum is shown in Fig. 7.40a and 7.40b, respectively. Detailed theories and approaches of laser device measurement is described in Chap. 8.

7.10 Burn In

Burn-in of diode lasers is the process by which diode lasers are exercised prior to being integrated into a system such as a solid state laser or a fiber laser or being placed in field service when they are directly used in the field. The intention is to detect those parts that would fail as a result of the initial, high-failure rate portion of the bathtub curve of device reliability or the infant failure and eliminate those parts from production. Diode lasers are generally detected on the basis of a change in one or more key operating parameters, such as power degradation or central wavelength drifting, which are measured before and after burn-in. In practical applications, the power degradation or current increment after burn-in should not be larger than 5 % of the initial value before burn-in. Burn-in temperature and operating current should be as high as possible to minimize burn-in time, yet not so high that a degradation mode is triggered [38]. Normally the applied current or power is 120 % of the normal operating value to accelerate the burn in. In theory, if the burn-in period is made sufficiently long and is even artificially stressful, the diode lasers can then be trusted to be mostly free of further early failures once the burn-in process is completed. In reality, the burn-in period should be balanced from early field failures (return materials), productivity and reduction of product lifetime. If the burn-in period is too short, early field failures will be high and high percentage of materials will be returned from customers. If the burn-in period is too long, the production cycle will be long. Also when the burn-in time is extended into the normal service portion of the bathtub curve, the effect of the burn-in is a reduction of product lifetime. In a mature production, the burn-in period is determined by failure time distribution through a very large number of device testing. The burn-in period is the time when the devices passed decreasing failure rate and reached a stable and very low percentage failure rate. Commonly, for 800-1000 nm diode lasers, burn-in period is 24–72 h at 120 % of operation condition depending on the application and device performance. Burn-in is to detect early failure parts and eliminate them from production, it does not improve the technology and manufacturing process. It is necessary to eliminate the root cause of early failures and improve the process with the various root causes for failures are identified and eliminated. The common failure modes and failure analysis are discussed in Chap. 9.

The operation flowchart of burn in consists of four steps: laser device installation, inspection of electrical short circuit, cooling and inspection, and laser device burn-in. The process is similar to the BBI and ABI test flow in Sect. 7.9 and it is shown in Fig. 7.42 [1].

All the electrical connections should be well cleaned to avoid the contaminations on the laser diode facet before installation. The material and dimensions of the

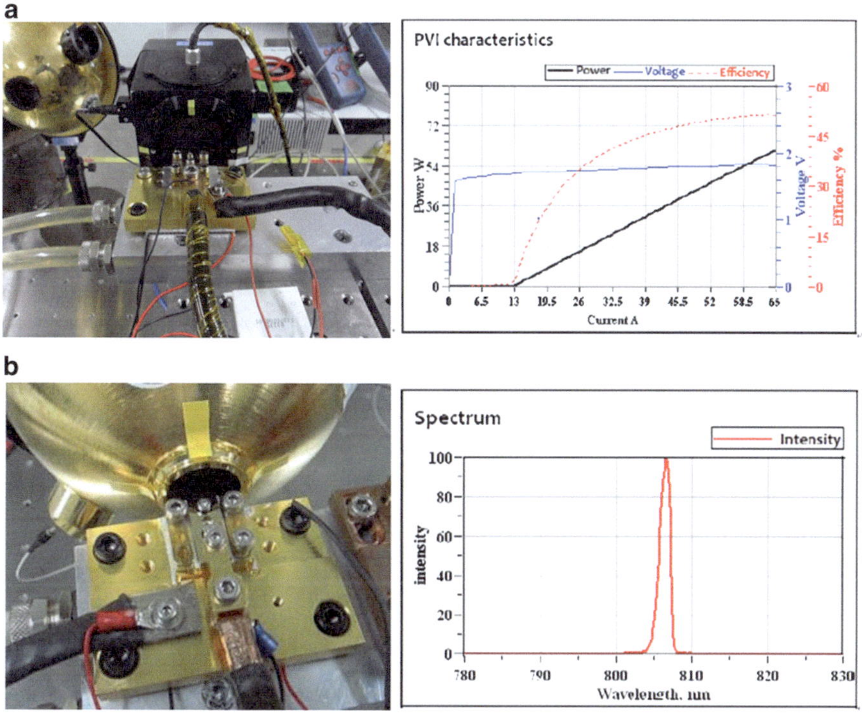

Fig. 7.41 Measurement of laser device for BBI and ABI test [1]. (**a**) Laser LIV measurement and typical measurement results [1]. (**b**) Laser spectrum and wavelength measurement and typical measurement results [1]

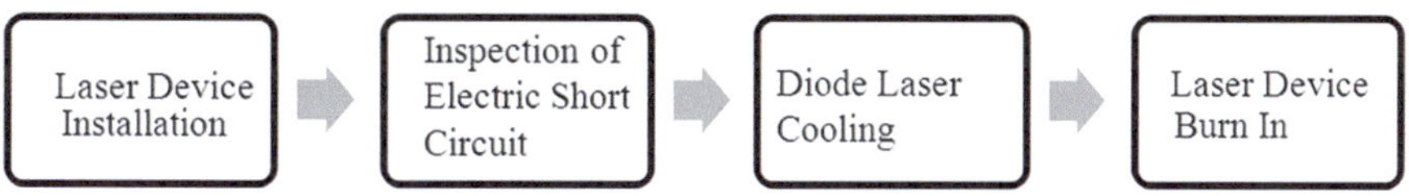

Fig. 7.42 Flow chart of diode laser burn-in process [1]

electrical connection should be considered to carry enough current and avoid over heating due to high resistance or ohmic contact. The water pipe and connections are inspected to make sure no water leaking exists. Cooling capability is calculated in advance to ensure that the generated heat can be removed. The laser devices are connected in series on the cooling plate for batch burn-in process. Figure 7.43 shows an example of CS burn-in setup which requires 80 A current handling capability and 2000 W heat dissipation capacity [1].

Temperature sensors such as thermistors are used on the product to monitor the temperature fluctuation and for safety protection, as shown in Fig. 7.44 [1]. The electrical short circuit of cathode and anode of laser is inspected using a multimeter before burn-in.

Fig. 7.43 An example of burn-in setup for CS devices [1]

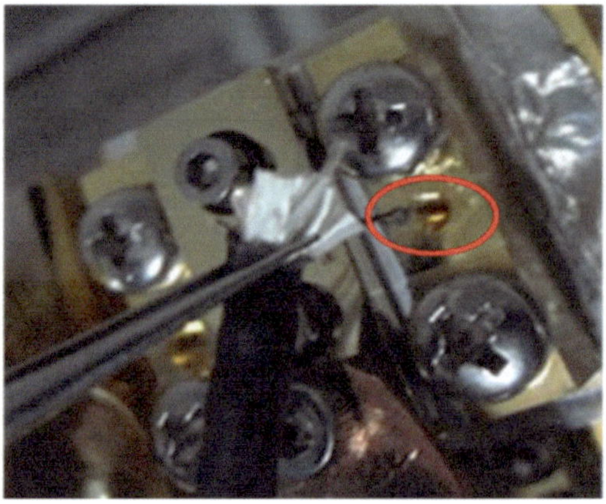

Fig. 7.44 Temperature sensor mounted on the laser device in burn-in for temperature monitoring and safety protection [1]

For safety consideration, burn-in is typically conducted in burn-in rack, as shown in Fig. 7.45 [1]. In the burn-in rack, a light absorber is placed in front of the laser devices to absorb the light emission and the light-turned-into-heat is removed from the light absorber. Temperature, water flow rate and electrical current and voltage are monitored simultaneously by integrated monitoring system. Abnormal change of these parameters will trigger warning and system automatic shutdown.

7.11 Final Inspection

Final inspection is performed after burn-in test (ABI test). Different from the screening inspection process, the final inspection is to identify more detailed defects ABI test to make a "decision" together with the BBI and ABI test results whether

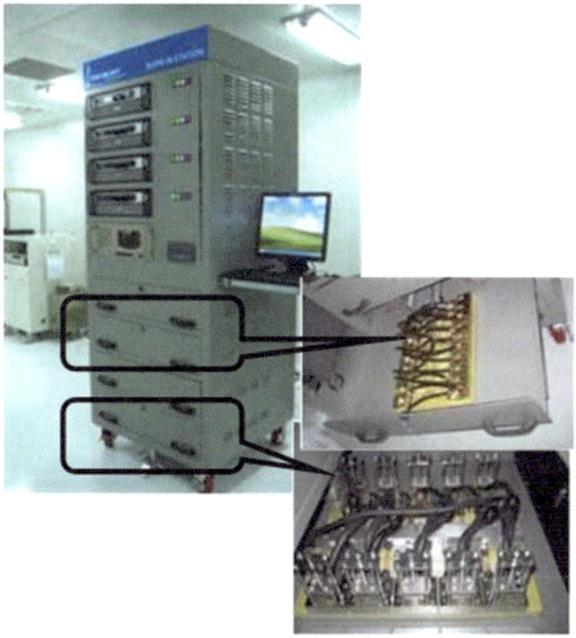

Fig. 7.45 Burn-in system [1]

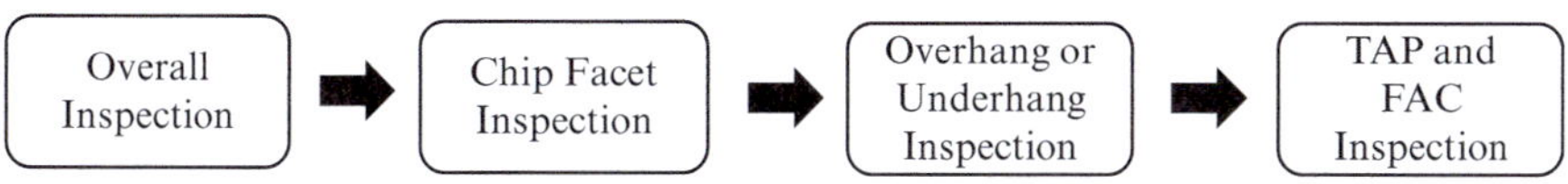

Fig. 7.46 Final inspection steps for diode lasers [1]

the laser devices are qualified products. Furthermore, different application or different customer may require different visual inspection specification. Therefore, the final inspection criteria may be tailored to a specific requirement. The final inspection generally includes following items, as shown in Fig. 7.46 [1]. The last two steps are optional depending on if the devices have fast axis collimation (FAC) lens and the tap used to fix the FAC lens.

7.11.1 Overall Visual Inspection

A finished product is inspected visually using naked eyes or a low magnification microscope. Any defects, such as solder overflow, chip damages, bump or pits, nicks, color deviation and oxidation should be recorded electronically or in production traveler sheet associated with the particular diode laser device. Some examples of the defects are shown in Fig. 7.47 [1].

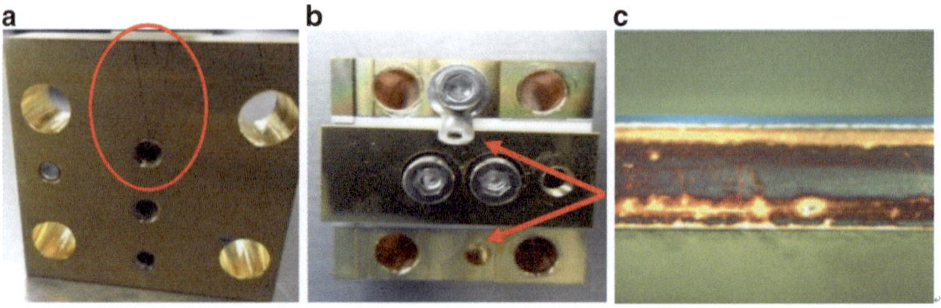

Fig. 7.47 Examples of defects in final overall visual inspection [1]. (**a**) Scratch on heat sink. (**b**) Color deviation. (**c**) Oxidation

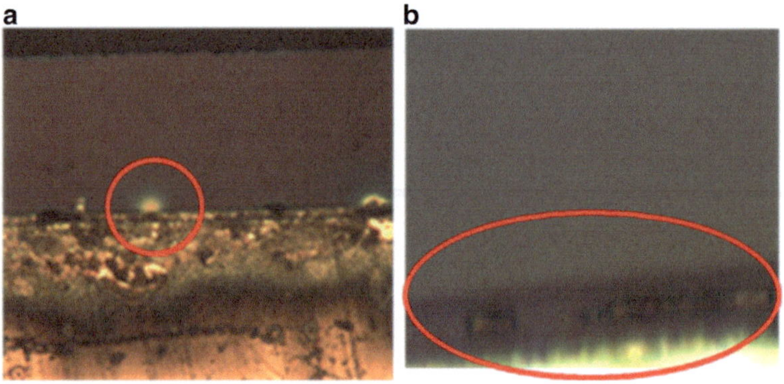

Fig. 7.48 Burn spot or COD caused by contamination or solder covering on the active area [1]. (**a**) Burn spot on the facet. (**b**) COD on the facet

7.11.2 Chip Facet Inspection

Facet inspection is the most critical step of final inspection. Defects on the chip (or bar) facet could affect the electrical performance and product reliability of the diode laser. The active area is the emitting laser area, and the inactive area is the nonradiative area. The size of active area is asymmetric, the thickness and width are about 2 μm and 100–200 μm, respectively. Generally, the cm bar consists of 19 emitters. The contamination or solder covering on the active area may lead to burn spot or COD after some duration of field service as shown in Fig. 7.48 [1]. Also, some other types of defects are illustrated in Fig. 7.49 [1]. The major types of defects performed in facet inspection are listed in Table 7.6 [1].

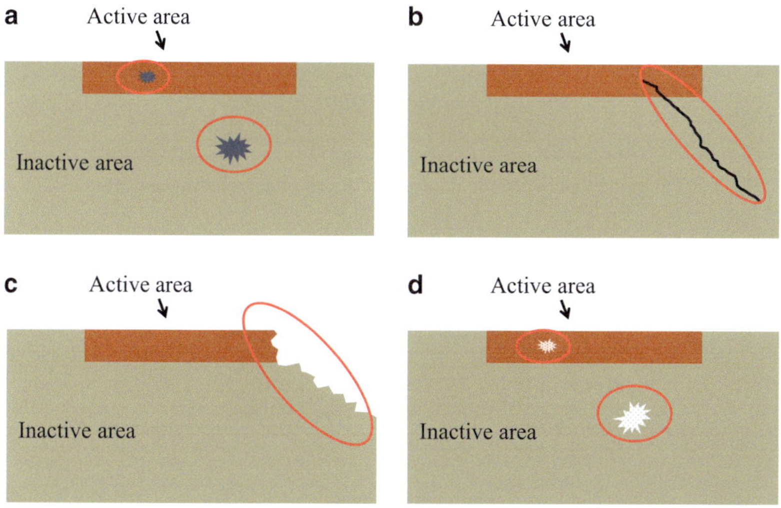

Fig. 7.49 Illustration of defects in active and inactive area [1]

Table 7.6 Major types of defects and inspection criteria performed in face inspection [1]

	Inspection criteria	
Defects	Active area	Inactive area
Contaminations	A contamination is a failure if the contamination is larger than 5 μm	A contamination is a failure if the contamination is larger than 1/3 thickness of the die (typically 100–150 μm)
Crack	A crack in the active area is a failure	Crack is a failure if the length of the defect is larger than 1/3 thickness of the die or the active area could be affected by crack propagation
Chipping	Chipping in the active is a failure	Chipping is a failure if the height of the defect is larger than 1/3 thickness of the die
Coating peeling	Coating peeling is a failure if the size of the defect is larger than 5 μm	Coating peeling is a failure if the size of the defect is larger than 1/3 thickness of the die
Mechanical damage	Mechanical damage in the active is a failure	Mechanical damage is a failure if the size of the defect is larger than 1/3 thickness of the die
Burn spots	A burn spot in the active is a failure	NA
Scratch	Scratch in the active is a failure	Scratch is a failure if the size of the defect is larger than 1/3 thickness of the die
COD	COD is a failure in the active area	NA
Solder overflow	Solder overflow is a failure if the size of the overflow is larger than 5 μm	Solder overflow is a failure if the size of the overflow is larger than 45 μm in any direction

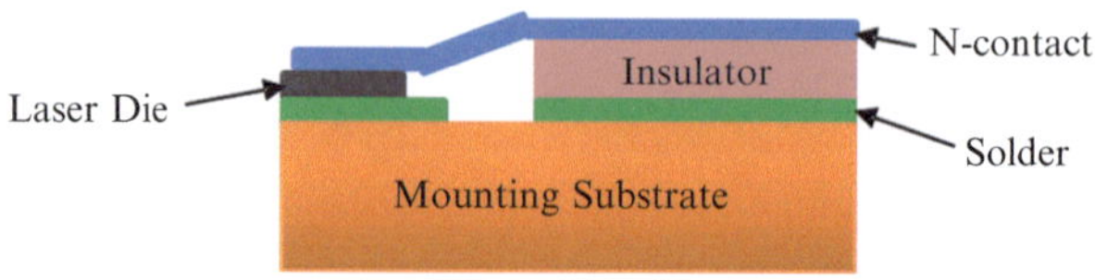

Fig. 7.50 Ideal alignment between the diode laser die and the mounting substrate [1]

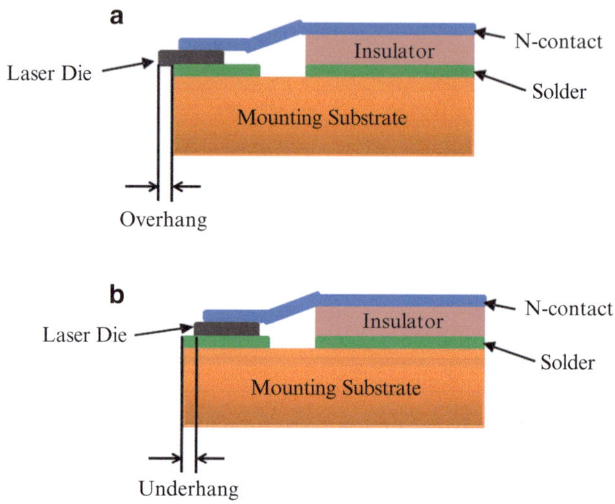

Fig. 7.51 Illustration of overhang (**a**) and underhang (**b**) of laser die on mounting substrate [1]. (**a**) Overhang of laser die on mounting substrate; (**b**) Underhang of laser die on mounting substrate

7.11.3 Overhang and Underhang Inspection

Overhang and underhang characterize the alignment between the diode laser die (could be a single emitter chip or a bar) and the mounting substrate. The consequence of overhang and underhang is ineffective heat conduction and blockage of light transmission, respectively. The inspection of overhang and underhang are conducted by using high magnification microscope.

Overhang and underhang are caused by inaccurate placement of the diode laser die or movement of the die during die bonding process. When laser chip stands out the boundary of submount, it is called overhang, with a positive value; when laser chip stands inside the boundary of submount, it is called underhang, with a negative value. Overhang and underhang are characterized by the value of the distance between laser chip facet and submount. Figure 7.50 shows a precise placement of laser chip on the submount, and examples of overhang and underhang are showed in Fig. 7.51 [1].

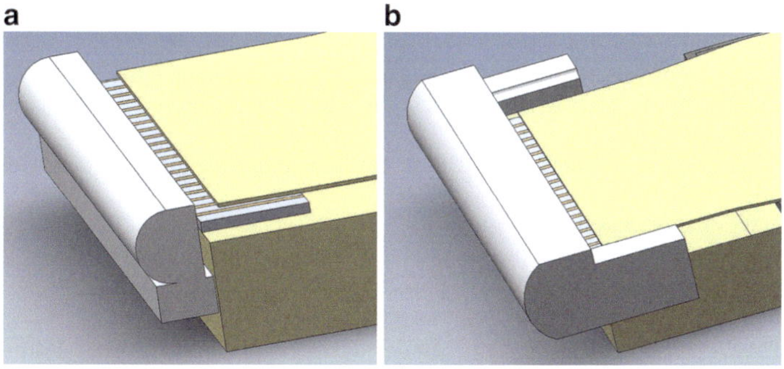

Fig. 7.52 Schematic illustration of FAC fixed with (**a**) bottom and (**b**) side TAP [1]

7.11.4 TAP/FAC Inspection

TAP and Fast Axis Collimation lens (FAC) inspection are necessary when the beam of a diode laser is collimated. When FAC is installed on a diode laser, a TAP needs to be fixed on the mounting substrate or a submount, and then the FAC is attached on the TAP using a UV curable adhesive. The inspection for TAP/FAC is mainly done visually with high magnification microscope. Figure 7.52a, b shows schematic illustration of FAC fixed with bottom and side TAP, respectively. Scratch, contamination, broken and adhesive residue are the main defects.

References

1. J.W. Wang, D. Hou, X.S. Liu, *Packaging Process of High Power Semiconductor Lasers*. (Internal Talk from Focuslight Technologies Co., Ltd., 2012), pp. 11–35
2. http://www.perkinelmer.com/CMSResources/Images/44-74542GDE_DSCBeginnersGuide.pdf
3. http://www.axic.com/images/thin-film-notes/AppNote2-CompositionThicknessMetalFilms.pdf
4. http://www.oxford-instruments.com/products/coating-thickness-measurement-tools/coating-thickness-analyser-x-strata
5. http://www.upcorp.com/monthlyspecial-bt2020.html
6. R.H. Todd, D.K. Allen, L. Alting, *Manufacturing Processes Reference Guide* (Industrial Press, Inc., New York, 1994)
7. http://nzic.org.nz/ChemProcesses/metals/
8. M.E.V. Shun'ko, V.S. Belkin, Cleaning properties of atomic oxygen excited to metastable state $2s^2 2p^4(^1S_0)$. J. Appl. Phys. **102**, 083304 (1–14) (2007)
9. S. Deiries, A. Silber, O. Iwert, E. Hummel, J.L. Lizon, *Plasma Cleaning: A New Method of Ultra-Cleaning Detector Cryostats* (European Southern Observatory, Garching, Germany, 2006)
10. http://plasmatreatment.co.uk/wp-content/uploads/2013/09/21.png
11. http://sindhu.ece.iisc.ernet.in/nanofab/twikii/pub/Main/E-beamEvaporation/E-beam%20manual1.pdf
12. http://www.jeol.co.jp/en/science/eb.html

13. K.S.S. Harsha, *Principles of Physical Vapor Deposition of Thin Films* (Elsevier, Great Britain, 2006), p. 400
14. http://www.oxford-vacuum.com/background/thin_film/evaporation.htm
15. http://sindhu.ece.iisc.ernet.in/nanofab/twiki/pub/Main/ThermalEvaporation/thermal_evaporation_procedure.pdf
16. https://biblio.ugent.be/input/download?func=downloadFile&recordOId=1095343&fileOId=1095356
17. http://www.directvacuum.com/sputter.asp
18. http://repository.tudelft.nl/view/ir/uuid:0058f840-4b34-41ff-bc73-a086bc797ce4/
19. K. Boucke, *Packaging of Diode Laser Bars* (Springer, Berlin, 2007)
20. http://www.micralyne.com/newslyne/ausnpaper.pdf
21. F. Bachmann, *High-Power Diode Lasers Technology and Applications* (Aachen, Germany, 2006)
22. X.S. Liu, R.W. Davis, L.C. Hughes, M.H. Rasmussen, R. Bhat, C.E. Zah, J. Stradling, A study on the reliability of indium solder die bonding of high power semiconductor lasers. J. Appl. Phys. **100**(1), 013104(1–11) (2006)
23. http://extra.ivf.se/ngl/documents/ChapterA/ChapterA1.pdf
24. http://www.microbonds.com/xwiretech/xwire_bkg.htm
25. http://extra.ivf.se/ngl/documents/ChapterA/ChapterA2.pdf
26. http://www.idt.com/document/power-systems-design-estimating-bond-wire-current-carrying-capacity
27. X.S. Liu, W. Zhao, *Technology Trend and Challenges in High Power Semiconductor Laser Packaging*. IEEE, 2009 Electronic Components and Technology Conference (2009), pp. 2106–2113
28. Y. Wang, L. Qin, Y. Zhang, Z.H. Tian, Y. Yang, Z.J. Li, C. Wang, D. Yao, H.H. Yin, Y. Liu, L.J. Wang, Packaging-induced strain measurement based on the degree of polarization in GaAsP-GaInP high-power diode laser bars. IEEE Photon. Technol. Lett. **21**(14), 963–965 (2009)
29. G.Q. Lu, J.N. Calata, Z.Y. Zhang, J.G. Bai, *A Lead-Free, Low-Temperature Sintering Die-Attachment Technique for High Performance and High-Temperature Packaging*. Proceeding of 6th IEEE CPMT Conference of High Density Microsystem Design and Packaging and Component Failure Analysis (2004), pp. 42–46
30. X.S. Liu, L.C. Hughes, M.H. Rasmussen, M.H. Hu, V.A Bhagavatula, R.W. Davis, S. Coleman, R. Bhat, C.E. Zah, *Packaging and Performance of 980 nm Broad Area Semiconductor Lasers*. IEEE 2005 6th International Conference on Electronic Packaging Technology (2005), pp. 67–73
31. http://www.finetech.de
32. http://www.atv-tech.de/en/index.html
33. S.A. Merritt, F. Seiferth, V. Vusirikala, M. Dagenais, Y.J. Chen, D.R. Stone, *A Rapid Flip Chip Die Bonding Method for Semiconductor Laser Diode Arrays*. IEEE 1997 Electronic Components and Technology Conference (1997), pp. 775–779
34. P. Zhalefar, A. Dadoo, M. Nazerian, A. Parniabaran, A.G. Mahani, M. Akhlaghifar, P. Abbasi, M.S. Zabhi, J. Sabbaghzadeh, Study on effects of solder fluxes on catastrophic mirror damages during laser diode packaging. IEEE Trans. Compon. Packag. Manuf. Technol. **3**(1), 46–51 (2013)
35. M. Hempel, M. Ziegler, S. Schwirzke-Schaaf, J.W. Tomm, D. Jankowski, D. Schröder, Spectroscopic analysis of packaging concepts for high-power diode laser bars. Appl. Phys. A **107**(2), 371–377 (2012)
36. Department of Defense, MIL-STD-883E, Test method standard, micro-circuits, Method 1014.9, March 14, 1995
37. http://assets.newport.com/webDocuments-EN/images/WP_Reduce_Test_Laser_Diode_IX.PDF
38. L.A. Johnson, Laser diode burn-in and reliability testing. IEEE Comm. Mag. **44**(2), 4–7 (2006)

Chapter 8
Testing and Characterization of High Power Semiconductor Lasers

High power semiconductor laser is a compact and precision optoelectronic device manufactured by a series of complicated fabrication processes. The performances of a semiconductor laser, such as its output power, wavelength, and spectrum are important for applications. Testing of the common parameters of a semiconductor laser is a route process for users for incoming material quality control and manufacturing procedure. Furthermore, testing is a part of semiconductor laser manufacturing process as well. Besides testing, characterization of some uncommon but important features is very important in the development of high power semiconductor lasers and failure analysis. By characterization of these features, such as spatial spectrum, smile and transient thermal impedance, one can have a deep understanding of the behavior of the diode laser, investigate the physics behind the features, and find out a solution to improve the performance and property of the devices.

In this chapter, the parameters to characterize high power semiconductor lasers is introduced, such as output power, threshold current, slope efficiency, electrical-to-optical conversion efficiency, series resistance, wavelength and spectrum, spatial spectrum, junction temperature, thermal resistance, near field, far field, smile (near-field linearity), and lifetime. The measurement methods and equipments used to characterize the above important parameters of the semiconductor lasers are presented.

8.1 Light Power–Current–Voltage

Light power–current–voltage (LIV) is one of the most important properties of high power semiconductor lasers. The LIV curves can be obtained by the measurement of output light power and voltage as a function of the driving current, as shown in Fig. 8.1 [1]. The output power and forward voltage is directly measured and can be

X. Liu et al., *Packaging of High Power Semiconductor Lasers*, Micro- and Opto-Electronic Materials, Structures, and Systems, DOI 10.1007/978-1-4614-9263-4_8,

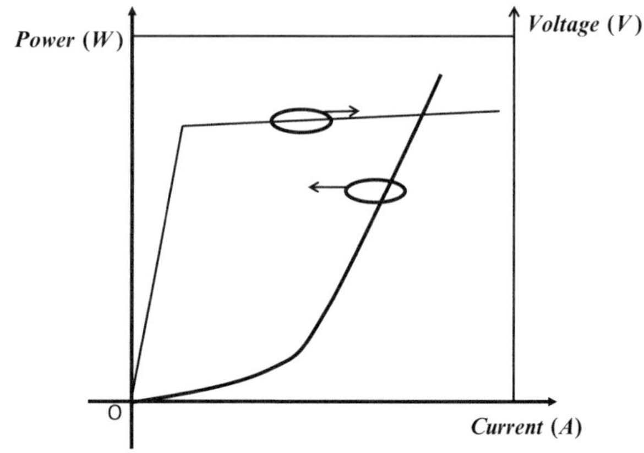

Fig. 8.1 Schematic diagram of LIV curve of a semiconductor laser [1]

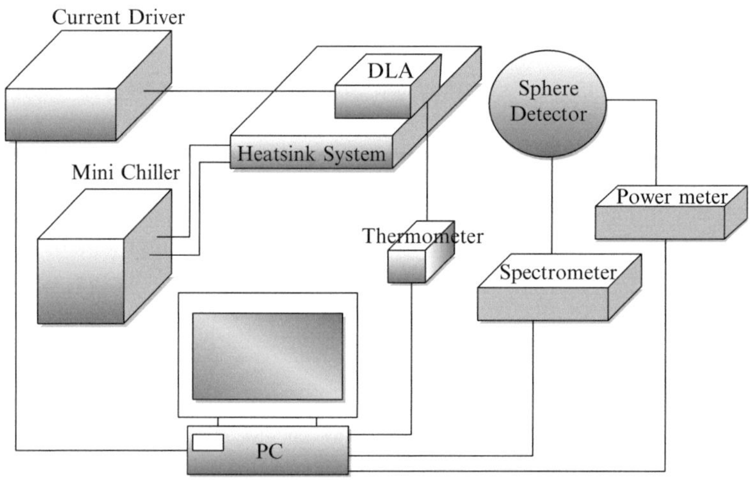

Fig. 8.2 Schematic diagram of a testing setup for LIV measurement of a diode laser array [2]

obtained from the LIV curve. The threshold current, slope efficiency, series resistance, and electrical-to-optical conversion efficiency can be calculated from the LIV curve.

Figure 8.2 shows the schematic diagram of a testing setup for LIV measurement of the diode laser [2]. The testing setup consists of sphere detector, power meter, spectrometer, current driver, controlling system by a computer, and cooling system which contain heat sink system, thermometer, and mini chiller. Data of LIV testing are sampled from the power meter, spectrometer, and driver. The computer analyzes the data and gives a standard LIV report.

Figure 8.3 shows the zoom-in of an actual LIV measurement station [3]. A diode laser array (DLA) is tightened on the top of heat spreader with fixtures. Generally an indium preformed foil or thermal grease is used as the thermal

Fig. 8.3 A zoom-in of an LIV measurement station of high power diode lasers [3]

interface material between the backside of the diode laser package and the top surface of the heat spreader to minimize the contact thermal resistance. The injection current of the diode laser is supplied by the current driver. The laser bar radiation is directly measured by a power meter or collected by an integrating sphere which has a photodiode detector and fiber port to detect the light. The photodiode detector is then connected to a power meter which converts the output current signal from the photodiode detector into optical power value. The fiber port in the integrating sphere is connected to a spectrometer which measures the wavelength and the spectrum. The measurement process is controlled by PC using proprietary software. The software displays and calculates all of the optical and electrical characteristics of laser such as output power, threshold current, operating voltage, wavelength, and spectrum. Figure 8.4 shows an example of a testing software interface [1], and a typical LIV test curve of a laser bar is shown in Fig. 8.5 [4].

The laser beam and the optical axis of the measuring system should be coaxial. Suitable optical alignment stages are used for this purpose. Any pointing variations of the beam during the measurement period shall be verified not to affect the accuracy required by the measurement. Optical elements such as beam splitters, attenuators, relay lenses shall be mounted such that the optical axis runs through their geometric centers. These optical elements are generally not required in LIV measurement, but for different purpose, they may be used. Care should be taken to avoid systematic errors. Reflections, external ambient light, thermal radiation, or air circulations are all potential sources of error [5].

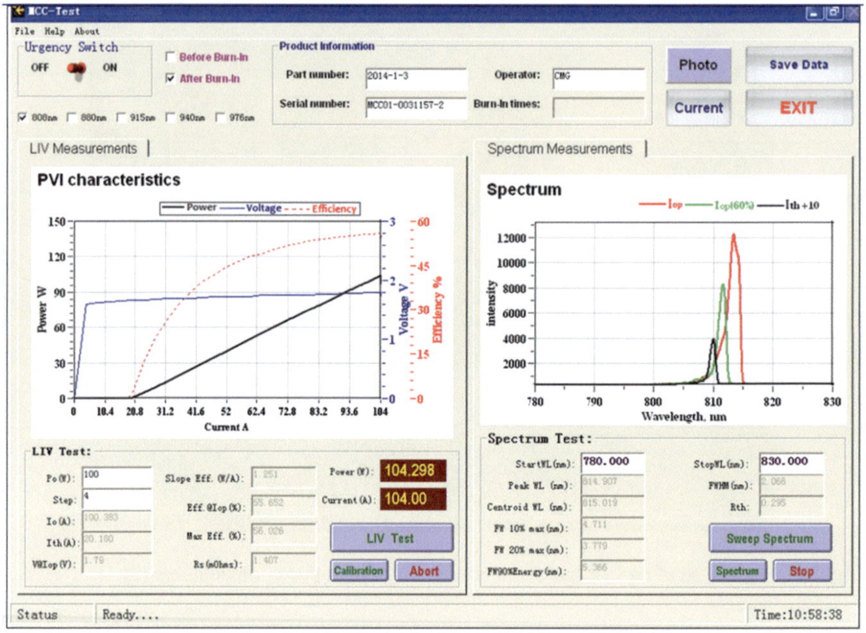

Fig. 8.4 An example of a testing software interface of high power diode lasers [1]

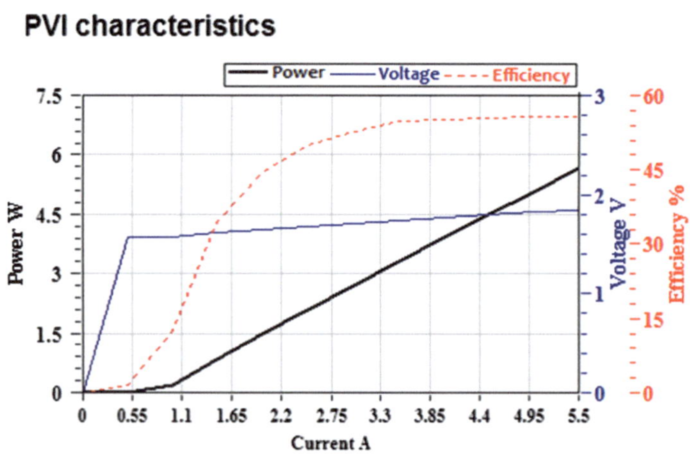

Fig. 8.5 A typical LIV test curve of a high power semiconductor laser bar [4]

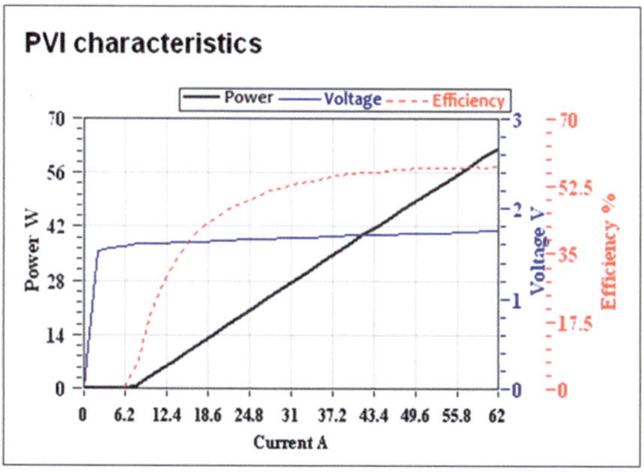

Pop(W)	60.00
Iop(A)	59.81
Ith(A)	7.17
Slope Eff.(W/A)	1.14
Eff.@Iop(%)	57.35
Max Eff.(%)	57.48
Rs(mOhms)	2.81
V @Iop(V)	1.75
TC	25.0

Fig. 8.6 A typical LI curve of a CW 60 W high power single bar laser (CS package) [1]

8.1.1 Output Power

Output power measurement is the most essential part in laser testing needed for laser safety classification, output power specifications, stability specifications, and other specific application requirements. Power measurement normally uses a calibrated power meter. Sometimes a calibrated attenuator is required in order to not saturate the photodiode detector. This section introduces power measurement including quasi-continuous wave (QCW) power measurement in details.

The optical power of diode laser is measured as a function of the driving current (I) and the corresponding curve, LI curve, is obtained by sweeping the current and measuring the resulting optical output power. The LI curve is a plot of the forward current against optical output power. To increase the testing accuracy, multiple times of measurement are needed [6]. However, to save manufacturing cost and improve testing efficiency, instead of sweeping the driving current and measuring the power many times, generally multiple power measurements are taken at each specific driving current point and the average of the readings is used in the output LI curve. Figure 8.6 shows a typical LIV of a commercial continuous wave (CW) 60 W conduction-cooled package single bar product [1]. The output power measurements at each specific driving current can be described as [1]:

$$P_{op} = \frac{1}{\tau_{\mathrm{j}}} \frac{1}{n} \sum_{i=1}^{n} P_i \tag{8.1}$$

Key:

P_{op}—output power, W;
τ_j—transmission ratio of attenuator, %;
n—number of testing;
P_i—the measured power at testing number i, W.

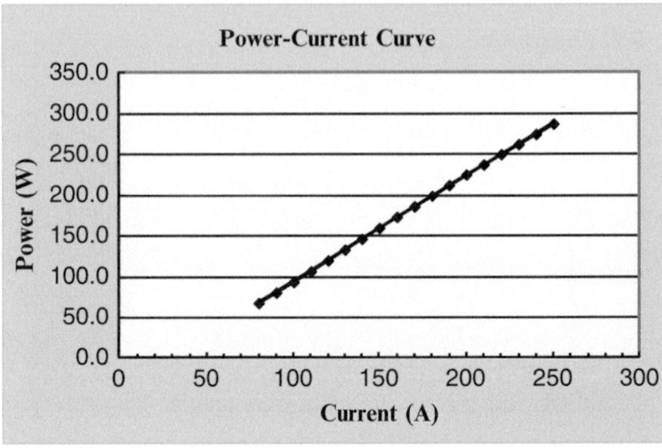

Fig. 8.7 A typical LI curve of a QCW (400 Hz, 200 μs) high power diode laser bar [7]

Besides CW operation mode, more and more applications require diode laser operating in QCW mode. For high power semiconductor lasers under QCW mode, the definition of QCW power is the maximum output power or peak power during the pulse width. There are two methods to test the power of high power semiconductor lasers in QCW mode [7]:

1. The pulse energy is measured directly with a calibrated energy meter. Sometimes a calibrated attenuator is required in order to not saturate the energy meter. The peak power is derived by calculating from the single pulse energy and pulse duration as below:

$$P_p = \frac{1}{\tau_j}\frac{1}{n}\sum_{i=1}^{n}\frac{E_i}{t} \tag{8.2}$$

   Key

   P_p—peak power, W
   τ_j—transmission ratio of attenuator, %
   n—number of testing
   E_i—measured pulse energy at the testing number i, J
   t—pulse duration, s

   Figure 8.7 shows a typical LI curve of a single bar high power semiconductor laser under QCW (400 Hz, 200 μs) condition [7].

2. The peak power is directly measured by synchronizing the measurement and the driving current. The measurement is typically done by a fast response photodiode that is placed into the laser beam. The photodiode provides a signal output proportional to the laser input. The signal output data from the photodiode is analyzed and calibrated by various analog and digital corrections such as compensating for the difference between the calibration wavelength and the wavelength of the laser, and then a calibrated measurement data is provided.

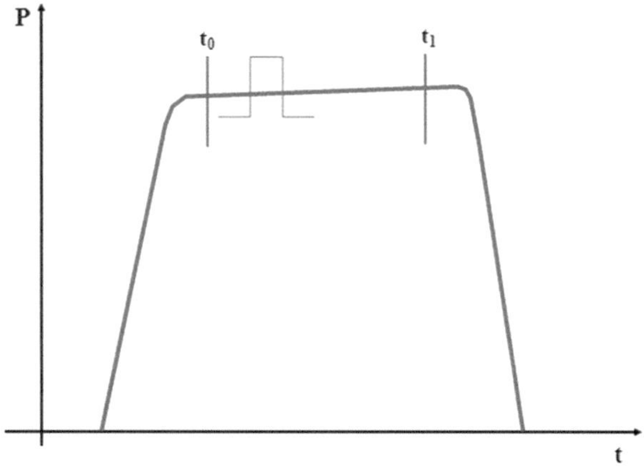

Fig. 8.8 Illustration of synchronization of peak power measurement and the driving current [1]

The synchronizing of measurement time and driving current in QCW operation mode is shown in Fig. 8.8. The measurement trigger time should be within the time range between t_0 and t_1. t_0 is a time when the driving current is stabilized after it is turned on and t_1 is a time that leave enough time for peak power to be measured before the driving current is turned off. Beyond the time range between t_0 and t_1, the diode laser may not work at stable state and the power measurement is unreliable.

The peak power is calculated as below:

$$P_p = \frac{1}{\tau_{\mathrm{j}}}\frac{1}{n}\sum_{i=1}^{n} P_{pi} \tag{8.3}$$

Key

P_p—peak power, W
τ_{j}—transmission ratio of attenuator, %
n—number of testing
P_{pi}—measured peak power at the testing number i, W

8.1.2 *Threshold Current*

Threshold current (I_{th}) is defined as the minimum current needed to achieve stimulated emission rather than spontaneous emission of radiation of a semiconductor laser. The I_{th} is described in the LI curve as shown in Fig. 8.9 [8]. At low current there is only spontaneous emission with a relatively broad spectrum. The spontaneous emission power rises slowly with the increasing current. At the threshold, the lasing gain equals to the sum of all the losses, such as material

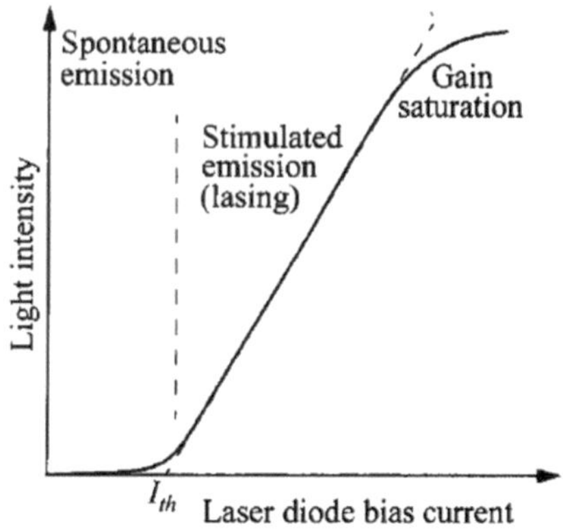

Fig. 8.9 Light power output versus laser bias current, showing the threshold current [8]

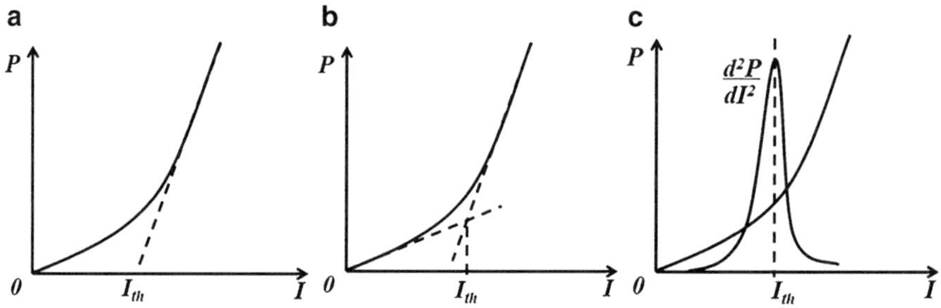

Fig. 8.10 Different methods of threshold current calculation; (**a**) Slope method; (**b**) Two-point method, (**c**) Differential method [1]

absorption loss and cavity loss. When the injection current is above the threshold value, the laser power increases much faster and laser diode starts lasing. For different types of high power semiconductor lasers, the threshold current level ranges from hundreds of milli-amperes for a single emitter to tens of amps for a laser bar.

The threshold current changes exponentially with the temperature [1]:

$$I_{th}(T_2) = I_{th}(T_1)\exp\left(\frac{T_2 - T_1}{T_0}\right) \tag{8.4}$$

where T_1 and T_2 are two different junction temperature. The threshold current can be measured at different temperatures. T_0 is the characteristic temperature of a diode laser design which specifies how sensitive I_{th} changes with temperature. The higher T_0, the less sensitive I_{th} changes with temperature and vice versa. For a typical GaAs-based 808 nm diode laser, the T_0 is approximately 155 K [9].

In order to obtain threshold current, the output power versus injection current is measured and plotted at a specific temperature. Figure 8.10 shows the methods of the threshold current calculation from the curve of the output power and injection current (P ~ I curve or LI curve) [1]. In general, there are three methods to calculate the threshold current as described below:

1. This method linearly fits the linear portion of the LI curve, and the *x*-intercept of the fitted line is defined as the threshold current, as shown in Fig. 8.10a.
2. This method linearly fits the linear portion of the LI curve before threshold and the linear portion after threshold; the threshold current is defined as the cross point of the two fitted lines, as shown in Fig. 8.10b.
3. This method first obtains the function of light power and current by curve fitting the LI curve testing data.

$$P = f(I) \tag{8.5}$$

where P is the output power and I is the injection current. Then the second derivative of the fitted function P $= f(I)$ is obtained.

$$\frac{d^2P}{dI^2} = G(I) \tag{8.6}$$

The extreme value of *G*(I) is calculated. The current at the corresponding extreme value of *G*(I) is defined as the threshold current I_{th}, as shown in Fig. 8.10c.

Among the above methods, the first method is widely used in industry due to its simplicity. The second method is more accurate than the first method one and easier in practice than the third method. The third method is rarely used in industry and it is occasionally used in scientific research and engineering.

The threshold of the conduction-cooled single bar product (CS package) is calculated based on the LI testing data shown in Fig. 8.6 using the above-mentioned three methods. By the first method, the experimental testing data is fitted and the result is shown in Fig. 8.11. The linear fitting function is expressed as

$$P(I) = 1.14I - 8.17 \tag{8.7}$$

The *x*-intercept of the fitted line is 7.17 A and thus the threshold current of this diode laser calculated by the first method is 7.17 A.

Using the second method, the experimental testing data is fitted and the result is shown in Fig. 8.12. The blue line, which is fitted to the pre-threshold portion of the LI curve, is expressed as

$$P(I) = 0.1I - 0.2 \tag{8.8}$$

The red line, which is fitted to the lasing portion in the same manner as the first method, is expressed as

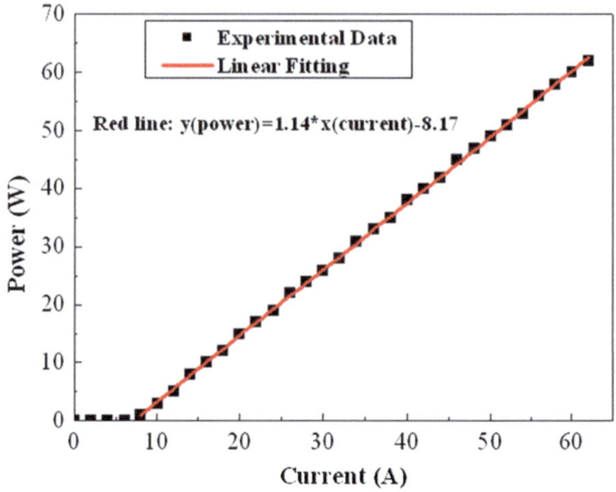

Fig. 8.11 Threshold current calculation using the first method indicated in Fig. 8.10(a) [1]

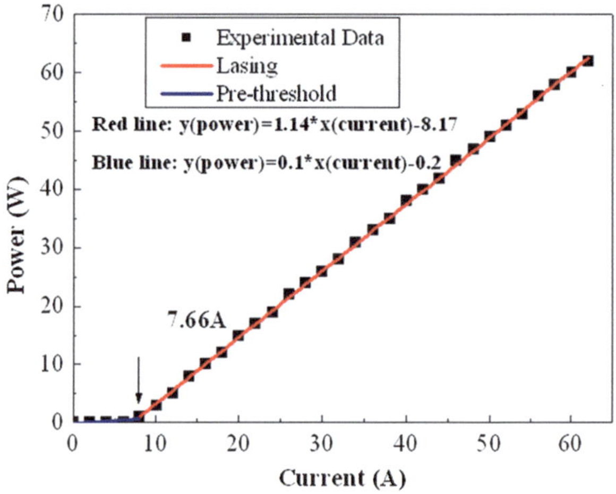

Fig. 8.12 Threshold current calculation using the second method indicated in Fig. 8.10(b) [1]

$$P(I) = 1.14I - 8.17 \tag{8.9}$$

According to two curve fitting function Eqs. (8.8) and (8.9), the threshold current can be calculated and the result is 7.66 A.

The threshold calculation by the third method is shown in Fig. 8.13. The d^2P/dI^2 can be obtained from measurement data by numerical solution, the result of which are marked as red dot in Fig. 8.13. The numerically calculated d^2P/dI^2 data can be fitted by Gaussian fitting. The fitting curve is shown in Fig. 8.13 and the fitting equation is

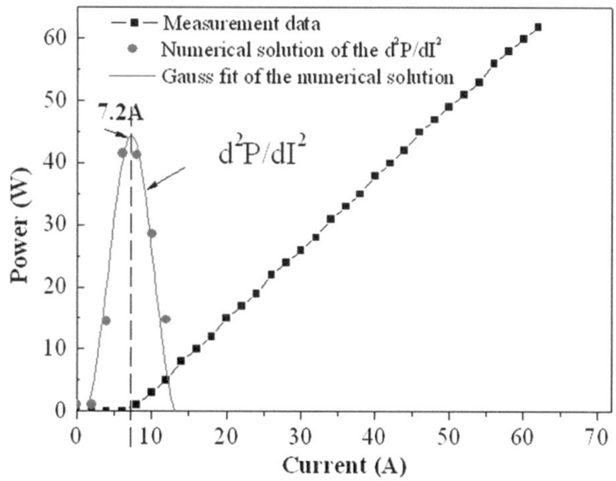

Fig. 8.13 Threshold current calculation using the third method indicated in Fig. 8.10(c) [1]

$$y = -0.04 + \frac{1.8}{5.87 \times \sqrt{0.244/2}} \times e^{-2\times((x-7.2)/5.87)^2} \tag{8.10}$$

According to Eq. (8.10), the maximum power P_m corresponds to the current of 7.2 A, and thus the threshold current I_{th} is 7.2 A.

8.1.3 Slope Efficiency

The slope efficiency is defined as the ratio of the difference of output optical power (ΔP) to the difference of corresponding input current (ΔI). It can be obtained from the LI curve by the double point method, as shown in Fig. 8.14 [1]. In practical calculation, the two points are selected as the ones corresponding to the output power of 30 and 90 % of the operating output power.

The slope efficiency (SE) of high power semiconductor laser is defined as

$$SE = \frac{\Delta P}{\Delta I} = \frac{P_2 - P_1}{I_2 - I_1} \tag{8.11}$$

where P_2 and P_1 are the output powers of a diode laser at the forward currents of I_2 and I_1, respectively. Figure 8.15 shows a typical L–I curve of a CW 60 W CS-packaged high power single bar laser [1]. To calculate slope efficiency of the CW 60 W CS-packaged single bar, the two points are selected as powers of 30 and 90 % of the operating output power which are 18 and 54 W, respectively. The currents corresponding to the selected powers are 23.14 and 54.71 A, respectively. According to Eq. (8.11), the slope efficiency is

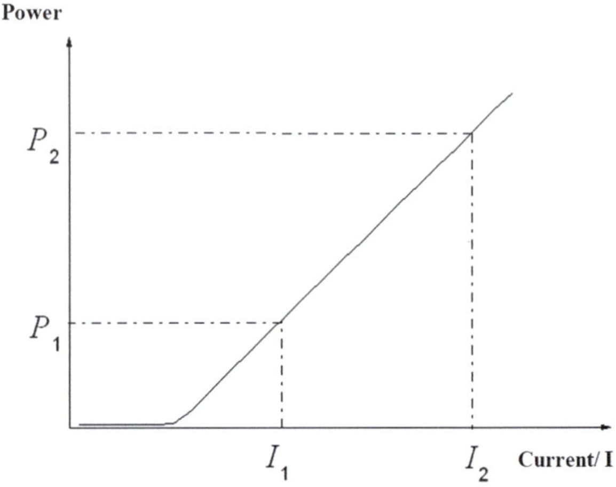

Fig. 8.14 Illustration of the calculation method of slope efficiency [1]

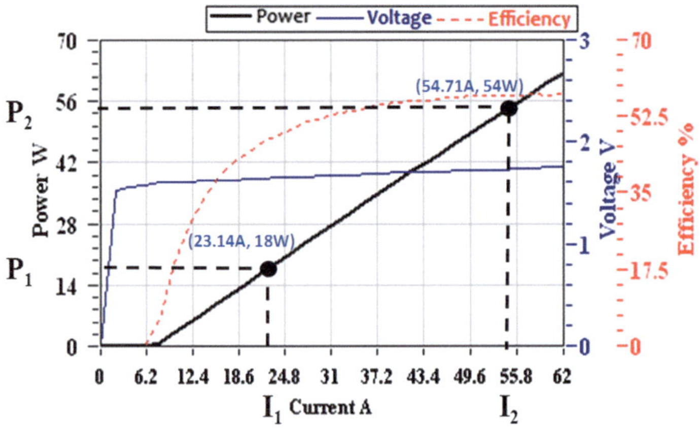

Fig. 8.15 Calculation of slope efficiency from a typical LI curve of a CW 60 W high power single bar laser (CS package) [1]

$$SE = \frac{P_2 - P_1}{I_2 - I_1} = \frac{54W - 18W}{54.71A - 23.14A} = 1.14W/A$$

8.1.4 Electrical-to-Optical Conversion Efficiency

Electrical-to-optical conversion efficiency (η_p) is defined as the ratio of output optical power (P_{op}) to input electrical power ($V_{op} \times I_{op}$) [1], that is, the conversion efficiency is

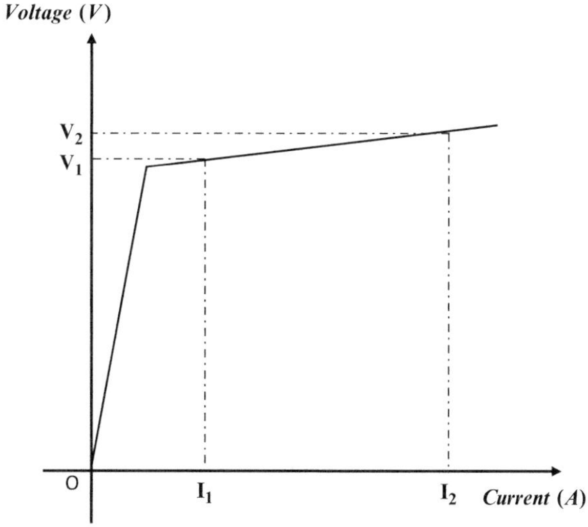

Fig. 8.16 Calculation methods of series resistance [1]

$$\eta_p = \frac{P_{op}}{I_{op} \times V_{op}} \times 100\% \tag{8.12}$$

Key:

η_p—Electrical-to-optical conversion efficiency, %;
I_{op}—operating current, A;
P_{op}—output power at operating current, W;
V_{op}—operating voltage, V.

For the LIV curve example shown in Fig. 8.6, when the Pop is at 60 W, the Iop and Vop are 59.81 A and 1.75 V, respectively. Thus the electrical-to-optical conversion efficiency is calculated as

$$\eta_p = \frac{P_{op}}{I_{op} \times V_{op}} \times 100\% = \frac{60W}{59.81A \times 1.75V} \times 100\% = 57.3\%$$

8.1.5 Series Resistance

The series resistance (Rs) is the sum of the bulk electrical resistance and the contacting electrical resistance of the materials and interfaces of the semiconductor laser device in which the driving current flows through. The higher Rs is, the more heat generated in the device and the lower electrical-to-optical conversion efficiency is. Rs is calculated from the LIV curve when the driving current is higher than the threshold current. It is defined as the ratio of the difference of forward voltage (ΔV) to the difference of corresponding input current (ΔI), as shown in Fig. 8.16 [1]. The series resistance can be described by

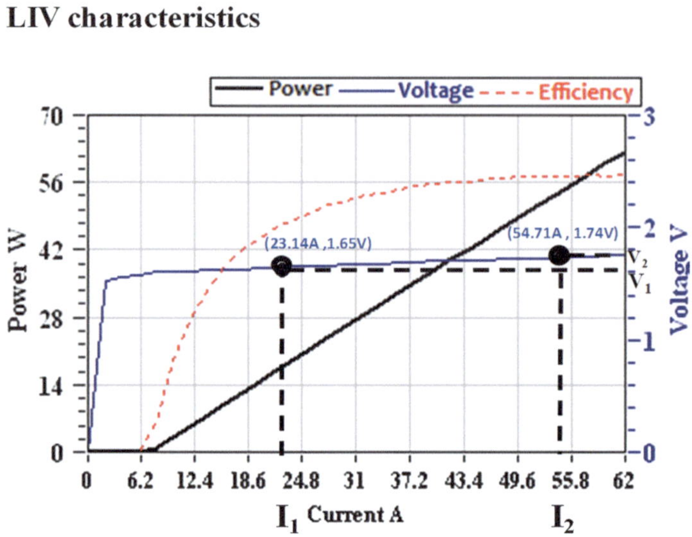

Fig. 8.17 Calculation of series resistance from a typical LIV curve of a CW 60 W high power single bar laser (CS package) [1]

$$Rs = \frac{\Delta V}{\Delta I} = \frac{V_2 - V_1}{I_2 - I_1} \tag{8.13}$$

where V_2, and V_1 are the forward voltage of a diode laser at the forward current of I_2 and I_1, respectively.

Figure 8.17 shows the typical V–I curve of a CW 60 W CS-packaged high power single bar laser [1]. To calculate series resistance, the two points are selected as ones which have the powers of 30 and 90 % of the operating output power. At the power of 30 % of the operating power which is 18 W, the voltage is 1.63 V and the current is 23.14 A, while at the power of 90 % of the operating power which is 54 W, the voltage is 1.74 V and the current is 54.71 A. According to Eq. (8.13), the series resistance is

$$Rs = \frac{V_2 - V_1}{I_2 - I_1} = \frac{1.74V - 1.65V}{54.71A - 23.14A} = 2.81m\Omega \tag{8.14}$$

8.2 Wavelength and Spectrum

The wavelength of a diode laser can be characterized by peak wavelength and centroid wavelength, and the spectrum is typically characterized by spectral width [10, 11]. The parameters used to characterize the spectral width are mainly full width at half maximum (FWHM) and full width at 90 % energy (FW90%E), with FWHM more commonly being used. For high power semiconductor lasers used for

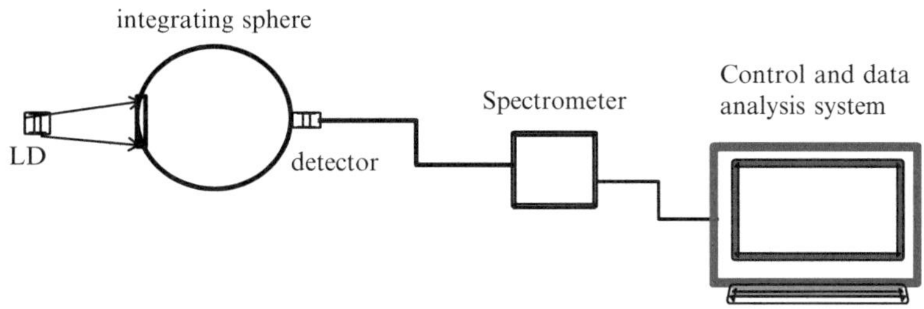

Fig. 8.18 Wavelength and spectrum measurement setup of high power semiconductor lasers [1]

pumping solid state lasers or fiber lasers, the spectral width is very important as the pumping source energy out of the absorption bandwidth of the active laser media is wasted and becomes heat. Therefore, spectral width of a diode laser as a pumping source significantly affects efficiency, power, beam quality, compactness, and thermal management of a laser system.

8.2.1 Measurement Method and Equipment

The wavelength and spectrum measurement system is schematically shown in Fig. 8.18 [1]. The measurement system mainly consists of four parts: (1) the light source, (2) the integrating sphere with signal port, (3) the spectrometer, and (4) control and data analysis software. Due to the high power output of the semiconductor laser, integrating sphere is usually used to homogenize the beam of the semiconductor laser. A small amount of light power is transferred to the spectrometer via the signal port in the integrating sphere. The spectrum information can be analyzed and displayed by the software. Figure 8.19 shows an example of the spectrum measurement software interface [1]. Figure 8.20 shows a typical measured spectrum of a laser diode array [1].

8.2.2 Wavelength and Spectrum Characterization

Peak Wavelength

The peak wavelength (or) peak-emission wavelength is defined as the wavelength at which the intensity of the laser emission is at the maximum value, as shown in Fig. 8.21 [1, 12, 13].

Figure 8.22 shows an example of a spectrum of 915 nm diode laser with a side peak on the left. The intensity distribution is normalized with wavelength. The maximum intensity I_{max} is marked in the figure. The wavelength corresponding to I_{max} is 913.33 nm and thus the peak wavelength of the diode laser is 913.33 nm.

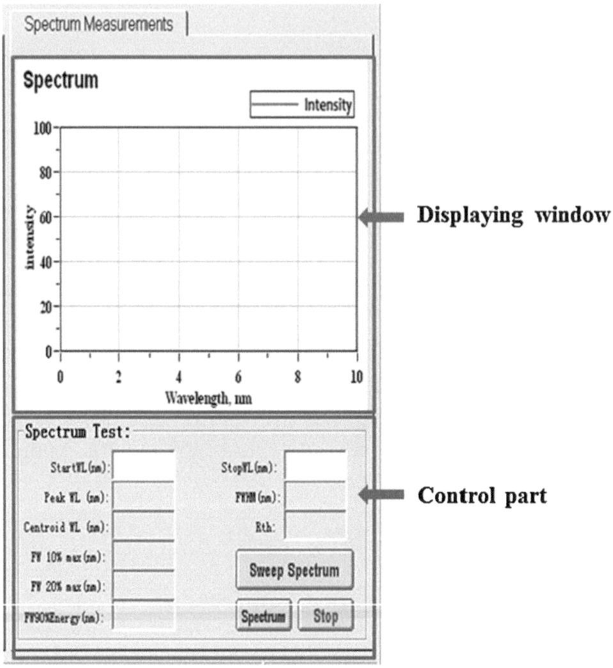

Fig. 8.19 An example of spectrum measurement software interface [1]

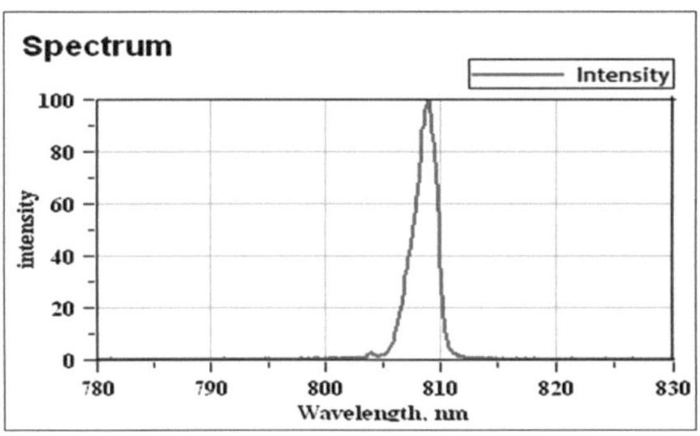

Peak wavelength(nm)	809.06
Centroid wavelength(nm)	808.82
FWHM(nm)	2.22
FW90% Energy(nm)	4.56

Fig. 8.20 A typical spectrum of a laser diode array [1]

Centroid Wavelength

The centroid wavelength is defined as the power weighted average wavelength of the intensity distribution as a function of wavelength of a diode laser [14]. It is calculated by

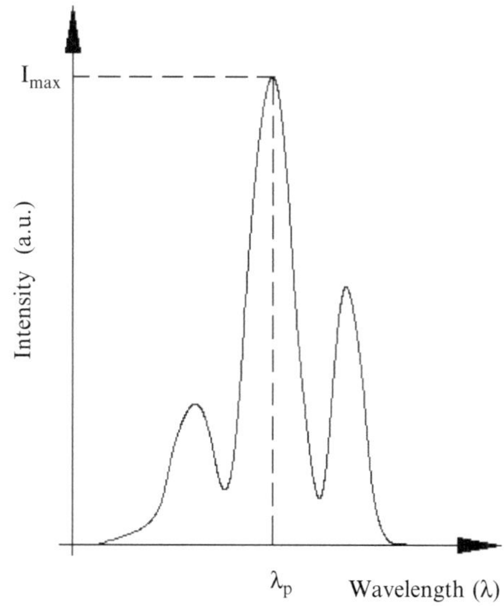

Fig. 8.21 Schematic diagram of peak wavelength of a diode laser [1]

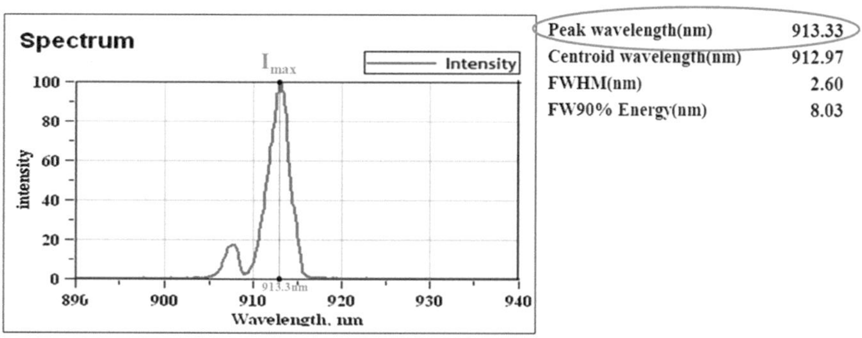

Fig. 8.22 An example of a spectrum of a 915 nm diode laser and its peak wavelength [1]

$$\lambda_c = \frac{1}{P_{total}} \int p(\lambda)\lambda d\lambda \tag{8.15}$$

the total power is

$$P_{total} = \int p(\lambda) d\lambda \tag{8.16}$$

where $p(\lambda)$ is the power spectral density. The above integrals theoretically extend over the entire wavelength range; however, it is usually sufficient to perform the integral over a certain wavelength range where the power spectral density $p(\lambda)$ is still meaningful. In practical, during measurement the intensity $I(\lambda)$ is sampled from

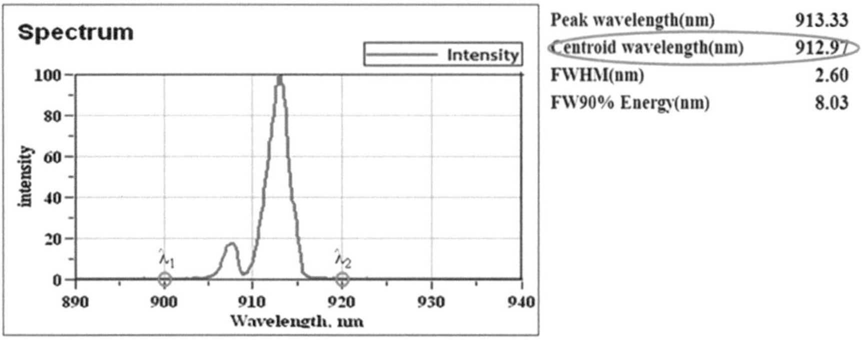

Fig. 8.23 An example of a spectrum of a 915 nm diode laser and its centroid wavelength [1]

discrete wavelength with same small interval. Thus, $I(\lambda_i)\Delta\lambda$ replaces $p(\lambda)d\lambda$, and Eq. (8.14) is changed into

$$\lambda_c = \frac{\sum_{i=1}^{i=N} I(\lambda_i)\lambda_i\Delta\lambda}{\sum_{i=1}^{i=N} I(\lambda_i)\Delta\lambda} = \frac{\sum_{i=1}^{i=N} I(\lambda_i)\lambda_i}{\sum_{i=1}^{i=N} I(\lambda_i)} \tag{8.17}$$

where N is the number of wavelength sampling during measurement.

Figure 8.23 shows the same spectrum of the 915 nm diode laser as the one shown in Fig. 8.22. As shown in the figure, the intensity at the marked λ_1 and λ_2 is almost zero. As shown in Fig. 8.23, λ_1 and λ_2 are 900 and 920 nm, respectively. It is noticeable that in order to obtain accurate centroid wavelength measurement, the number of wavelength sampling N should be large enough or the wavelength sampling interval should be small enough. Here, the number of the wavelength sampling N from λ_1 to λ_2 is 170, and $\Delta\lambda$ is 0.1176 nm. According to Eq. (8.17), the centroid wavelength λ_c is calculated based on the measurement data as

$$\lambda_c = \frac{\sum_{i=1}^{i=170} I(\lambda_i)\lambda_i}{\sum_{i=1}^{i=170} I(\lambda_i)} = 912.97nm \tag{8.18}$$

Full Width at Half Maximum (FWHM)

Spectral width of FWHM is defined as the difference between the wavelengths at which intensity is half of its peak value in the intensity distribution as a function of wavelength curve. Figure 8.24 shows schematic diagram of FWHM [1, 12].

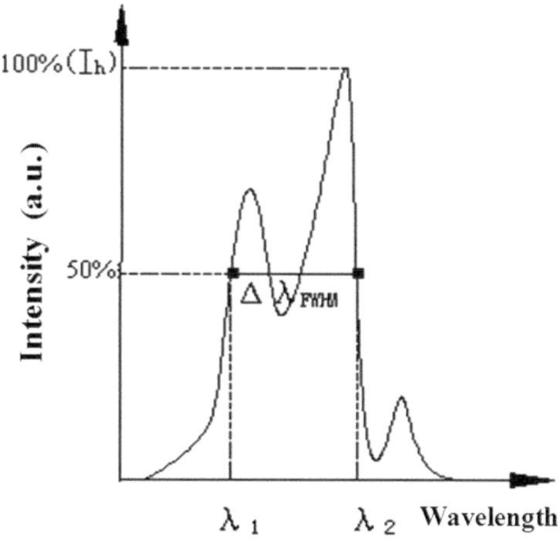

Fig. 8.24 Schematic diagram of full width at half maximum (FWHM) [1]

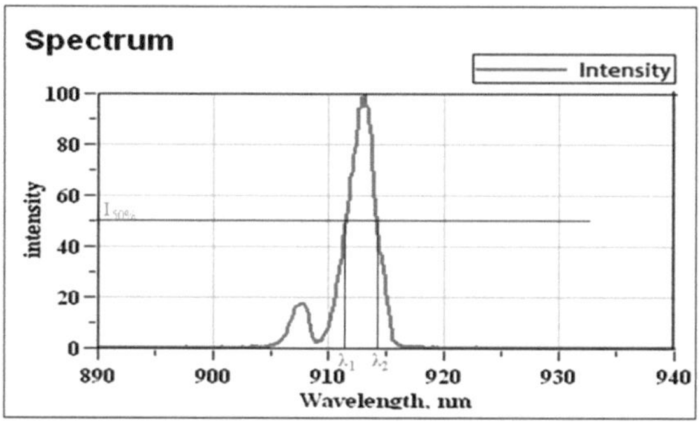

Peak wavelength(nm)	913.33
Centroid wavelength(nm)	912.97
FWHM(nm)	2.60
FW90% Energy(nm)	8.03

Fig. 8.25 An example of a spectrum of a 915 nm diode laser and its FWHM [1]

The spectral width of FWHM can be calculated as

$$\Delta\lambda_{FWHM} = \lambda_2 - \lambda_1 \tag{8.19}$$

where $\Delta\lambda_{FWHM}$ is FWHM spectral width, λ_1 and λ_2 are the wavelengths at which the corresponding intensity is 50 % of maximum intensity.

The spectrum shown in Fig. 8.22 is also used as an example to calculate the FWHM. The half maximum intensity $I_{50\%}$ and λ_1 and λ_2, at which the intensity $I(\lambda)$ is equal to $I_{50\%}$, are marked in Fig. 8.25. From testing data it is known that $\lambda_1 = 911.52$ nm and $\lambda_2 = 914.12$ nm, and the spectral width of FWHM is calculated according to Eq. (8.19).

$$\Delta\lambda_{\mathrm{FWHM}} = \lambda_2 - \lambda_1 = 2.6nm$$

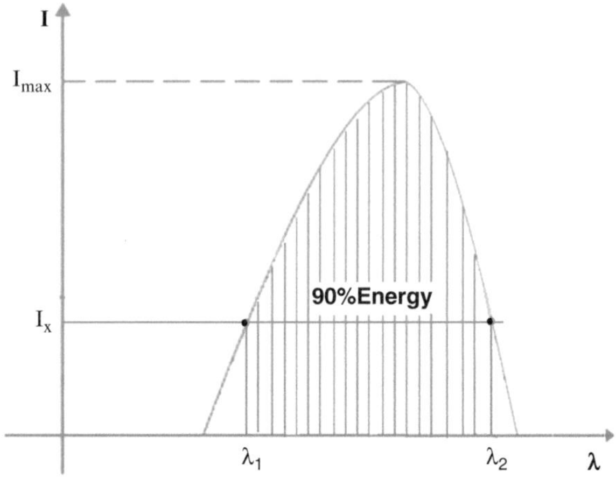

Fig. 8.26 Schematic diagram of full width at 90 % energy (FW90%E) [1]

FW90%E

The spectrum width of FW90%E is defined as the difference of the wavelengths between which the energy is 90 % of the total energy in the intensity distribution as a function of the wavelength curve. Figure 8.26 shows schematic diagram of FW90%E [1, 13]. In the figure, the energy marked by shaded area is 90 % of the total energy, and the difference of the two corresponding wavelengths λ_1 and λ_2 is the full spectrum width of 90 % energy. The FW90%E is calculated as

$$\Delta\lambda_{FW90\%E} = \lambda_2 - \lambda_1 \tag{8.20}$$

where $\Delta\lambda_{FW90\%E}$ is FW90%E, λ_1 and λ_2 are the wavelengths between which the energy is 90 % of the total energy.

Figure 8.27 illustrates the calculation of FW90%E. λ_3 and λ_4 are the wavelengths beyond which the intensity is negligible. In this case, λ_3 and λ_4 are 900 and 920 nm respectively. The total energy E_{total} is

$$E_{\text{total}} = \int_{\lambda_3}^{\lambda_4} I(\lambda)d\lambda \tag{8.21}$$

λ_1 and λ_2 are the wavelengths between which the energy is 90 % of the total energy. At λ_1 and λ_2, the intensity is equal which is presented in Eq. (8.22). The energy calculation between λ_1 and λ_2 is expressed as Eq. (8.23).

$$I(\lambda_1) = I(\lambda_2), \tag{8.22}$$

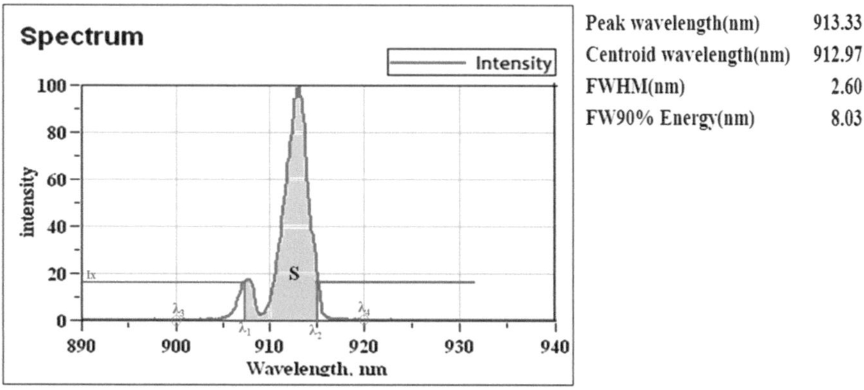

Fig. 8.27 An example of a spectrum of a 915 nm diode laser and its FW90%E calculation [15]

$$\int_{\lambda_1}^{\lambda_2} I(\lambda)d\lambda = 90\% E_{total}. \tag{8.23}$$

Same as centroid wavelength calculation, in practical, during measurement the intensity I(λ) is sampled from discrete wavelength with same small interval. λ_1 and λ_2 can be calculated numerically from the measurement data using software. For the spectrum shown in Fig. 8.27, the calculated λ_1 and λ_2 are 907.53 and 915.5 nm, respectively. The $\Delta\lambda_{FW90\%E}$ can be obtained by $\lambda_2 - \lambda_1$ which is 7.97 nm.

8.3 Spatial Spectrum

The spatial spectrum of a diode laser bar is defined as the wavelength distribution of each individual emitter in spatial dimension [16, 17]. Generally, a semiconductor laser bar consists of 19 ~ 62 emitters. The emitting wavelength from individual emitters is affected by wafer uniformity as well as packaging-related voids and thermal stress effects. With the advancement of epitaxy growth technology and wafer fabrication technology, the wavelength uniformity within a wafer is much improved and the variation of the wavelength from the individual emitters within a diode laser bar is negligible. The difference in emitting wavelength from the individual emitters of a diode laser bar is mainly affected by packaging-related effects. The spectral broadening of laser arrays is a result of nonuniform emitting wavelength from individual emitters. The spatial spectrum is an important parameter to characterize the mechanism of spectral broadening in high power semiconductor lasers.

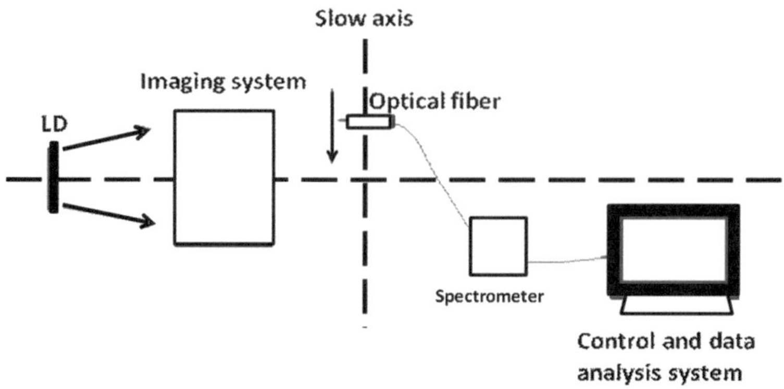

Fig. 8.28 The principle of spatial spectrum measurement [1]

8.3.1 Measurement Method and Equipment

In spatial spectrum measurement, the wavelength of each individual emitter should be measured accurately. Therefore, the radiation crosstalk from neighboring emitters should be avoided as the spacing of two neighboring emitter is small, which generally ranges from 150 to 500 μm. Also the beam in the slow axis is diverged. The wavelength of each emitter should be tested before the beam overlaps with neighboring emitters [17]. In order to avoid the interference from other emitters, the beams of the emitters should be separated in space and then detected one by one. An optical imaging system is used to carry out the separation of the beams of each emitter. Figure 8.28 shows the principle of the measurement method [1]. The imaging system is installed in front of the diode laser. The amplified image of each emitter is located in imaging plane, and they are divided spatially. An optical fiber, which is installed in the imaging plane, is used to transmit the optical signal of each emitter. The other end of the fiber is connected to a spectrometer which measures the wavelength and outputs to the control and data analysis system. In a diode laser bar, the pitch of the emitters is pretty accurate. After the imaging system, the distance interval of the images of the emitters should be the same. In this way, the fiber can be moved with same distance interval to measure the wavelength of each emitter automatically along the slow axis.

The key part of this method is the design of the optical imaging system. At the image plane, the image system should have low aberration to avoid the beam superposition with other emitters and reduces the influences of other neighboring emitters. A four-lens optical imaging system is designed. The emitters of the laser bar are magnified by the imaging system. The schematic diagram of the four-lens optical imaging system is shown Fig. 8.29 [1].

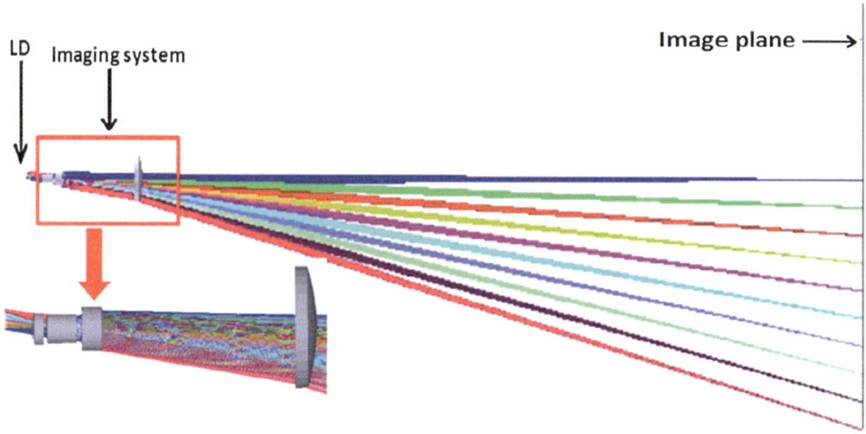

Fig. 8.29 Schematic diagram of a four-lens optical imaging system designed for acquiring spatial spectrum [1]

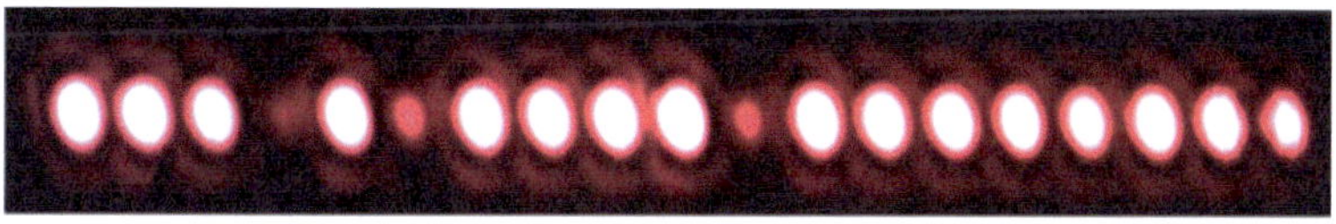

Fig. 8.30 The near-field image of a 19 emitter laser bar [1]

8.3.2 Typical Testing Results and Data Analysis

A semiconductor laser bar with 19 emitters is chosen and the spatial spectrum testing and spatial spectral imaging is introduced. The near-field image of the laser bar is shown in Fig. 8.30 [1]. The image was taken after filters. It can be seen that 3 of the 19 emitters have weaker intensity than the others. Figure 8.31 shows the output spectrum of the laser bar [1]. There is a left shoulder in the spectrum. The spectrum of the diode laser bar is an overlapping of the spectrum of the 19 emitters in the bar. The spectrum of the 19 individual emitters in the laser bar was tested, and the result is shown in Fig. 8.32. As shown in Fig. 8.32, the emitters from No. 8 to No. 15 obviously have blue shift relative to the other emitters, and the left shoulder in the spectrum of the laser bar is generated due to the blue shift of these emitters.

The average centroid wavelength of the laser emitters can be calculated by

$$\bar{\lambda} = \frac{\sum_{n=1}^{N} \lambda_n}{N} \tag{8.24}$$

where $\bar{\lambda}$ is the average centroid wavelength of the emitters; N is the emitters number of the laser bar; and λ_n is the centroid wavelength for the nth emitter. Based on the

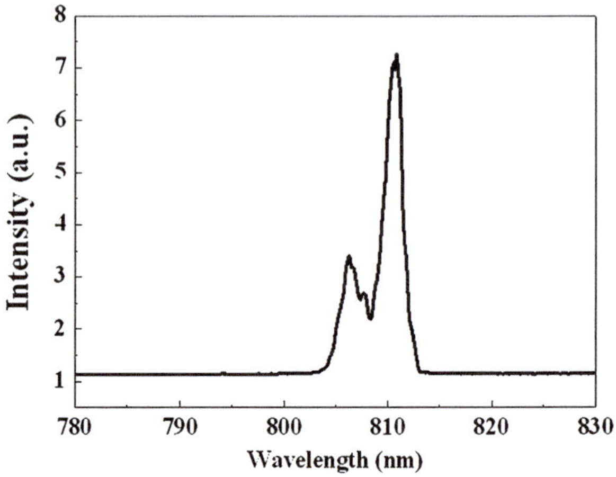

Fig. 8.31 The spectrum of the laser bar [1]

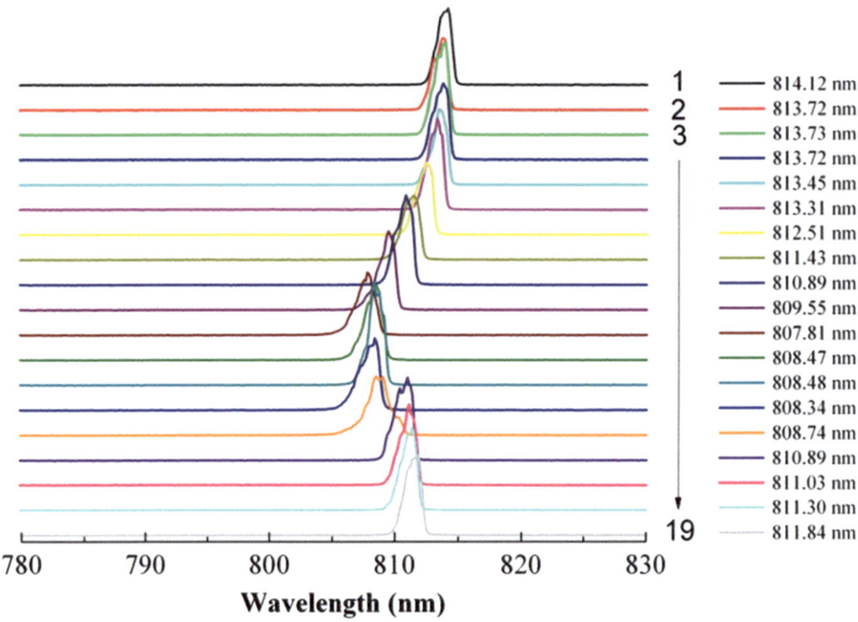

Fig. 8.32 The spectrum of the 19 emitters in laser bar [1]

testing results shown in Fig. 8.32, the average centroid wavelength is 811.23 nm calculated by Eq. (8.24). The centroid wavelength distribution of the individual emitters can be expressed visually by displacing the near-field image of the emitters spatially in proportional to the difference between the centroid wavelength of the individual emitters and the average centroid wavelength. Figure 8.33 shows the spatial spectral imaging of the laser bar [1]. As can be seen from Fig. 8.33, spatial

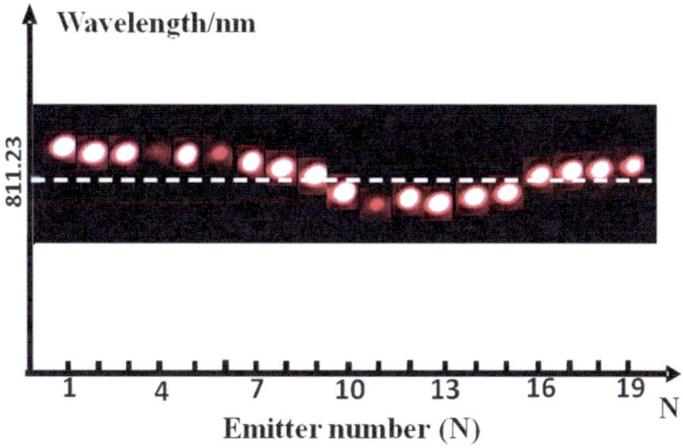

Fig. 8.33 The spatial spectral imaging of the laser bar with 19 emitters [1]

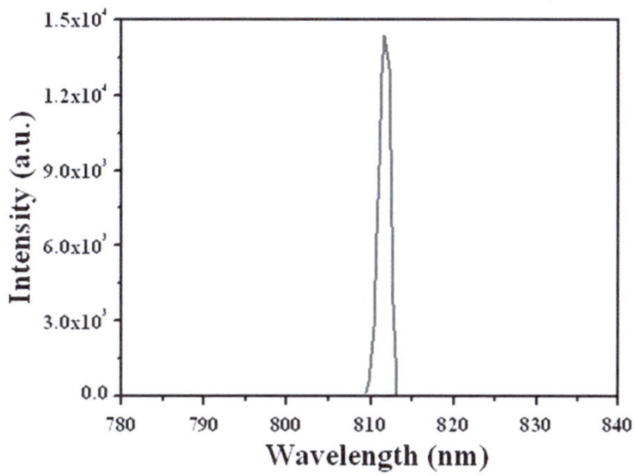

Fig. 8.34 A typical spectrum of a good diode laser bar [17]

spectral imaging can display the wavelength distribution of the emitters vividly and effectively. It can also indicate the relative intensity distribution among the emitters in the laser bar.

Spatial spectrum is a very useful tool in studying the characteristics of a diode laser bar, especially on packaging-related thermal and thermal stress effects. It is also effective in studying the mechanism of spectral broadening in a diode laser bar. The spectral broadening of laser bars is a result of nonuniform emitting wavelength from individual emitters. The spectral broadening of a laser bar as a pumping source significantly affects compactness, efficiency, power, beam quality, and thermal management of a laser system such as diode pumped solid state lasers and fiber lasers. All the pumping source energy out of the absorption bandwidth of the active laser media is wasted and becomes heat. Figure 8.34 shows a typical spectrum of a good diode laser bar. The spectrum is in Gaussian shape. The possible spectral broadening includes a shoulder, tail or secondary peak on the right side or left side or both sides the spectrum, or double or multiple peaks, which are demonstrated in Fig. 8.35a–d, respectively [17].

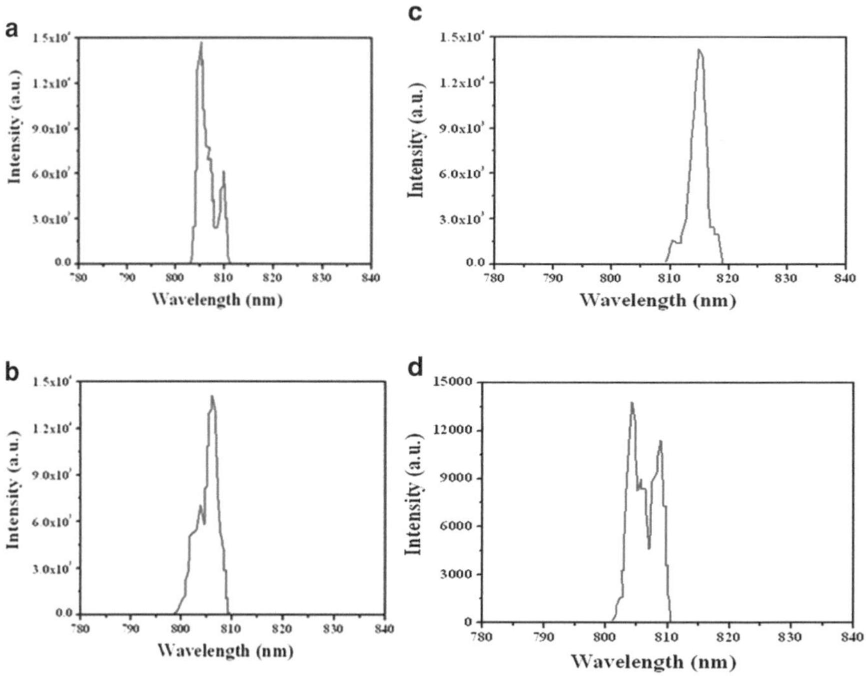

Fig. 8.35 Typical examples of broadened spectra of diode laser bars; (**a**) right shoulder; (**b**) left shoulder; (**c**) secondary peak on the both sides; (**d**) double peaks [17]

Figure 8.36 shows the spatial spectral images of the laser bar samples whose spectra were shown in Figs. 8.34 and 8.35. The red-dotted line is the main peak wavelength location in their respective spectrum shown in Figs. 8.34 and 8.35. Figure 8.36a shows the spatial spectral image of the good diode laser bar shown in Fig. 8.34, which has a narrow and Gaussian-shaped spectrum. As expected, the spatial spectral images of the emitters are pretty linear which indicates that the wavelength of the individual emitters in the laser bar is pretty uniform. Figure 8.36b–e shows the spatial spectral images of the laser bar samples with broadened spectra shown in Fig. 8.35a–d, respectively. It can be seen from Fig. 8.36b that the five emitters on the right side in the image have relatively longer wavelengths than majority of the emitters in the array. The light emission from these longer wavelength emitters forms a separate peak from that of the rest of the emitters corresponding to the secondary peak on the right side of the spectrum in Fig. 8.35a. In the same manner, the light emission from the emitters at the center of the image in Fig. 8.36c creates the shoulder on the left side of the spectrum in Fig. 8.35b. Similarly, from the images in Fig. 8.36d, e, one can find out which emitters are responsible for the tails on both left and right sides of the spectrum and the double peaks shown in Fig. 8.35c, d, respectively.

Figure 8.37 shows the overlap of spatial spectral images with the SAM images obtained from the same samples shown in Fig. 8.36 [17]. When the solder interface

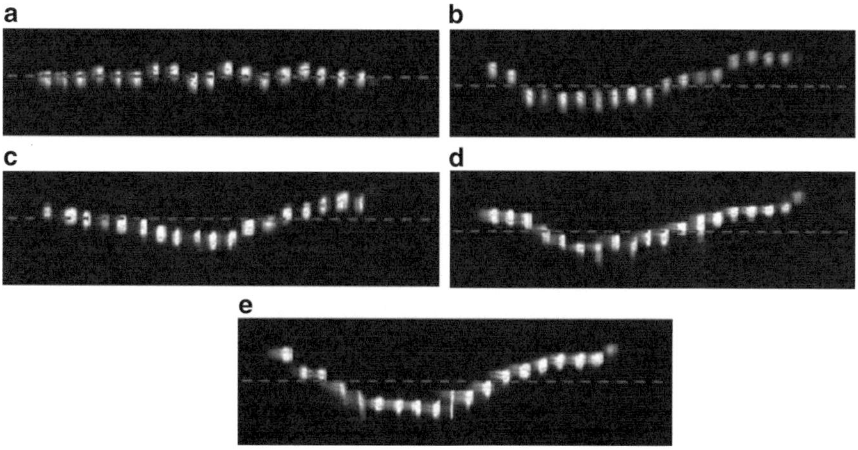

Fig. 8.36 Spatial spectral imaging of the diode laser bar samples with different spectral shapes; (**a**) good spectrum as shown in Fig. 8.34; (**b**) spectrum having a secondary peak/shoulder on the right side as shown in Fig. 8.35(a); (**c**) spectrum having a secondary peak/shoulder on the left side as shown in Fig. 8.35(b); (**d**) spectrum having tails on both left and right sides as shown in Fig. 8.35 (c); (**e**) spectrum having double peaks as shown in Fig. 8.35(d) [17]

is uniform and free of large voids, the wavelength of the individual emitters are uniform and thus the laser array has narrow and Gaussian-shaped spectrum. The spatial spectral imaging is linear as shown in Fig. 8.37a. From Fig. 8.37b and scanning electron microscope (SEM) results, it can be seen that the first five emitters on the right side of the array have large solder voids and correspondingly these five emitters have longer wavelength [17]. To verify the white-colored regions in the SAM images have solder voids, thc samples were cross-sectioned and examined using SEM. Figure 8.38 is the SEM image of the cross-section of the solder interface under the second right emitter in Fig. 8.37b, where the SAM image is the brightest which indicates large and deep solder voids [17]. Same explanation can be applied to the sample in Fig. 8.37e, only that the number of emitter having solder voids is greater than the sample in Fig. 8.37b.

The color at the two ends of the SAM image in Fig. 8.37c is lighter than the central region. It indicates that the two ends either have shallow solder voids or different acoustic impedance from the central region. The SEM results showed that the two ends do not have significant voids, but have thicker solder interface than the central region. Figure 8.39a, b is typical SEM images of the cross-section of the solder interface under the emitters in the light-colored and dark-colored regions in Fig. 8.37c [17]. It can be seen that the solder interface in the light-colored region has no significant voids and it is about 4–5 μm thick. Also, in Fig. 8.37c, the emitters at the lighter-colored region in the SAM image have longer wavelength than those at the dark region in the SAM image although SAM imaging and SEM results showed there were no significant solder voids in under the emitters at the lighter-colored region. Therefore, the spectral broadening of

a

b

c

d

e

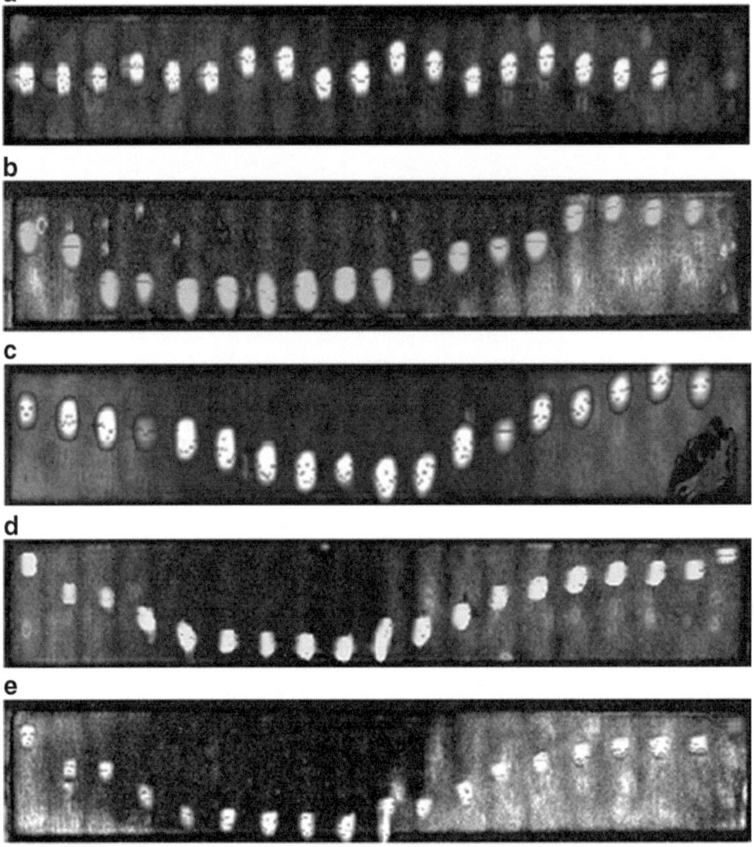

Fig. 8.37 Overlap of spatial spectral images with the scanning acoustic microscopy; (**a**) a typical spectrum of a good laser diode array; (**b**) spectrum having a secondary peak/shoulder on the right side as shown in Fig. 8.35(b); (**c**) spectrum having a secondary peak/shoulder on the left side as shown in Fig. 8.35(c); (**d**) spectrum having tails on both left and right sides as shown in Fig. 8.35 (d); (**e**) spectrum having double peaks as shown in Fig. 8.35(e) [17]

this sample could not be explained by thermal effect. The fact that the spectral broadening occurred at the short wavelength direction and it is known that tensile or compressive stress in the epitaxial material of a laser affects the emitting wavelength with a coefficient on the order of ~$1*10^{-5}$ eV/bar (or ~0.005 nm/bar), with tensile stress causing red-shift and compressive stress having blue-shift, the spectral broadening of this type can be attributed to thermal stress effect. Thermal stress is an inherent problem with the use of copper heat sink since copper has much larger coefficient of thermal expansion (CTE) than the laser bar which is essentially made of GaAs material. The SAM result shown in Fig. 8.37c and SEM results shown in Fig. 8.39 revealed that the two ends have different acoustic impedance from the central region and the two ends have thicker solder interface

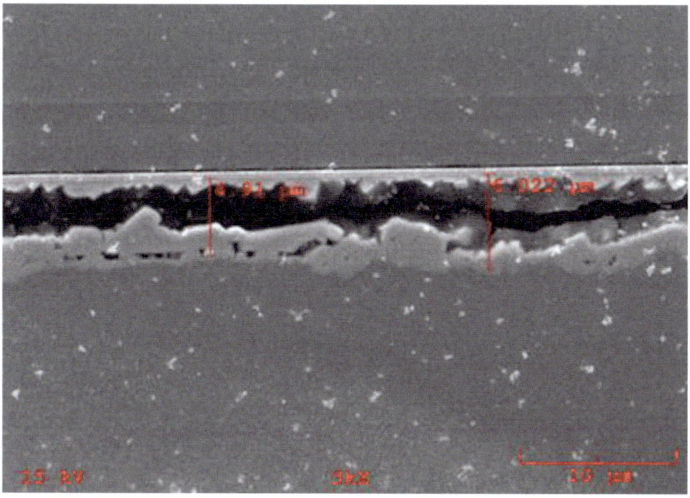

Fig. 8.38 SEM image of the cross-section of the solder interface under the second right emitter in Fig. 8.37(b) [17]

than the central region. The central region no longer has pure indium left as all the indium was consumed during the soldering process to form intermetallics with the Cu heat sink which has Au finish and the metallization on the laser bar while there remains a layer of pure indium although the indium solder forms intermetallics with the Cu heat sink and the metallization on the laser bar on the lower and upper interfaces. Material analysis results confirmed that the porous layer in the solder interface under the emitters at the ends was pure indium material while the indium solder was completely consumed to during soldering process form AuIn intermetallics which become a hard material and can no longer relieve the thermal stress [17]. The compressive thermal stress can be easily transferred to the laser bar when there is no soft indium material left which leads to blue-shift of wavelength at the central region. For Fig. 8.37d, the thermal effect caused by void and thermal stress are both included in the sample, which generated the spectrum having tails on both left and right sides [17].

8.4 Junction Temperature

The junction temperature is defined as the temperature at the center of the active region in a semiconductor laser when it operates at a given input current. It is the maximum temperature inside a laser. The junction temperature is mainly caused by the waste heat of the high power semiconductor laser. Diode laser junction temperature T_j can be calculated by the following relationship:

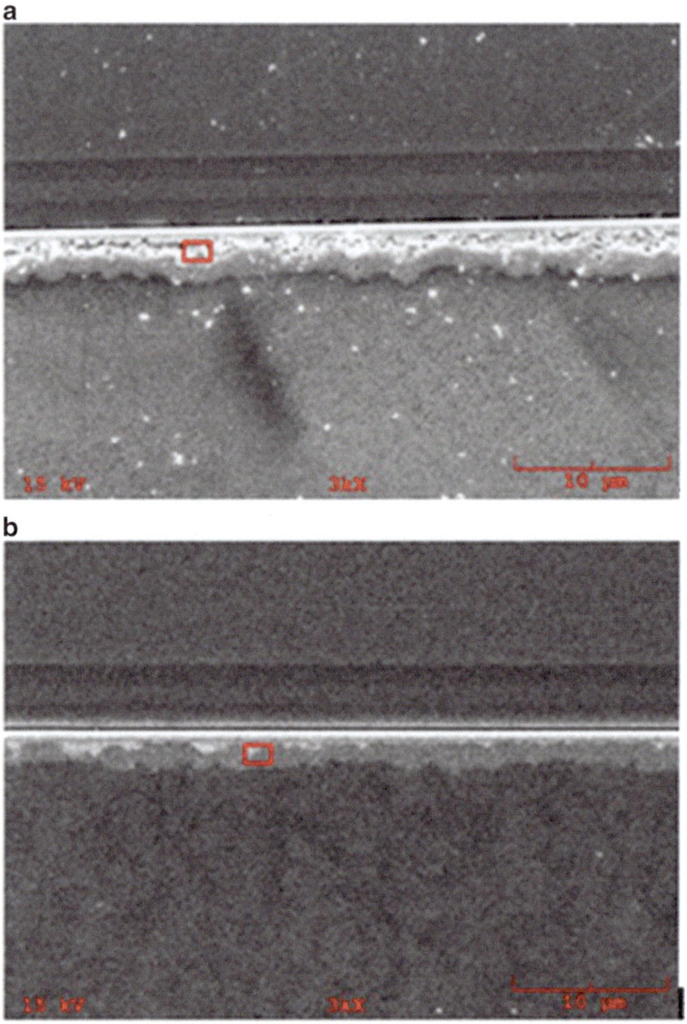

Fig. 8.39 Typical SEM images of the cross-section of the solder interface under the emitters in the light-colored and dark-colored regions in Fig. 8.37(c); (**a**) corresponds to light-colored region and (**b**) corresponds to dark-colored region [17]

$$T_j = T_{heat\sin k} + \Delta T = T_{heat\sin k} + R_{th} \times Q_{waste-heat} \tag{8.25}$$

where $T_{heat\ \sin\ k}$ is the heat sink temperature, ΔT is junction temperature rise, R_{th} is the thermal resistance, $Q_{waste\ -\ heat}$ is the waste heat generated in the laser junction. Experimentally, the junction temperature T_j can be measured through forward voltage method (FVM), power output method (POM), and wavelength shift method (WSM). Generally, these methods are based on a change in the measured physical

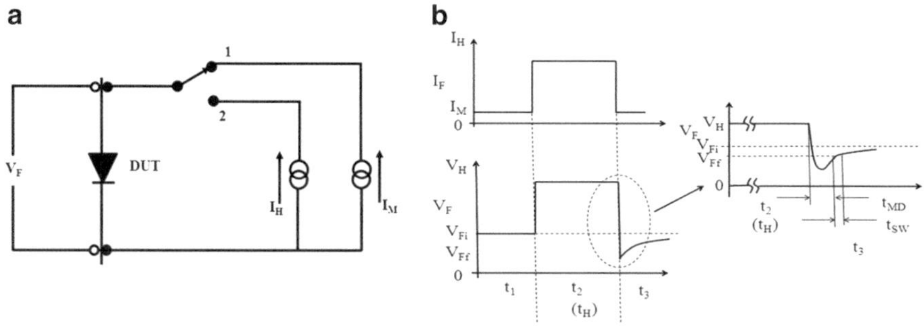

Fig. 8.40 (**a**) Junction temperature change testing setup for diode laser (*DUT* device-under-test, V_F forward voltage, I_H heating current, I_M measurement current); (**b**) principle of the forward voltage method (V_{Fi} initial V_F value before application of heating power, V_{Ff} final V_F value after application of heating power, I_H the current applied to the DUT during the heating time in order to cause power dissipation, V_H the heating voltage resulting from the application of I_H to the DUT, t_H the duration of V_H applied to the DUT, I_M the measurement current used to forward bias the temperature sensing diode junction for measurement of V_F, t_{SW} sample window time during, t_{MD} measurement delay time) [19–21]

property between pulsed and continuous wave operation of the laser diode. When operated with short pulses and low duty cycle, there is essentially no heating in the semiconductor junction and the temperature of the junction is equal to that of the heat sink that the packaged laser is mounted to [18]. Based on the description above, the junction temperature at certain current can be obtained by the change in optical output power, forward voltage, or wavelength.

8.4.1 Measurement Method and Equipment

The three junction temperature measurement methods including (1) FVM, (2) POM, and (3) WSM are introduced in this section.

Forward Voltage Method

The FVM, which is applied to measure the junction temperature change of a diode laser, uses a three-step sequence of applied current levels to determine a change in junction voltage (ΔV_F) under measurement current (I_M) conditions [19–21]. The setup for the measurement is shown in Fig. 8.40a [19–21]. The test begins with the adjustment of I_M and I_H to the desired values. I_M is set to a small value that does not generate self-heating within the diode laser junction region while it is large enough for measurement accuracy and repeatability. The value of heating current (I_H) is usually at least 50 times greater than the value of I_M. It normally is the operating current of the diode laser. When operating junction temperature is being measurement the I_H heat current is the operating current. First, I_M is applied and the junction voltage of the diode laser is measured. The measurement value is referred to as V_{Fi}. Second, I_M is replaced with the desired current I_H for a time duration (t_H) consistent with the steady-state or a transient condition required. During this time the voltage (V_H) of diode laser is measured for determining the amount of heat power (P_H) being generated in the diode laser. Finally, at the conclusion of t_H, the switch is again moved to position 1 and the current is changed from I_H to I_M. The forward voltage is quickly measured at the measurement current I_M within a time period defined by sampling time window t_{SW}. This voltage is refered to as V_{Ff}. The value of t_{SW} should be small. In theory, the sampling time window should be as close to the termination of the heating current as possible. However, due to inductive effects of the measurement setup which would cause voltage overshoot and oscillation, V_{Ff} should be measured after a measurement delay time t_{MD}. Measurement delay time is defined as the time from the start of heating power (P_H) removal to the start of the V_{Ff} measurement time. The two current sources are then turned off at the completion of the test. The three-step operation is shown graphically in Fig. 8.40b [19].

Once this three-step measurement process has been completed and the appropriate data collected, the next step is to use the data to compute ΔT_J:

$$\Delta T_J = K \times \Delta V_F \tag{8.26}$$

$$\Delta V_F = \left| V_{Fi} - V_{Ff} \right| \tag{8.27}$$

where K is correlation factor for the change in forward voltage and the corresponding change in junction temperature; ΔT_J and ΔV_F are the change of junction temperature and forward voltage, respectively.

The junction temperature of laser device can be derived by following equation:

$$T_j = K \times \Delta V_F + T_A \tag{8.28}$$

Power Output Method

The light power output of a semiconductor laser is linearly proportional to its junction temperature T_j in a certain range. Thus ΔT_j can be derived by the following equation [21]:

$$\Delta T_j = (P_o - P_{oi})/K_p \tag{8.29}$$

where K_p is the power-temperature coefficient of high power semiconductor laser. Bring above equation into Eq. (8.25), then T_j can be calculated by

$$T_j = T_A + (P_o - P_{oi})/K_p \tag{8.30}$$

P_{oi} is the output power operating with short pulses and low duty cycle; P_o is the output power at measured current.

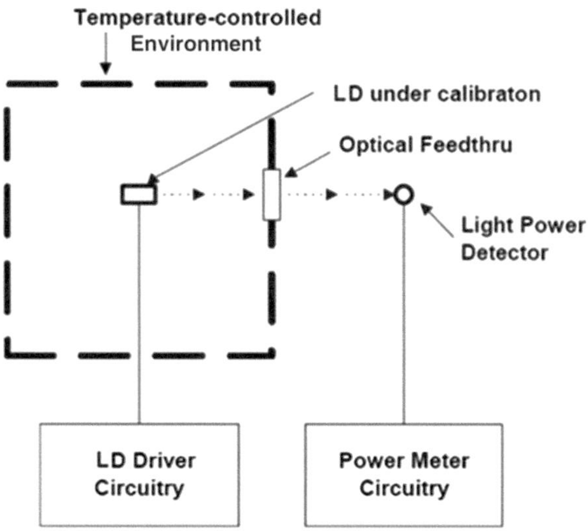

Fig. 8.41 The junction temperature measurement setup for power output method [20]

Implementing this method requires a measurement setup as shown in Fig. 8.41 [20]. The laser diode (LD) is contained in a temperature controlled environment and pulsed with a very short current pulse at a very low duty cycle to minimize self-heating. Under these conditions the junction temperature can reasonably be assumed to equal the environment temperature. The output power is measured each time when the environment temperature is changed and allowed to reach a steady-state condition. The data can then be plotted as P_o versus T_j and the value of K_p is determined from taking the reciprocal of the plotted line. After K_p is acquired, the junction temperature of laser device can be derived by Eq. (8.30).

Wavelength Shift Method

The other optical method makes use of the fact that the light output wavelength, primarily the center wavelength (λ_0), shifts linearly with device junction temperature with a positive slope. This method is particularly useful for narrow spectral laser devices. Similar to the previous method, a calibration procedure is necessary to obtain the wavelength–temperature coefficient K_λ. The value of K_λ is dependent on the diode construction. Once K_λ is known, ΔT_j can be derived by the following equation:

$$\Delta T_j = (\lambda_o - \lambda_{oi})/K_\lambda \tag{8.31}$$

Bring above equation to Eq. (8.25), then T_j can be calculated by

$$T_j = T_{heat\sin k} + (\lambda_o - \lambda_{oi})/K_\lambda \tag{8.32}$$

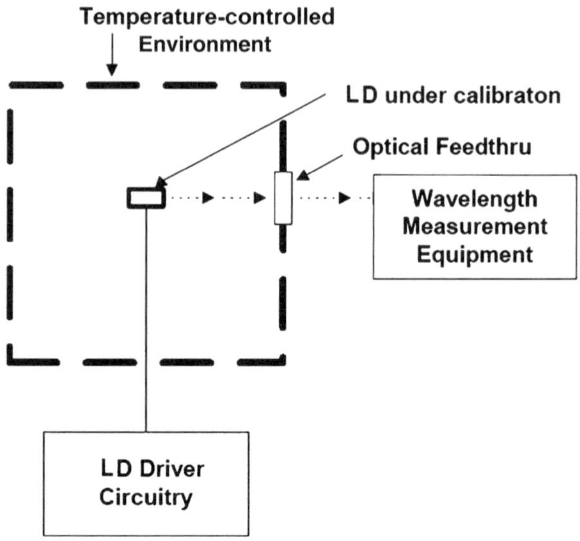

Fig. 8.42 The junction temperature measurement setup for wavelength shift method [20]

where λ_{0i} is the output wavelength operated under short pulses and low duty cycle at the temperature of $T_{heatsink}$; λ_0 is the output wavelength at measured current and condition.

The measurement setup and procedure for this method, shown in Fig. 8.42, is very similar to that for the POM [20]. The only difference is the light measurement instrumentation. The procedure of obtaining K_λ consists of applying short, low duty cycle pulses and measuring the wavelength as the environment temperature is changed. The measurement data is used to generate a plot of wavelength versus junction temperature from which the coefficient value of K_λ can be obtained. Junction temperature measurement is made by multiplying the K_λ by the wavelength shift from initial turn-on to the steady-state condition under operating conditions.

8.4.2 *Typical Testing Results and Analysis*

In this section, an example will be presented to measure the junction temperature by WSM. Operating with short pulses (100 μs) and low duty cycle (0.1 %) at a current just above threshold current, the wavelength of the MCC semiconductor laser is measured and the output wavelength is 803.5 nm as shown in Fig. 8.43 [1]. Under this testing conduction, no heat generation in chip is assumed and the junction temperature is equal to that of the heat sink (25 °C) that the packaged laser is mounted to. When the MCC semiconductor laser operates at 270 A with pulse width 200 μs and 400 Hz, the output wavelength is 807.5 nm. The wavelength–temperature coefficient of K_λ for the output wavelength between 790 and 825 nm is 0.28 nm/°C [22].

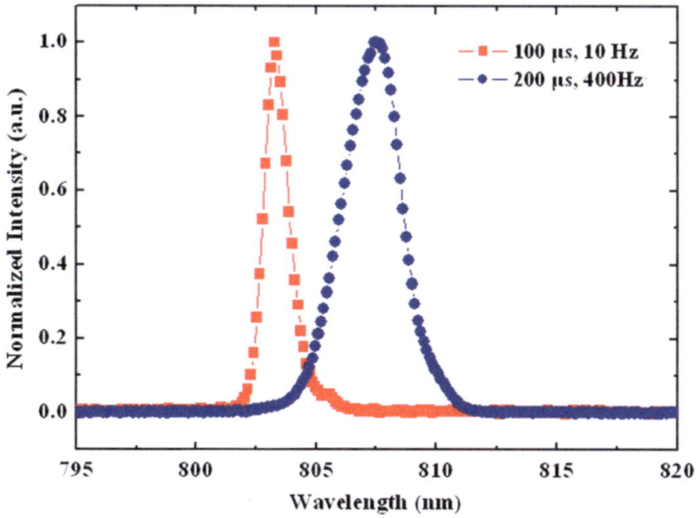

Fig. 8.43 Wavelength measurement under different pulse testing condition (*red square*: 100 μs, 10 Hz; *blue circle*: 200 μs, 400 Hz) [1]

According to Eq. (8.32), for the junction temperature of the MCC semiconductor laser operating at 270 A with pulse width 200 μs and 400 Hz, the junction temperature can be calculated as

$$T_j = 25°C + (807.5 - 803.5)/0.28 = 39.3°C \tag{8.33}$$

8.5 Thermal Resistance

Thermal resistance is a property of an object and a measurement of a temperature difference by which an object or material resists a heat flow (heat per time unit or thermal resistance) [23]. Thermal resistance is the reciprocal of thermal conductance. Thermal resistance has the units K/W. From Fourier's Law for heat conduction, the following equation can be derived, and it is valid as long as all of the parameters (x and k) are constant throughout the sample [23]:

$$R_{th} = \frac{x}{A^* k}, \tag{8.34}$$

where R_{th} is the absolute thermal resistance (across the length of the material) (K/W), x is the length of the material (measured on a path parallel to the heat flow) (m), k is the thermal conductivity of the material (W/(K · m)), A is the cross-sectional area (perpendicular to the path of heat flow) (m^2).

For high power semiconductor lasers, thermal resistance can be measured and calculated by the temperature difference across a heat-flow path divided by the heat generation that caused the temperature difference [21]. Thermal resistance of a high power semiconductor laser indicates the temperature difference between junction and atmosphere temperature when the device is operating. The thermal resistance evaluates the ability of the heat dissipation of a semiconductor laser.

Thermal resistance can be calculated as

$$R_{th} = \frac{\Delta T}{\Delta Q} \tag{8.35}$$

where R_{th} is the thermal resistance, ΔT is the junction temperature rise, ΔQ is heat generation of a semiconductor laser.

8.5.1 Measurement Method and Equipment

In order to calculate thermal resistance, ΔT and ΔQ should be obtained. The approaches to measure ΔT were introduced in Sect. 8.4 and they will not be repeated here. ΔQ can be obtained by the equation $\Delta Q = (IV - P_o)$, where I is the operating current and V is the operating voltage. Po is the output optical power of the device. All these three parameters of I, V, and Po can be directly measured, as discussed in Sect. 8.1.

In practice, thermal resistance can be more directly obtained by wavelength method and FVM. The principles are the same as measuring junction temperature rise ΔT. Wavelength method is based on the correlation between wavelength and temperature. FVM is based on the correlation between forward voltage and temperature.

According to the wavelength method, the thermal resistance of the formula (8.35) can be rewritten as

$$R_{th} = \Delta T/\Delta Q = \frac{1}{\Delta\lambda/\Delta T} \times \frac{\Delta\lambda}{\Delta Q} \tag{8.36}$$

where $\Delta\lambda/\Delta T$ is the coefficient of wavelength shift of the diode laser. Coefficient of wavelength shift of a particular diode laser is inherited from the epitaxy structure design. It can be theoretically calculated and experimentally measured. For 808 nm diode lasers, the generally accepted coefficient of wavelength shift is 0.28 nm/°C, while it is 0.32 nm/°C for 980 nm diode lasers. $\Delta\lambda/\Delta Q$ is the wavelength–heat generation power coefficient. It can be calculated from the curve fitting of measured wavelength corresponding to various heat generation powers.

According to the FVM, thermal resistance is described by [21]

$$R_{th} = \frac{\Delta T}{\Delta Q} = \frac{K \times \Delta V_F}{(IV - P_o)} \tag{8.37}$$

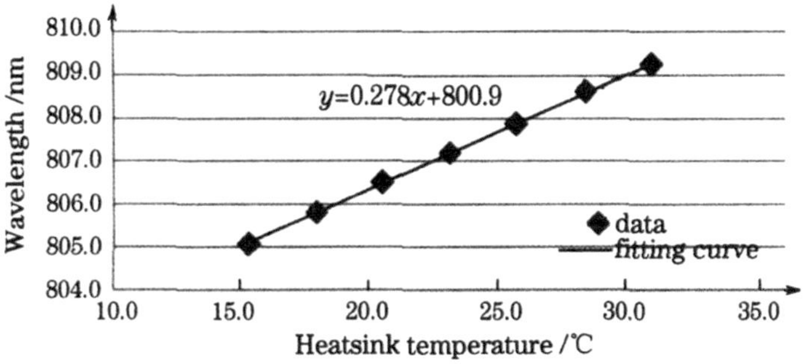

Fig. 8.44 Measured wavelength as a function of temperature [24]

generally I_M is much smaller than I_H, then $\Delta Q \approx IV - P$. In Eq. (8.37), K factor is defined as the reciprocal slope of the $V_f \sim T_j$ line (see Sect. 8.4). Again the K factor of a particular diode laser is inherited from the epitaxy structure design. It can be theoretically calculated and experimentally measured. ΔV_F is the change in forward junction voltage under measurement current (I_M) conditions before application of heating current and after application of heating current, which generates heating in the junction. It can be calculated from the measured initial forward voltage V_{Fi} and final forward voltage V_{Ff} by $\Delta V_F = V_{Fi} - V_{Ff}$. The heating current I_H, heating voltage V_H, the measurement current I_M, measurement voltage V_M and output optical power Po at heating current can be directly measured, as discussed in Sect. 8.1.

8.5.2 *Typical Testing Results and Analysis*

We illustrate the wavelength method of measuring thermal resistance by using a C-mount-packaged single emitter high power semiconductor laser as an example [24]. Figure 8.44 gives the measurement curve of wavelength versus temperature for the C-mount-packaged semiconductor laser. From this measurement curve, the wavelength–temperature coefficient $\Delta\lambda/\Delta T$ can be obtained as 0.278 nm/K [24]. For a specific type of diode laser, the wavelength–temperature coefficient $\Delta\lambda/\Delta T$ does not have to be measured since it is generally known.

Figure 8.45 shows the measured wavelength versus power dissipation (heat generation power) for the C-mount-packaged semiconductor laser. From this measurement curve, the wavelength–power dissipation coefficient $\Delta\lambda/\Delta Q$ can be calculated to be 1.09 nm/W [24]. According to the Eq. (8.36), the thermal resistance of the C-mount-packaged single emitter high power semiconductor laser is $1.09/0.278 = 3.92$ K/W.

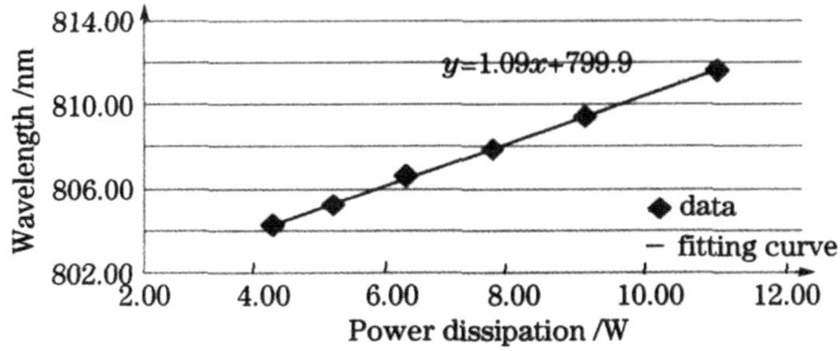

Fig. 8.45 Measured wavelength as a function of heat power dissipation [24]

8.6 Near Field

The near field of a laser beam is the region around the beam waist. For diode lasers, it is the region around the front facet. Near-field intensity is defined as the intensity distribution of the diode laser irradiation near diode laser front facet. The near-field intensity profile (irradiance profile) measurement is very important. Some diode laser beams may exhibit "hot spots", i.e., regions with higher intensity, which may play a role in catastrophic optical damage phenomena. There can also be satellite structures, halos, and other deviations from beam uniformity which are not desired for applications. The shape of the beam intensity profile may change during propagation, and is also not always stable over time. Generally, the near-field intensity measurement distance is several times of wavelength of the tested laser. Two measurement methods, including scanning near-field optical microscopy and direct imaging method, will be addressed in this section.

8.6.1 Scanning Near-Field Optical Microscope

Initially, the scanning near-field optical microscope (SNOM) is designed for measuring the non-radiating material [25, 26]. In order to measure the near-field light emission of semiconductor lasers, SNOM needs to have some important changes. The one change is to remove the illuminating light source of He–Ne laser. A reflective microscopic system was designed and added for the rough adjustment of the distance between the diode facet and the fiber tip. The other is that the original transmitive objectives are not suitable for positioning the fiber tip at the facet of the laser diode, and the signal feedback system of shear force is added to fine adjustment [25]. The principle of modified SNOM is shown in Fig. 8.46 and the changed parts are marked by dashed box in the figure [1, 25].

The distance between the fiber tip and facet of the diode laser is roughly controlled by the rough adjustment system by movement of the platform marked below in Fig. 8.46. The charge-coupled device (CCD) and computer in rough

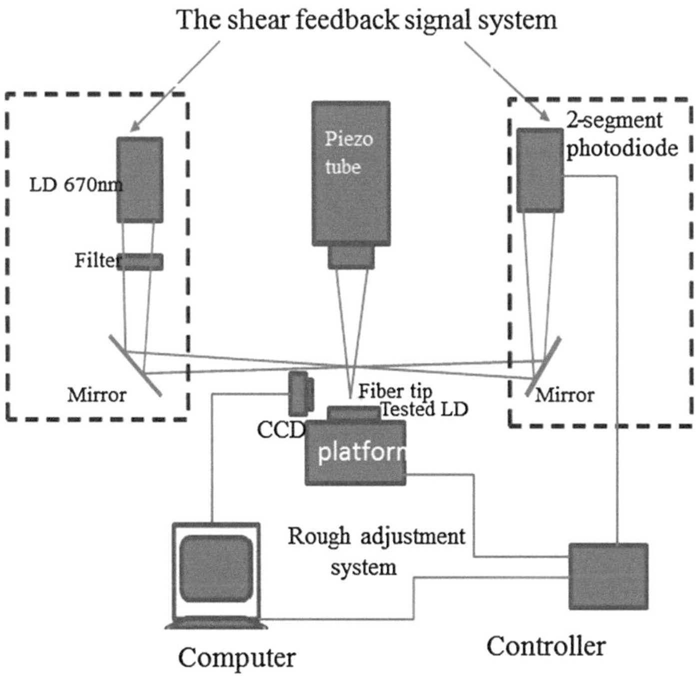

Fig. 8.46 The measurement setup of modified SNOM for semiconductor laser near-field intensity measurement [1, 25]

adjustment system are used to observe the distance of fiber tip and LD. The controller is used to adjust the distance. The separation distance between fiber tip and diode laser can be controlled at several micrometers by the rough adjustment system.

After rough adjustment, the distance should be fine adjusted by the shear feedback signal system. As shown in the figure, the detection of the motion of the excited tip is achieved by illuminating the fiber tip from one side with a focused laser beam of 670 nm and observing the movement of the diffraction pattern on the opposite side with a photodiode placed some centimeters away [25]. By controlling the shear force, the fiber tip can be driven to the position just 20 nm above the facet of laser diode.

After distance adjustment, the fiber tip is directly illuminated by the laser beam emitted from the semiconductor laser, and it collects the beam intensity and transfers into the piezo tube.

8.6.2 Direct Imaging Method

Near-field intensity distribution can also be tested by a near field microscope. This method is more effective in measuring the relative intensities of individual emitters in a semiconductor laser bar. The test system has four parts as shown in

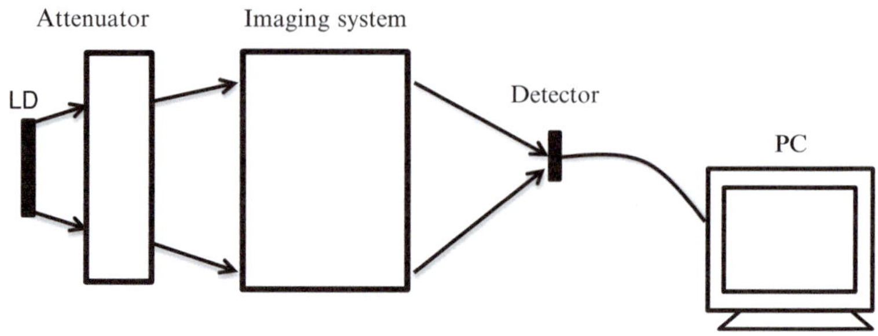

Fig. 8.47 Schematic diagram of near-field intensity profile measurement by direct imaging method [1]

Fig. 8.48 The near-field beam of a laser bar [1]

Fig. 8.47 [1]. The beam generated from the active region transmits through the attenuator to the imaging system. The attenuator is used to reduce the intensity of the laser. A CCD is placed behind the imaging system to detect the intensity distribution. The PC is used for analyzing and presenting measurement results.

8.6.3 Typical Testing Results and Analysis

The direct imaging method is used to measure and screen the near-field intensity behavior of high power diode lasers. Based on this method, the beam relative intensity is received and recorded by the PC, and the beam intensity of each emitter of a laser bar is recorded and presented. It is know, that when the beam intensity of the emitters in a diode laser bar is non-uniform, the emitters are more likely to fail if the intensity is too strong or too weak. The near field intensity testing is important finding out the satisfied laser beam, especially for pumping applications.

The near field patterns of semiconductor laser bars in CS package shown in Fig. 8.48 are illustrated and analyzed [1]. The relative intensity of each emitters of a

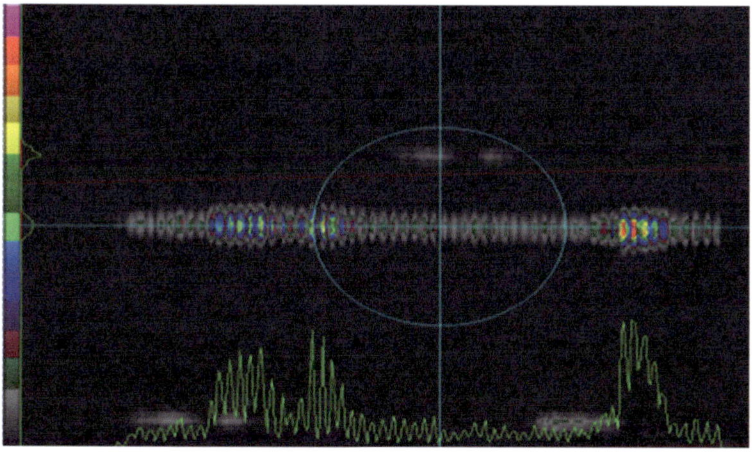

Fig. 8.49 Example of a near field pattern and profile [1]

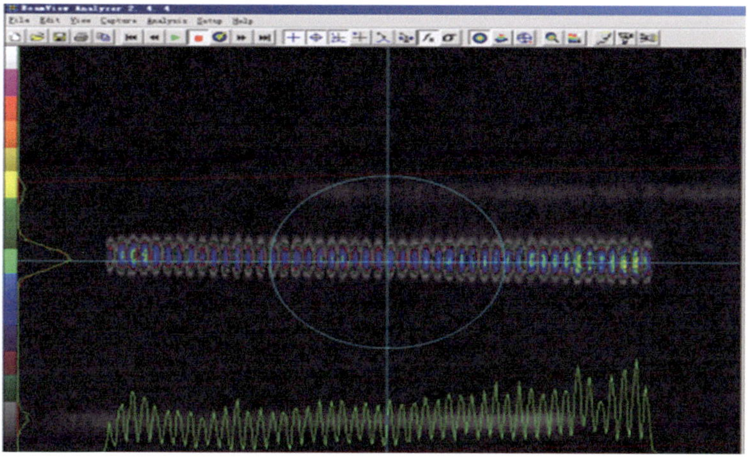

Fig. 8.50 Example of a near field pattern and profile of a diode laser bar [1]

laser bar can be obtained by near field test, and the beam uniformity of a laser bar is analyzed and presented. Two typical test results are shown in Fig. 8.49 and Fig. 8.50, respectively.

Figure 8.49 shows the near-field test result of CS-packaged single bar high power semiconductor laser with poor near field pattern [1]. The image exhibits the near field pattern of every emitter in the laser bar and the profile gives the relative intensity distribution quantitatively. It can be seen that the intensity of the emitters in the laser bar is not uniform. The intensity in center region and the two ends are relatively low while the intensity at left-middle and right-middle region has peaks. During operation, the emitters with high intensity peaks would be more vulnerable than the ones with low intensity due to COMD and other failures. From the near field pattern and especially the intensity profile, it is obvious to find the difference among the emitters and it is a good approach to evaluate the property of multi-emitter laser bar products, screen out the fragile products failure analysis.

Figure 8.50 shows the near-field test result of CS-packaged single bar high power semiconductor laser good near field pattern [1]. It can be seen from both the image and the intensity profile that the intensity distribution of the emitters in this diode laser bar is more uniform. During field application, the output power from each emitter over time is more likely to be uniform and the "burden" of optical power and heat is shared more equally by the emitters and therefore the reliability of this device would be better than the one shown in Fig. 8.49.

8.7 Far Field

Far field indicates the radiation field of a laser at a distance which is much greater than the Rayleigh length. For edge emitting semiconductor lasers, far-field divergence angle and intensity distribution are important parameters which are used to describe the characteristics of far-field property.

The fast axis is the direction perpendicular to the p–n junction plane of semiconductor laser, and the slow axis is the direction parallel to the p–n junction plane of semiconductor laser. The fast axis divergence angle with the symbol of $\theta_\perp$ is the maximum value of the divergence angle in the fast axis direction. Similarly, the slow axis divergence angle with the symbol of $\theta_{||}$ is the maximum value of the divergence angle in the slow axis direction, as shown in Fig. 8.51 [1].

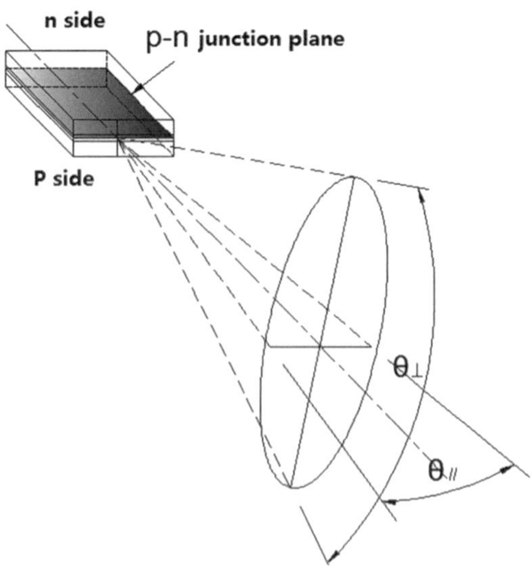

Fig. 8.51 Schematic diagram of fast and slow axis divergence angle [1]

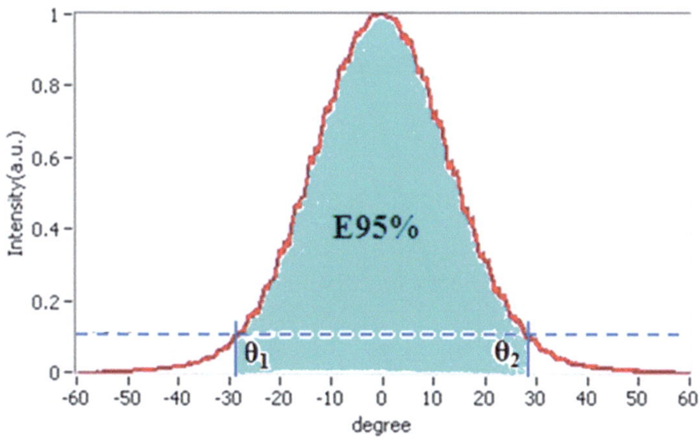

Fig. 8.52 Definition of divergence angles of full width at FW95%E [1]

Three types of the divergence angles are defined and calculated: (1) full width at half maximum intensity; (2) full width at 95 % energy (FW95%E); (3) full width at $1/e^2$ intensity.

The divergence angles of FW95%E are defined as the difference between the angles in which the energy is 95 % of the whole energy. Figure 8.52 shows schematic diagram of FW95%E [1]. In the figure, the energy marked by shade area is 95 % of the whole energy, and the difference of the two corresponding angles of θ_1 and θ_2 are the full angles of 95 % energy. The FW95%E can be calculated by

$$\theta_{95\%} = \theta_2 - \theta_1. \tag{8.38}$$

where $\theta_{95\%}$ is the divergence angle of full width at 95 % energy, θ_1 and θ_2 are the minimum and maximum angles of the shade area in the figure.

Divergence angles of FWHM or $1/e^2$ of maximum intensity are defined as the difference between the angels for which intensity is half or $1/e^2$ of its peak value. Figure 8.53 shows schematic diagram of FWHM and full width at $1/e^2$ intensity [1, 13].

The divergence angles can be calculated by

$$\theta_{1/2} = \theta_2 - \theta_1 \tag{8.39}$$

$$\theta_{1/e^2} = \theta_4 - \theta_3 \tag{8.40}$$

where $\theta_{1/2}$ and θ_{1/e^2} are divergence angles of full width at half maximum and full width at $1/e^2$ intensity, θ_1, θ_2, θ_3 and θ_4 are the minimize and maximum angles at which the corresponding intensity are 50 % and $1/e^2$ maximum intensity, as shown in the figure.

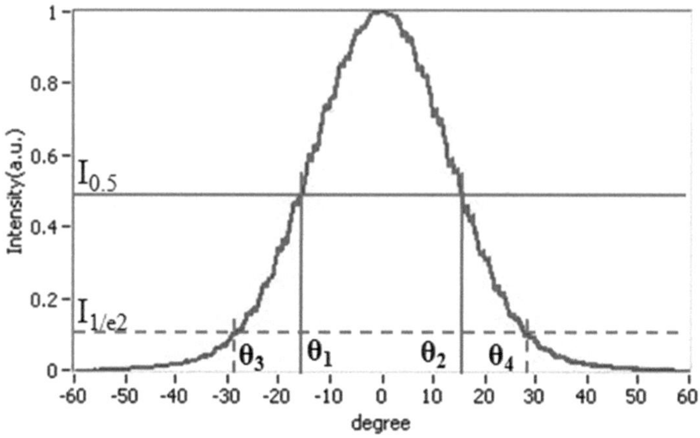

Fig. 8.53 Definition of divergence angles of FWHM and full width at $1/e^2$ intensity [1]

Fig. 8.54 The schematic position of the pinhole scanning method [1]

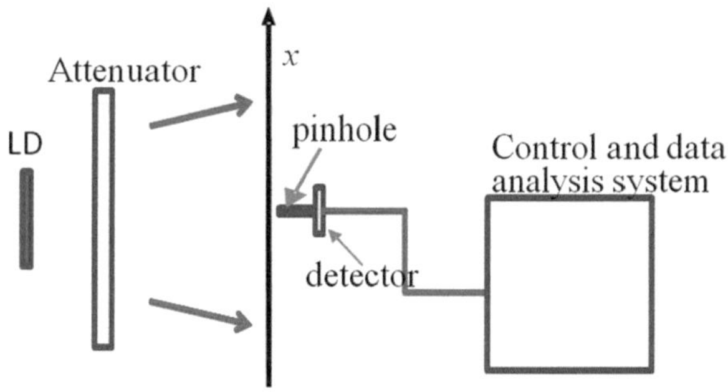

8.7.1 Measurement Method and Equipment

There are several methods to measure the far-field intensity, such as knife-edge test, CCD optical imaging, and optical focusing methods [27]. However, due to the large divergence angles and high power densities, some measurement methods are difficult to use for high power semiconductor lasers. Some other methods, such as pinhole scanning and avalanche diode methods, have been proposed in testing far field of diode lasers [28–31]. The pinhole scanning method is commonly used to test the far field of high power semiconductor lasers [28, 29].

The pinhole scanning method has five components: (1) light source, (2) attenuator, (3) pinhole with a moving stage, (4) detector behind the pinhole, (5) control and data analysis system. The arrangement of these five components is shown in Fig. 8.54 [1]. The detector is used to receive and transform the light emitting intensity information. It is placed behind the pinhole. Generally, the diameter of the pinhole is very small within 0.5 mm or preferably within 0.2 mm. The pinhole is fixed on a moving frame and is able to scan along fast and slow axis. If light power intensity is too strong, an attenuator is needed to be placed between the diode laser and the pinhole to decrease the light intensity and prevent the detector from saturation.

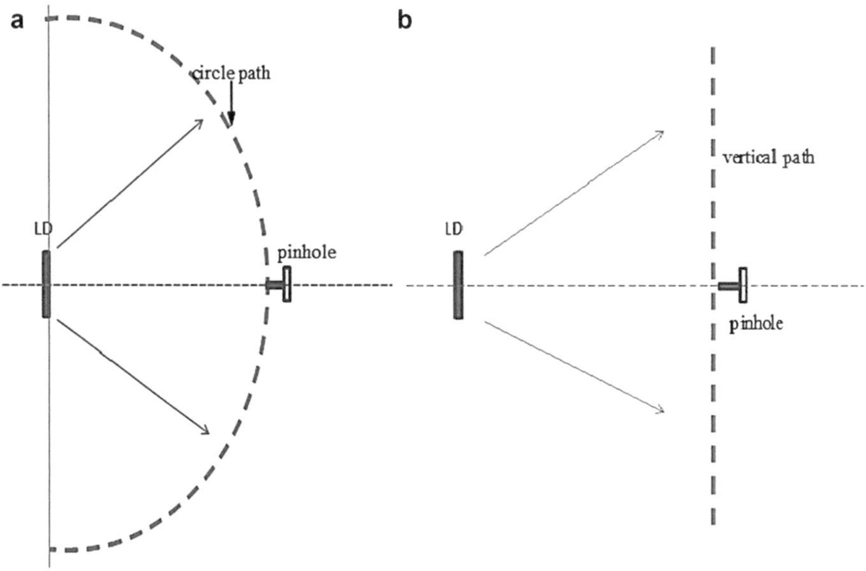

Fig. 8.55 The schematic diagram of two movement paths (**a**) angular path and (**b**) planar path [1]

During measurement, the pinhole and the detector is mounted on a moving stage, and move the pinhole and the detector around to generate the far-field pattern. The moving stage could have two movement paths and two options, which are angular path and planar path as shown in Fig. 8.55, respectively [1]. The movement path in Fig. 8.55a is an angular path around the LD being tested which is a semicircle path denoted by a dotted line, and the radius of the semicircle is the test distance from the LD active region to the pinhole and detector. The irradiance pattern and intensity distribution along the pinhole movement path is measured and recorded, typically in the fast axis (vertical) and slow axis (horizontal) directions. The far field divergence angles can be obtained from the intensity curves directly.

Figure 8.55b shows the planar path, and the test distance from the LD active region to the pinhole and detector is z. The pinhole and detector scans the intensity of the beam in the plane, typically along the vertical and horizontal axis in the plane, and the far-field divergence angles can be calculated.

A measurement system based on angular movement path shown in Fig. 8.55a can be designed as shown in Fig. 8.56 [1]. For both the fast axis and slow axis, the centers of the active region of a diode laser being tested and the pinhole with detector should be coaxial. In Fig. 8.56, two pinholes and detectors fixed on the slow and fast axis movement frames and the movement of the frames are controlled by the PC system. The test distances for the fast and slow axis are a constant value with R = 150 mm. The far-field intensity distributions and divergence angles along fast and slow axis can be obtained and the PC system will analyze the measurement data and give a detailed report.

Fig. 8.56 An example of a far-field testing equipment of high power semiconductor laser [1]

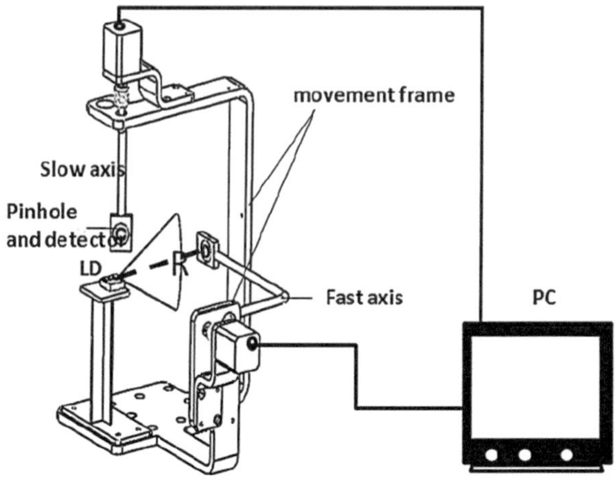

Figure 8.57 shows an example of a far-field intensity distributions along fast and slow axes of a CW 50 W CS-packaged single bar high power semiconductor laser with 19 emitters [1]. The divergence angles of full width 95 % energy $\theta_{95\%}$ can be calculated. When $\theta_{95\%}$ is E_x and E_y are equal to 95 % of whole energy, respectively. The equations are:

$$E_{\text{total}} = \int_{\theta_{i3}}^{\theta_{i4}} I(\theta)d\theta, \tag{8.41}$$

$$\int_{\theta_{i1}}^{\theta_{i2}} I(\theta)d\theta = 95\% E_{total}. \tag{8.42}$$

where, *i* in the equations denotes x and y, representing fast axis and slow axis. When calculation of $\theta_{x95\%}$ is calculated, i is x, and when slow axis $\theta_{y95\%}$ is calculated *i* is y. As we know, the intensity distribution as a function of angle $I(\theta)$ is obtained by measurement. As $I(\theta)$ and θ are discrete values, θ_{x1}, θ_{x2}, θ_{y1}, and θ_{y2} can be numerically calculated from Eqs. 8.41 to 8.42. Based on Eq. (8.38), one can obtaine $\theta_{x95\%} = \theta_{x2} - \theta_{x1} = 51.19^o$ and $\theta_{y95\%} = \theta_{y2} - \theta_{y1} = 9.84^o$.

Fig. 8.58 shows an example of the calculation of divergence angles of, full width half maximum intensity $I_{50\%}$ and full width $1/e^2$ intensity I_{1/e^2}. θ_{i1}, θ_{i2}, θ_{i3}, and θ_{i4} which makes the intensity $I(\theta)$ equaling to $I_{50\%}$ and I_{1/e^2} along fast and slow axes, are marked, respectively. When the divergence angles along fast axis is calculated, i is x, and when the divergence angles along slow axis is calculated, i is y. Based on Eqs. (8.39) and (8.40), full width half maximum intensity divergence angle is $\theta_{x1/2} = \theta_{x2} - \theta_{x1} = 30.47^o$, $\theta_{y1/2} = \theta_{y2} - \theta_{y1} = 6.66^o$, and full width $1/e^2$ intensity divergence angle is $\theta_{x1/e^2} = \theta_{x4} - \theta_{x3} = 53.44^o$, and $\theta_{y1/e^2} = \theta_{y4} - \theta_{y3} = 9.28^o$.

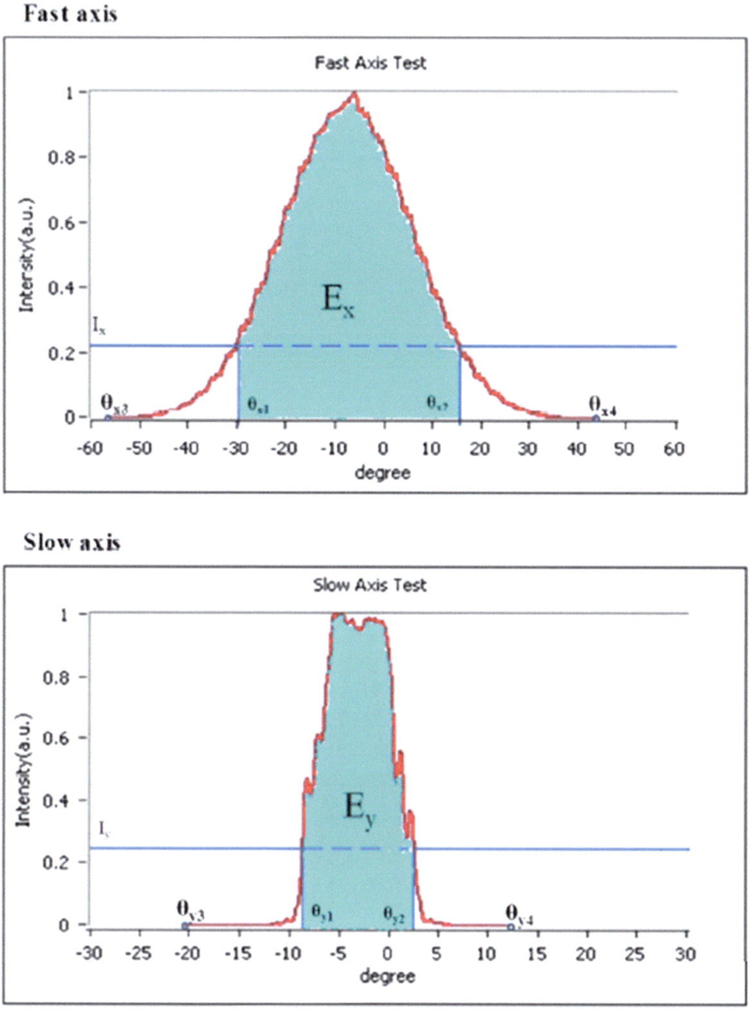

Fig. 8.57 An example of far-field testing and $\theta_{95\%}$ calculation [1]

8.7.2 *Far-Field Testing Data Analysis*

The far field characterizes not only the divergence angles, but also the beam quality of high power broad area diode lasers, in which far field beam shape and profile could cause facet damage and reliability concerns. Here far-field analysis on a 980 nm single emitter broad area lasers have been investigated under CW and pulsed current conditions. Figure 8.59 shows the typical far-field patterns of a 980 nm single emitter diode laser package under CW operation from 1 to 12 A [32]. Figure 8.60 shows the

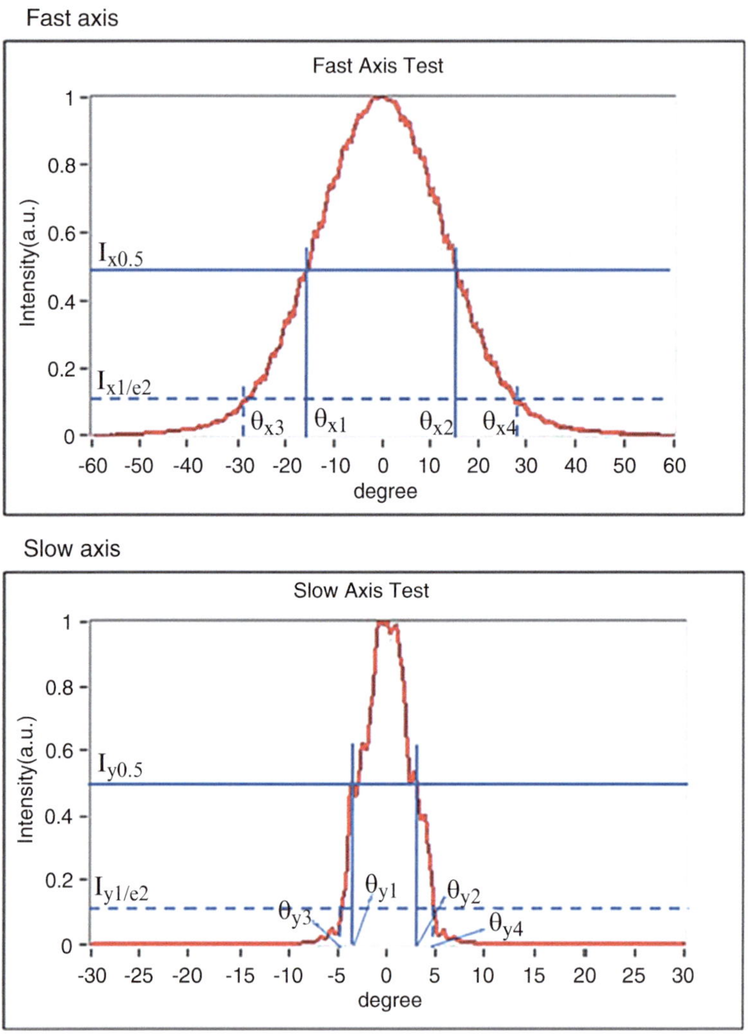

Fig. 8.58 An example of far-field testing and $\theta_{1/2}$ and θ_{1/e^2} calculation [1]

far-field patterns of the same device under pulsed operation condition with pulse width of 1 μs and frequency of 2 kHz from 1 to 20 A [32].

The fast axis far-field patterns of CW and pulsed operation are similar. For both the CW operation results shown in Fig. 8.59 and the pulsed operation results shown in Fig. 8.60, the intensity profiles in the fast axis almost have the same distribution profiles as a function of injection current, that is, the far field patter in the fast axis does not change with injection current.

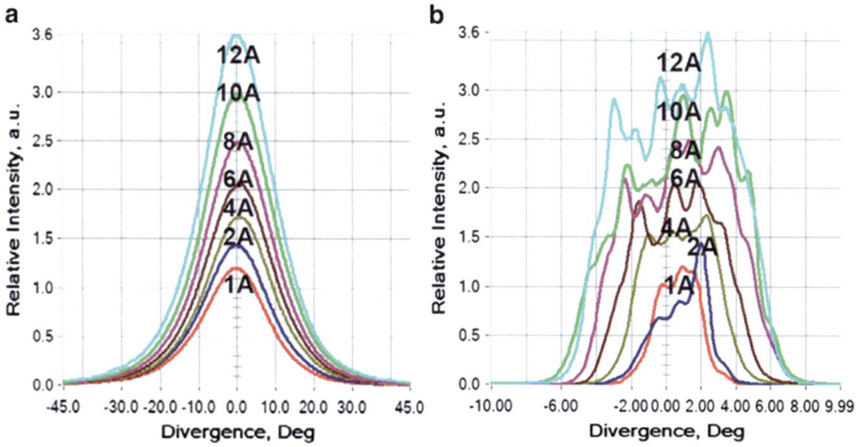

Fig. 8.59 Typical far-field patterns of a 980 nm single emitter diode laser package measured under CW condition; (**a**) fast axis and (**b**) slow axis [32]

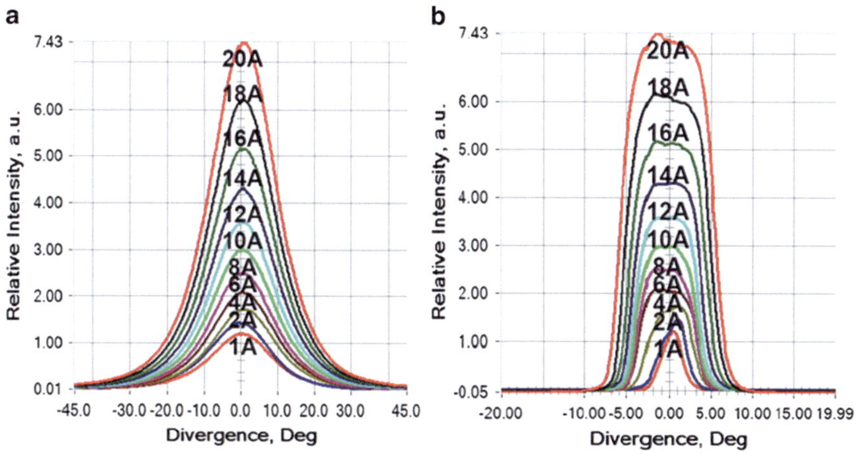

Fig. 8.60 The far-field patterns of the same device shown in Fig. 8.59 measured under pulsed current condition (pulse width = 1 μs, frequency = 2 kHz); (**a**) fast axis and (**b**) slow axis [32]

For the slow axis far field, the patterns of CW and pulsed operation are very different. The CW slow axis far-field patterns have a lot of spikes with the increasing of the injection current, and the divergence angle increased obviously with the increasing of the injection current. As shown in Fig. 8.60, the divergence angle in the slow axis also increases gradually with the increasing current, but at a much slower rate. Furthermore, the profile in slow axis is not changed much with the increasing current under QCW. It is believed that the elevated junction temperature due to heat generation in the laser greatly enhance the lateral index confinement which leads to existence of many multimodes in the lateral direction. The overlap of these modes creates a nonuniform light pattern at the output facet which could cause facet damage if a high amount of light is concentrated on a small spot in the facet [32]. In CW operation, more heat is generated and the junction temperature is much higher than that in QCW operation, thus the effect is more obvious in CW operation than in QCW operation.

8.8 Smile

The high power semiconductor laser array or bar consists of many emitters. In most of the cases, the emitters in the laser array are not perfectly aligned because of mechanical and/or thermal-mechanical stresses generated in the process of epitaxy material growth, chip fabrication and packaging [33, 34]. Packaging induced stress caused by CTE mismatch between the diode laser bar and the mounting substrate during solder bonding is the primary source of smile. The smile is defined as the bending of the line of emitters in transverse direction. More commonly the smile is expressed as the nonlinearity of the near field of the emitters and measurement techniques could be developed based on this principle. Smile in diode laser bars can prevent the fast axis collimation (FAC) lens being correctly positioned for all emitters in the bar and can have detrimental effects on the ability to focus beams from diode bars. Figure 8.61 shows the "smile" of high power semiconductor laser arrays with 62 emitters which are concave, convex, or waved [33]. The individual emitters form a curvature, which can be concave or convex, corresponding to "smile" (Fig. 8.61a), "cry" (Fig. 8.61b), or other types (Fig. 8.61c, d). Both "smile", "cry", or other shapes are commonly called "smile" due to historical reason.

There are mainly two test methods to characterize the smile: imaging method (IM) and the interferometric method [33, 34]. The imaging test method is the most commonly employed method for the characterization of the emitter linearity of a laser array.

8.8.1 *Imaging Test Methods*

Figure 8.62 shows the schematic setup of IM for smile measurement [1]. In this smile measurement setup, the beam emitted from the semiconductor laser is firstly collimated in fast axis and slow axis position by a collimator system. After this, the beam is splitted into two parts by a splitting mirror. One part of the beam transmits into the CCD through an optical system and the emitters are imaged onto CCD. The signal of beam curvature is digitized, processed, and displayed in a personal computer (PC). The other part of the beam transmits into

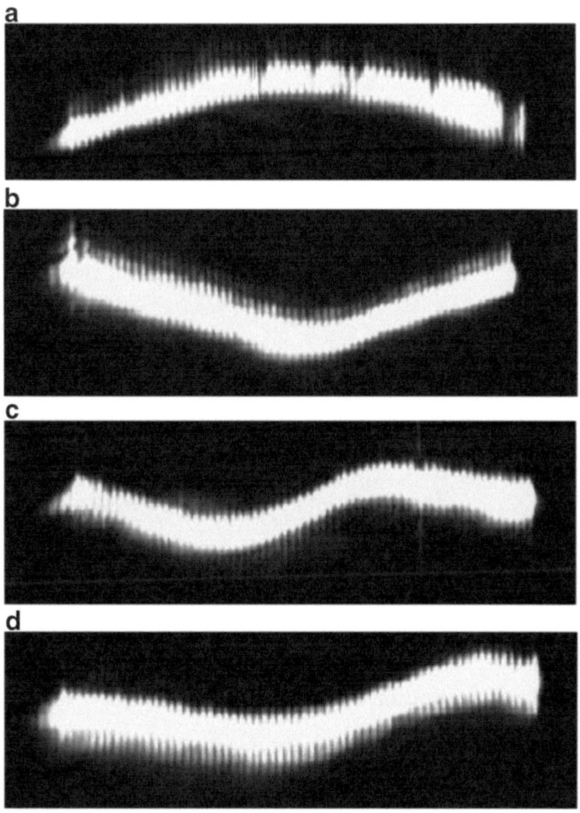

Fig. 8.61 Images of diode laser bars with various "smile" (**a**) 'cry' of approx. 2 μm; (**b**) 'smile' of approx. 2.5 μm, or others (**c**),(**d**) [33]

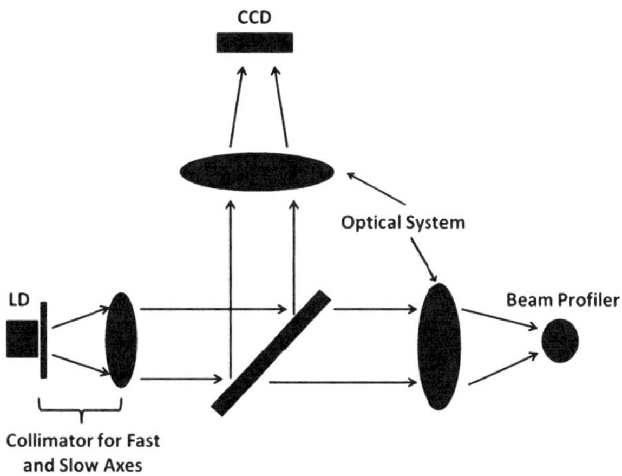

Fig. 8.62 The smile measurement setup of imaging method [1]

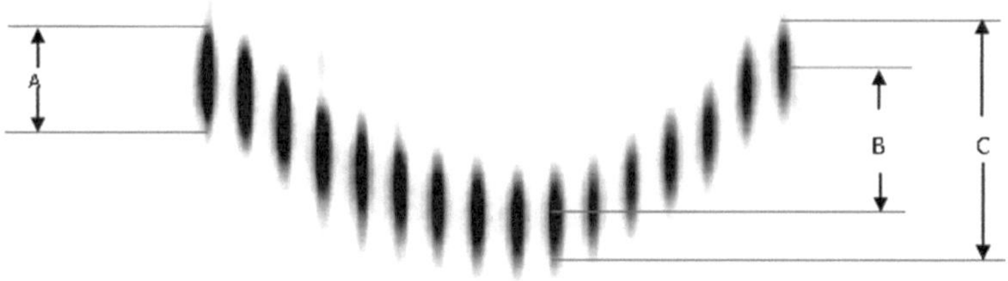

Fig. 8.63 Simplified calculation model of smile for semiconductor laser bars [1]

the beam profiler. The function of beam profiler is to monitor the quality of beam collimation along fast axis.

To quantify the smile of a laser array or bar effectively, a simplified calculation model has been developed [33]. Figure 8.63 shows the simplified model to calculate the smile of a typical centimeter bar.

The smile of a laser bar is described by

$$\text{Smile} = B\mu m = (C - A)\mu m. \tag{8.43}$$

where the definition of A, B, and C in Eq. (8.43) is described below:

A—the width of a typical emitter in a laser bar in fast axis direction;
B—the central distance between the highest emitter and the lowest emitter of the bended laser bar
C—the distance between the upper end of the highest emitter and the lower end of the lowest emitter of the bended laser bar

In addition, the values of A and C can be directly measured from the CCD image. In calculation of smile, the values of A and C is divided by the magnification number of the optical imaging system.

8.8.2 Typical Testing Results and Analysis

Figure 8.64 shows a smile test result of a CS packaged diode laser bar with 19 emitters by IM test method. In the figure, D and E are the lowest and highest intensity position of emitters in the bar, respectively [1]. D is set as 0 and as the origin of the abscissa. The C is calculated by C = E − D, and according to Eq. (8.43), smile = C − A = (E − D) − A = 1.13 μm.

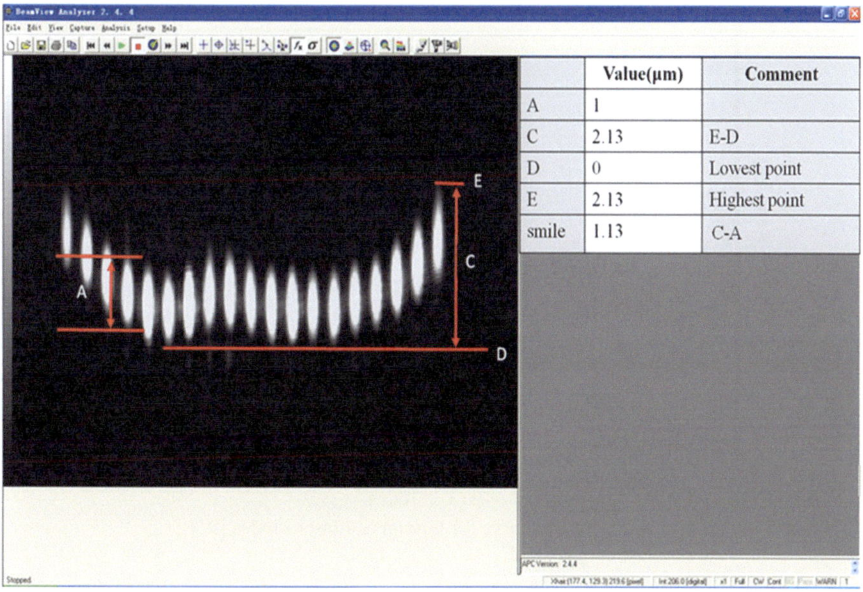

Fig. 8.64 An example of smile measurement and the test result of a semiconductor laser bar [1]

8.9 Burn-in and Lifetime Testing

Burn-in and lifetime testings are used to screen out the defective devices and to evaluate the reliability of high power semiconductor laser. For the high power semiconductor lasers, normally all devices are required to pass the procedure of burn in. In this section, test methods and equipments of burn-in and lifetime testing are described [35].

8.9.1 Burn in

Principles of Burn in

Infant mortality failures are often caused by the defects introduced during the manufacturing process or intrinsic semiconductor defects [36–38]. In contrast to lifetime test, burn-in test is applied to all lasers right after their completion of manufacturing in order to identify and eliminate defective devices to avoid infant mortality failure. From a laser user's point of view, many of the issues related to laser diode reliability are revealed by the hazard rate characteristic curve for a population of lasers as shown in Fig. 8.65 [37]. Due to the consideration of effective

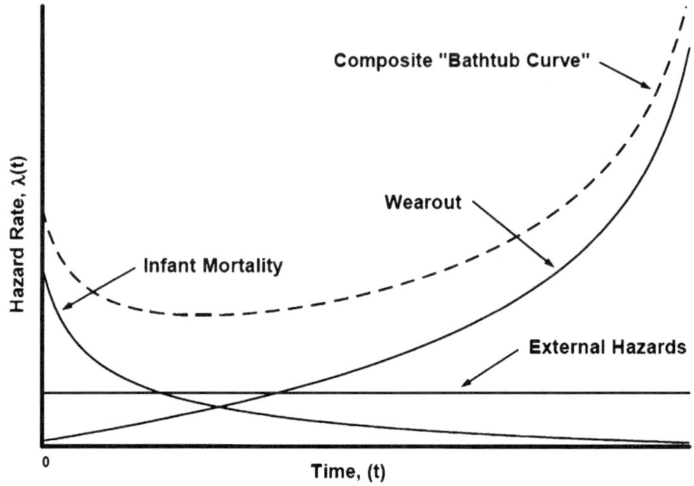

Fig. 8.65 Hazard rate characteristic curve for unscreened laser diodes [37]

cost and cycle time, burn-in time is commonly controlled to be less than 100 h for mass production. The screening criteria are developed through a series of engineering tests to determine the most effective burn-in conditions.

In order to shorten the burn-in duration and save the cost, accelerated burn in is normally applied to verify the quality of laser product in short time. Accelerated burn in may be implemented through high temperature, injection current, or optical power, in which temperature acceleration is most commonly used in industrial to screen out the devices with unacceptable short lives.

The burn-in conditions required to reduce the early failure rate after shipment to the target value can be determined using the failure distribution function F(t) obtained from burn-in study [39]. Labeling the burn-in time as t_0 and the coefficient of acceleration for the burn-in conditions as K, the cumulative early failure rate that can be eliminated by burn in is given as F(K · t0), and the new cumulative early failure rate F(t) up to time t after burn in can be obtained by the following formula [39]:

$$F(t) = F(K * t_0 + t) - F(K * t_0) \tag{8.44}$$

This relationship can be expressed in graph form as shown in Fig. 8.66 [39]. The burn-in conditions are selected according to the combination of the acceleration conditions and time that will reduce this value to the target early failure rate or lower. Normally, initial defects that are the causes of early failures occur at the highest rate in the initial stages of process development, and then decrease thereafter due to process improvements and process mastery. The early failure rate decreases in proportion to these initial defects, so the burn-in time is reviewed as appropriate in accordance with process improvements [39].

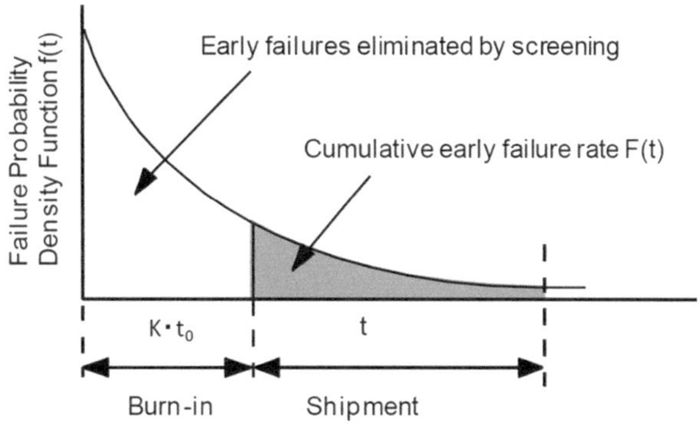

Fig. 8.66 Early failure screening by burn in [39]

Burn-in Test Methods

The two most common test strategies for burn-in screening of production are (1) Burn in with In Situ Test and (2) Burn in with Separated Test.

Burn in with in situ test: In the case of low production volumes or when the same system is used for both engineering evaluations and burn in, it is often cost effective to perform parametric testing in the same system that is used for burn in. In this case parametric data may be taken continuously or at the beginning and end of a burn-in cycle.

Burn in with separate test: In almost all other cases it is more cost effective to use simple constant current, constant temperature burn-in chambers which are separated from the parametric test system. A semi-automated parametric test system can easily provide the throughput required to process over 1,000 lasers in an 8 h shift [35]. This approach also has the advantage that the parametric test system can also be designed to incorporate spectral measurement which is difficult to implement in an in situ test system.

8.9.2 Lifetime Test

The lifetime test is important to both manufacturers and users of laser device to have an estimation of the product work duration through statistical data analysis.

Lifetime test is defined as the interval (time or number of pulses) over which a laser device or a laser assembly maintains the performance characteristics specified by the manufacturer. The criterion of the end-of-life is commonly considered as a 20 % drop in laser performance, such as the power degradation with the same current.

Lifetime test is used to collect laser lifetime data under carefully controlled operating conditions in order to develop statistical models that can then be used to predict laser lifetime under intended operating conditions. In order to obtain statistically meaningful data, lifetime test normally involve dozens of lasers monitored for periods of at least 1,000 h and often these test studies extend to over a year or years in length [40, 41].

Method and Equipment

Depending on the type and application of the laser diode, lifetime test involves the periodic measurement of a variety of device parameters including operating current, optical output power, threshold current, and forward voltage under accelerated aging conditions. Accelerated lifetime test may be implemented through high temperature, injection current, or optical power, in which temperature acceleration is the most common applied approach.

Lifetime test is conducted in one of the following three modes of operation: (1) Constant current aging, (2) Constant power aging, and (3) Periodic sample testing.

Constant current aging: Often referred to as ACC mode (automatic current control). In this mode, the laser current is held constant for the duration of the test.

Constant power aging: Often referred to as APC mode (automatic power control). In this mode, the laser output power is held constant by continuously adjusting current as required to maintain constant output power. The optical output power is measured either with an external photo-detector or by using an internal monitor photodiode if one is available within the laser package.

Periodic sample testing: In periodic sample testing, measurements may either be made in situ within the test system or at a separate test stand. In situ testing gives the most repeatable measurement results and reduces the hazard of laser damage due to handling. The use of a separate test stand reduces overall cost, especially when several thousand lasers are involved in long-term lifetime test studies.

Figure 8.67 shows over 5,000 h long lifetime test under ACC mode, or the constant current aging mode [42]. These tests consist of a series of shorter tests that typically last for 720 h. After each shorter lifetime session was completed, the electrical and optical characteristics of the device were measured.

A typical laser diode array semi-automated lifetime test system with a PC has been developed, as shown in Fig. 8.68 [43]. With the specially designed software, the PC sets operational and environmental parameters, acquires and archives data, feed back anomalous readings, and generates a number of warning status and alert messages when necessary. All the laser performance parameter data are continuously monitored and recorded using a set of instruments for analysis and evaluation.

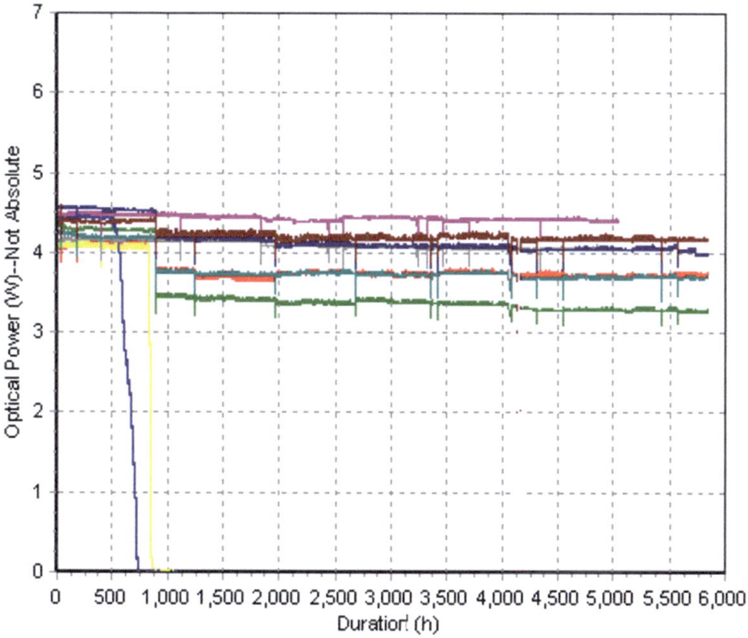

Fig. 8.67 An example of life time test results under ACC mode [42]

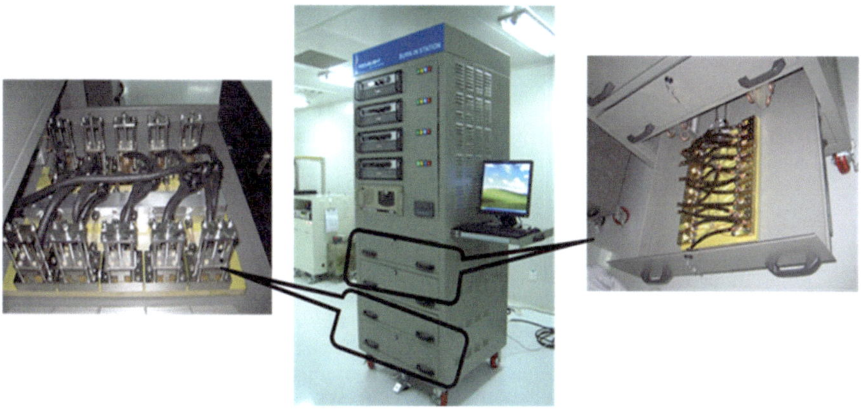

Fig. 8.68 An example of lifetime test system [43]

Typical Testing Results and Data Analysis

One of the most common equations used to analyze laser diode lifetime is the Arrhenius model. The Arrhenius model uses temperature and activation energy to predict time to fail [44].

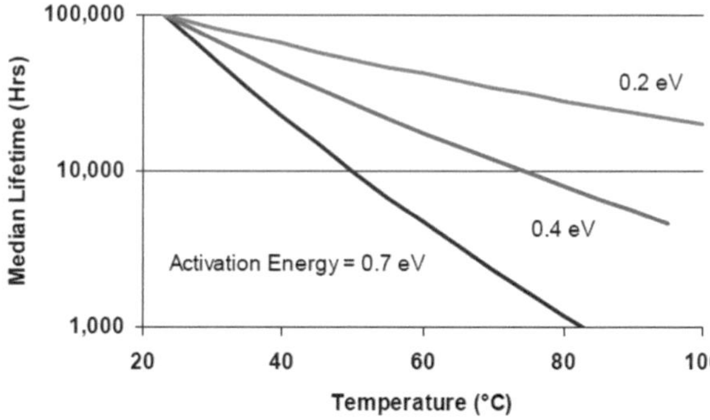

Fig. 8.69 Laser lifetime varies with operation temperature [35]

$$Lifetime = A_t \, exp\left(\frac{E_a}{kT}\right) \tag{8.45}$$

where At is a constant, Ea is the activation energy, and k and T are Boltzman's constant and temperature respectively. Depending on the type of laser, typical activation energies range from 0.2 to 0.7 eV. Laser lifetime test can be significantly accelerated at high temperatures as shown in Fig. 8.69 [35]. As can be seen in the figure, a laser diode with activation energy of 0.7 eV and median lifetime of 100,000 h at room temperature has a time of only 2,300 h when the device is operated at 70 °C [35].

References

1. J.W. Wang, D. Hou, Y.L. Liu, X.S. Liu et al., Test and characterization of high power semiconductor lasers, Technique Report of Focuslight Technologies, Inc.
2. E. Farsad, S.P. Abbasi, A. Goodarzi, M.S. Zabihi, Experimental parametric investigation of temperature effects on 60W-QCW diode laser. World Acad. Sci. Eng. Technol. **59**, 1190–1196 (2011)
3. K. Boucke, J. Jandeleit, W. Brandenburg, A. Ostlender, P. Loosen, R. Poprawe, Packaging and characterization equipment for high-power diode laser bars and VCSELs. Testing, reliability, and applications of optoelectronic devices, in *Proceedings of SPIE*, vol. 4285 (2001), pp. 165–172
4. X.N. Li, Y.X. Zhang, J.W. Wang, L.L. Xiong, P. Zhang, Z.Q. Nie, Z.F. Wang, H. Liu, X.S. Liu, Influence of package structure on the performance of the single emitter diode laser. IEEE Trans. Compon. Packaging Manuf. Technol. **2**(10), 1592–1600 (2012)
5. http://www.iso.org/iso/home/store/catalogue_ics/catalogue_detail_ics.htm?ics1=31&ics2=260&ics3=&csnumber=22264
6. http://std265.infoeach.com/view-MjY1fDIyMTM3Mw==.html
7. J.W. Wang, Z.B. Yuan, Y.X. Zhang, E.T. Zhang, D. Wu, X.S. Liu, 250W QCW conduction cooled high power semiconductor laser, in *Proceedings of 10th on Electronic Packaging Technology & High Density Packaging (ICEPT-HDP)*, 2009, pp. 451–455
8. H.Y. Sun, *Laser Diode Beam Basics, Manipulations and Characterizations* (Springer, New York, 2012)
9. G. Erbert, A. Bärwolff, J. Sebastian, J. Tomm, *High-power Broad-area Diode Lasers and Laser Bars*, vol. 78 (Springer, New York, 2000), pp. 173–223

10. Z.N. Yang, H.Y. Wang, Y.D. Li, Q.S. Lu, W.H. Hua, X.J. Xu, J.B. Chen, A smile insensitive method for spectral line width narrowing on high power laser diode arrays. Opt. Commun. **284** (21), 5189–5191 (2011)
11. G. Giuliani, M. Norgia, Laser diode linewidth measurement by means of self-mixing interferometry. IEEE Photonics Technol. Lett. **12**(8), 1028–1030 (2000)
12. BS EN ISO 13695-2004, Optics and photonics—Lasers and laser related equipment—Test methods for the spectral characteristics of lasers
13. BS-EN-ISO-11145-2006, Optics and photonics—Lasers and laser related equipment—Vocabulary and symbols
14. http://www.measurement.gov.au/Publications/. . ./NMI%20TR%203.doc
15. J.W. Wang, Z.B. Yuan, L. Guo, L.L. Xiong, Y.X. Zhang, C.H. Peng, X.N. Li, X.S. Liu, Packaging of high power semiconductor laser arrays using a novel macro-channel cooler, in *Proceedings of 11th on Electronic Packaging Technology & High Density Packaging (ICEPT-HDP)*, 2010, pp. 92–97
16. X.C. Ding, P. Zhang, L.L. Xiong, X. Ou, X.N. Li, Z.F. Xu, J.W. Wang, X.S. Liu, Thermal reaction of high power semiconductor laser with voids in solder layer. Chin. J. Lasers **38**, 0902006-1 (2011)
17. X.S. Liu, J.W. Wang, P.Y. Wei, Study of the mechanisms of spectral broadening in high power semiconductor laser arrays, in *Proceedings of 58th Electronic Components and Technology Conference (ECTC)*, 2008, pp. 1005–1010
18. http://assets.newport.com/webDocuments-EN/images/AN30_Measure_Laser_Diode_IX.PDF
19. http://www.thermengr.net/PDF/MilStd750M3101_Diode.pdf
20. B. Siegal, Practical considerations in high power LED junction temperature measurements, in *Electronics Manufacturing and Technology*, 2007, pp. 62–66
21. B. Siegal, Measurement of junction temperature confirms package thermal design. Laser Focus World **39**, 12–14 (2003)
22. http://www.laserlines.co.uk/htm/la/nlight_cascade_Hor.htm
23. http://en.wikipedia.org/wiki/Thermal_resistance
24. Y.X. Zhang, J.W. Wang, D. Wu, K. Yang, Y.L. Ma, X.S. Liu, A new package structure for high power single emitter semiconductor laser and performance analysis. Chin. J. Lasers **37**(5), 1186–1191 (2010)
25. C.Q. Gao, *Characterization and Transformation of Astigmatic Laser Beams* (Wissenschaft Und Technik, Berlin, 1999)
26. S.H. Chen, D.P. Tsai, Y.F. Chen, P.M. Ong, True near-field optical characters of a GaAlAs semiconductor laser diode. Rev. Sci. Instrum. **70**(12), 4463–4466 (1999)
27. B. Lv, *Laser Optical: Beam Description, Propagation and Optical Cavity Technology in Physics* (Higher Education Press, Beijing, China, 2003)
28. S. Nemoto, Experimental evaluation of a new expression for the far field of a diode laser beam. Appl. Optics **33**(27), 6387–6392 (1994)
29. Y.J. Li, J. Katz, Nonparaxial analysis of the far-field radiation patterns of double-heterostructure laser. Appl. Optics **35**(9), 1442–1451 (1996)
30. M.H. Hu, X.S. Liu, C. Caneau, Y.B. Li, R. Bhat, K. Song, C.E. Zah, Testing of high-power semiconductor laser bars. J. Lightwave Technol. **23**(2), 573–580 (2005)
31. C. Borgentun, J. Bengtsson, A. Larsson, Full characterization of a high-power semiconductor disk laser beam with simultaneous capture of optimally sized focus and farfield. Appl. Optics **50**(12), 1640–1650 (2011)
32. X. Liu, L.C. Hughes, M.H. Rasmussen, M.H. Hu, V.A. Bhagavatula, R.W. Davis, S. Coleman, R. Bhat, C.-E. Zah, Packaging and performance of 980nm broad area semiconductor lasers, IEEE, in *2005 6th International Conference on Electronic Packaging Technology*, pp. 67–73
33. J.W. Wang, Z.B. Yuan, L.J. Kang, K. Yang, Y.X. Zhang, X.S. Liu, Study of the mechanism of "Smile" in high power diode laser arrays and strategies in improving near-field linearity, in *IEEE 2009 Electronic Components and Technology Conference*, 2009, pp. 837–842

34. L. Mart-Lopez, J.A. Ramos-de-Campos, R.A. Martnez-Celorio, Interferometric method for characterizing the smile of laser diode bars. Opt. Commun. **275**(2), 359–371 (2007)
35. L.A. Johnson, Laser diode burn-in and reliability testing. IEEE Commun. Mag. **44**(2), 4–7 (2006)
36. H.Y. Li, L.Y. Qi, J.W. Shi, E.S. Jin, Z.T. Li, D.S. Gao, J.Z. Yu, L. Guo, Effective method for evaluation of semiconductor laser quality. Microelectron. Reliab. **40**(2), 333–337 (2000)
37. F.R. Nash, *Estimating Device Reliability: Assessment of Credibility* (Academic Publishers, Boston, 1992)
38. M. Fukuda, Optical source reliability in recent optical fiber transmission systems and consumer electronics. Microelectron. Reliab. **46**(2–4), 263–269 (2006)
39. *Sony Semiconductor Quality and Reliability Handbook* (Sony Corporation, October 2000)
40. GRE-468-CORE, Generic reliability assurance requirements for optoelectronic devices used in telecommunications equipment, Bellcore (now available from Telcordia Technologies Inc.), 1998
41. GRE-3013-CORE, Generic reliability assurance requirements for optoelectronic devices used in short-life, information-handling products and equipment, Telcordia Technologies Inc., 1999
42. Y. Sin, B. Foran, N. Presser, M. Mason, S.C. Moss, Reliability and failure mode investigation of high power multi-mode InGaAs strained quantum well single emitters, in *Proceedings of SPIE*, vol. 6456, 2007, pp. 645605(1–13)
43. J.W. Wang, L.J. Kang, P. Zhang, Y.X. Zhang, X.S. Liu, High power semiconductor laser array packaged on micro-channel cooler using gold-tin soldering technology, in *Proceedings of SPIE*, vol. 8241 (2012), pp. 82410H(1–9).
44. O. Ueda, *Reliability and Degradation of III-V Optical Devices* (Artech House, London, 1996)

Chapter 9
Failure Analysis and Reliability Assessment in High Power Semiconductor Laser Packaging

High reliability and durability are two of the important requirements for a commercially used semiconductor laser. There are multiple causes for the semiconductor laser to fail during operation. Therefore, it is important to analyze and identify the root causes for laser failures and provide effective solutions to improve the reliability and durability of the semiconductor laser. In this chapter, different failure modes of semiconductor lasers are introduced. By analyzing these failure modes, some approaches to improve reliability of the semiconductor laser are demonstrated. The lifetime prediction of the semiconductor lasers is also presented.

9.1 Failure Modes

Based on the decreasing rate of output power when failure occurs, the failure modes associated with laser diodes can be classified into three categories: rapid, catastrophic, gradual as illustrated in Fig. 9.1 [1]. If a diode laser is operated at a constant driving current, rapid failure can be detected by a very fast decrease in output power. A sudden decrease in output power is typical for catastrophic failure. The typical characteristic of gradual failure is a gradual decrease of output power at a constant driving current. Gradual failure occurs over a long period, and it determines the ultimate lifetime of the diode laser.

Failure phenomena in the semiconductor laser can also be classified as external and internal failure, depending on the failure location relative to the chip region. Internal failure, also called bulk failure, is related to the internal degradation within the bulk chip region. External failure takes place outside the chip, and can be divided into solder-related failure due to packaging, facet-related failure, and package structure failure such as micro-channel cooler corrosion. In addition, performance instability due to intermetallic formation in the die bonding solder layer and optical feedback could deteriorate the performance of a laser device.

X. Liu et al., *Packaging of High Power Semiconductor Lasers*, Micro- and Opto-Electronic Materials, Structures, and Systems, DOI 10.1007/978-1-4614-9263-4_9,

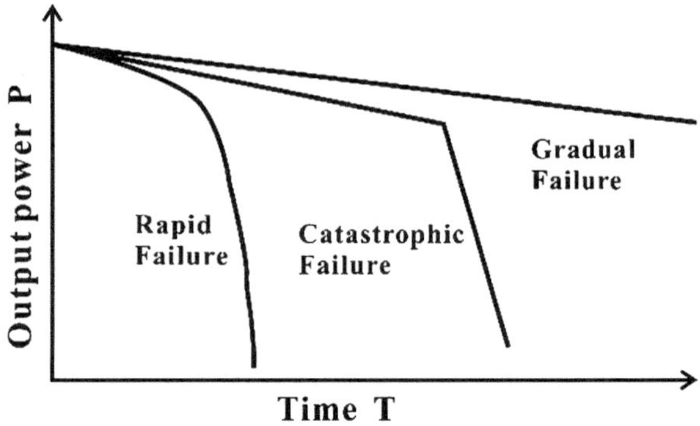

Fig. 9.1 Failure modes of high power diode lasers at constant current [1]

9.1.1 Bulk Failure

Bulk failure (wear-out failure) is generally the result of the junction temperature increase in the active region of the laser. This failure may be caused by the growth of defects at the region, and is exhibited by a slow degradation in the performance of the laser. Figure 9.2 shows the slow degrease of output power of a high power 808 nm gold–tin mounted laser bars with 1.6 × P_{nom} at $T_{heatsink}$ = 20 °C operated in CW mode [2]. Generally, the lifetime of a diode laser exhibiting a slow degradation can be approximated by the following relationship:

$$P_{out}(t) = P_0 e^{-t/\tau} \tag{9.1}$$

where P_{out} is the output power at time t, P_0 is the initial power output of the device, and τ is the lifetime of the laser. Assuming that for a given time t, the power output of the device has dropped to a percentage from the initial power level such that the power ratio is $P_R = P_{out}/P_0$. Thus the lifetime of τ of a laser device can be expressed as

$$\tau = -t/\ln(P_R) \tag{9.2}$$

9.1.2 Facet Failure

The optical power density at the laser facet can be greater than 10 MW/cm^2. The optical absorption due to high power density and nonradiative recombination at a laser diode facet can result in a large amount of localized heat at the facet. The generated heat will also cause the bandgap to shrink and as a result the current concentration will increase which creates more heat and eventual thermal runaway and catastrophic optical mirror damage (COMD). The scanning electron microscope (SEM) image of a laser diode facet after COMD failure is shown in Figs. 9.3 [3] and 9.4 [4]. The SEM image in Fig. 9.3 clearly demonstrates a blister of the melted

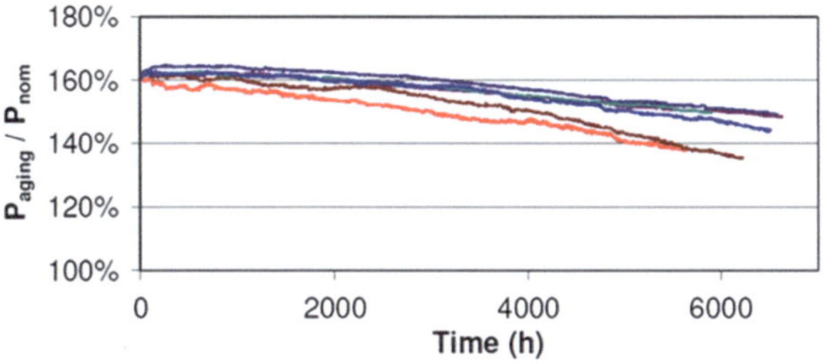

Fig. 9.2 Slow degradation of output optical power of gold–tin mounted laser bars with 1.6 × P_{nom} at $T_{heatsink}$ = 20 °C operated in CW mode [2]

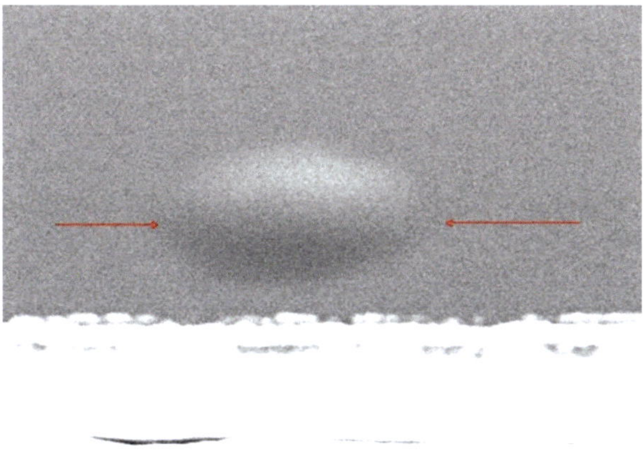

Fig. 9.3 SEM image of a laser diode facet after COMD failure [3]

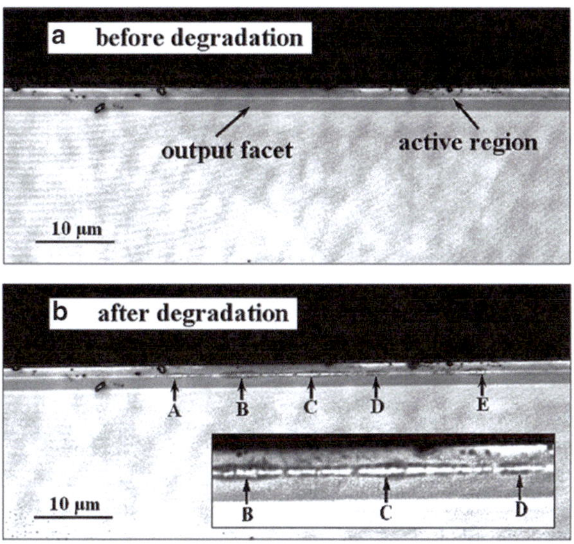

Fig. 9.4 The laser scanning confocal microscopy (LSCM) images of the output facet (**a**) before degradation (**b**) after degradation. The *inset* of (**b**) shows the profiles of melting spots B, C, and D [4]

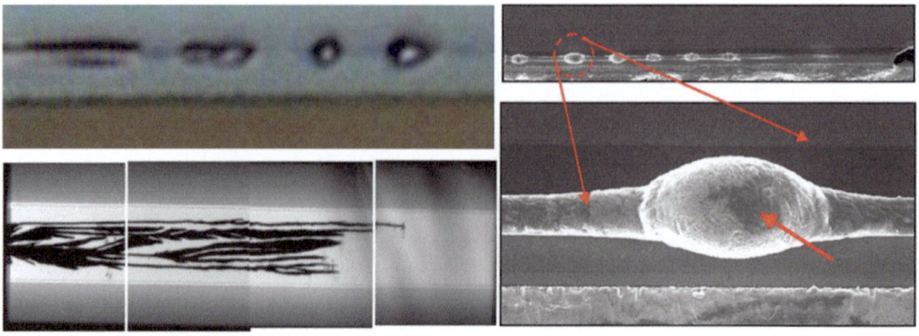

Fig. 9.5 Optical (*top left*) and SEM (*right*) images of COMD on the facet. *Bottom left* shows top view of COMD propagated throughout the emitter, as seen through an n-side window in the metallization [5]

materials. In Fig. 9.4, some equally spaced melting spots can be seen in the output facets after degradation. These melting spots are caused by the rise of the local temperature at the output facet. Figure 9.5 shows a facet with COMD and a top view of the effects of COMD within the laser diode cavity, seen through a window in the n-side metallization [5]. COMD limits the maximum output power and is one of the major failure mechanisms of laser diodes. Therefore, an optimized laser package structure design to reduce the front facet temperature is crucial to increase the COMD level for higher output power and improved reliability. It is known that the COMD level is mainly determined by the facet design and process such as facet passivation of the laser chip or bar.

9.1.3 Solder Joint Failure

The solder joint failure may be caused by different mechanisms: thermal fatigue, electromigration, thermo-electromigration, and oxidation. In semiconductor laser packaging, Indium is typically used as the bonding solder for its high ductility in order to reduce the thermal stress caused by large coefficient of thermal expansion (CTE) mismatch between copper and GaAs. However, similar to some metallic conductors, electromigration and thermo-electro-migration phenomenon can occur in indium solder at the current density of ~10^3 A/cm^2 or higher. Indium atoms are pushed downwards, as the electrons mainly flow downwards under high current density when the laser is under operation. As a result of electromigration, voids are created and gradually enlarged in the Indium solder layer causing local heating near the facets of the laser during the operation. Cross-sectional images of many samples confirm the existence of voids in the indium solder layer after 600 h testing. In some samples, widespread, large voids have been observed, as shown in Fig. 9.6. As the testing time gets longer, the voids become more extensive and larger, which indicates that most of the isolated voids are already connected together [6].

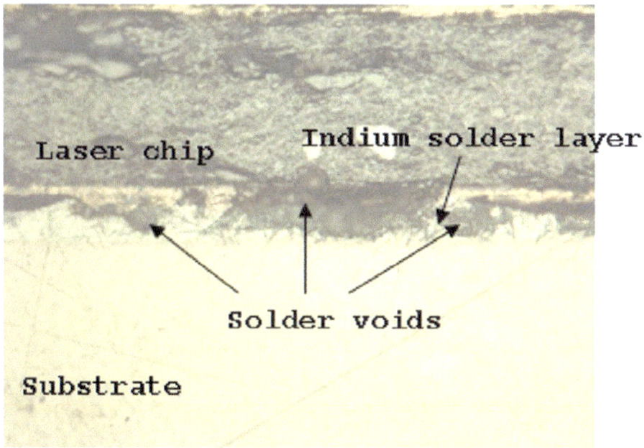

Fig. 9.6 An optical image of a typical cross-sectioned sample after 600 h testing at 7 A and 40 °C showing large voids formation in the indium solder layer [6]

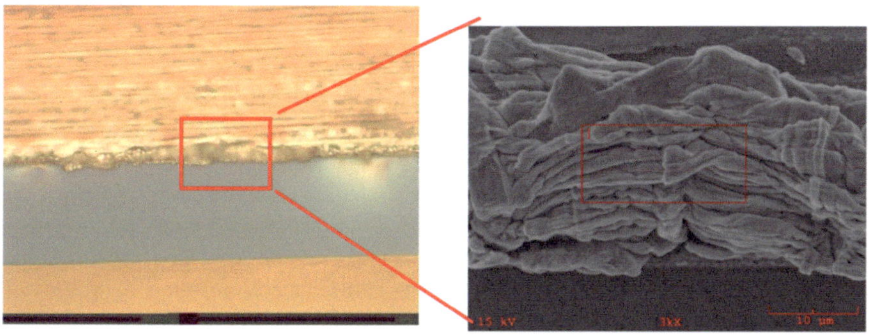

Fig. 9.7 (*Left*) Microscope image of a laser bar facet on an indium-based CCP (Conduction-Cooled Single Bar Package) package after 264 h of 1 s on/1 s off cycling. (*Right*) Close-up image from a scanning electron microscope [7]

Moreover, Indium-bonded conduction-cooled semiconductor laser package usually fails in 1 s on/1 s off cycling operation mode, and exhibits catastrophic failure modes of burned or blown-out bars. This is because the quasi-CW conditions of on-off cycling construct the thermal gradients in the solder joint, and elevate the junction temperature and generate the thermal fatigue. Figure 9.7 shows a device that has degraded by 25 % after 264 h of on-off cycling [7]. Using a SEM under high magnification, one can see indium extruding in the front of the laser facet. The material is verified as >96 % indium using energy dispersion X-ray fluorescence (EDX). Scanning acoustic microscopy (SAM) analysis on this device shows corresponding voiding near the facet where the extruded indium is located. The folds, seen in Fig. 9.7, indicate that the indium may be pushed out by a mechanical force from the on-off cycling [7].

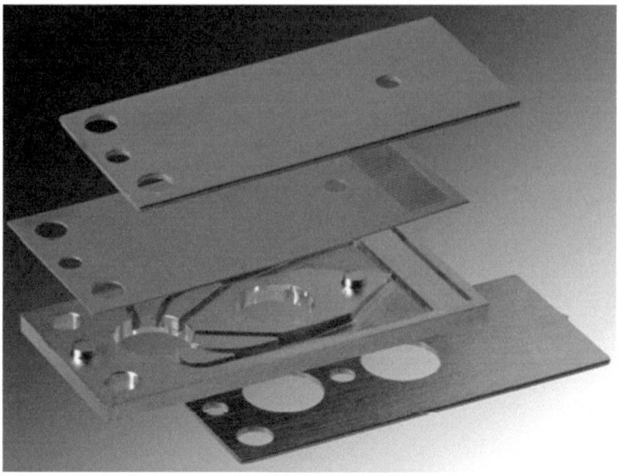

Fig. 9.8 Assembly of a micro-channel heat sink [8]

9.1.4 Micro-channel Cooler Corrosion

The most common method of removing large amounts of waste heat in laser diode arrays is through the use of copper-based micro-channel coolers in which a cooling liquid is pumped through the micro-channels to remove the heat. In the majority of commercially available coolers as shown in Fig. 9.8, the coolant is in electrical contact with the diode bars [8]. This requires the use of deionized water. However, even the use of the deionized water as the cooling fluid, a number of corrosions can still occur and the lifetime of copper micro-channel heat sinks can be shortened. Although the application of appropriate protective coatings can reduce corrosion, the process of erosion can eventually damage the protective coatings, and at these damaged areas the corrosion may even be accelerated by electrochemical mechanisms.

9.1.5 Performance Instability

The hard solder AuSn can overcome the disadvantageous properties of thermal fatigue and creep rupture of the soft solders, such as Indium, by staying within the elastic deformation zone. However, microstructure analysis of the as-reflowed samples as well as the samples subjected to thermal aging show that the thick Au layer in the traditional Ti/Pt/Au system is mostly consumed due to the interactions between the Sn in the solder and the Au layer [9]. The metallurgical interaction of AuSn solder with Au layer is to form the Au_5Sn, $AuSn_4$, and other intermetallic compounds (shown in Fig. 9.9) [9]. The Au layer might even be completely consumed during aging or upon reflow, which is undesirable since the Au is necessary for thermal management. Furthermore, the next metallization layers Pt and Ti may start to react with Sn due to the diffusion of the Sn in the AuSn solder to

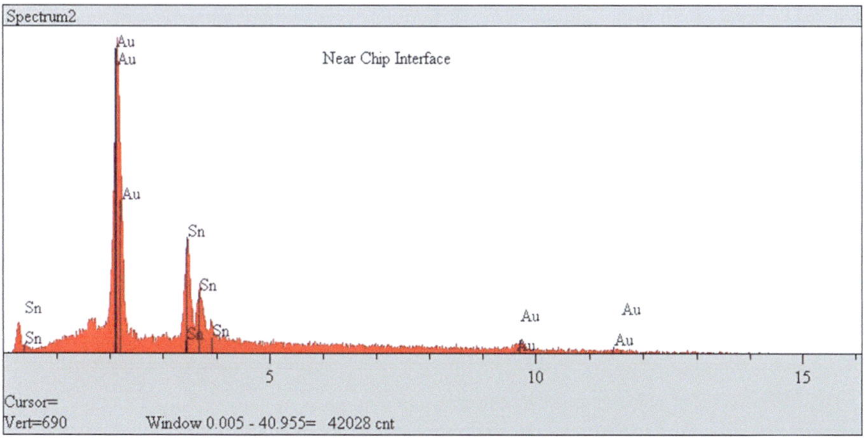

Fig. 9.9 Energy dispersion X-ray fluorescence (EDX) spectra of the material close to the chip and solder interface (inside the epi-side Au layer) aged for 3 months at 150 °C [9]

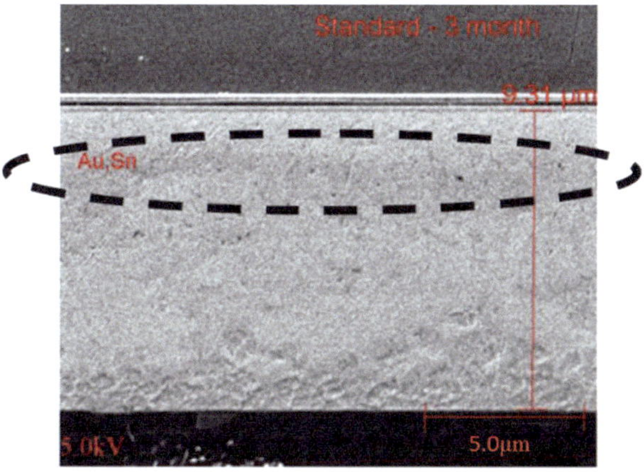

Fig. 9.10 Typical cross-sectional SEM images of the die bonding solder interface of a laser chip having the traditional Ti/Pt/Au (thick Au of 2–3 μm) metallization system aged for 3 months at 150 °C [9]

form various intermetallics. These metallurgical interactions would severely degrade the thermal performance of the laser as the thermal conductivity of the Au and Sn intermetallic compounds are much smaller than that of the Au. Also, as can be seen from the SEM image in Fig. 9.10 (circled), one dark region, which mainly consists of δ-phase (AuSn), is formed close to the chip interface where the epi-side Au layer is originally located [9]. Due to the severe degradation of the thermal performance of the laser, the wavelength shift and the output spectrum

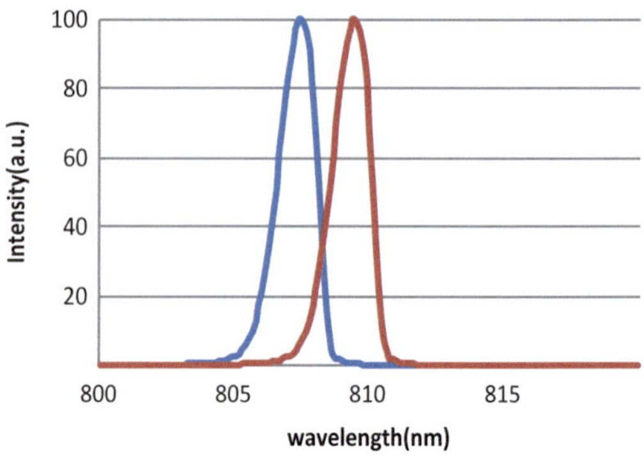

Fig. 9.11 Wavelength shift of the diode laser before (*blue line*) and after (*red line*) aged for 3 months [1]

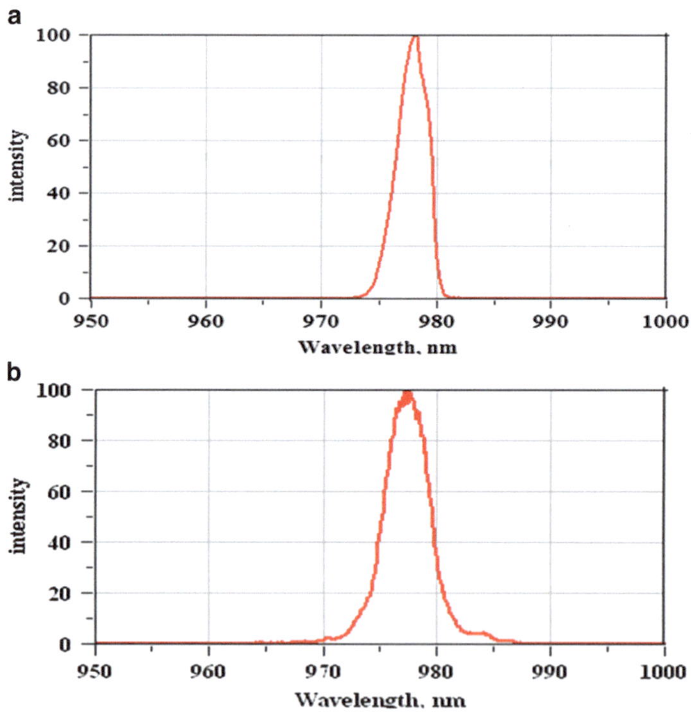

Fig. 9.12 Spectrum broadening of the diode laser before (**a**) and after (**b**) aged for 1 month [1]

shape change (spectrum broadening) may happen during the lifetime, which are shown in Figs. 9.11 and 9.12, respectively [1]. Therefore, good diffusion barrier metallization scheme needs to be selected to preserve the desired Au layer thickness and to achieve thermal stability.

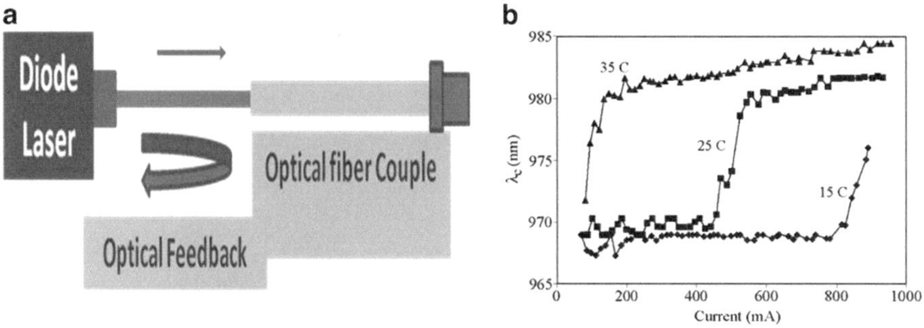

Fig. 9.13 (**a**) Schematic of optical feedback in optical fiber coupling; (**b**) dependence of output wavelength on current under different submount temperatures of the high-power fiber-coupled 980-nm pump laser [1, 10]

9.1.6 Optical Feedback

It is well known that lasers, including semiconductor lasers, are susceptible to optical feedback. Feedback affects the laser spectral property and causes unstable laser outputs. Figure 9.13a gives the schematic description of optical feedback in optical fiber coupling. The direct fiber coupling schemes usually can increase the probability of occurrence of these stray reflections. As can be seen in Fig. 9.13b, below 800 mA and at 15 °C, the device lases consistently close to 969 nm but the output wavelength sharply increases over 800 mA [10]. At 25 °C, the device lases consistently close to 969.5 nm with a current up to 450 mA, at which the wavelength jumps to 980 nm and then slowly increases as the current goes up. Finally, at 35 °C, lasing starts near 970 nm, but rapidly increases to 980 nm as the current increases to 50 mA. The wavelength then increases slowly as the current continues to go up. These detrimental effects due to optical feedback can occur even when minimal reflectivity antireflection (AR) coatings are used on the fiber lens surface.

The optical feedback of the semiconductor lasers also happens in direct diode laser head for material surface treatment applications such as laser cladding. Figure 9.14a shows the feedback from the machined workpiece in laser cladding [1]. Figure 9.14b, c shows that the shape of the output spectrum clearly deviates from the symmetry under the optical feedback in cladding system [1]. The instability is manifested in the appearance of many external cavity modes in the power spectrum and in the strong coupling between the delayed feedback effect and the relaxation oscillation. Good understanding of the optical feedback effect is important not only for understanding laser dynamics but also for applications of lasers in various systems.

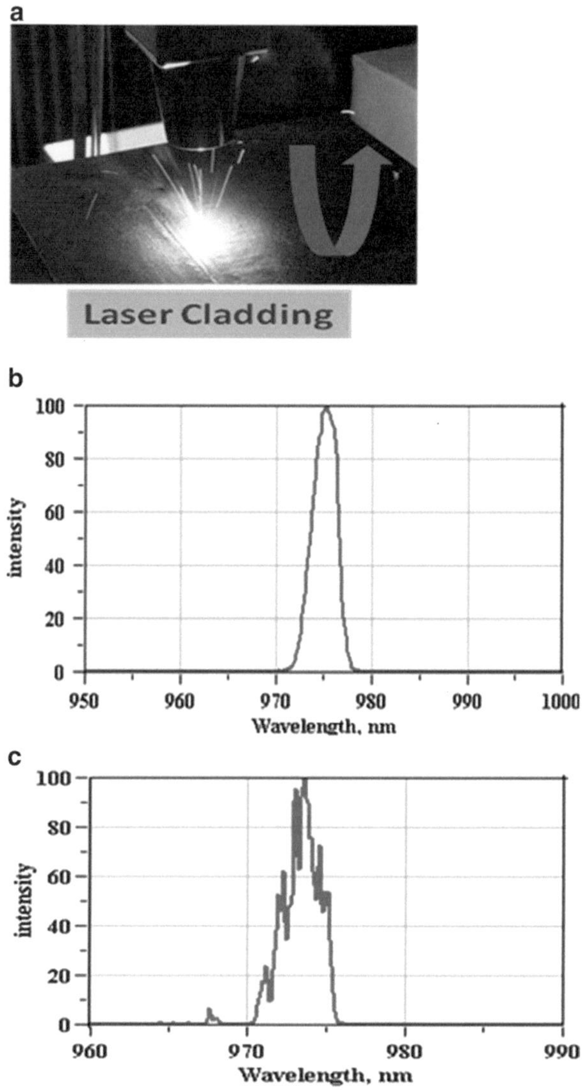

Fig. 9.14 (**a**) Optical feedback from the machined workpiece in laser cladding; output spectrum without (**b**) and with (**c**) optical feedback [1]

9.2 Approaches to Improve Reliability

9.2.1 Thermal Management

Junction temperature has great effect on the lifetime of a diode laser. Thermal management is very crucial in controlling the junction temperature and has become one of the major obstacles for the higher output power of a semiconductor laser.

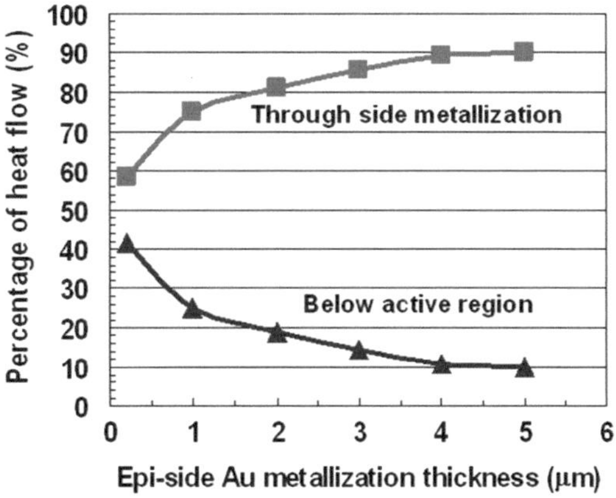

Fig. 9.15 Percentage of heat flow through the Au metallization and solder interface right below the active region (4.2 μm) and through side metallization as a function of the epi-side Au metallization thickness [9]

Thermal management of high power semiconductor lasers mainly focuses on the thermal resistance reduction, thus reducing bulk temperature rise in the laser diode chip. Some methods used to reduce the thermal resistance are: (1) optimization of laser chip and mounting substrate; (2) optimization of package structure.

Optimization of laser chip and mounting substrate: The commonly used epi-side metallization scheme for semiconductor lasers consists of Ti, Pt, and Au. As the Ti and Pt layers are relatively thin and the thermal conductivity of Ti film (~22 W m^{-1} K^{-1}) and Pt film (73 W m^{-1} K^{-1}) is low compared with that of Au layer, Ti and Pt layers have much less impact on the thermal performance of the laser diode than the Au layer. Therefore, the design and optimization are essentially done on the Au layer [9].

The thermal conductivity of the die attach solder is not very high (57 W/m K for AuSn solder), while that of the epi-side metal is high (315 W/m K for Au). The p-metal is close to the active region; the epi-side metallization can laterally spread the heat and enlarge the heat flow area, thus having a significant effect on the thermal behavior of the laser diode. Figure 9.15 shows the percentage of heat flow through the Au metallization and solder interface right below the active region (4.2 μm) and through side metallization as a function of the epi-side Au metallization thickness [9]. The curves suggest that as the Au layer gets thicker, the percentage of the heat flow through the side metallization increases, while the percentage of the heat flow dissipated directly below the active region through the solder interface decreases. After the Au thickness is beyond 3 μm, the return of thermal resistance reduction and heat spreading effect diminishes [9].

The thermal performance of a junction-down-bonded laser could also be very sensitive to die bonding solder voids since the die attach interface is so close to the active region. The existence of voids in the solder interface undoubtedly reduces the reliability of the laser [11, 12]. Modeling results show that the local junction temperature of the laser chip with 0.2-μm-thick Au increases much more than that of the laser with 3-μm Au metallization when voids exist below the active region. Figure 9.16 summaries the local junction temperature rise and temperature

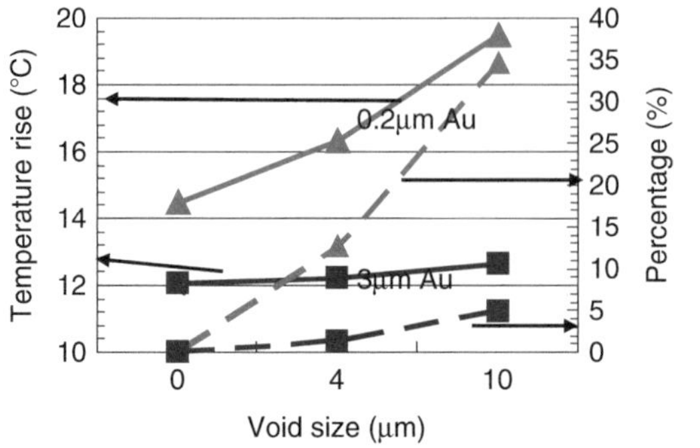

Fig. 9.16 Local junction temperature rise and temperature increase percentage of a junction-down-bonded laser chip having 0.2- and 3-μm-thick Au metallization at the epi-side under different situations: no voids, voids of 4 μm wide and 1 μm thick, and voids of 10 μm wide and 1 μm thick below the active region [9]

increase percentage of a junction-down-bonded laser chip, which has 0.2- and 3-μm-thick Au metallization at the epi-side under different situations [9]. The figure clearly shows that a laser chip with thin Au metallization is much more sensitive to die bonding solder voids than a laser with thick Au metallization. Considering the cost and technical issues of plating very thick Au as well as the diminishing returns, it is concluded that a thick Au layer of ~3 μm is preferred to achieve effective heat dissipation for an epi-down-bonded GaAs-based high power semiconductor laser.

Optimization of package structure: The package structure critically influences the thermal resistance of semiconductor lasers. Single emitter diode laser F-mount package structure and the traditional package structure C-mount were studied as an example to demonstrate the influence of package structure on thermal resistance. C-mount package is one of the typical single emitter laser packages available commercially. F-mount is another package style for single emitter semiconductor lasers, which has been described in Chap. 2. Compared to C-mount package, in F-mount package, the heat flow path is shorter and there is no direct heat crosstalk. The heat generated from chip can be easily dissipated downward to the environment through ceramic submount (AlN) and the heat sink. As discussed in Chap. 3, the thermal resistance of F-mount and C-mount laser is 2.24 and 3.02 °C/W, respectively. This shows that F-mount has better thermal management than C-mount [13].

9.2.2 Facet Protection

The absorption of the emitted light and the nonradiative recombination dominate the local facet heating when the operation power is high. The local facet heating can easily form COMD and largely reduces the reliability of the device. The heating leads to a reduction of the semiconductor bandgap, which can cause more

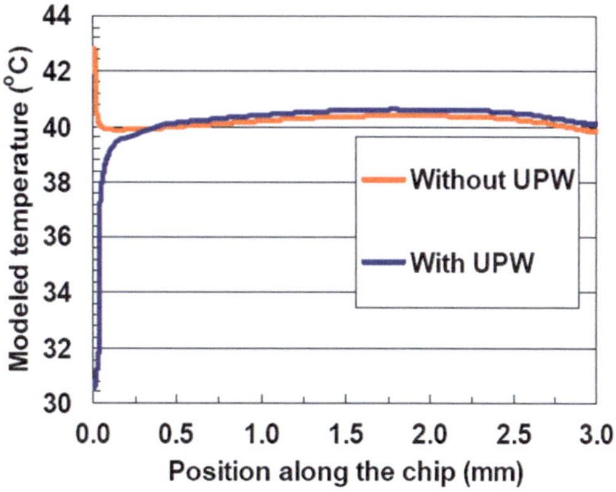

Fig. 9.17 Temperature profiles along the active region for laser chips with and without UPW [3]

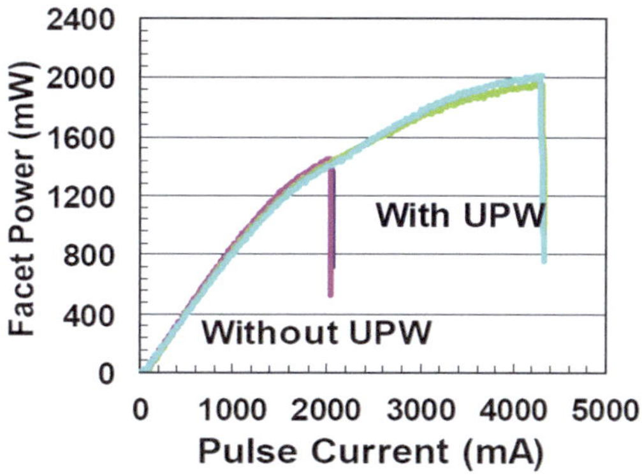

Fig. 9.18 Output power vs. current characteristics of 980 nm laser diodes with and without UPW under pulse operation [3]

absorption of the laser beam at the facet in return. Generally, unpumped window, facet passivation, and quantum well intermix are the three main methods to reduce the local facet heating and enhance the threshold of COMD.

1. *Unpumped window technique*:
   The facet heating of a high-power pump laser can be reduced by the introduction of an unpumped window (UPW), which is a current blocking region located at the front facet of the laser [3]. The blocking layer effectively prevents current injection next to the facet and thus reduces the carrier density and the surface nonradiative recombination.

   The UPW can also reduce the bulk temperature at the facet region as there is less or no heat generation in the active region under the blocking layer. Figure 9.17 gives the modeling results on temperature profiles along the active region for laser chips with and without UPW [3]. One can see that the bulk temperature at the front facet is lower for the chip with UPW design.

   The effect of the UPW on reducing the facet temperature and improving the COMD level of the 980 nm laser is demonstrated experimentally as shown in Fig. 9.18 [3]. The COMD currents are about 2 and 4.3 A, respectively, for the

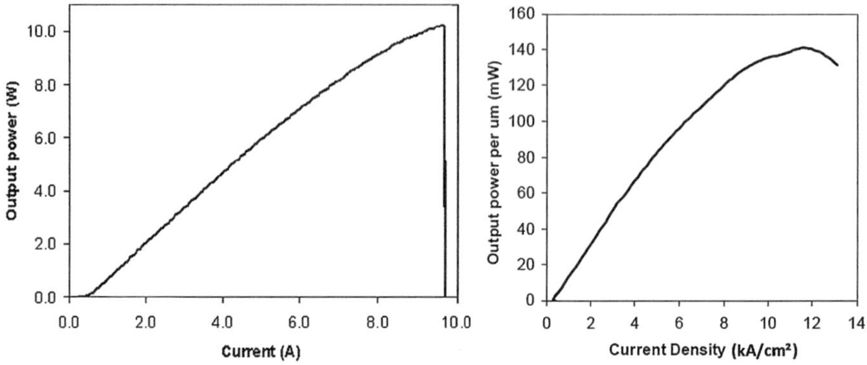

Fig. 9.19 808 nm single emitter COMD test without (*left*) and with (*right*) facet passivation [5]

chips without and with UPW, corresponding to 1.4 and 1.9 W optical power. Measurement results clearly suggest that UPW can reduce the facet temperature and improve the COMD level of the 980 nm lasers.

2. *Facet passivation*:
   Facet passivation technology has been developed to substantially increase COMD levels and in some cases virtually eliminate COMD. Facet passivation can effectively eliminate the oxides and dislocations on the facet, and leave no initiation sites for COMD. Figure 9.19 shows the results of COMD testing of a standard 808 nm single emitter device with and without facet passivation [5]. A device's COMD level is determined by ramping the current until sudden failure when the facet is melted due to high local temperature. Sometimes, thermal rollover is observed when the driving current is ramped up instead of sudden failure. This indicates the COMD level is higher than the thermal rollover power. In this case, COMD failure is of less concern. As shown in the figures, the device fails at a linear power density (LPD) of about 50 mW/μm without facet passivation; while with facet passivation, a similar device reaches a LPD over 140 mW/μm right before thermal rollover. No COMD is observed throughout the testing.
3. *Quantum well intermixing*:
   Quantum well intermixing (QWI) can bring considerable benefits to the reliability and performance of high power laser diodes by intermixing the facet regions of the device. The QWI can increase the bandgap and hence eliminate absorption, thus avoiding catastrophic optical damage. The effect of intermixing on the band structure of the quantum well is illustrated in Fig. 9.20. The red lines show the conduction and valence bands of the as-grown structure. Because the central GaAs layer is a QW, the electron and hole levels are quantized, so the effective bandgap is wider than that of bulk GaAs. Intermixing causes Al to diffuse into and Ga out of the well, widening the well and increasing its average bandgap (shown in blue). The bandgap of the intermixed QW is wider than that of the original well allowing the intermixed structure to be used as a passive waveguide [14].

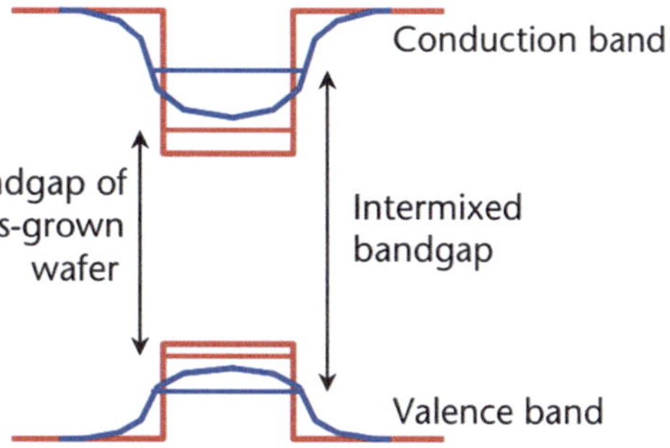

Fig. 9.20 Conduction and valence band structure of a GaAs QW with AlAs barriers before intermixing (*red*) and after QWI (*blue*) [14]

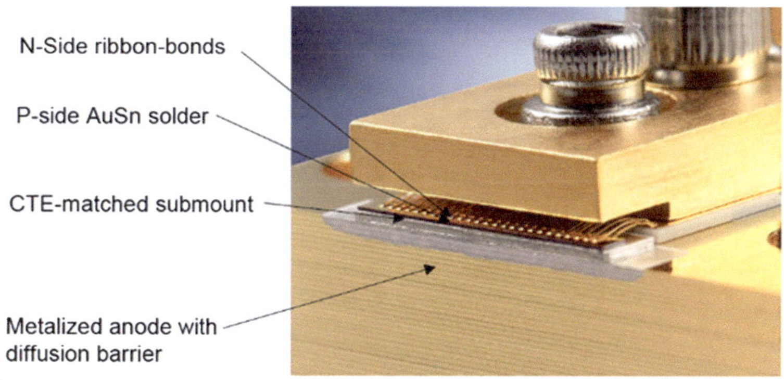

Fig. 9.21 A typical view of a hard-soldered conduction-cooled package [7]

QWI has two main benefits for the fabrication of laser diodes. Firstly, the non-absorbing mirror regions dramatically reduce catastrophic optical damage of the laser, which is the main failure mechanism of a laser diode, enabling reliability to be significantly improved. Secondly, the non-absorbing mirror regions relax cleaving tolerances for device manufacture, thus facet coating robustness is less critical compared to a more conventional semiconductor laser [15].

9.2.3 *Indium-Free Packaging*

As a replacement of the soft solder, the hard solder conduction-cooled package is to mount semiconductor diodes on a CTE matched submount with Indium-free solder such as AuSn solder. Figure 9.21 shows the packaging design in detail [7].

Figure 9.22 compares the relative optical power vs. aging time of the 980 nm broad area laser CT mount assemblies bonded with indium and AuSn solders [6]. One can see that the AuSn solder-bonded lasers have much longer lifetime than the indium

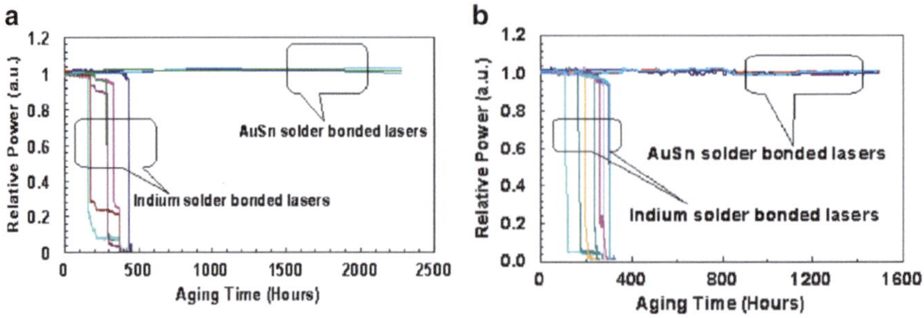

Fig. 9.22 Relative optical power vs. aging time of the 980 nm CT mount laser assemblies bonded with indium and AuSn solders under the aging conditions of (**a**) 6 A and 35 °C and (**b**) 7 A and 40 °C [6]

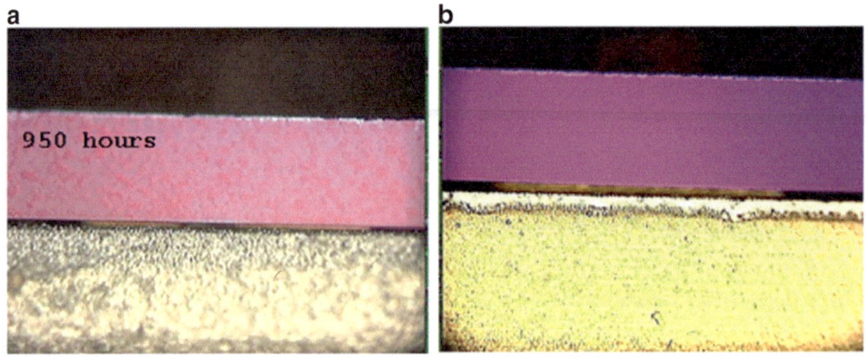

Fig. 9.23 Images the front surface of the (**a**) indium and (**b**) AuSn-bonded laser CT mount during lifetime test under the testing condition of 6 A and 35 °C [6]

solder-bonded lasers under both test conditions. The relative power curves in Fig. 9.22 show that the indium solder-bonded devices exhibit sudden failure at a quite short aging time, while most of the AuSn solder-bonded devices keep quite stable optical power at a much longer aging time.

From Fig. 9.23a, one can see that after 950 h in indium solder-bonded laser samples, the bright patterns spread to the whole front facet and the temperature at the front part of the laser chip is so high that the chip is burned [6]. Compared with indium solder-bonded laser samples, AuSn-bonded laser samples do not have evident material growth on the front or top surface of the CT mount after 2,000 h of testing at 6 A and 35 °C, as shown in Fig. 9.23b. Also no blisters and light-colored spots are observed at the laser front facet after 2,000 h [6].

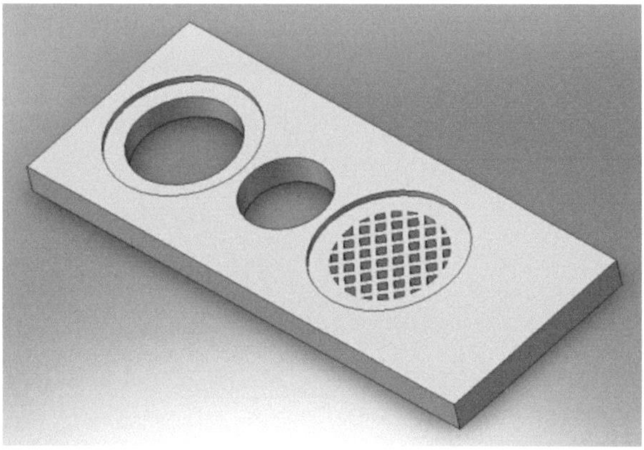

Fig. 9.24 Structure of the MaCC cooler [1]

Fig. 9.25 Indium-free vertical MaCC-packaged stacks (40 bar-808 nm-QCW-8.8 kW) [1]

9.2.4 New Cooler Design

Erosion and corrosion in copper-based micro-channel coolers are serious issues, which can shorten the lifetime of the coolers. Optimizing the cooler structure is one approach to resolve this problem. The other approach to mitigate corrosion is to coat the copper with a protective layer.

1. *New cooler structure*:
   A macro-channel cooler (MaCC) has been developed for packaging of high power semiconductor laser arrays, as shown in Fig. 9.24 [1]. Compared with conventional copper micro-channel cooler, the new design is more tolerable to particles and corrosion debris. The damage from electric-chemical erosion is less severe. Moreover, the maintenance for the MaCC is much easier than conventional micro-channel coolers.

   Based on the MaCC laser array unit, the QCW Indium-free laser stack and array are produced, as shown in Fig. 9.25 [1]. The test results indicate that the performance of a MaCC-packaged laser arrays in QCW mode is similar to that of copper micro-channel cooler. More importantly, MaCC enables the indium-free

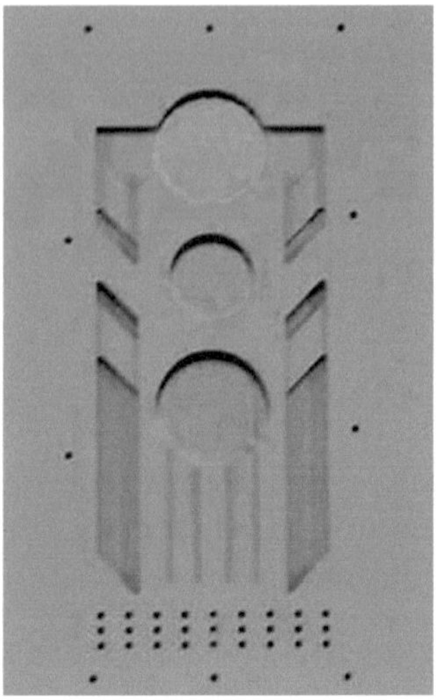

Fig. 9.26 Interior structure of ceramic MCC [16]

packaging process in fabricating high power semiconductor laser array, multiple-bar vertical stack and area array laser. Therefore, the MaCC is a good candidate in high power semiconductor laser array/bar, stack and area array for QCW applications.

Another new type of ceramic-based micro-channel cooler for high-power laser diode arrays has been developed that works better than existing copper-based designs. The interior structure of the micro-channel cooler made from ceramic material is shown in Fig. 9.26 [16]. The cooler design uses layers of ceramic material for its ability to operate in harsh conditions without causing internal corrosion and for a longer lifetime. The thermal performance of the ceramic-based micro-channel cooler packages is comparable to that of standard copper micro-channel-cooled packages. Furthermore, because the coolant is electrically isolated from the current path in an optoelectronic device, it is not necessary to use deionized water. Instead, standard filtered water is adequate.

The ceramic-based micro-channel cooler design leverages existing technology to create a low-cost, high-performance alternative to copper-based micro-channel coolers. This approach offers the greatest promise for future development due to the vast assortment of existing capabilities that have already been developed for similar ceramic structures used in the electronics industry.

2. *Protective coating on the copper*:
   Coating conventional micro-channel coolers could help alleviate some of the erosion/corrosion issues [17]. With a passivation process a pinhole-free protective coating for nickel–gold (Ni/Au) plated copper micro-channel coolers is made. In this process, atomic layer deposition (ALD) is used to deposit highly uniform ceramic thin films on commercially available coolers. The ALD coating can provide corrosion resistance and erosion protection for the micro-channel, thereby

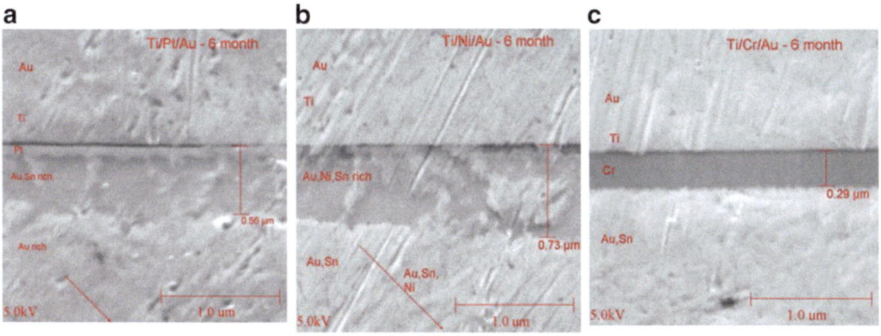

Fig. 9.27 High magnification SEM images of the die bonding solder interface of a laser chip with different metallization system in samples subjected to various aging times at 150 °C. (**a**) Ti/Pt/Au/Ti/Pt/Au; (**b**) Ti/Pt/Au/Ti/Ni/Au; (**c**) Ti/Pt/Au/Ti/Cr/Au [9]

increasing reliability. To increase the strain tolerance of the ceramic coating and eliminate any pinholes in the coating, the micro-channel surface is pretreated with an organic self-assembled monolayer. This layer can provide an ideal surface for ALD bonding, improve uniformity by masking surface defects in the metal, and provide compliance for any mismatch in thermal expansion coefficient.

9.2.5 Diffusion Barrier Design

The thick Au layer in the traditional Ti/Pt/Au system is mostly consumed due to the interactions between the Sn in the solder and the Au layer. Therefore, an effective diffusion barrier is needed to preserve the thick Au layer in order to achieve the desired thermal performance. The metallization structures with different diffusion barriers, namely Ti/Pt/Au/Ti/Pt/Au, Ti/Pt/Au/Ti/Ni/Au, and Ti/Pt/Au/Ti/Cr/Au, are implemented and the metallurgical stability of these structures is studied to preserve the integrity of the metallization system. For the metallization structure with Pt and Ni as a diffusion barrier, the solder–Pt and the solder–Ni intermetallic interface grow significantly with aging time, no Sn could be detected inside the thick Au layer after 6 months aging, as shown in Fig. 9.27a, b, respectively [9].

Unlike the samples with Pt and Ni as diffusion barriers, the Cr layer has clear definition in the samples subjected to thermal aging, as shown in the cross-sectional SEM images of the samples having the Ti/Pt/Au/Ti/Cr/Au metallization system in Fig. 9.27c [9].

Die shear test results show that the bonding strength of a laser with Ti–Pt–thick Au–Ti–Cr–Au metallization is higher than that of a chip with Ti–Pt–thick Au metallization. It is revealed that a laser with both metallization structures can have good die bonding and solder wettability. These data above suggest that the transition metal Cr can act as an effective diffusion barrier metal for AuSn solder and preserve the integrity of the metallization system. It is promising to use Cr as the diffusion barrier in the thermal management of laser devices and improve their reliability.

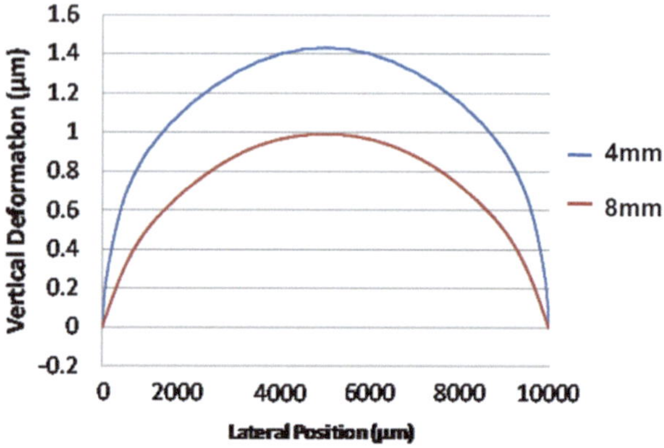

Fig. 9.28 The deformation curve of laser array in vertical orientation using copper heat sink with different thickness [18]

9.2.6 Thermal Stress Management

Thermal stress occurring during the packaging and operating processes influences the reliability of high power diode laser bars. Thermal stress is one of the most critical problems for the packaging technique of high power diode laser. The thermal stress can have a significant effect on the device performance: "smile," threshold, wavelength, polarization. Also, thermal stress may cause the formation of the "crack" on the facet to damage the laser chips and reduce the lifetime of the devices. According to the discussions in Chap. 4, thermal stress is a function of the mismatch of CTE between the heat sink and laser bar material, as well as the freezing point of the solder. Optimization of the heat sink material and die bonding process are two effective ways to reduce the thermal stress.

Optimization of the heat sink structure: The thickness of heat sink may have significant effects on deformation. Figure 9.28 shows the simulation results on the deformations of two 808 nm CS high power semiconductor laser arrays with different thickness (4 and 8 mm) of copper anode block [18]. The laser bar deformation in vertical orientation using thicker mounting copper heat sink is lower than that using thinner copper block. If the thickness of copper block heat sink is increased from 4 to 8 mm, the laser bar deformation in vertical position is reduced from 1.41 to 1 μm, which is a reduction of 29 %.

If the thickness of CS copper anode block is thicker, it can release the bulky thermal stress to some extent, which is caused by the large CTE mismatch between laser chip and CS mounting block heat sink. This makes the curvature smaller. If the thickness of CS copper anode block is thinner, it is very easy to generate warpage during the reflow process, bending together with the bar/array. In this way, a resultant force will be imposed on the bar/array. This leads to bar bending severely. With such thin and mechanically sensitive heat sinks the beam may be seriously degraded due to mechanical deformations which, in turn, lead to large smile [18].

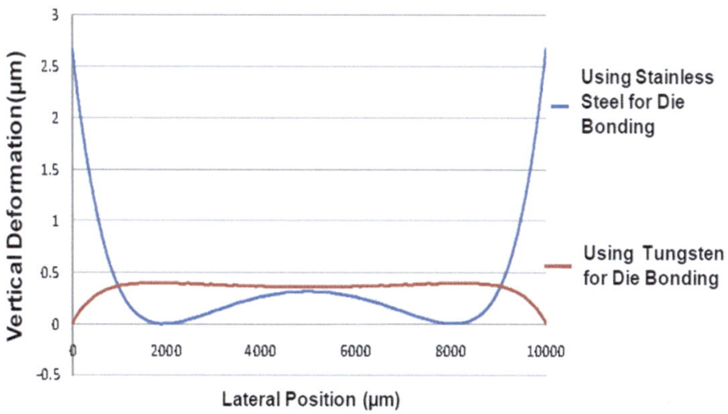

Fig. 9.29 The deformation curve of laser array in vertical orientation using different pickup tool for die bonding process [18]

Optimization of pickup tool for die bonding process: The laser bar deformation as a function of different pickup tool for die bonding process is simulated. Figure 9.29 shows the laser bar deformation in vertical orientation as a function of different materials of pickup tool for die bonding process [18].

Results from Fig. 9.29 indicate that the laser bar deformation in vertical orientation is significantly affected by the pickup tool material. The laser array is concave if Tungsten is used during the die bonding process, and the smile observed is less than 0.5 μm. In contrary to the tungsten, stainless steel pickup tool makes the smile larger than tungsten pickup tool. The smile is convex and exceeds 2.5 μm. These results suggest that tungsten is a better option for the pickup tool in die bonding process.

9.2.7 Approaches to Reduce Optical Feedback

1. *Optical isolator*:
   A solution for the diode failure from facet catastrophic optical damage COMD, which can be caused by the feedback of the high power laser beams, is the use of optical isolator. A low-cost isolator integrated within the module package has been developed and presented in Fig. 9.30 [19]. In this approach, an isolator is added in between the diode and the fiber laser. The isolator transmits the 9xx nm pump light and reflects the fiber laser light at 1,060 nm, thereby protecting the diode from potential optical damage. Normal modules with no feedback protection can be damaged at fiber laser powers of 10 W peak power. With the isolator,

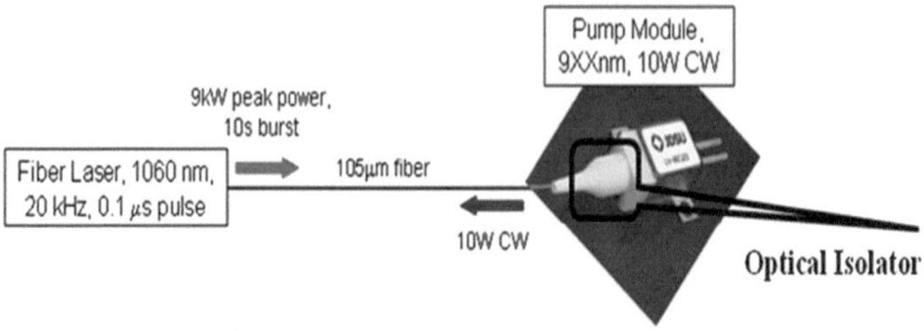

Fig. 9.30 Schematic of experiment demonstrating feedback isolation to pulsed fiber laser light at 1,060 nm [19]

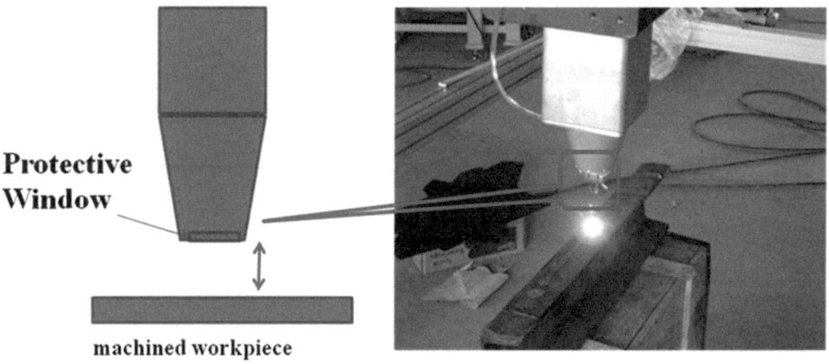

Fig. 9.31 Protective window designed for the laser processing [1]

no damage to the pump module is observed after being exposed to 10 s of fiber laser pulses at a peak power of 9 kW. It is clear that the feedback isolation of the pump module offers a convenient means to eliminate a common failure mode associated with damage to the laser diode from high-peak-power fiber laser pulses.

2. *Protective window*:
   The other solution for the diode failure, which is caused by the feedback from the workpiece in laser processing, is to add a protective window as shown in Fig. 9.31. The protective window is usually made of quartz glass, which is fixed in front of the laser head and coated with antireflection film. The protective window is an effective device to prevent the light reflection from the workpiece being processed. Also, the dust generated during laser processing can be kept out from the laser diode. As the protective window can be damaged under bad work environment, the protective window is usually designed such that it can be replaced with ease.

9.3 Lifetime Prediction

A laser diode lifetime can be divided into three periods: early failure period, random failure period, and wear-out failure period. The curve used to show the lifetime period is called bathtub curve, the failure in the early failure period is typically due to assembly and major semiconductor defects, which can be screened out with a well-designed burn-in process. The wear-out failure period is characterized by an increasing failure rate at the end of a laser's life. In between early failure period and wear-out failure period is the random failure period, which represents a diode's operation life with a relatively stable failure rate.

9.3.1 *Distribution Functions in Reliability Analysis*

Typical distribution functions used to analyze reliability data of semiconductor lasers are described below [20].

1. *Normal distribution*:
   The normal distribution is a typical continuous distribution used for quality control. In reliability analysis, the normal distribution is often applied to wear-out life. The probability failure density function *f*(t) and failure distribution function F(t) are expressed by the following equations.

$$f(t) = \frac{1}{\sqrt{2\pi}\sigma} \exp\left\{ -\frac{(t-\mu)^2}{2\sigma^2} \right\} \tag{9.3}$$

$$F(t) = \frac{1}{\sqrt{2\pi}\sigma} \int_{-\infty}^{t} \exp\left\{ -\frac{(x-\mu)^2}{2\sigma^2} \right\} dx \tag{9.4}$$

   This distribution is given by the mean parameter μ and the dispersion (variance) parameter σ. As shown in Fig. 9.32, the normal distribution has a symmetrical bell shape centering on μ [20].

2. *Exponential distribution*:
   The exponential distribution represents the life distribution (failure distribution function) in the random failure region where the failure rate λ is constant over time, and the probability failure density function f(t) and failure distribution function F(t) are expressed by the following equations. This distribution corresponds to the case when the shape parameter m = 1 in the Weibull distribution described hereafter (Fig. 9.33).

$$f(t) = \lambda e^{-\lambda t} \tag{9.5}$$

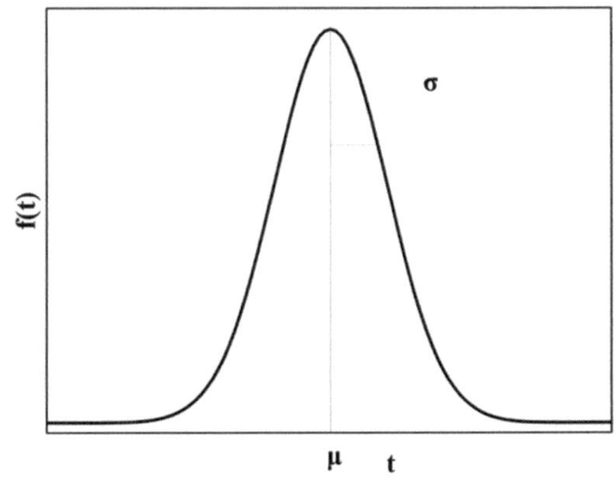

Fig. 9.32 Normal distribution [20]

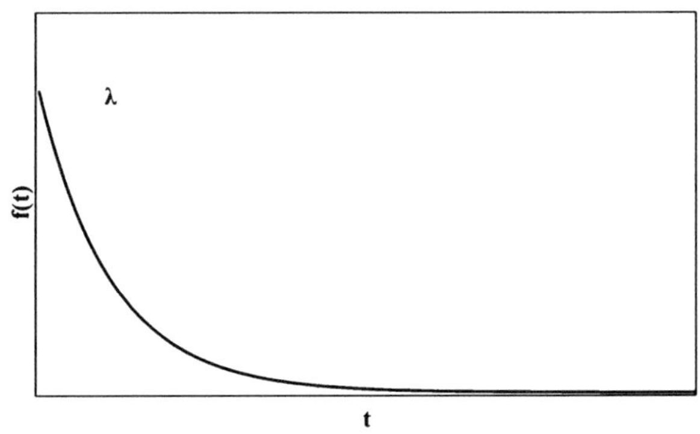

Fig. 9.33 Exponential distribution [20]

$$F(t) = \lambda e^{-\lambda t} \tag{9.6}$$

Note that as shown in the following Eq. (9.7), the MTTF is given from t_0, which is the inverse of the failure rate λ.

$$1/\lambda = t_0 = \text{MTTF} \tag{9.7}$$

3. *Logarithmic normal distribution*:
   The logarithmic normal distribution is a distribution function where ln t, which is the logarithm of the life time t, follows the above-mentioned normal distribution. The probability failure density function *f*(t) and failure distribution function *F*(t) are expressed by the following equations (Fig. 9.34).

$$f(t) = \frac{1}{\sqrt{2\pi}\sigma t}\exp\left\{-\frac{1}{2}\left(\frac{\ln t - \mu}{\sigma}\right)^2\right\} \tag{9.8}$$

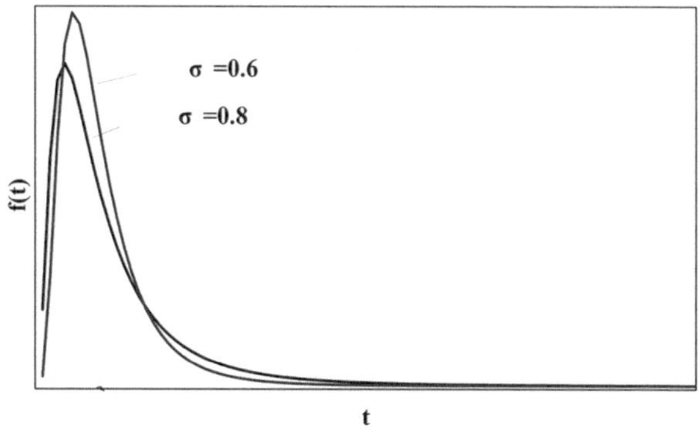

Fig. 9.34 Logarithmic normal distribution [20]

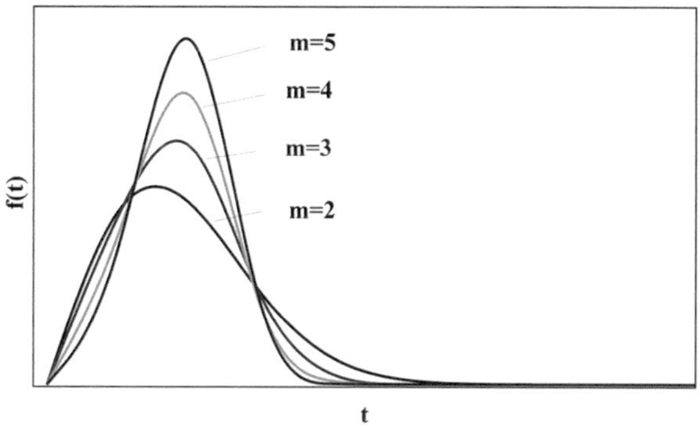

Fig. 9.35 Weibull distribution [20]

$$F(t) = \frac{1}{\sqrt{2\pi}\sigma}\int_{-\infty}^{t} \frac{1}{x}\exp\left\{-\frac{1}{2}\left(\frac{\ln x - \mu}{\sigma}\right)^2\right\}dx \tag{9.9}$$

4. *Weibull distribution*:
   The Weibull distribution is a weakest link model proposed by W. Weibull (Sweden) in 1939 as a mechanical breakdown strength distribution. This model was applied by J.H.K. Kao in 1955 to analyze the life of vacuum tubes, and has often been used since then to model life distributions in the analysis of semiconductor device reliability.

   The probability failure density function *f*(t) and failure distribution function *F*(t) are expressed by the following equations (Fig. 9.35).

$$f(t) = \frac{m}{\eta}\left(\frac{t-\gamma}{\eta}\right)^{m-1}\exp\left\{-\left(\frac{t-\gamma}{\eta}\right)^{m}\right\} \tag{9.10}$$

$$F(t) = 1 - \exp\left\{-\left(\frac{t-\gamma}{\eta}\right)^m\right\} \tag{9.11}$$

Here, m is called the form parameter, η is the measure parameter (characteristic life), and γ is the position parameter.

In addition, assuming $t_0 = \eta^m$, the failure rate λ(t) is expressed by the following equation.

$$\lambda(t) = \frac{m}{\eta}\left(\frac{t-\gamma}{\eta}\right)^{m-1} = \frac{m}{t_0}(t-\gamma)^{m-1} \tag{9.12}$$

The failure pattern can be defined from the value of the form parameter m.

$0 < m < 1$: Early failure pattern where the failure rate decreases over time;
$m = 1$: Random failure pattern where the failure rate is constant (matches with the exponential distribution);
$m > 1$: Wear-out failure pattern where the failure rate increases over time.

9.3.2 Life Estimation Method

To estimate the failure rate under normal operating conditions, a multi-cell test is performed under highly accelerated conditions. The failure rate is accelerated by high temperature, current, and optical power. Multi-cell tests are designed with different temperature and current/power combinations to derive the sensitivity of device reliability to each parameter. Analysis of the observed failure rate at each condition produces a model from which failure rates at normal operating conditions are extrapolated.

Life test data are analyzed with the maximum likelihood estimation method assuming a Weibull failure distribution and the dependence of the Weibull scale parameter η on junction temperature and optical power [21]:

$$F(t) = 1 - \exp\left(-\left(\frac{t}{\eta}\right)^\beta\right) \tag{9.13}$$

$$\eta(T_j, P) = \eta_{op} \bullet \exp\left(\frac{E_A}{k_B}\left(\frac{1}{T_j} - \frac{1}{T_{op}}\right)\right) \bullet \left(\frac{P}{P_{op}}\right)^{-n} \tag{9.14}$$

where F(t) is a cumulative failure density function of time t, β is the Weibull shape parameter, E_A is thermal activation energy, n is exponent of power acceleration, η_{op}, T_{op}, and P_{op} are scale parameters, junction temperature, and optical power at operating conditions. In ref. [21], at 35 °C heat sink temperature ($T_{op} = 51$ °C, $P_{op} = 6.5$ W), the following model parameters are given: $\beta = 0.84$; $E_A = 0.47$ eV;

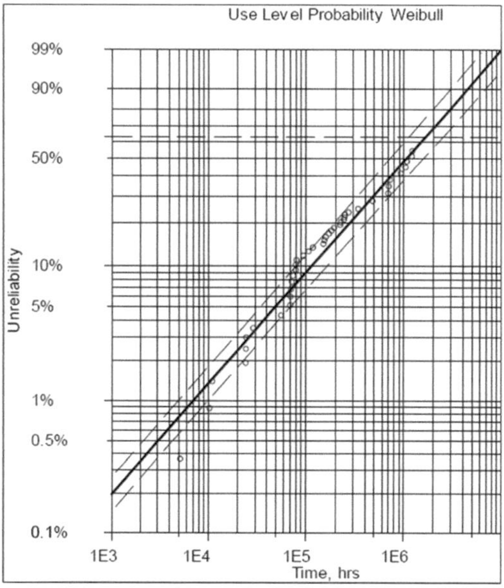

Fig. 9.36 Use level Weibull probability plot of the multi-cell test failures. *Dashed lines* represent 60 % one-sided confidence bounds. Projection is for 6.5 W, 35 °C heat sink temperature [21]

$n = 6.6$; $\eta_{op} = 1.7 \times 10^6$ h. Figure 9.36 presents a Weibull plot of all multi-cell test failures on a use level time axis where times are calculated using the acceleration model (9.13) at operating conditions.

References

1. B. Hao, D. Hou, T. Song, X. Yin, X.S. Liu, *Reliability Analysis of the High Power Semiconductor Laser in Application* (Internal Talk from Focuslight Technologies Co., Ltd., 2012), pp. 1–15
2. H. Kissel, G. Seibold, J. Biesenbach, G. Groenninger, G. Herrmann, U. Strauss, A comprehensive reliability study of high-power 808 nm laser diodes mounted with AuSn and indium. Proc. SPIE **6876**, 687618(1–10) (2008)
3. X.S. Liu, M.H. Hu, C.G. Caneau, R. Bhat, L.C. Hughes, C.E. Zah, in *Thermal Management Strategies for High Power Semiconductor Pump Lasers*. IEEE 2004 Inter Society Conference on Thermal Phenomena, vol. 2 (2004), pp. 493–500
4. Y.B. Qiao, S.W. Feng, C. Xiong, X.Y. Ma, H. Zhu, C.S. Guo, G.H. Wei, Spatial hole burning degradation of AlGaAs/GaAs laser diodes. Appl. Phys. Lett. **99**(10), 103506(1–3) (2011)
5. A. Hodges, J. Wang, M. DeFranza, X.S. Liu, B. Vivian, C. Johnson, P. Crump, P. Leisher, M. DeVito, R. Martinsen, J. Bell, A CTE matched, hard solder, passively cooled laser diode package combined with nXLT™ facet passivation enables high power, high reliability operation. Proc. SPIE **6552**, 65521E(1–9) (2007)
6. X.S. Liu, R.W. Davis, L.C. Hughes, M.H. Rasmussen, R. Bhat, C.E. Zah, J. Stradling, A study on the reliability of indium solder die bonding of high power semiconductor lasers. J. Appl. Phys. **100**(1), 013104(1–11) (2006).

7. D. Schleuning, M. Griffin, P. James, J. McNulty, D. Mendoza, J. Morales, D. Nabors, M. Peters, H. Zhou, M. Reed, Robust hard-solder packaging of conduction cooled laser diode bars. Proc. SPIE **6456**, 645604(1–11) (2007).
8. M. Leers, K. Boucke, C. Scholz, T. Westphalen, Next generation of cooling approaches for diode laser bars. Proc. SPIE **6456**, 64561A(1–10) (2007)
9. X.S. Liu, K. Song, R.W. Davis, M.H. Hu, C.E. Zah, in *Design and Implementation of Metallization Structures for Epi-down Bonded High Power Semiconductor Lasers*. IEEE 2004 Electronic Components and Technology Conference, vol. 1 (2004), pp. 798–806
10. M.K. Davis, A. Kussmaul, G.W. Yang, M.H. Hu, X.S. Liu, Y.Q. Zhu, D.A.S. Loeber, C.E. Zah, Impact of near-end residual reflectivity on the spectral performance of high-power pump lasers. IEEE J. Quantum Electron. **40**(4), 354–363 (2004)
11. M.T. Sheen, C.M. Chang, H.C. Teng, J.H. Kuang, K.C. Hsieh, W.H. Cheng, The influence of thermal aging on joint strength and fracture surface of Pb/Sn and Au/Sn solders in laser diode packages. J. Electron. Mater. **31**(8), 895–902 (2002)
12. E. Zakel, H. Reichl, Au-Sn bonding metallurgy of TAB contacts and its influence on the Kirk end all effect in the ternary Cu-Au-Sn. IEEE Trans. Compon. Hybrids Manuf. Technol. **16**(3), 323–332 (1993)
13. X.N. Li, Y.X. Zhang, J.W. Wang, L.L. Xiong, P. Zhang, Z.Q. Nie, Z.F. Wang, H. Liu, X.S. Liu, Influence of package structure on the performance of the single emitter diode laser. IEEE Trans. Compon. Packag. Manuf. Technol. **2**(10), 1592–1599 (2012)
14. J. Marsh, Quantum well intermixing revolutionizes high power laser diodes: monolithically integrated systems drive applications. Laser Technik J. **4**(5), 32–35 (2007)
15. S.P. Najda, G. Bacchin, B. Qiu, X. Liu, O.P. Kowalski, M. Silver, S.D. McDougall, C.J. Hamilton, J.H. Marsh, Benefits of quantum well intermixing in high power diode lasers. Proc. SPIE **5365**, 1–13 (2004)
16. R. Feeler, S. Colemanb, J. Levya, E. Stephensa, Elimination of deionized cooling water requirement for microchannel-cooled laser diode arrays. Proc. SPIE **6456**, 645617(1–8) (2007)
17. http://www.nadya-anscombe.com/downloadlibrary/EOOct1122-26.pdf
18. J.W. Wang, Z.B. Yuan, L.J. Kang, K. Yang, Y.X. Zhang, X.S. Liu, in *Study of the Mechanism of "Smile" in High Power Diode Laser Arrays and Strategies in Improving Near-field Linearity*. IEEE 2009 Electronic Components and Technology Conference (2009), pp. 837–842.
19. V. Wong, V. Rossin, J. Skidmore, P. Yalamanchili, X.D. Qiu, R. Duesterberg, P. Doussiere, D. Venables, R. Raju, J. Guo, M. Au, L. Zavala, M. Peters, G.W. Yang, Y.Z. Dai, E. Zucker, Recent progress in fiber-coupled multi-mode pump module and broad area laser-diode performance from 800–1500 nm. Proc. SPIE **7198**, 71980S(1–8) (2009)
20. *Sony Semiconductor Quality and Reliability Handbook* (Sony Corporation, October 2000), pp. 232–236
21. V. Rossin, M. Peters, E. Zucker, B. Acklin, Highly reliable high-power broad area laser diodes. Proc. SPIE **6104**, 610407(1–10) (2006)

Chapter 10
Applications of High Power Semiconductor Lasers

At early stage, high power semiconductor lasers have been mainly applied in pumping of solid-state lasers and fiber lasers. However, with the development of chip/bar technology and packaging technology in recent years, semiconductor lasers as light sources have been directly used in many new fields, such as medical and cosmetic, IR illumination, and material surface processing which includes cladding, hardening, and annealing [1–3]. The main applications of semiconductor lasers are shown in Fig. 10.1 [4]. In this chapter, we discuss the applications of high power semiconductor lasers in pumping of solid-state and fiber lasers, in material surface processing as well as medical and cosmetic fields. For pumping application, we mainly discuss the advanced pumping structure and the requirement on semiconductor laser pumping source. For material surface processing application, we introduce a high power semiconductor laser head and discuss its key technologies and applications. For medical and cosmetic, we mainly use laser hair removal as an example to discuss the direct semiconductor laser application in this field.

10.1 Pumping Applications

The major purpose of using light source as a pump source for lasers is to transform electric energy to radiation energy efficiently. Today only flash lamps, cw arc lamps, and semiconductor lasers are of practical applications.

Compared with lamps as the pump source, the semiconductor laser is the most efficient pump source for lasers, the advantages of the semiconductor laser pumping can be summarized as below [5]:

1. *Increased system efficiency*: A semiconductor laser pump source has a very narrow emission bandwidth. Although lamps have a higher efficiency (70 %), only a small fraction of the radiation is absorbed by the laser crystal due to its wide spectrum width. In contrast, the wavelength of a semiconductor laser can be chosen to fall completely within an absorption band of a particular gain medium.

X. Liu et al., *Packaging of High Power Semiconductor Lasers*, Micro- and Opto-Electronic Materials, Structures, and Systems, DOI 10.1007/978-1-4614-9263-4_10,

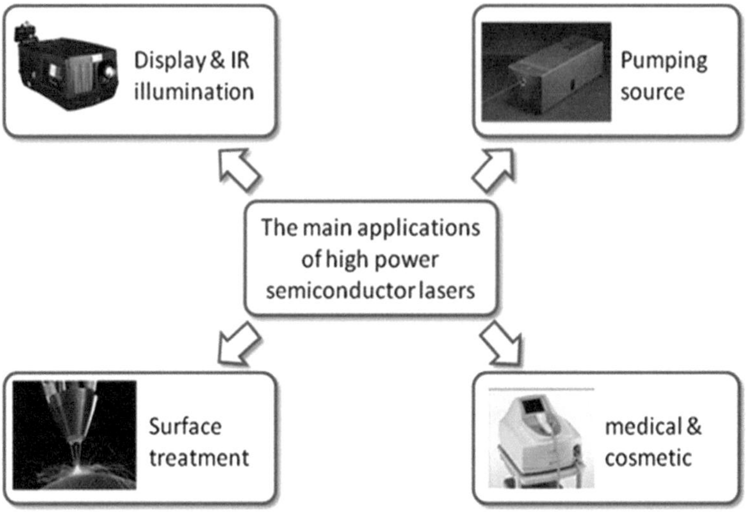

Fig. 10.1 The main application fields of high power semiconductor lasers [4]

2. *Improved beam quality*: A concomitant advantage derived from the spectral match between the semiconductor laser emission and the gain medium absorption band is a reduction in the amount of heat which is generated in the laser material. This reduces thermo-optic effects and therefore leads to better beam quality. In addition, the directionality of semiconductor radiation allows designs with good spatial overlap between pump radiation and low-order modes in the resonator, which in turn leads to a laser output with high beam quality.
3. *Enabling technology for new laser materials*: The most prominent laser materials which are pumped with semiconductor lasers can also be pumped with lamps. However, a number of very useful materials such as Nd:YVO_4, Yb:YAG, and Tm:YAG have reached prominence only as a result of semiconductor laser pumps.
4. *Increased component lifetime*: System lifetime and reliability is higher in semiconductor laser-pumped lasers compared to lamp-based system. Semiconductor lasers exhibit lifetimes on the order of 10^4 h in CW operation and 10^9 shots in the pulsed mode. Lamp life is on the order of 10^8 shots, and about 500 h for CW operation.
5. *Benign operating benefits*: The absence of high-voltage pulses, high temperatures, and UV radiation encountered with lamps lead to much more benign operation features of semiconductor laser-pumped systems. Furthermore, the high pump flux combined with a substantial UV content in lamp-pumped systems cause material degradation in the pump cavity. Such problems are virtually eliminated with semiconductor laser-pumped sources.

6. *Increased pulse repetition rate*: Besides CW operation, quasi-cw semiconductor lasers permit pulsed operation of lasers in the regime from a few hundred hertz to a few kilohertz.

10.1.1 Pumping for Solid-State Lasers

In the 1960s, the first solid-state laser pumped by semiconductor lasers was produced by Lincoln Laboratory in the USA [5]. The performances and pumping technologies of solid-state lasers have been continuously improved due to the performance improvement of semiconductor lasers over the years. In recent years, high power solid-state lasers with the output power of 10 kW or even up to100 kW have been widely studied. Ultra-high power solid-state lasers are the trend of the development of solid-state lasers.

The key criteria for semiconductor laser pumping in solid-state lasers are (1) efficient transport of pump power to the gain medium, (2) efficient absorption of pump radiation, and (3) high uniformity of absorbed pump power density. Therefore, one should select diode pump source with particular spectral irradiance, geometry, and temporal characteristics.

Materials for laser operation must possess strong absorption bands, and the spectral irradiance of the selected semiconductor laser source must match the absorption peak of the laser medium. So the center wavelength and the spectral width of diode lasers are very critical for the pumping of solid-state lasers. For example, the absorption peak of the Nd:YAG laser which is by far the most commonly used solid-state laser is 808 nm, and its absorption bandwidth at full width and half maximum (FWHM) is about 3 nm. Figure 10.2 shows the absorption spectrum of Nd:YAG which is expanded around the wavelength of 808 nm [5]. Therefore, the semiconductor laser with peak wavelength of 808 nm and spectral width of 3 nm is recommended [5]. Table 10.1 presents the absorption peak and band width of some common laser materials [5].

Temperature gradients and stress nonuniformity within a diode laser array lead to a broader spectral output for the whole array as compared to a single emitter device. In diode-pumped lasers that have only a short optical absorption path, it is important to have a narrow spectral emission from the semiconductor laser array in order to absorb most of the pump radiation. However, generally speaking, in optically thick materials, such as lager laser rod or slab with dimensions on the order of 10–15 mm, spectral width becomes less critical [5].

Usually the size of a laser crystal is small, while the beam size of the semiconductor laser, especially the semiconductor laser stacks, is too large to efficiently transport the pump power to the gain medium. Hence, the beam shaping and focus optical system should be designed to improve beam quality of semiconductor lasers and satisfy the requirement of pump beam. The beam shaping optical system includes three parts which are collimation lenses, beam transformation system, and beam focus system. Light duck and other complex focus systems have been

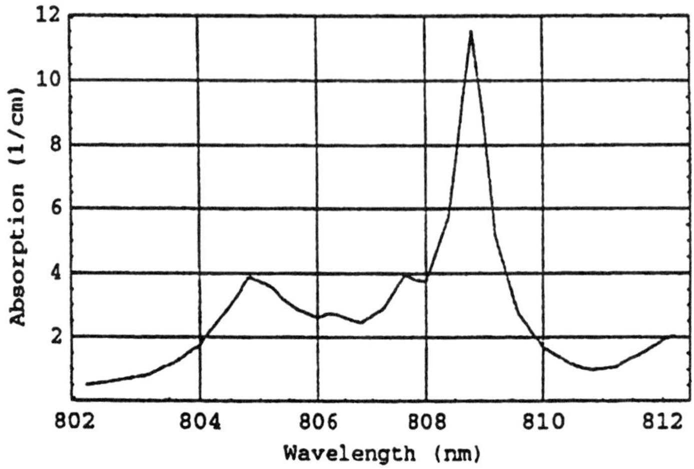

Fig. 10.2 Absorption spectrum of Nd:YAG around 808 nm [5]

Table 10.1 Absorption peak and bandwidth of some laser materials [5]

Laser material	Absorption peak (nm)	Absorption bandwidth (nm)
Nd:YAG	808	3
Nd:YVO4	808	16
Yb:YAG	940	18
Er:YAG	963	–
Tm:YAG	783	5

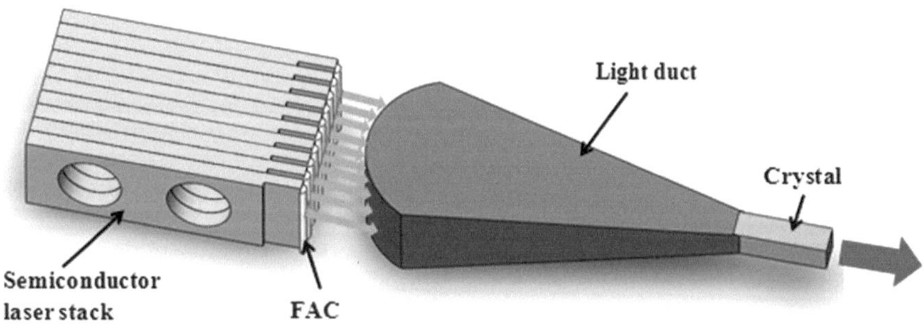

Fig. 10.3 A typical pump system with light duct [4]

designed for beam shaping and compressing [6]. Figure 10.3 shows the principle of beam shaping system with light duct [4]. The collimation of fast axis of semiconductor laser beam is generally needed for most of the beam shaping systems. The light duct is designed in such a way that total reflection of the pimp beam is achieved. In transmission, the pump beam is reflected many times in the light duck, beam size is compressed gradually and meanwhile the beam intensity becomes

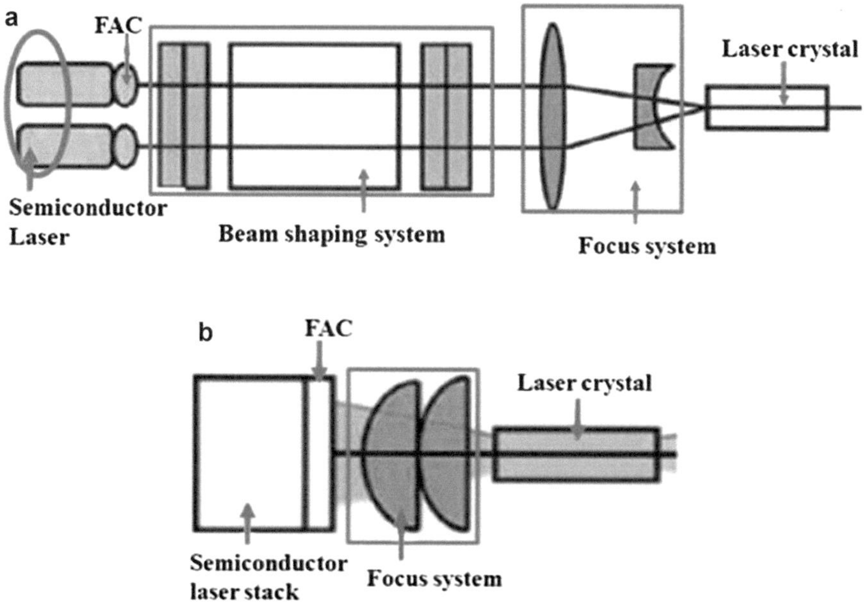

Fig. 10.4 The optical system of pump source of semiconductor lasers [4]. (**a**) Pump system with shaping and focus optical system. (**b**) Pump system with focus optical system

homogenize at output end. If smaller beam size of laser stacks are needed, the beam compressing system becomes complex, and beam shape and focus optical system should be designed. Figure 10.4a, b shows another two designs of beam shape and focus system [4]. In order to obtain smaller beam size, slow axis collimation lens also should be used to collimate slow axis beam. After that, the collimated beam should be shaped to make beam parameter product (BPP) of fast and slow axes symmetrical, and then the shaped beam is focused by focusing system.

In the following, the pumping structures of the most common diode-pumped solid-state lasers are introduced. Based to the shape of gain media, the solid-state lasers are categorized into rod lasers, slab lasers, and disk lasers. The requirements of pumping source for these three different lasers are described and the common pumping sources for different pumping schemes are presented.

10.1.2 Rod Lasers

The rod laser is the most mature laser type and it has been widely used. This geometry is most easily manufactured and gives the output beam a generic rotational symmetry, which is often favorable in applications. The rod gain media mainly includes Yb:YAG and Nd:YAG. The most common pumping schemes of rod lasers are side pumping and end pumping.

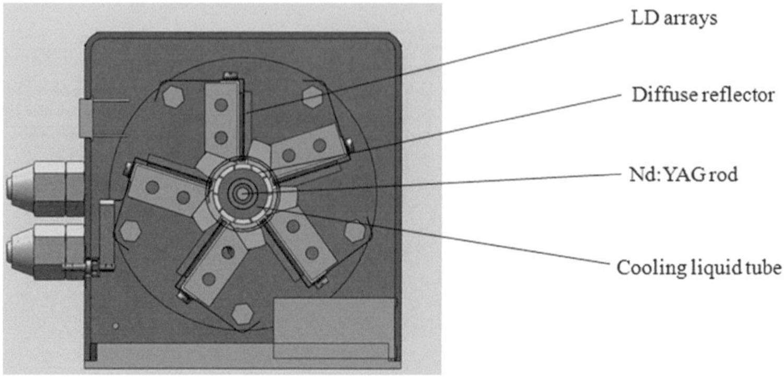

Fig. 10.5 Schematic structure of a side-pumped rod laser [4]

Side pumping: Side pumping is shown in Fig. 10.5 [4]. The laser rod is surrounded by pump sources from five sides which are well-distributed in five directions, each containing several bars arranged horizontally. This side-pumped geometry results in excellent gain uniformity. In addition, three or seven direction-pumped structures are also employed regularly.

Depending on the structure of the side-pumped rod lasers, and for high output power, the LD bars even bar stacks are often chosen as pump sources for side-pumped rod lasers. For high output power or high quality, pumping uniformity along the rod and from different sides need to be maintained. Uniform pumping can cancel bi-focusing of thermal lens in a rod-geometry medium and is necessary for high-power, high-quality beam generation. Hence, the LD bars of every stack array should be with the same wavelength which is matched with the absorption spectral of the gain medium. In addition, the beam profile of every array should be uniform.

As shown in Fig. 10.6, the horizontal arrays are chosen as pump sources for side-pumped rod lasers [4]. Three, five, or seven of the H-arrays are arranged around the gain medium depending on three-side, five-side, or seven-side pumping. The output power of horizontal lasers with four laser bars can reach 400 W in CW operation, and the output power of a few hundred watts for solid-state laser can be obtained.

In order to achieve higher output power, more semiconductor lasers should be arranged around crystal rod and laser stacks have been used as pump source. Figure 10.7 shows four 3-bar vertical stacks are arranged horizontally on a liquid cooler [4]. Due to the small pitch of the vertical stacks, these pump sources are often used only in QCW mode.

In order to further improve the output power, the rod is pumped in high density from more directions. Actually the pump source forms a half circular shape or polygonal shape, as shown in Fig. 10.8 [4]. The output power of a pump module is 8 kW in QCW with 4 % duty cycle. One or two of such high density pump sources are used depending on the design.

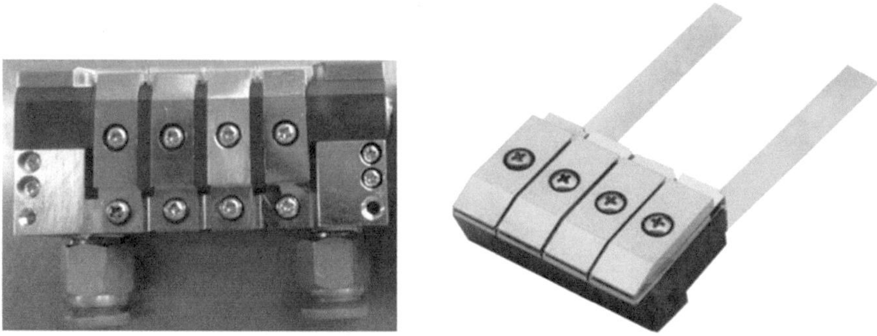

Fig. 10.6 Horizontal array semiconductor lasers for side pumping of a rod laser [4]

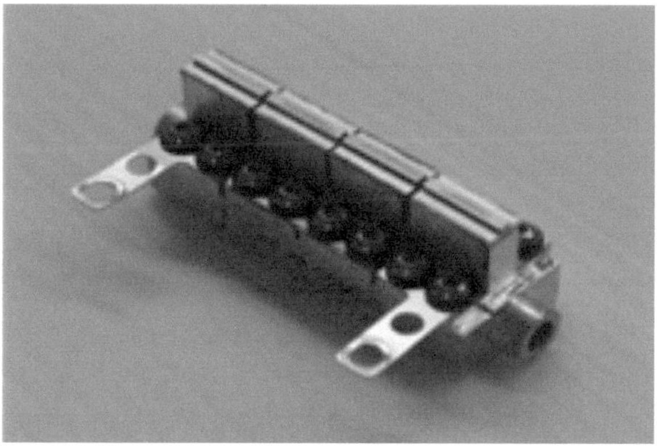

Fig. 10.7 Horizontally arranged bar stacks for side pumping of a rod laser [4]

Fig. 10.8 Polygonal shaped bar stacks for side pumping of a rod laser [4]

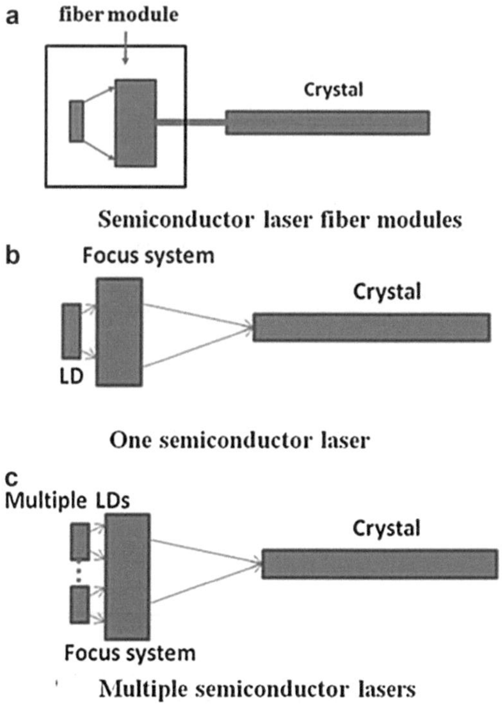

Fig. 10.9 Schematic structures of the end-pumped rod lasers [4]

End pumping: End-pumped rod lasers give the optimum beam quality and operate in the lowest-order transverse resonator mode. In addition, this architecture offers higher pump intensities, which are important in quasi-three-level lasers, such as Yb:YAG.

Usually the diameter of the laser rod is less than 1 cm. The semiconductor laser as the pump source of the end-pumped rod laser can be focused into the rod through an imaging system. Figure 10.9a shows a typical pumping structure which uses a fiber-coupled LD module as the pump source [4, 7]. The beam out of the fiber is focused into the rod through an imaging system. When an open package diode laser is used for a pumping source, a focus system is needed to focus the laser beam into the rod, as shown in Fig. 10.9b [4, 7]. For higher output power, multiple diode lasers with complex beam shape and focus optical system are used as pump source as shown in Fig. 10.9b [7]. As mentioned above, the diameter of the rod is usually small, so the maximum pump power depends on the damage threshold of the gain media.

Figure 10.9a shows an end-pumped rod laser which employs fiber-coupled diode as pump source. Using fiber-coupled semiconductor laser as pump source has advantages: the beam at the fiber output has a circular distribution and ability to

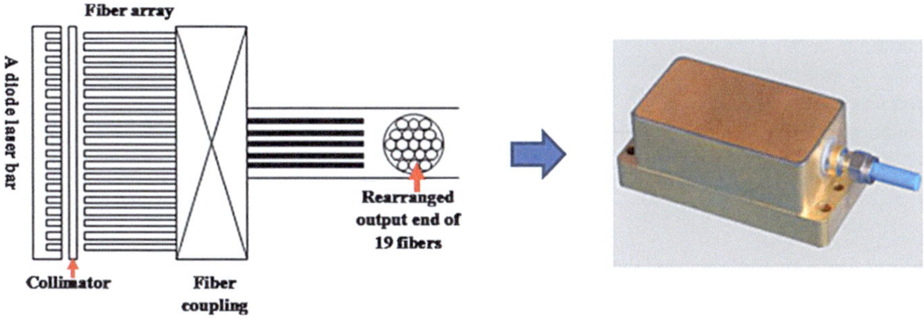

Fig. 10.10 Structure of a fiber-coupled module used for end pumping of a rod laser [4]

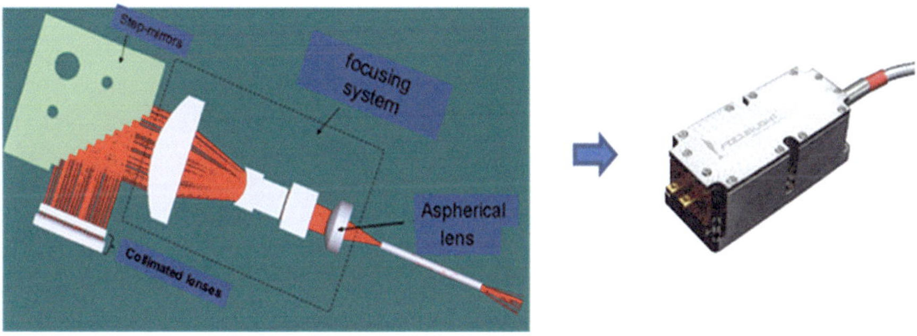

Fig. 10.11 Structure of a high brightness fiber-coupled module used for end pumping of a rod laser [4]

remove heat from the diode to the optical components of the laser; the pump source can be replaced relatively easily without disturbing the alignment. By far most of the low power solid-state lasers employ fiber-coupled semiconductor lasers as their pump sources. Figure 10.10 shows one structure of the fiber-coupled modules [4]. This type of pump source offers high output power up to 60 W, high reliability and durability. However, the brightness is relatively low. For higher pump brightness and power density, fiber-coupled modules which employ advanced beam shaping and coupling technology are used. Figure 10.11 shows one typical example [4]. For these types of module, an output power up to 50 W from a 400 or 200 μm fiber can be achieved.

For the pumping configuration shown in Fig. 10.9b, a conduction-cooled laser bar as shown in Fig. 10.12 would be a good choice and it is commonly used [4]. This type of laser bar can provide up to 80 W CW power and is cost-effective as compared to fiber-coupled modules. For the pumping structure shown in Fig. 10.9c, a V-stack diode laser as shown in Fig. 10.13 can be chosen as pump source [4]. V-stacks generally can provide a few hundred watts or even a few thousand watts of power. Due to the poor beam quality of these types of pump sources, a beam shaping and focusing optical system has to be used.

Fig. 10.12 A conduction-cooled laser bar commonly used for end pumping of a rod laser with a beam shaping and focusing optical system [4]

Fig. 10.13 An example of a V-stack diode laser used for end pumping of a rod laser with a beam shaping and focusing optical system [4]

10.1.3 Slab Lasers

Owning to the large area for pumping and heat dissipation, slab laser can achieve high output power while maintaining good beam quality and efficiency. The invention of zigzag slabs in the early 1970s by Bill Martin and Joe Chernock launched a new paradigm in the development of solid-state lasers [8]. The idea of propagating laser beams in a direction that averages the temperature gradients in the gain medium has been the cornerstone of power scaling of solid-state lasers. The pumping structures can also be divided into end-pumped and side-pumped structures.

Side Pumping

As shown in Fig. 10.14, the laser slab is sandwiched between a diode array and a heat sink [5, 9]. The heat sink contains a reflective coating to return unused pump

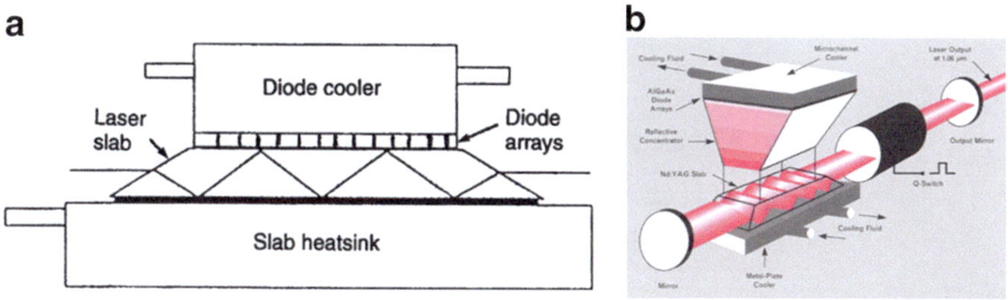

Fig. 10.14 Structures of side-pumped slab laser (**a**) with and (**b**) without intervening optics [5, 9]

radiation back to the slab for a second pass to reduce coupling losses. Generally, no intervening optics is used as shown in Fig. 10.14a. In order to get higher pump power density, as shown in Fig. 10.14b, then a reflective concentrator is employed. The purpose of the zigzag optical path is to mitigate the effects of thermal lens caused by thermal gradients in the slab.

Obviously, the pump source would be a diode array or two dimensional arrays. The laser bars in the full array should have close center wavelength for high absorption and good beam quality. The beam profile of the pump source on the slab surface should be uniform. A laser crystal slab typically has large size, therefore the pump laser arrays generally only need fast axis collimation to reduce divergence angle in fast axis. However, for super high power applications, more pumping power is required and the number of laser bars used is in the high tens of even hundreds. In these cases, the size of laser crystal slabs is smaller than the beam size of pump laser area array and beam shaping system should be considered to reduce pump beam size. Some examples of pump sources used for side-pumped slab lasers are shown in Fig. 10.15 [4]. The pump sources can deliver a few thousands or even tens of thousands of power in CW or QCW mode. It was reported that the average output power of 67 kW was achieved based on this structure [10].

End Pumping

End-pumped slab architectures decouple the slab absorption length from the traditional cooling geometry of the slabs, thus providing scalability that comes with using thinner slabs. In addition, the same with the end-pumped rod lasers, end-pumped slab laser offers high pump intensities, which are important in quasi-three-level lasers, such as Yb:YAG. Figure 10.16 shows an example of a conduction-cooled, end-pumped slab laser architecture [8, 11, 12].

In end-pumped slab laser, the pump source must have high brightness along the thickness direction of the slab to allow efficient optical coupling into the slab

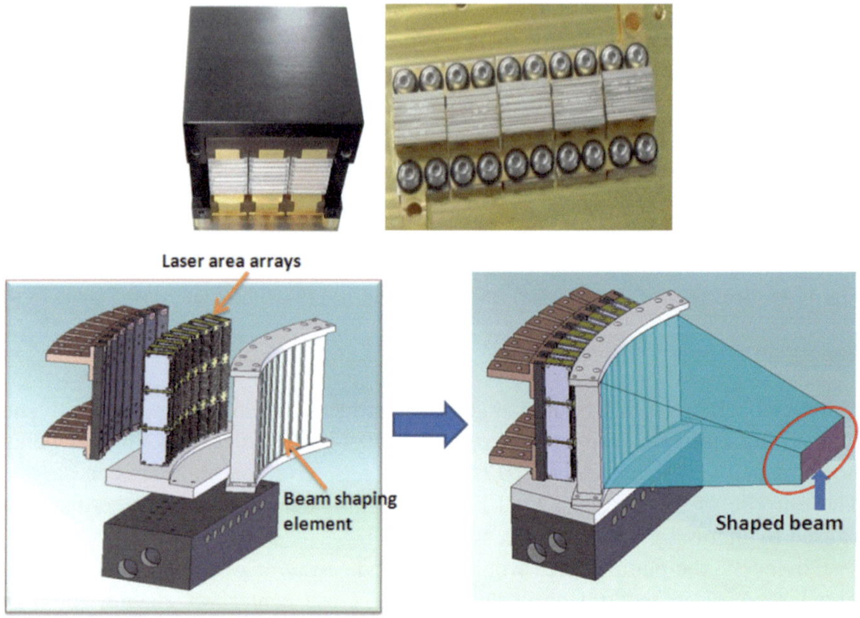

Fig. 10.15 Examples of pump sources used for side pumping of slab lasers [4]

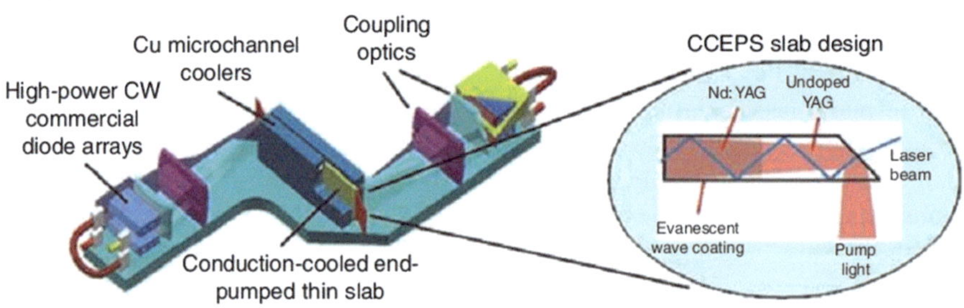

Fig. 10.16 Structures of conduction-cooled, end-pumped slab laser [8]

aperture. With the requirement of high output power and high brightness, diode laser vertical stack or stack arrays are the preferred pumping sources. The pump sources introduced in side pumping of slab lasers could be used for end pumping as well, but more sophisticated beam shaping and focusing technologies are required. In order to achieve high brightness, the diode bars used in the stacks are fast axis collimated by using micro-lenses. Figure 10.17 shows a diode bar with fast axis collimator (FACs), which is the building block of the pump source for end pumping of slab lasers [4].

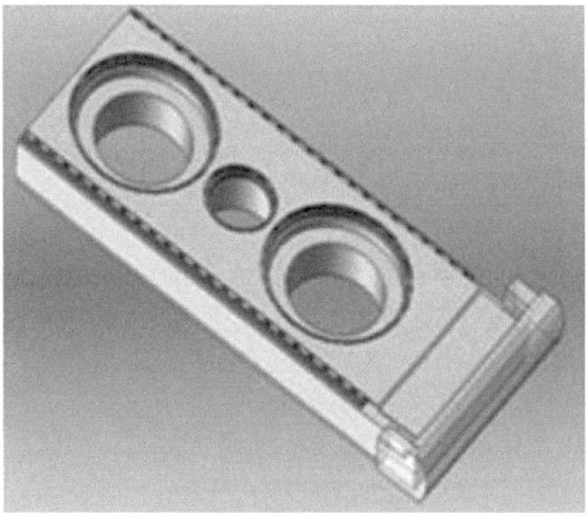

Fig. 10.17 A laser bar with a FAC [4]

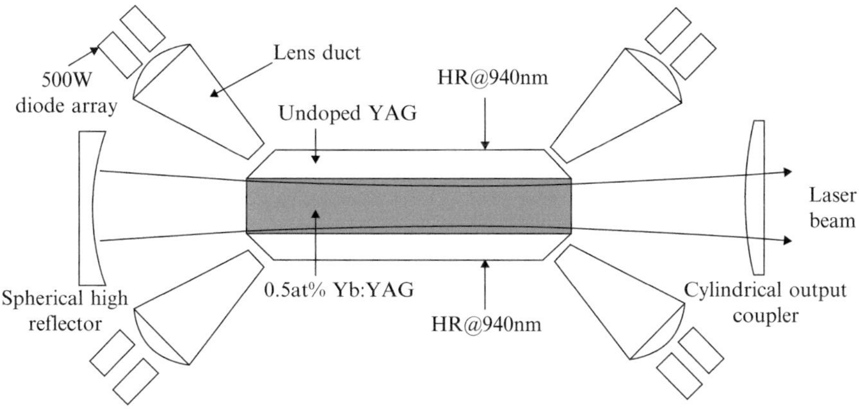

Fig. 10.18 The structure of the corner-pumped slab laser [13]

Corner Pumping

Another pump method for high-average-power slab laser which has particular application to high-average-power quasi-three-level lasers is corner pumping. Figure 10.18 shows the structure of this pumping method [13]. As can be seen, diode arrays with lens ducts are used as the pump source.

10.1.4 Disk Laser

The design of disk laser makes high power solid-state lasers feasible which yield high efficiency and good beam quality. In addition, this design also shows excellent results with quasi-three-level laser crystals like Yb:YAG, which cannot be operated with good results using thick rods or slabs [11].

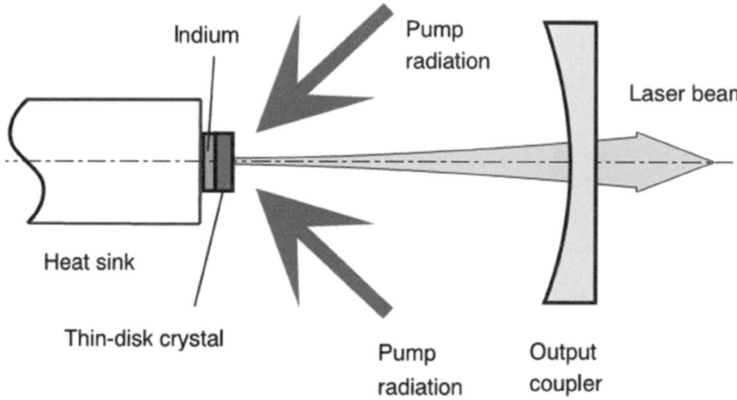

Fig. 10.19 Fundamental structure of a face-pumped disk laser [11]

Face Pumping

Figure 10.19 shows the face-pumped disk laser which is the fundamental scheme of disk laser design [11]. The disk itself is antireflection coated for both pumping laser wavelengths at the front side and high-reflection coated for both wavelengths at the rear side which is mounted on the cooling device. The laser resonator is built by the disk itself and an output coupler in front of the thin disk. The thickness of the crystal disk is small compared to the diameter of the pumped area. So the heat is effectively conducted into the heat sink. The output power of such a thin-disk laser can be scaled easily by increasing the diameter of the pumped area while keeping the pump power density constant, thus the beam quality remains unaffected.

Depending on the design and diameter of the pumping area, proper pump source, such as fiber-coupled modules, H-array, V-stack, or area array with certain beam shaping and focus optical system should be chosen [14, 15].

End Pumping

Another way for increasing the output power is to increase the absorption efficiency of the pump power and the use of several disks in one resonator [11]. As shown in Fig. 10.20a, 16 passes in the disk gain medium can be realized by the pumping scheme. In this way, the optical to optical efficiency of up to 48 % can be achieved [11]. Figure 10.20b shows the resonator design with two disks. In this pumping scheme, fiber-coupled modules are preferred as its pump source, so that the size of beam profile would be reflected enough times by the parabolic mirror.

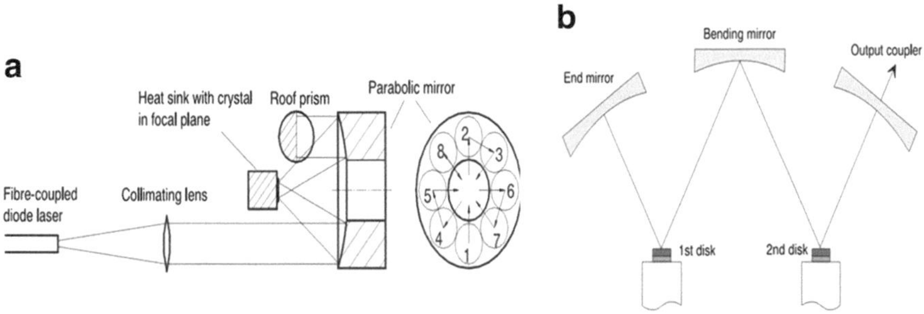

Fig. 10.20 Structures of disk laser design for the scale of the output power by the use of (**a**) 16 pass of the pump radiation and (**b**) two disks in one resonator [11]

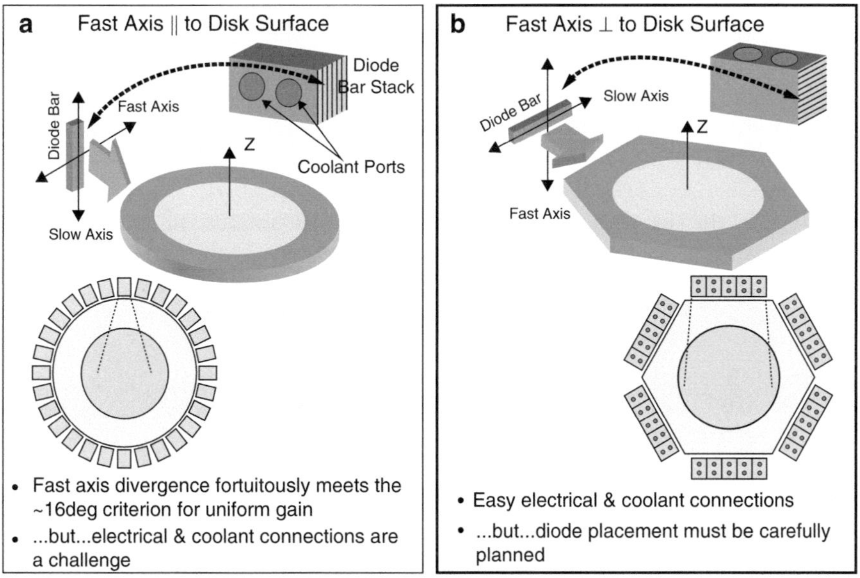

Fig. 10.21 Examples of configurations for edge pumping of disk lasers. (**a**) Diode fast axis parallel to disk face; (**b**) diode fast axis perpendicular to disk face [10]

Edge Pumping

The disk lasers can also be edge pumped. In the edge-pumped disk laser, the disk is usually of composite construction with a laser-active central portion and a peripheral edge which can offer good coupling between the pump sources and the gain medium. The pump lasers are arranged around the circumference of the composite disk and generally point toward its center. Figure 10.21 shows two approaches of edge pumping of disk lasers that yield uniform pumping [10]. In the former case, in Fig. 10.21a, the fast axis is parallel to the disk surface. And in the latter case in

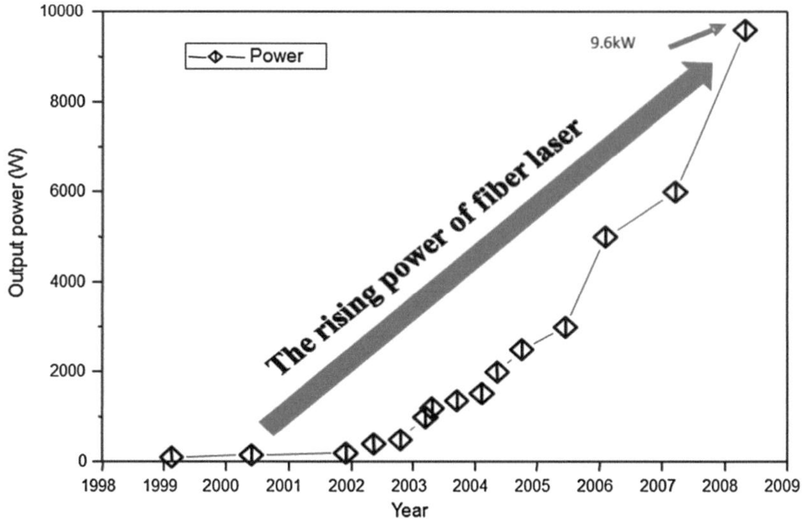

Fig. 10.22 The development of the output power of fiber lasers from a single fiber [4]

Fig. 10.21b, the fast axis is perpendicular to the disk surface. In these pumping schemes, vertical stacks are chosen as pump sources. Typically beam shaping and focusing optics are required to make the pump beam smaller and easier to couple into the edge of the disks.

10.1.5 Pumping for Fiber Lasers

The first fiber laser pumped by a laser diode laser was born in 1987, and the gain fiber is an Nd-doped silica fiber and the pump source is GaAs laser diodes [16]. After 30 years, the fiber lasers were developed rapidly and applied in many fields. As shown in Fig. 10.22, from 2002 to 2009 the output power of fiber lasers had been improved fast, and the output power of 9.6 kW from a single fiber had been realized at 2009 by IPG of USA [4, 10, 17–23].

Pumping Structures of Fiber Lasers

There are two kinds of resonant cavity structures of fiber lasers. One is space-coupled resonant cavity, and the other is all-fiber resonant cavity with all-fiber components. The most important property of space-coupled resonant cavity is that the resonant cavity is composed by two coated mirrors which are similar as solid-state lasers [24]. The optical focus system is added to compress pump beam into the doped fiber. However, in actual applications, the disadvantage of the space-coupled

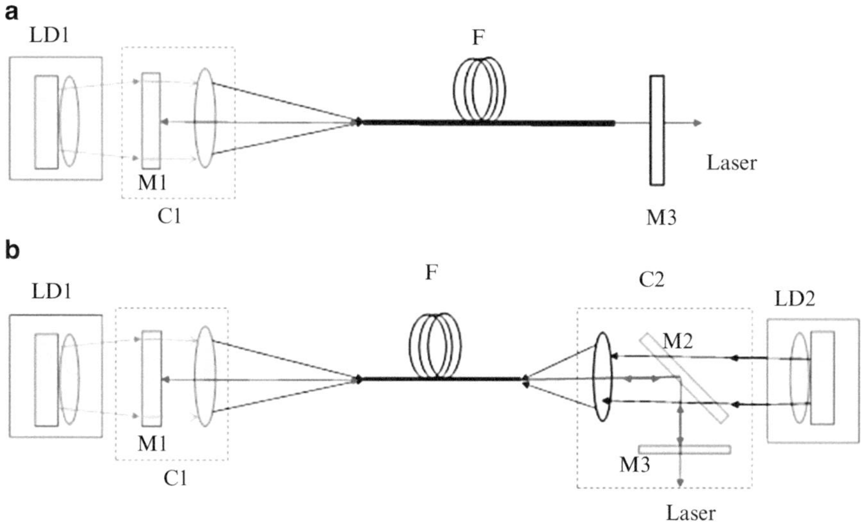

Fig. 10.23 Schemes of end pumping using optical focus systems [4, 26]

resonant cavity of fiber lasers is obvious. Space-coupled fiber lasers have high requirements on the work environment. The space-coupled resonant cavity is easily disturbed by the change of ambient vibration and temperature conditions during operation, and in some case, the laser beam cannot be generated. It is difficult to design a sealed package for space-coupled fiber lasers and it is difficult to manufacture this structure commercially. In addition, space-coupled fiber lasers have low wall-plug efficiency due to complex optical system.

In order to overcome the disadvantages of space-coupled fiber lasers, an all-fiber resonant cavity with all-fiber components has been proposed in recent years [25]. With the development of the all-fiber technology, it replaced the space-coupled resonant technology. Besides its simple structure, it has high pumping efficiency and it is easy to maintain. The design of all-fiber lasers is flexible and versatile, and many new structures have been proposed and applied to achieve higher output power. In an all-fiber laser system, the two coated resonant cavity mirrors is replaced by Fiber Bragg Gratings (FBG) and the pump light source from fiber-coupled modules is coupled into the doped fiber directly without other optical elements [25]. According to the relative position between pump source and doped fiber of fiber lasers, the pump structures can be classified into end-pumped and side-pumped structures with end pumping being dominant in applications.

End Pumping with Optical Focus Systems

End pumping is used early in space coupling pump structure for fiber lasers. In the scheme of end pumping, as shown in Fig. 10.23, C1 and C2 are collimation and

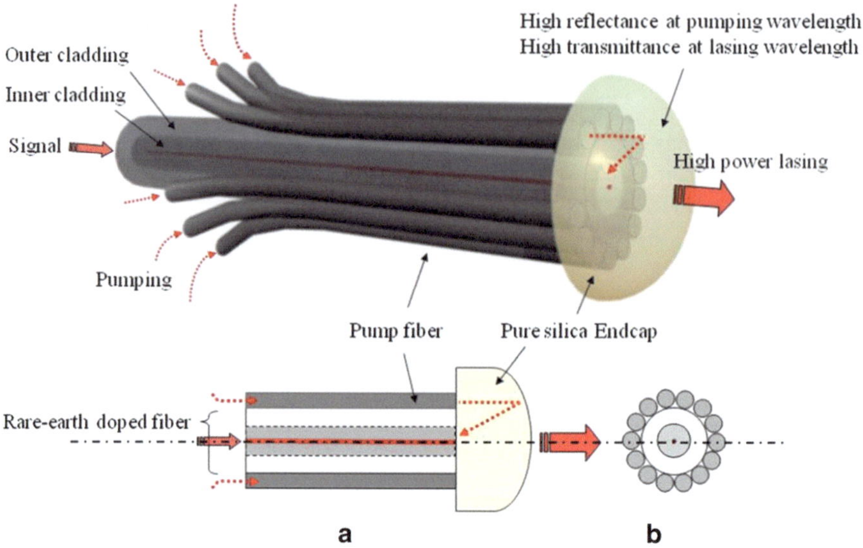

Fig. 10.24 Plano-convex two-color mirror of end-side pump technology [27]

focus systems and both of them include two lenses and a dichroic mirror, respectively [4, 26]. The function of C1 and C2 is to compress and focus pump beam of semiconductor laser LD1 and LD2 first, and then the focused beam is coupled into doped fiber at focal point. In addition, the dichroic mirrors of M1 and M2 and M3 in C1 and C2 are used to form the resonant cavity.

Plano-convex two-color mirror end pumping technology: This pump structure, shown in Fig. 10.24, is one of all-fiber resonant cavity of fiber lasers [4, 27]. As presented in Fig. 10.24, the pump sources are composed of multiple fiber-coupled diode laser modules. The fibers of the modules are arranged around the doped fiber of the fiber laser. In order to couple the pump beam into doped fiber, the plano-convex two-color mirror is used to compress the pump beam and deliver the pump light into the inner cladding. The surface of plano-convex two-color mirror is a concavity with high-reflection film as shown in Fig. 10.24. The pumping beam are reflected and focused at focus point. The doped fiber is put at focus point of plano-convex two-color mirror and pump beam is coupled into the inner cladding. However, due to the fact that the beam quality of diode laser fiber-coupled modules is not very good, there is large beam loss after reflection by the plano-convex two-color mirror, especially when the diameter of fiber lasers is small.

Taper-fused fiber combiner end pumping technologies: This technology is to fuse the fibers of pump sources and doped active fiber together. After fusing, the end side of pump fibers and doped fiber are drawn to be one fiber which is the taper-fused fiber combiner. There is N + 1 fibers to be fused, in which N denotes number of pump fibers and "1" denotes one doped fiber. Before fusing, the coating layers of pump

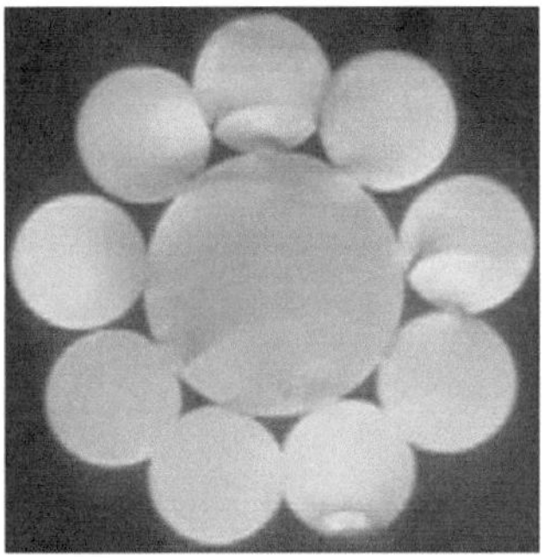

Fig. 10.25 The arrangement structure of pump laser fibers and doped active fiber of a fiber laser [28]

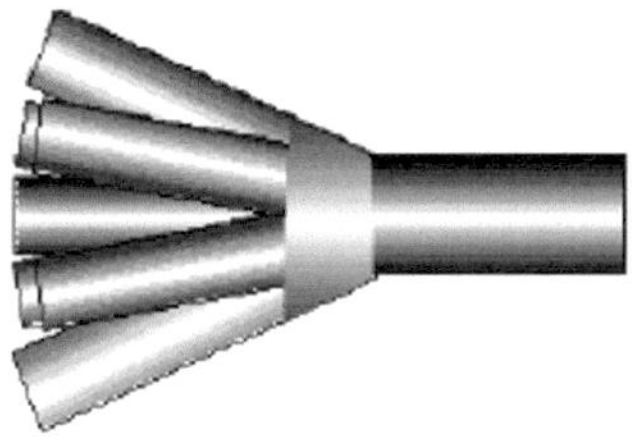

Fig. 10.26 Illustration of taper-fused fiber combiner used in fiber lasers [28]

fibers and the doped fiber should be stripped. Then the N fibers of pump sources are arranged around doped fiber as shown in Fig. 10.25. Finally the arranged pump fibers and the doped active fiber are fused together under high temperature [28]. As shown in Fig. 10.26, the taper-fused fiber combiner is formed [28]. In order to achieve high beam coupling efficiency, the diameter and numerical aperture of pump lasers and doped fiber should meet the following relationship [28]:

$$\sqrt{N}D_iNA_i \leq D_0NA_0 \tag{10.1}$$

where N is the number of pump lasers in taper-fused fiber combiner; D_i and D_0 are the fiber diameters of pump laser and doped fiber, respectively; NA_i and NA_0 are the fiber numerical aperture of pump laser and doped fiber, respectively. Besides pump lasers and doped fiber are fused together, a signal fiber also can be fused with them to compose the taper-fused fiber combiner.

Based on this coupling technology, the output power of fiber lasers from a single fiber can be over 3,000 W, and the coupling efficiency is 81.7 % [29]. Due to the high coupling efficiency and simple design structure, the taper-fused fiber combiner technology is wildly used in fiber lasers today.

V-grooves end pumping technology: V-grooves technology is to erode V-grooves on doped fiber [30, 31]. The process is fairly complex and need high precision control. The process steps are: (1) the coating layer of double-clad active fiber is removed; (2) the V-grooves are etched in the exposed inner cladding. The depth must be controlled accuracy to avoid the damage of inner core of the active fiber; (3) the surface of the V-grooves is optically polished and the reflecting film is coated onto

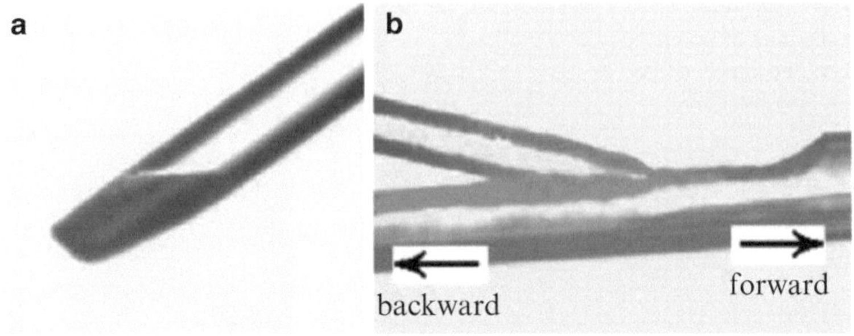

Fig. 10.27 The microscopic photographs of (**a**) the polished fiber end and (**b**) the fiber coupler [34]

V-grooves; (4) antireflection film for pump laser wavelength is coated on the reception side of V-grooves; and (5) the pump beam is focused into the inner core by optical lenses and reflected by the V-grooves to pump the doped fiber. This technology has good power expansion ability. It was reported that the output power of 1 kW was achieved from a single fiber by this method. The coupling efficiency is above 80 %. However, the inner cladding layer of doped fibers is easy to be damaged during the fabrication process of V-grooves. This technology has yet to be improved and optimized to obtain high yield and manufacturability.

Angle-polished side pumping technology: The principle of angle-polished method is shown in Fig. 10.27 [33, 34]. It consists of multi-mode fibers of pump sources and a double-clad doped fiber. The output end of fibers of pump sources is polished at angle γ, and the outer coating of doped fiber is removed. The microscopic photographs of the polished fiber end and the fiber couplers are shown in Fig. 10.27. The coupling angle (the angle between the axis of the pumping fiber and the double-clad fiber) is determined mainly by the polished angle.

The coupling efficiency of this method is mainly determined by two factors which are the polished angle and the fiber interconnection process. The value of polished angle should be optimized. If it is too large, the pumped beam will have a large incidence angle and some of the beam will be reflected at the boundary interfaces at the inner and outer cladding in doped fiber. If the angle is too small, the light leakage at both interfaces will be large [33]. The optimum angle of polished angle is 15° and the coupling efficiency of fiber laser is 90 % [34]. The index-matching material is added between pump fibers and doped fiber to strengthen the connection of them.

Pump Sources for Fiber Lasers

Same as pumping for solid-state lasers, one of the key criteria for semiconductor laser pumping in fiber laser is efficient absorption of pump radiation. Therefore, particular pump wavelength and good beam quality are required.

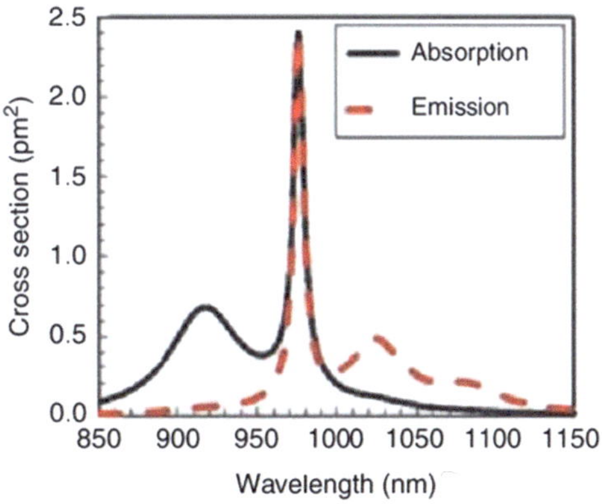

Fig. 10.28 The absorption and emission cross section of Yb^{3+} ions in aluminosilicate fibers [8]

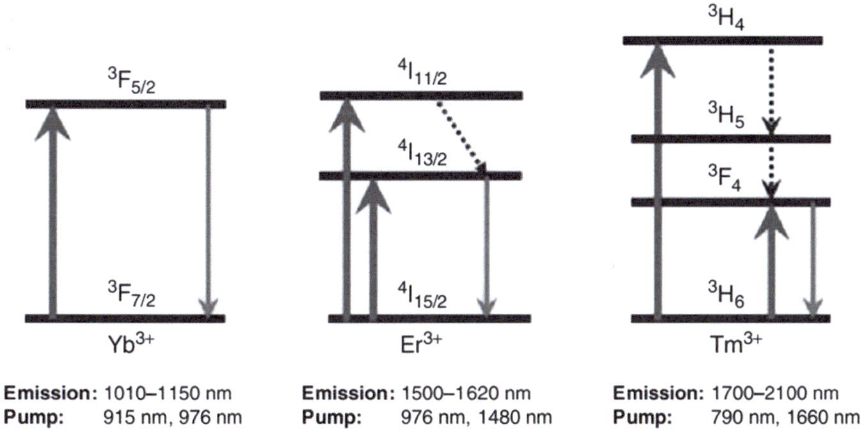

Fig. 10.29 The lower-level transition and wavelength of pump and emission for Yb^{3+}, Er^{3+}, Tm^{3+} [8]

Wavelengths of the Pump Sources

There are various gain medium with different doping, and the wavelength of pump light sources should match the absorption of the gain medium [33]. As shown in Fig. 10.28, the peak absorption wavelength of Yb^{3+} doped fiber is 915 and 976 nm [8]. Semiconductor lasers have wide wavelength range and the right wavelength should be selected to meet the wavelength requirements. Erbium-doped optical fibers are the most studied due to their ability to amplify around the telecommunication application window of 1.55 μm. Because of their low quantum defect, Yb^{3+} ions have recently become the dopant of choice for high-power fiber lasers. Tm^{3+}-doped optical fiber lasers have recently attracted much interest for applications around 2 μm. Figure 10.29 illustrates the lower-level transitions and wavelength of pump and emission for Yb^{3+}, Er^{3+}, Tm^{3+}. The semiconductor lasers with wavelength of 790, 915, 976, 1,480, and 1,660 nm can be used as their pump sources [8].

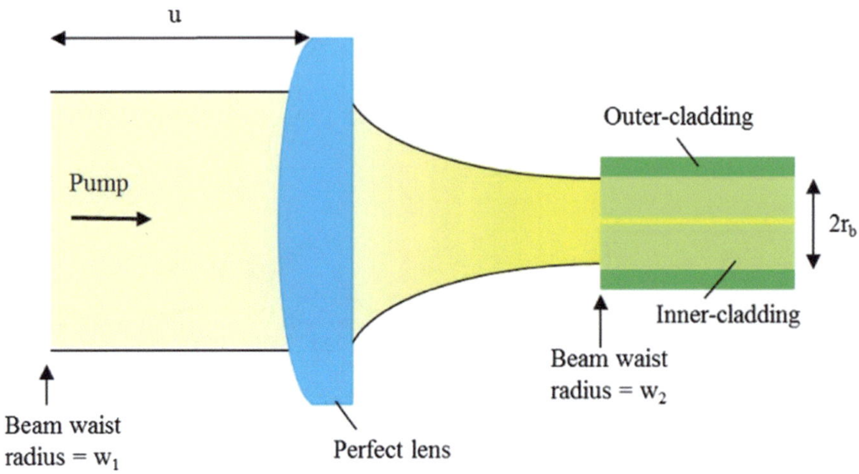

Fig. 10.30 Coupling of pump light into double-clad fibers [4]

Beam Parameter Considerations of Pump Sources

Figure 10.30 shows the basics structure of coupling pumps light into a double-clad fiber and the pump light must meet the pumping conditions of fiber lasers [4]. As presented, w_1 and w_2 are the beam waist radiuses of pump light before and after the perfect beam shaping lenses, respectively; r_b is the diameter of the doped fiber of fiber laser; u is the distance between w_1 and perfect lens; and f is focus distance of the perfect lens. According to beam propagation principle, the relationship of w_1 and w_2 meets the following equation:

$$\frac{w_2}{w_1} = \frac{1}{\sqrt{z_{01}^2 + (\text{u-f})^2}} \tag{10.2}$$

where $z_{01} = \frac{\pi w_1^2}{M^2 \lambda}$ and the square of M is the beam quality factor. For efficient coupling of pump light into the inner cladding, r_b and M_p which denote the beam propagation factor of pump light meet the relationship of Eq. (10.3)

$$r_b \geq \frac{\lambda_p M_p^2}{\pi \theta \gamma} \tag{10.3}$$

where $\theta = \arcsin(\text{NA})$ and γ is a factor which takes into account the needs to underfill the inner cladding and inner cladding's NA to avoid pump-induced damage to the outer coating. The value for γ depends on the situation and, in particular, the pump power. Based on Eq. (10.3), the beam propagation factor of M_p must satisfy the equation:

Table 10.2 The M_p^2 of diode laser pump beam for different diameters of double-clad fiber

$2r_b$ (μm)	M_p^2
125	66
200	105
400	210
600	315

Table 10.3 The M^2 of diode lasers

Semiconductor laser	Divergences [90 % E half angle (rmad)]	Beam size (μm^2)	Output power (W)	M^2
Single emitter (multi-mode)	θ_x=35°; θ_y=7°	1 × 100–200	7–10	$M_x^2 \approx 1$; $M_y^2 \approx 39$
Diode laser bar	θ_x=35°; θ_y=7°	1 × 10,000	40–120	$M_x^2 \approx 1$; $M_y^2 \approx 1,956$
[a]Diode laser stack	θ_x=35°; θ_y=7°	N × bar spacing × 10,000	40 × N–120 × N	$M_x^2 \approx 10,564$; $M_y^2 \approx 1,956$

[a]The typical bar pitch of a stack is 1.8 mm; N is 6 used in calculations

$$M_p^2 \leq \frac{\pi \theta r_b \gamma}{\lambda_p} \tag{10.4}$$

Theoretical upper limit P_p which denotes the theoretical maximum power of fiber laser pumped by a pump source (i.e., for a "perfect" pump launching scheme) is given by

$$P_{p\max} \approx \frac{P_s}{M_x^2 M_y^2} \left(\frac{\pi r_b \theta \gamma}{\lambda_p} \right)^2 \tag{10.5}$$

where P_s is the power of a single (constituent) emitter, and M_x and M_y are its beam propagation factors along orthogonal and parallel directions, respectively. According to Eq. (10.4), the beam quality M_p^2 of semiconductor lasers should be less than $\frac{\pi \theta r_b \gamma}{\lambda_p}$, and the typical values of double-clad fiber are NA $= 0.4$, $\lambda_p = 980$ nm, $\gamma_{uf} = 0.8$, and $\theta_{na} = 0.4$. Based on Eq. (10.4), the values of M_p^2 of diode laser pump beam for different diameters of double-clad fibers can be calculated and presented in Table 10.2.

As shown in Table 10.2, for the smaller diameter of clad fiber, the smaller M_p^2 of diode laser pump beam is required. However, due to poor beam quality, most of the diode lasers cannot meet the condition before beam shaping. The typical beam parameters of diode lasers, including divergence angles and M^2, are shown in Table 10.3.

Fig. 10.31 A typical fiber-coupled semiconductor laser module with multiple emitters [4]

It is known from Table 10.3 that only the beam propagation factors of single emitter lasers can meet the requirement of pump source. Naturally, single emitters have been used as pump sources for fiber laser in recent years and most of the time the beam is coupled into fiber. However, the output power of single emitters cannot meet the demand of the fiber laser development which requires higher power and higher brightness pump sources. In recent years, the beam shaping and fiber coupling technologies of semiconductor lasers for pumping fiber lasers have been progressed rapidly. The output power of pump sources has been improved very fast. These technologies mainly use spatial beam combining and sometimes even combined with polarization coupling techniques to integrate the output beams of multiple single emitters to achieve high power and high brightness. Fiber-coupled modules with output power of 30–40 W out of 100 μm fiber which consists of 3–4 single emitter diode lasers are common commercial products. Figure 10.31 shows a typical fiber-coupled module with multiple emitters [4]. Commercial products with output powers up to 100 W from 100 μm fiber are also available and higher power and higher brightness products are being developed based on these technologies.

For semiconductor laser bars, the beam shaping and fiber coupling designs are more complex. The principle of beam shaping and fiber coupling technologies for laser bars were introduced in Chap. 5 in detail. However, fiber-coupled single laser bar products are currently not popularly used in fiber lasers due to the fact that it is very difficult to couple the light out of a laser bar into a 100 μm fiber. If it is larger core size, such as 200 μm, the brightness is not competitive enough compared with the multiple single emitter approaches. Multiple laser bars are used to obtain high output powers. Spatial and polarization coupling technologies have been employed to achieve higher output power and higher brightness. Output power of 200 W with fiber diameter of 200 μm is commercially available and they are widely used as pump sources for fiber lasers. Figure 10.32 shows such a fiber-coupled diode which has multiple bars [36]. However, there is still not an efficient beam shaping and fiber coupling method for diode laser stacks to couple the output beams into a small size fibers. Some technologies are being developed and stack-based fiber coupling could offer the best cost value for hundreds of watts to kilowatt level high power fiber laser applications.

Fig. 10.32 Fiber-coupled semiconductor laser with multiple bars [36]

10.2 Material Surface Treatment

Lasers have been introduced to material surface treatment industry for many years. It can provide a well-controlled heat source, both in terms of the power delivered and the spatial precision of the directed energy. The interaction of the laser beam with the material being processed is substantially different than that of other thermal processing methods, and lasers offer distinct advantages over competing thermal processing methods used for surface treatment [37]. Currently, the CO_2, solid-state and fiber lasers dominate the laser material processing industry. However, with the recent development of packaging and beam shaping technologies, semiconductor lasers as light sources have been directly applied in material surface treatment [38–44].

There are two approaches to apply the semiconductor laser in surface treatment application as shown in Fig. 10.33, one is pumping source for solid-state laser and fiber laser as described in Sect. 10.1, and the other is direct application by coupling laser light to fiber or shaping output beam by optical lens system [4]. In this section, we will discuss direct shaped beam output for material surface treatment applications.

10.2.1 Characteristics of High Power Semiconductor Laser (HPSL) in Material Surface Treatment

HPSL have many inherent properties which make them particularly suitable for surface treatment applications. Some of their main characteristics are as follows [45]:

Wavelength: The wavelength of the radiation emitted from most practical HPSLs is between 800 and 1,000 nm. A major concern in laser materials processing is the

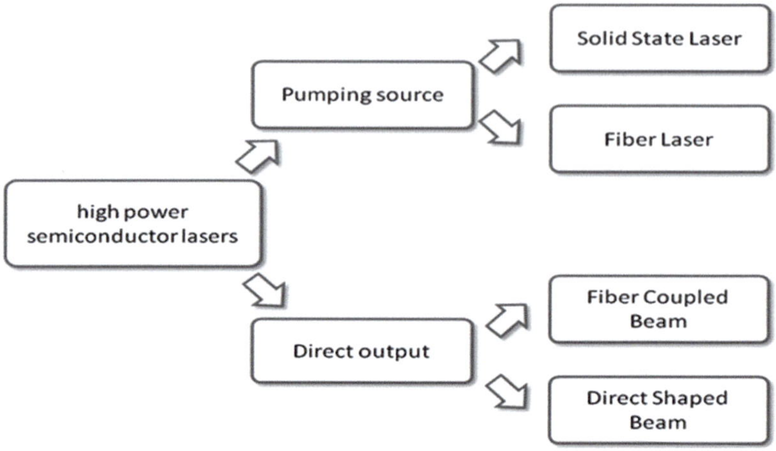

Fig. 10.33 Approaches to apply semiconductor laser in material surface treatment [4]

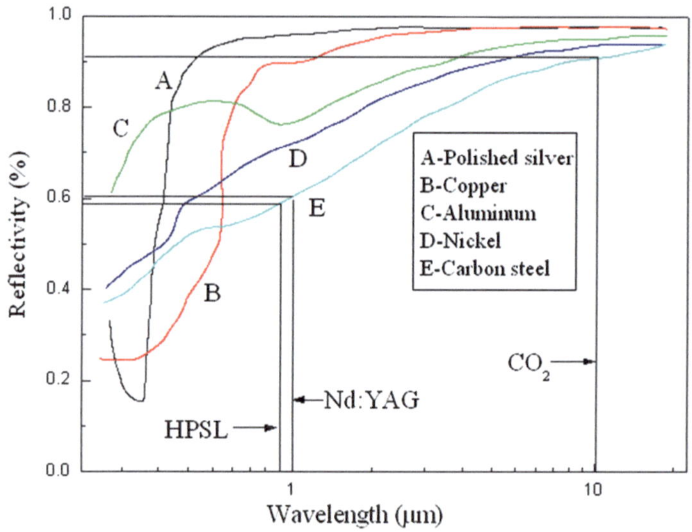

Fig. 10.34 The reflectivity of different metal material as a function of wavelength of the incident light beam [45]

fraction of incident laser radiation which is reflected from the surface of the work piece. This is characterized by the work piece reflectivity, defined as the ratio of the radiant power reflected from the surface to the radiant power incident on the surface. The relationship between the reflectivity of various metals and the wavelength of the incident light is shown in Fig. 10.34 [45]. It can be seen that due to its shorter wavelength, HPSL radiation has a considerably higher degree of absorption

Table 10.4 Summary of materials processing lasers characteristics [46]

	Laser type				
	HPSL	DPSS	LPSS	Fiber	CO_2
Wavelength(μm)	0.98	1.06	1.06	1.07	10.6
Power range	10s of watts to 10 kW	100s of watts to a few kW	100s of watts to few kW	100s of watts to 10s of kW	100s of watts to 10s of kW
Size	Very small	Large	Large	Medium	Large
Electrical efficiency	40 %	15 %	5 %	25 %	10 %
Maintenance interval	2 years	1 year	3 months	2 years	6 months
Initial capital cost	Low	High	Medium	High	Medium
Cost of ownership[a]		\$53/h	\$51/h	\$43/h	\$49/h

[a]Including capital and operating cost

into metallic surfaces than CO_2 lasers which generally emit radiation at 10.6 μm and a slightly higher degree of absorption than Nd:YAG lasers which generally emit radiation at 1.06 μm.

Efficiency, costs, and size: Table 10.4 shows the comparison of the characteristics of HPSL, diode-pumped solid-state (DPSS) laser, lamp-pumped solid-state (LPSS) laser, fiber laser, and CO_2 laser [46]. Maintenance and running costs of HPSLs, however, are the lowest compared to those of competing laser types. The semiconductor laser is the most efficient of all laser types with a typical electrical to optical efficiency of 30–40 % compared with CO_2 and Nd:YAG (LPSS) lasers which have typical efficiencies of 10–15 % and 2–10 % respectively.

Because of the nature of the HPSL construction and the reduced cooling requirements of semiconductor lasers compared with other laser types, HPSLs are considerably more compact than both CO_2 and Nd:YAG lasers. The size of the laser optical head is dependent on the power output and the manufacturer's design, with a volume of approximately 5–12 L being typical for a 3 kW direct application HPSL [42]. The HPSL optical head size can be up to three orders of magnitude smaller than competing laser types and this allows them to be relatively easily integrated into conventional machining systems or onto robot arms.

Beam profile, quality, and stability: The beam profile of a HPSL is generally top-hat in the slow axis direction and Gaussian in the fast axis direction with a rectangular shaped spot due to the nature of the beam formation process, as shown in Fig. 10.35 [4]. This profile is beneficial for many applications where uniform heating of a surface is required, such as laser hardening, laser alloy, and laser cladding.

Compare to 1–5 mrad CO_2 and 1–10 mrad Nd:YAG laser beam, the quality of HPSL is poor with typical values of 85 mrad × 200 mrad [47, 48]. For material surface treatment application, the beam quality of HPSL is good enough. One of the major differences between HPSL and other laser sources is the number of laser

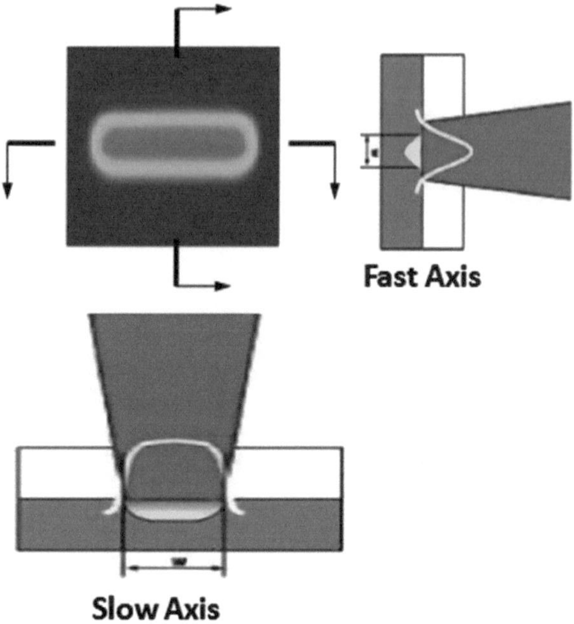

Fig. 10.35 A typical beam profile of a HPSL [35]

beams generated in the system (CO_2 and Nd:YAG lasers generating just a single beam). With single beam lasers, small variations during the excitement of the laser-active medium result in intensity fluctuations, apparent as irregular spikes in the beam profile. The great number of superimposed laser beams of a HPSL system compensates fluctuations within individual beams, giving the combined beam an outstanding modal stability [49].

10.2.2 The HPSL System and Optical Technology

Overview of Direct HPSL System

As shown in Fig. 10.36, the typical direct HPSL material surface processing system mainly consists of laser head, controller, power supply, and chiller [4]. The gantry type machine or robot is usually selected to locate laser head, as shown in Fig. 10.36. Powder feeder or wire feeder is optional according to the processing type. The power supply in HPSL is DC and high current source working in continuous or modulated mode. The chiller provides cooling water to laser head to keep laser working at constant temperature. The laser head body and lens may also be cooled by the water from chiller. The laser head, the most important component in the system, delivers the light to work piece.

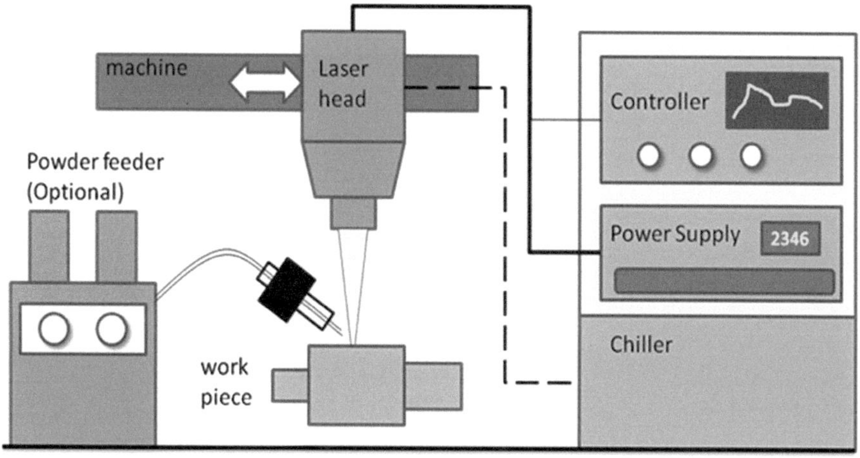

Fig. 10.36 Schematic diagram of a typical direct HPSL system for material surface treatment [4]

Fig. 10.37 Laser device used in direct HPSL system [4]

The Optical Technologies of Direct HPSL

The typical power of single semiconductor emitter is just several watts, which is far away from the kilowatts power level for surface treatment applications. A single bar can output several tens to hundred watts. Laser stacks consist of multiple bars can achieve kilowatts output, which is typically the power source for directly shaped beam output HPSL (Fig. 10.37).

Although the assembled laser stack can reach thousands watts, the output beam is large and the intensity is not high enough for most of the material surface treatment applications [4]. An optical system should be introduced to focus these beams and obtain high power density. Spatial beam combining, polarization beam combining, and wavelength beam combining are the most popular technologies in laser head design. Furthermore, these methods can be combined to achieve very high level output.

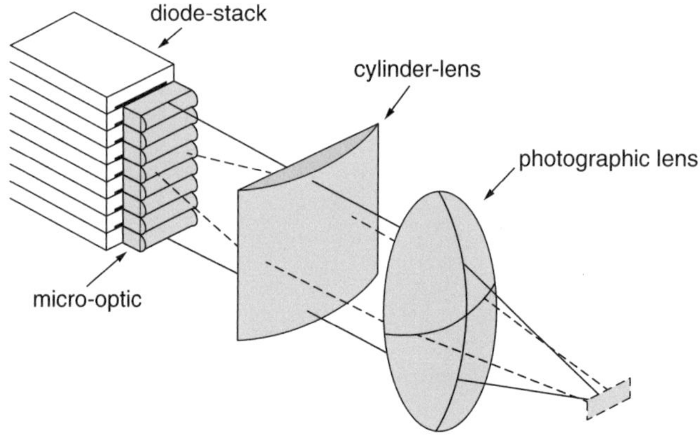

Fig. 10.38 Visualization of focusing of light from a laser stack [31]

Spatial beam combining: The simplest spatial combining is to use a lens system in front of single laser stack, as shown in Fig. 10.38 [31]. The emission from a single semiconductor laser bar is confined to the narrow junction region (typically 1 μm), diffraction of the light results in a large beam divergence of up to 45° half angle in the direction perpendicular to the emission line ("fast" axis) and up to 10° half angle in the direction parallel to the emission line ("slow" axis) [50]. The fast axis is collimated by micro-optics. The output from the micro-optics is approximately parallel beams. A lens then focuses the beam to a spot. For the slow axis, a cylinder-lens is used to reduce the divergence angle so that the output beam at the working plane is not too large. However, it is relatively difficult to achieve high power and high power density using this approach. Furthermore, the focal plane of this design is generally small which results in short working distance. This in turn will limit the applications.

To increase the output power density, spatial beam shaping may be applied to the stack to reduce the output beam size. One of approaches is to apply two triangular prisms as shown in Fig. 10.39. The collimated beam of laser stack is divided into three parts by two prisms [4]. The beam in the middle passes through directly, and the other two beams are redirected by two triangular prisms. The angle of the triangular prism is designed to redirect the laser beam and makes complete superposition of the three parts of beam at the design focal plane. After superposition, the beam size is reduced as one third of the whole beam size of the laser stack.

Figure 10.40 presents another approach to combine single stack beam with strip mirror and anamorphic prisms. The beam in the fast axis direction is divided into two parts, each one contains several sub-beams. The top part passes through a reflective stripe mirror and reflects from a high-reflection mirror on the lower part of the stripe mirror in such a way that fills the dead space between the five lower beams. In order to further decrease the spot size on the fast axis, a pair of anamorphic prisms is used. This technique compresses the beam size significantly [49].

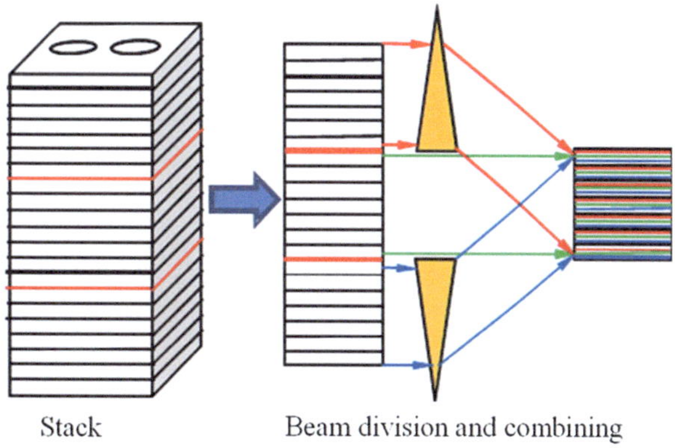

Fig. 10.39 Beam combining with two triangular prisms [4]

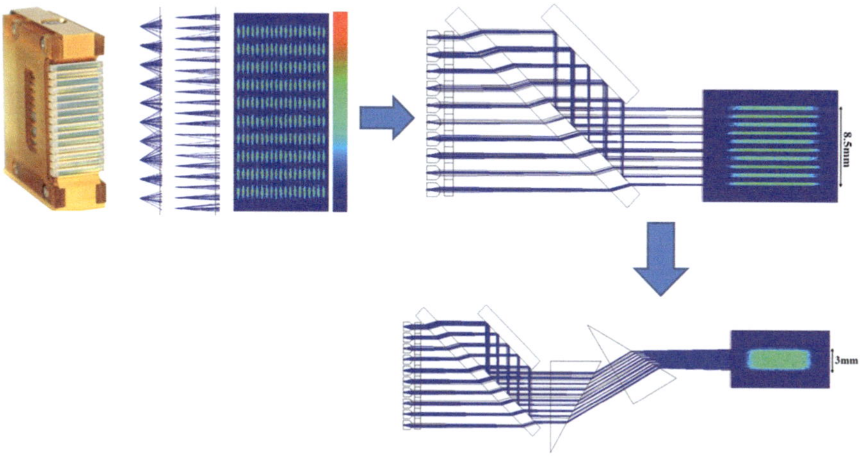

Fig. 10.40 The output beam profile of laser stack after strip mirror and anamorphic prisms [49]

Polarization beam combining: The basic principle of polarization beam combining is represented in Sect. 5.3.1. Figure 10.41 shows how the polarization beam combining technology is applied in the laser head design. To achieve high output power, two stacks are used. All the specifications of the stacks are the same and the polarization of one of the stack is changed by a wave plate. In combining the two beams, one beam from a stack is reflected by a mirror and then reflected again by the polarization coupler which totally reflects the beam but is of transparent to the other beam which has different polarization. The combined beam is focused on the working plane by a focusing system. This approach is widely used in high power laser head design [50].

Wavelength beam combining: Wavelength coupling is to couple several beams with different wavelengths through a narrow pass filter, as shown in Fig. 10.42. This method can make the output power increase several times, while maintaining the spot size and the quality of the beam [51]. This approach has high requirements on

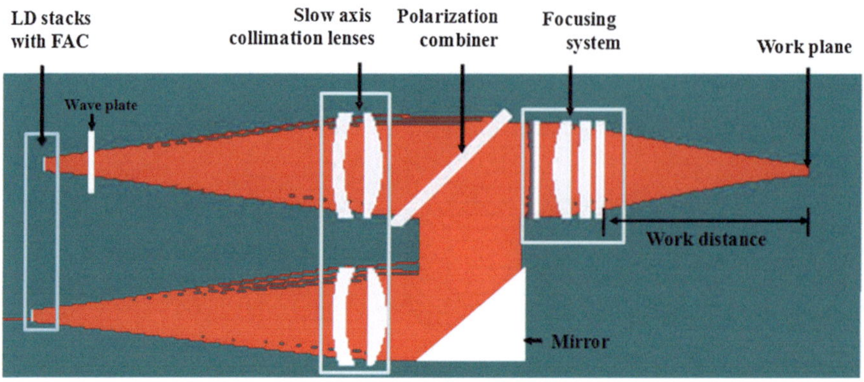

Fig. 10.41 Beam combining of two stacks with polarization combiner [50]

Fig. 10.42 Schematic diagram of wavelength coupling [51]

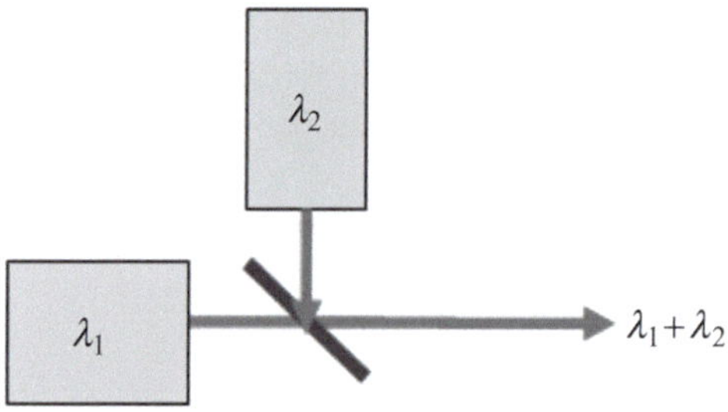

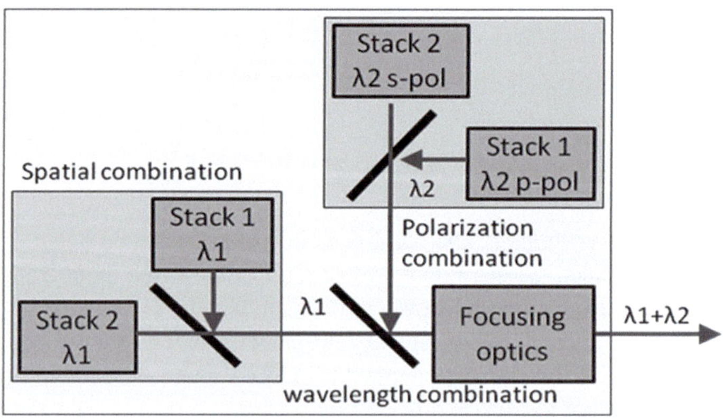

Fig. 10.43 Beam combination with different combining technologies [4]

the thin film coupling devices especially if the wavelength difference of the beams is small and the power density is high [52].

To obtain even higher output power, different combining methods can be used in one system simultaneously. One example is shown in Fig. 10.43 [4].

Fig. 10.44 Cladding with direct HPSL [53]

10.2.3 Applications of Direct HPSL in Surface Treatment

As we discussed, the beam profile of direct HPSL is beneficial to surface treatment, especially for laser cladding, laser hardening, and laser alloying.

Laser Cladding

Laser cladding uses a high energy density laser heat source to deposit selected materials (such as corrosion-resistance or wear-resistant materials) on structural material substrates. Historically, the development of high-power gas lasers led to the initial development of laser cladding technology in the late 1970s. In the 1980s, the process became more refined with improved system development and process understanding. Laser cladding is now widely used in a range of industries, including automotive, aerospace, energy production, medical, and marine industries [37]. Direct semiconductor laser cladding has significant market potential due to its mobility (low weight and small size) and high efficiency.

Because the laser beam is generally concentrated and small, the energy provided by the laser can be directed at a specific location on the substrate. This offers great flexibility and minimizes the impact of the process on the substrate material such as distortion. Laser cladding typically produces a high-quality clad layer with extremely low dilution, low porosity, and good surface uniformity. Figure 10.44 shows cladding with direct HPSL [53].

In laser cladding, the clad material can be in the format of powder or wire. Figure 10.45a, b illustrates powder-based and wire-based cladding process, respectively. In both cases, the laser beam is directed at the powder or wire, which is melted on the substrate surface. Each material introduction approach, however, has unique properties, along with its characteristic advantages and disadvantages [37].

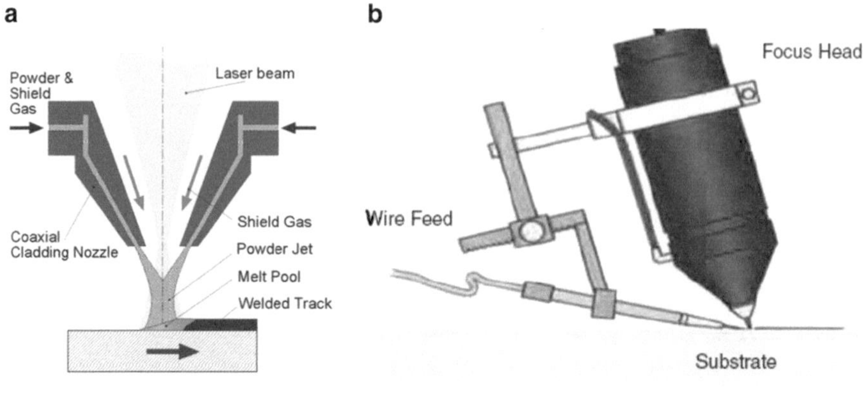

Fig. 10.45 Illustration showing (**a**) coaxially fed powder laser cladding [53] and (**b**) wire-fed laser cladding [37]

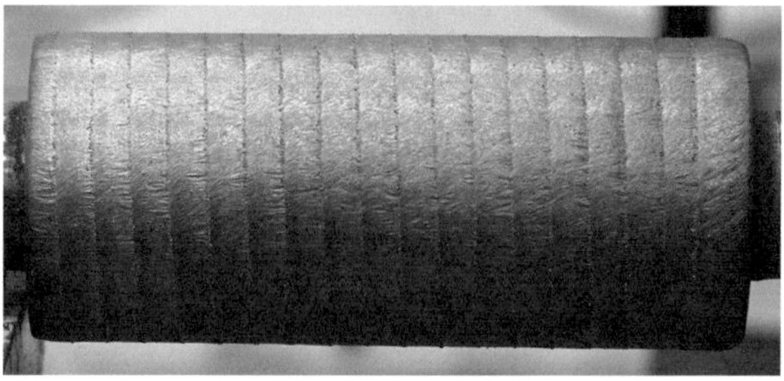

Fig. 10.46 Print roller shaft clad with Hastelloy C-22 [44]

Figure 10.46 shows a repaired roller shaft, which is made of mild carbon steel clad with 0.5 mm Hastelloy C-22 by a 4 kW direct HPSL system [44]. The clad layer is corrosion and wear-resistant material which extends the lifetime of roller significantly. From the composition analysis shown in Fig. 10.47, the trace of chromium changes very sharp at the interface between cladding layer and substrate. The dilution rate of cladding material is extremely low, that means, the property of cladding material is depredated by direct HPSL cladding.

Laser Hardening

Laser hardening of ferrous materials is an established process used to enhance the mechanical properties of highly stressed machine parts, such as gears and bearings. Surface hardening increases the wear resistance of the material, and under favorable circumstances, increases the fatigue strength caused by residual compressive stresses that are induced in the work piece surface by the transformation hardening process. The laser surface hardening process is not fundamentally different from conventional

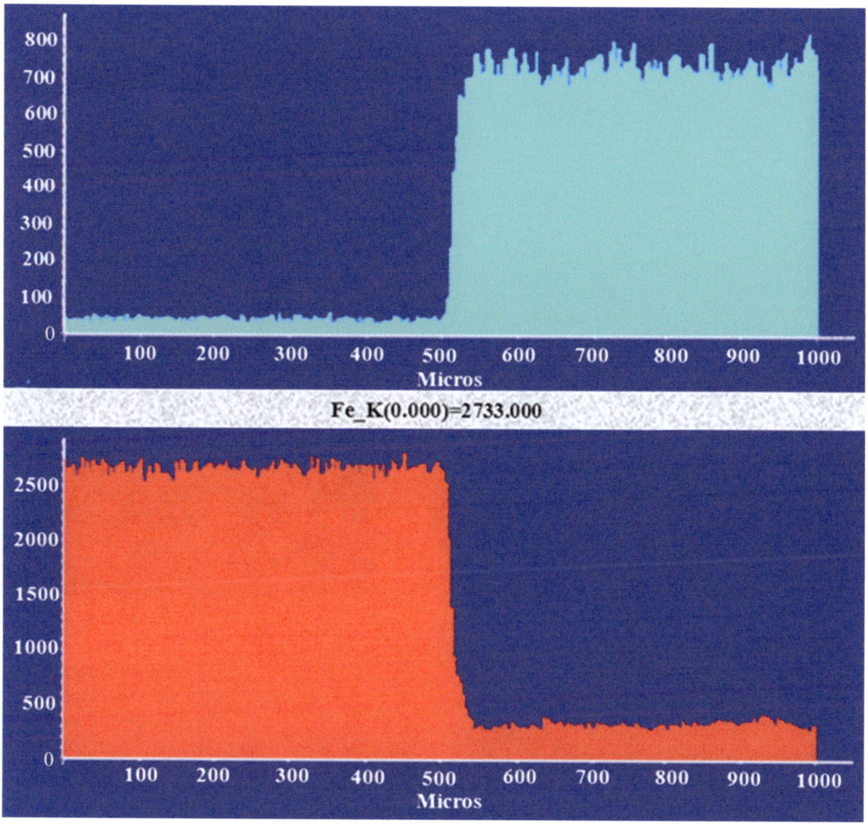

Fig. 10.47 SEM line trace of C-22 clad with a high power diode laser at 4 kW and 0.5 m/min [44]

hardening process of ferrous materials. Laser surface hardening differs from conventional hardening in that only a thin surface layer is heated to austenitization temperatures prior to quenching, leaving the interior of the workpiece essentially unaffected [54]. Figure 10.48 shows an example of laser hardening on the surface of gear shaft [4].

The principle of laser hardening is a phase change by rapid heating and cooling from the laser beam. When a laser beam impinges on a surface, part of its optical energy is absorbed as heat at the surface. Since the power density of the laser beam is sufficiently high, heat will be generated at the surface at a rate higher than heat lost through conduction. This results in rapid temperature increase at the surface layer. In a very short time, a thin surface layer will have reached austenitizing temperatures, whereas the interior of the workpiece remains in its original state. Even with a relatively moderate power density of 500 W/cm^2, a temperature gradient of 500 °C/mm can be obtained. By moving the laser beam over the surface of the workpiece, the area within the path of the beam is rapidly heated as the beam passes,

Fig. 10.48 Laser hardening on the surface of a gear shaft [4]

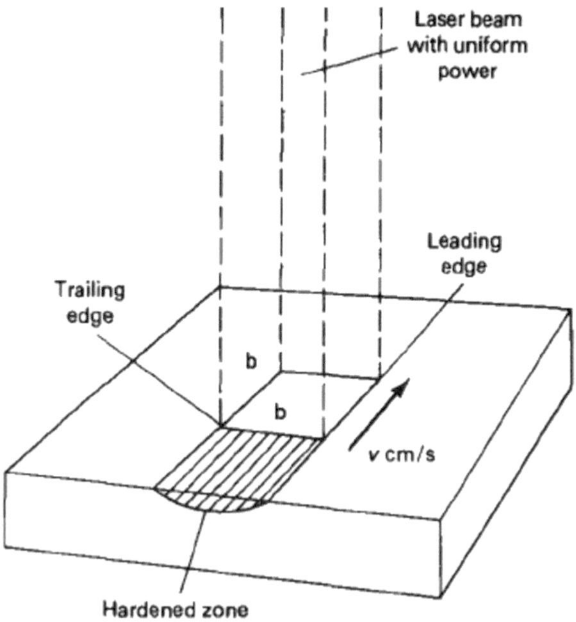

Fig. 10.49 Illustration of laser hardening process [54]

as shown in Fig. 10.49 [54]. This area is subsequently cooled rapidly through heat conduction to the surrounding areas after the beam has passed. By selecting the appropriate power density and the relative moving speed of the laser beam to the workpiece being processed, the material can be hardened to the desired depth.

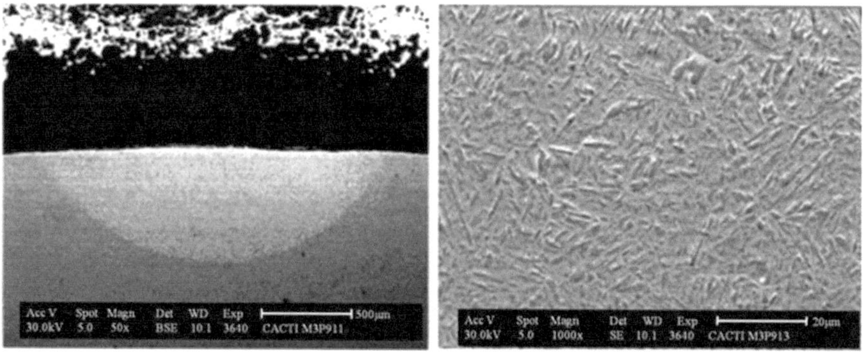

Fig. 10.50 SEM micrographs of the hardened zone of the sample processed at $P=470$ W, $v=5$ mm/s and *T*aus = 820 °C [38]

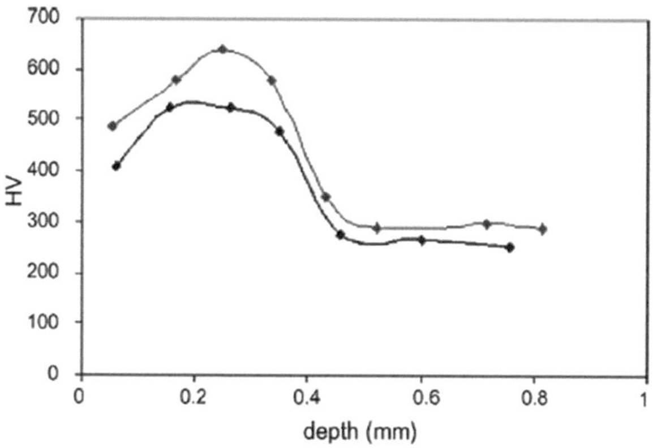

Fig. 10.51 Influence of the austenization temperature on the superficial hardness (upper curve: *T*aus = 800 °C, lower curve: *T*aus = 750 °C) [38]

Figure 10.50 shows the SEM micrographs of hardened zone produced on AISI 1045 steel by direct HPSL [38]. The hardness from surface to substrate as shown in Fig. 10.51, is increased compare to non-hardened zone [38].

Laser Alloying

Laser alloying is a material processing method which utilizes the high power density laser sources to rapidly melt the alloying elements and a portion of the underlying substrate simultaneously. These alloying elements diffuse rapidly into the melt pool, and the desired depth of alloying can be achieved in a short period of time. Since the melting occurs in a very short time and only at the surface, the bulk of the material remains cool, thus serving as an intimate heat sink. Large temperature gradients exist across the boundary between the melted surface region and the underlying solid substrate. This results in rapid self-quenching and resolidification. By this approach, a desired alloy chemistry and microstructure can be generated on the substrate. The degree of microstructure refinement will depend on the solidification rate. The surface of a low-cost alloy, such as mild steel, can be selectively alloyed to enhance properties, such as resistance to wear, in such a way that only the locally modified surface possesses properties typical of tribological alloys. This results in substantial cost savings and reduces the dependence on special materials [55].

Laser alloying has been primarily applied to improve corrosion resistance. The laser alloying has been applied on surface treatment of magnesium and its alloys [56]. The experiment has been performed on MCMgAl12Zn1 by high power diode laser with feeding of hard silicon carbide particles under an argon shielding gas. Argon was used during laser re-melting to prevent oxidation of the surface layer and the substrate. During metallographic examinations of the MCMgAl12Zn1 alloy, a uniform distribution was observed of the employed SiC particles in the entire alloyed zone. The result shows that it is possible to make surface layers on cast magnesium alloys with ceramic particles in the microstructure. This surface has better mechanical and abrasiveness properties than cast magnesium alloys in initial state.

10.3 Hair Removal

Laser hair removal is the process of removing unwanted hair by exposure to pulsed laser light that damages the hair follicle. Laser hair removal has been commercially available since the 1990s. Diode laser hair removal has become the trend in recent years replacing IPL (Intense Pulsed Light) technology [57]. Compare to Nd:YAG laser (1,064 nm) and Alexandrite laser (755 nm) as well as ruby laser (694 nm), diode lasers at 810 nm are considered the most efficient and most cost-effective laser light sources available. Diode lasers also have less adverse effects and are well suited for clinical hair removal applications [57].

10.3.1 The Principle of Laser Hair Removal

The fundamental principle of laser hair removal is selective photothermolysis found by Anderson and Parrish, which uses a specific wavelength of light and pulse duration to obtain optimal effect on a targeted tissue with minimal effect on surrounding tissue [57]. Laser hair removal focuses laser light on the endogenous chromophore melanin, which is mainly found in the hair shaft, with a small amount present in the upper third of the follicular epithelium, as shown in Fig. 10.52. When appropriate laser energy is directed at the skin, light is primarily absorbed in the hair shaft melanin. When the laser light is absorbed, heat is generated and it diffuses to

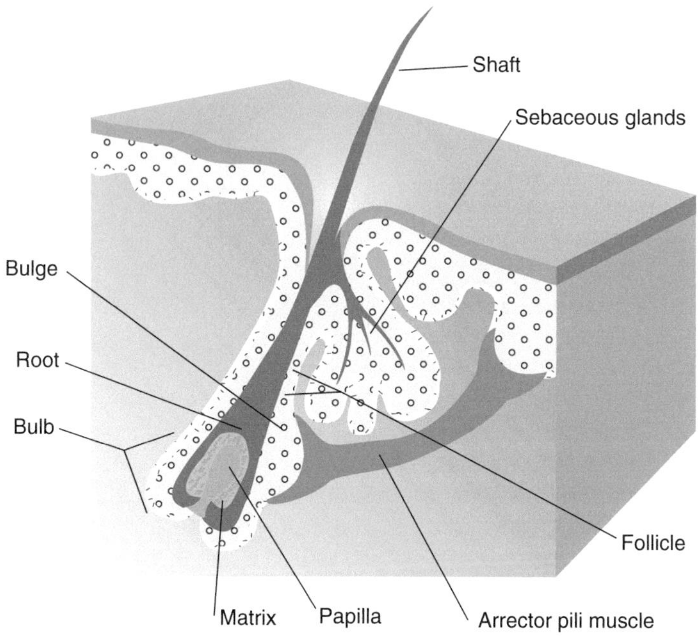

Fig. 10.52 The anatomy of a typical terminal hair. Laser hair removal targets the melanin in the hair shaft [58]

the surrounding follicular epithelium [58]. It is necessary to heat up the follicles enough to disable them from producing hair.

The light absorption of chromophores at specific wavelength is different in the range from red to near-infrared, as shown in Fig. 10.53. Wavelengths between approximately 700 and 1,000 nm are selectively absorbed by melanin and oxyhemoglobin and water absorb less energy at these wavelengths. Selective tissue destruction occurs with optimal parameters of wavelength, fluence, pulse duration, and spot size which confine heating and subsequently damage to the desired chromophore without affecting the surrounding tissues [4, 58, 59].

As shown in Fig. 10.54, the auxiliary hair was removed after 6 months by long-pulse laser in darker skin types with no signs of dyspigmentation or burns [60].

Semiconductor laser is an ideal light source for hair removal. Compared to other light sources, the advantages of diode lasers are as follows:

1. *Broad wavelength*: The wavelength range of semiconductor lasers is very broad; it is very easy to find the device to fulfill requirements for hair removal; currently, 810 nm semiconductor lasers are widely used. 1,060 nm is also commonly available.
2. *Adjustable pulse width and frequency*: According to selective photothermolysis theory, pulse width must be equal to or shorter than the thermal relaxation time (TRT) of the targeted follicles in human skin to confine thermal damage. The TRT

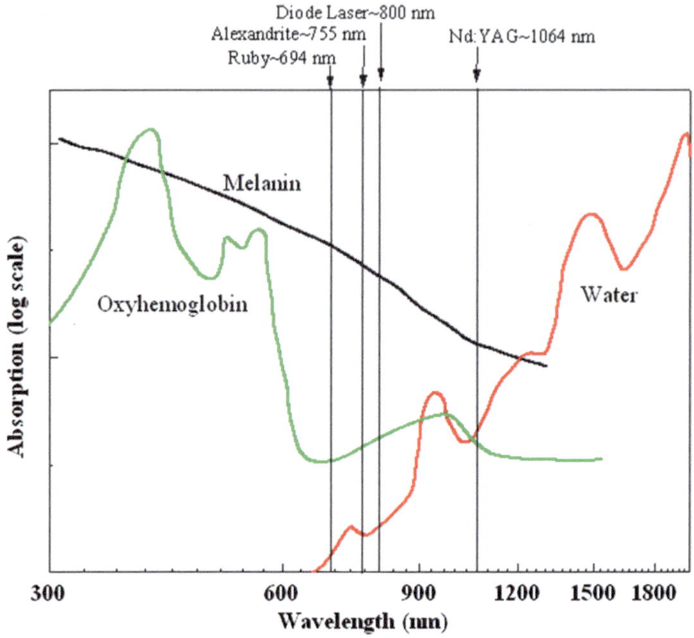

Fig. 10.53 The absorption rate of various chromophores as a function of the wavelength of light [4, 58, 59]

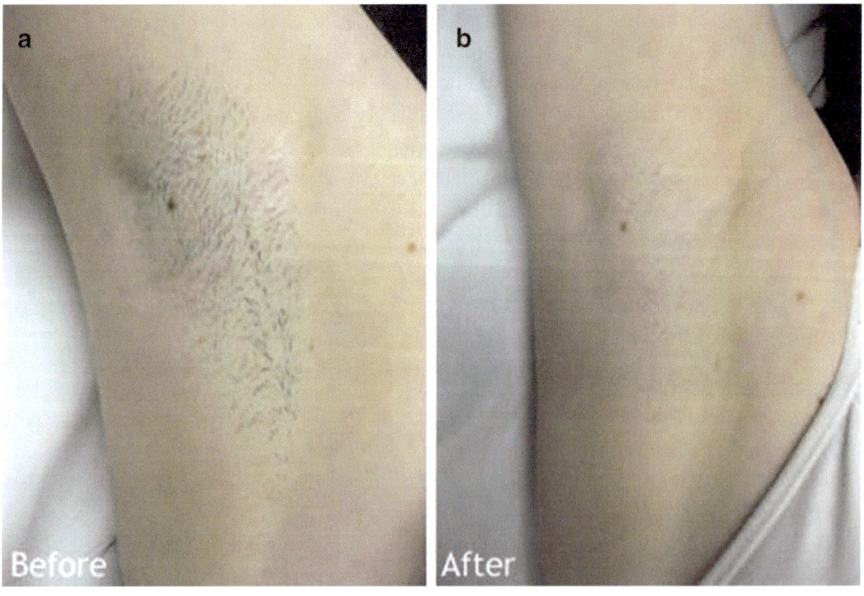

Fig. 10.54 Auxiliary hair removal with darker skin before (**a**) and after (**b**) treatment [60]

of the whole follicular structure depends on its diameter and is normally 10–100 ms. Repetition rate or frequency is believed to have a cumulative effect. Shooting multiple pulses at the same target with a specific delay between pulses can cause a slight improvement in the heating of an area. It was also reported that longer pulse widths are needed for specific skin type, e.g., pulse width expanded to 400 ms can achieve safety and efficacy in long-term hair reduction and pulse durations longer than 100 ms may be required to safely and effectively treat Skin ECT VI patients [61, 62]. Consequently, the laser source must have a range of pulse widths and frequencies. The pulse width and frequency of semiconductor laser can be easily tuned to meet any requirement of hair removal process.

3. *Beam spot*: The beam spot size affects treatment. Theoretically, the size of the ideal laser beam should be about four times as large as the treatment target since the target is deep in the skin. Generally hair removal lasers have a spot size about the size of a fingertip (3–18 mm). Larger spot sizes, such as 20–30 mm acting on the legs, back, and other large body parts, help laser light penetrate deeper, effectively enhance the efficiency, and speed and improve comfort of patient [63]. The spot size of semiconductor laser can be easily beam shaped to range of 8–30 mm with uniform intensity distribution.
4. *Wide power range*: Although available frequency varies, clinically effective lasers should be capable of delivering in excess of 30–40 J/cm^2 [64]. Due to its modular design concept,the output power of a typical semiconductor laser vertical stack or stack array can be from 500 to 3,000 W. It can meet different power requirements in the hair removal application.
5. *Compact size*: The semiconductor diode lasers have small footprint and light weight, which make it possible to be integrated in hand pieces. Also, hand piece can be integrated with sapphire contact cooling [65]. As a result, the systems are simple and compact.
6. *Proven lifetime and reliability*: Semiconductor lasers have high electro-optical efficiency and long lifetime. For hair removal application, due to its application conditions, the lifetime of diode lasers is not expected to be as long as those used for industrial applications which is typically longer than 20,000 h. The proven lifetime in hair removal is in the order of ten million shots in the long-pulsed operation mode. For shorter pulse width operation, longer lifetime is expected. The stability of the output power is generally within 2 % during operation.

10.3.2 Semiconductor Laser Hair Removal System and Optical Design

Figure 10.55 shows a typical semiconductor laser hair removal system [4]. The system consists of light generation and treatment unit (hand piece), diode laser driver and controller, cooling system for diode laser light source and skin contact, touch screen control unit, and safety protection and detection system. The hand piece is the most important part of the system. Figure 10.55b shows a hand piece.

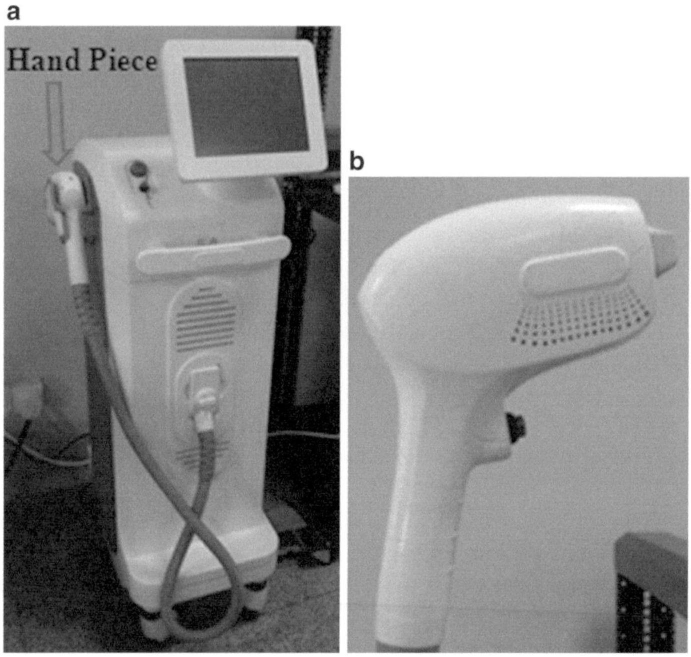

Fig. 10.55 Hair removal system (a) entire system (b) hand piece [4]

The detailed structure and design is discussed below. The driver and controller provide adjustable current to the diode laser source and it is integrated with the touch screen human control interface. The pulse width, repetition rate, and current magnitude can be controlled and adjusted. The diode laser is generally cooled by water. Depending on the design, the skin contact cooling is achieved by thermoelectric cooler or cold water condenser. The thermoelectric cooler is then cooled by water. Therefore, in the cooling system, water tank, pump, and pipelines are required. The water is cooled by chiller or heat spreader. Touch screen is the brain of the system and is used by the operator to control the system. Some sensors and detectors are built in detect the output power, skin contact temperature, cooling water quality, cooling water flow rate, and other parameters for safety protection and machine operation protection.

Hand piece is the core part of the hair removal system. Figure 10.56 shows an example of construction of the hand piece [66]. The hand piece mainly consists of diode laser light source, light guide, and treatment head. The light source is typically a laser bar stack or stack arrays. The light guide is used to transmit the light from the light generation to the treatment head. Most of the time, the light guide also serves as a laser beam shaper to reshape the beam into a spot size designed for treatment. The treatment head delivers the light to the skin. Most of the

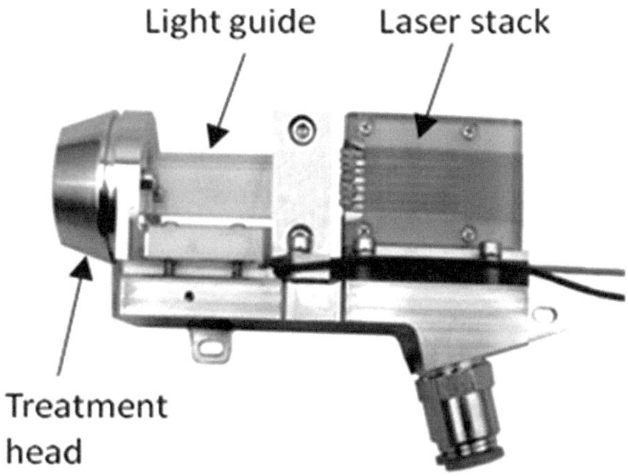

Fig. 10.56 An example of a hair removal hand piece construction [66]

time, there is a cooled sapphire contact window in treatment head to reduce the skin temperature for patient comfort and better treatment results [66].

Optical design is one of the key aspects in diode laser hair removal system. Optical design should be optimized to achieve high transmission efficiency, that is, the energy from the diode laser should be transmitted to the treatment head with minimum loss. Less light loss also improves the reliability of the hair removal hand piece since the lost optical energy would generate heat inside the hand piece. Optical design should ensure low light loss at the end surface of the waveguide or at the interface between the waveguide and the sapphire window to keep the skin contact temperature cool for patient comfort and effective hair removal. Furthermore, a uniform output beam is required for hair removal application and optical design should be optimized for intensity distribution.

An example of waveguide design is shown in Fig. 10.57 [66]. As shown in the figure, the important parameters of optical waveguide are the length L, the angle of θ, and the sizes of input and output surfaces.

In order to improve beam transmission efficiency, the design of waveguide should meet the total reflection condition. The propagation of incidence beams in a waveguide is shown in Fig. 10.58. The maximum incidence angle is denoted by α_0, and α_1 is the refraction angle by incidence surface, and N is the number of reflections of an incidence beam in waveguide [66]. In order to avoid optical leakage, the propagation of laser beam should be meet total reflection law in waveguide. As shown in Fig. 10.58, and n_0 and n_1 are the refractive index of air and waveguide, respectively. θ and N have the following relationship to avoid light loss from waveguide:

$$\theta \leq [\arcsin(n_1/n_0) - \arcsin(n_0 \sin \alpha_0/n_1)]/[2(N-1)] \tag{10.6}$$

The laser stack without collimator is located at the input end of the waveguide. After propagation in the waveguide, the beam is reshaped. A compressed and homogenized beam is achieved at the output end. Figure 10.59 shows the beam

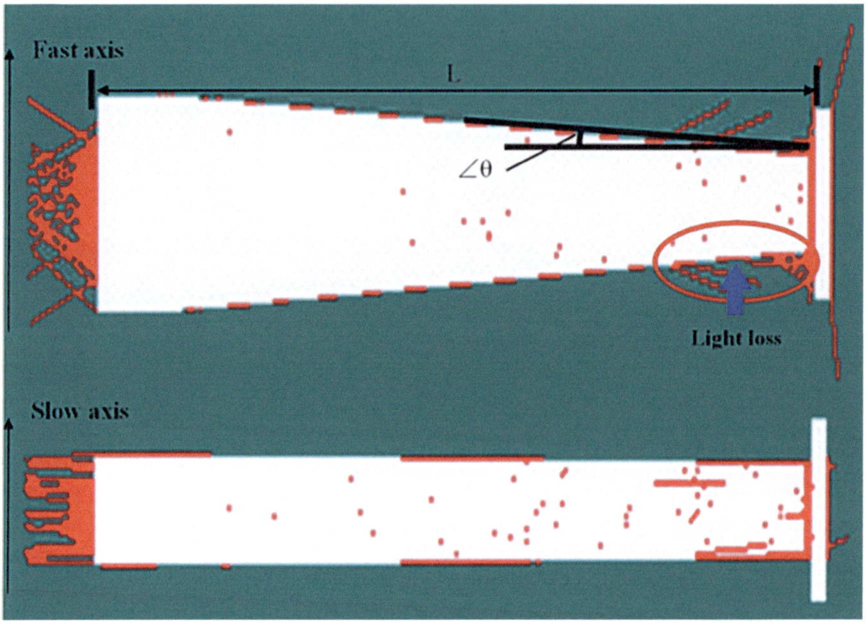

Fig. 10.57 An example of waveguide design of a hand piece in a diode laser hair removal system [66]

Fig. 10.58 The propagation of the incidence beam in waveguide [66]

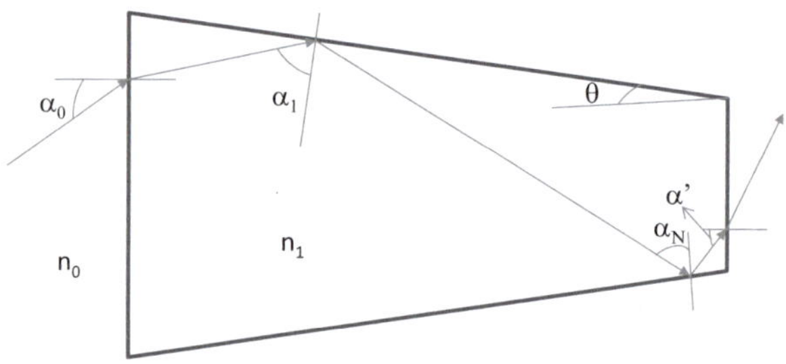

transmission inside of the hand piece [66]. The waveguide is usually either a tapered or a cavity-shaped structure, made by optical or metallic materials [67]. Figure 10.60 shows examples of light guide structures [66].

10.3.3 Semiconductor Lasers for Hair Removal

There are mainly three types of semiconductor laser light sources for hair removal application: (1) High duty cycle light sources; (2) High peak power light sources; and (3) Industrial water-cooled light sources.

Fig. 10.59 The beam transmission inside of the hand piece; (**a**) a vertical stack diode laser; (**b**) hand piece and (**c**) homogenized beam at output end [66]

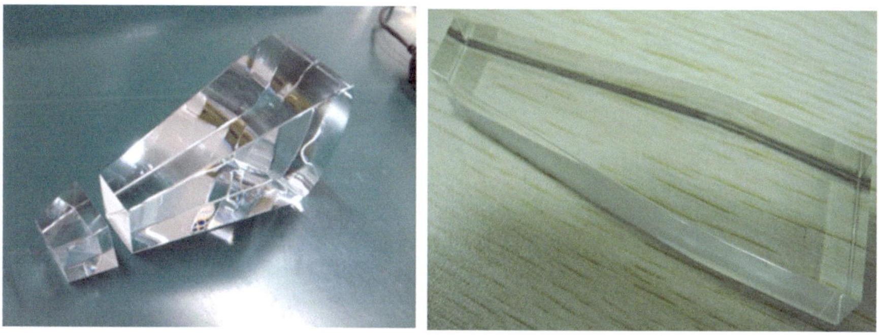

Fig. 10.60 Examples of light guide structures [66]

1. *High duty cycle light sources*
   Micro-channel-cooled (MCC) vertical stack lasers as high duty cycle light sources are the most commonly used light sources in diode laser hair removal application. The pulse width can be up to 400 ms and duty cycle can be up to 50 %. The basic unit of vertical stack is MCC single bar. The output power per bar is generally 100 W with wavelength of 808 nm. The number of bars in the stack can be up to 12 bars for hair removal application. For higher power, two stacks in parallel can be designed. Figure 10.61 shows examples of vertical stack high duty cycle light sources [68].
2. *High peak power light sources*
   G-Stack-based stack arrays can offer high peak power and compact light emission area. From a 350 mm^2 light emission area, a peak power of 3 kW can be reached. Typically, the pulse width can be up to 30 ms and duty cycle can be up to 10 % for the G-stack-based high peak power light sources. The output power for per bar in G-stack can be up to 100 W. The pitch of the G-stacks is generally

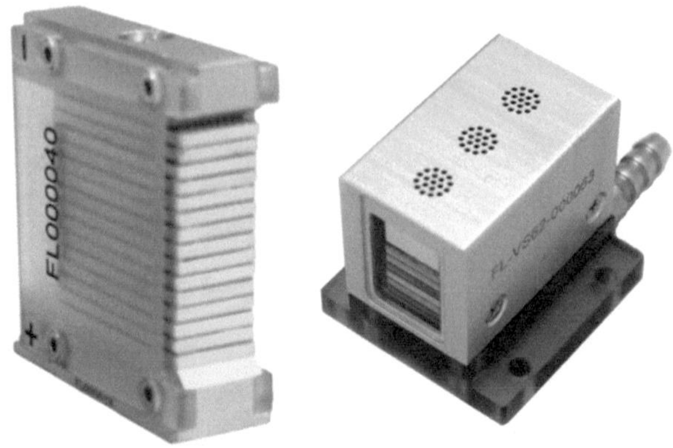

Fig. 10.61 Vertical stack high duty cycle light sources for hair removal applications [68]

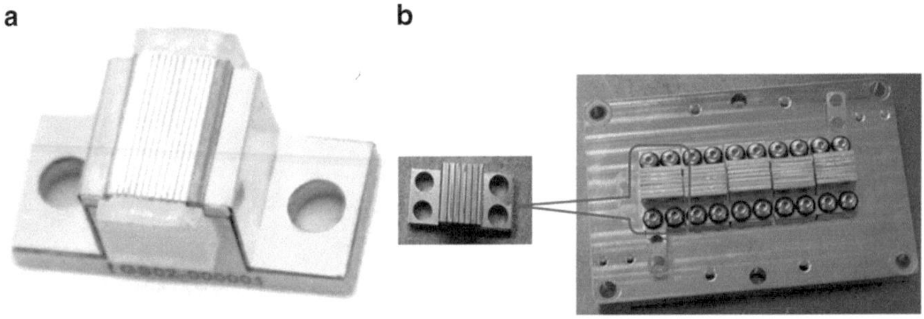

Fig. 10.62 High peak power light sources for hair removal applications; (**a**) a single G-stack unit and (**b**) a G-stack array [4]

either 0.75 or 1.15 mm. Figure 10.62 shows two examples of G-stack lasers used in hair removal application [4]. Figure 10.62a shows a G-stack composed of many bars. The number of bars in the G-stack can be up to 20 bars and the power can be up to 2 kW. Figure 10.62b shows a G-stack array which consists of five "building blocks," with each of the "building block" having 6 bars. Total there are 30 bars and the power can be up to 3 kW. The "building blocks" may be arranged differently based on the application requirement and design.

3. *Industrial water-cooled light sources*

   Industrial water-cooled light sources offer longer lifetime, user convenience and low maintenance cost for diode laser hair removal application because of the avoidance of the deionized water cooling. The lifetime of industrial water-cooled light sources can reach up to 100 M shots. The end user does not need DI water and change water frequently. Usually, the industrial water-cooled light sources use a cooling plate with large liquid flow channels or macro-channel coolers (MaCC) for heat dissipation. Figure 10.63 shows examples of light sources with industrial water cooled using cooling plate and MaCC [66]. For the cooling plate-cooled light sources, the output power per bar can be from 40 to 100 W/bar depending on

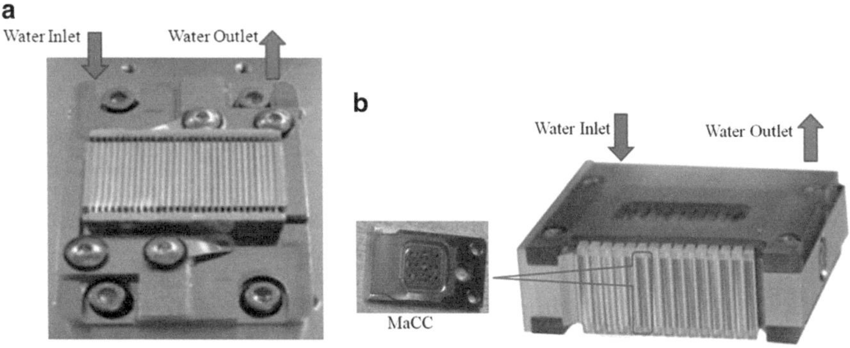

Fig. 10.63 Industrial water-cooled light sources for hair removal applications; (**a**) cooling plate cooled; (**b**) macro-channel cooled (MaCC) [66]

the pulse width and duty cycle [66]. The MaCC V-stacks are assembled using the basic unit of MaCC single bar, which is shown in Fig. 10.63b. The pitch of MaCC V-stack is usually 1.8 mm. The output power can reach 80 W per bar and rated pulse width can be up to 100 ms with 20 % duty cycle.

References

1. R. Diehl, *High-Power Diode Laser* (Springer, Berlin, 2000)
2. N. Lichtenstein, B. Schmidt, A. Fily, S. Weiß, S. Arlt, S. Pawlik, B. Sverdlov, J. Müller, C. Harder, DPSSL and FL pumps based on 980nm-telecom pump laser technology: changing the industry. Proc. SPIE **5336**, 77–83 (2004)
3. B. Faircloth, High-brightness high-power fiber coupled diode laser system for material processing and laser pumping. Proc. SPIE **4973**, 34–41 (2003)
4. T. Song, Y. Wang, M. Wang, X.S. Liu, *Application of High Power Semiconductor Lasers.* Internal Talk from Focuslight Technologies Co., Ltd. (2012), pp. 20–27
5. W. Koechner, *Solid State Laser Engineering* (Springer, Berlin, 1966)
6. C. Orth, R. Beach, C. Bibeau, E. Honea, K. Jancaitis, J. Lawson, C. Marshall, R. Sacks, K. Schaffers, J. Skidmore, S. Sutton, Design modeling of the 100-J diode-pumped solid-state laser for project mercury. Proc. SPIE **3265**, 114–129 (2008)
7. R.J. Beach, E.C. Honea, S.B. Sutton, C.M. Bibeau, J.A. Skidmore, M.A. Emanuel, S.A. Payne, P.V. Avizonis, R.S. Monroe, D.G. Harris, High-average-power diode-pumped Yb:YAG lasers. Proc. SPIE **3889**, 246–260 (2000)
8. H. Injeyan, G.D. Goodno, *High Power Laser Handbook* (The McGaw-Hill Companies, New York, 2011)
9. http://www.ll.mit.edu/publications/journal/pdf/vol03_no3/3.3.5.diodepumpedlaser.pdf
10. J. Vetrovec, R.S. Shah, T. Endo, A. Koumvakalis, K. Masters, W. Wooster, K. Widen, S. Lassovsky, Progress in the development of solid state disk laser. Proc. SPIE **5332**, 235–243 (2004)
11. R. Diehl (ed.), *High-Power Diode Lasers: Fundamentals, Technology, Applications* (Springer, Berlin, 2000)
12. H. Injeyan, C.S. Hoefer, End pumped zigzag slab laser gain medium, U.S. Patent 6,094,297, 5 July 2000

13. Q. Liu, M. Gong, F. Lu, W. Gong, C. Li, D. Ma, Corner-pumped Yb: yttrium aluminum garnet slab laser emitted up to 1 kW, Appl. Phys. Let. **88**(10), 101113(1–3) (2006)
14. N. Hempler, J.M. Hopkins, A.J. Kemp, N. Schulz, M. Rattunde, J. Wagner, M.D. Dawson, D. Burns, Pulsed pumping of semiconductor disk lasers. Opt. Exp. **15**(6), 3247–3256 (2007)
15. N. Kugler, T. Brand, I. Schmidt, C. Gao, H. Weber, Yb:YAG disk laser pumped by a stacked diode array. Proc. SPIE **3682**, 36–46 (1998)
16. M. Shimizu, H. Suda, M. Horiguchi, High-efficiency Nd-doped fibre lasers using direct-coated dielectric mirrors. Electron. Lett. **23**(15), 768–769 (1987)
17. J. Limpert, A. Liem, H. Zellmer, A. Tunnermann, 500W continuous wave fibre laser with excellent beam quality. Electron. Lett. **39**, 645–647 (2003)
18. C.H. Liu, A. Galvanauskas, B. Ehlers, F. Doerfel, S. Heinemann, A. Carter, J. Farroni, *810-W Single Transverse Mode b-Doped Fiber Laser, OSA/ASSP*, Post Deadline Paper, 1-2 (2004)
19. G. Bonati, H. Voelckel, T. Gabler, U. Krause, A. Tünnermann, J. Limpert, H. Zellmer, in *Photonics West: 1.53 kW From a Single Yb-Doped Photonic Crystal Fiber Laser.* San Jose, Late Breaking Developments, Session 5709-2a (2005)
20. V. Gapontsev, D. Gapontsev, N. Platonov, O. Shkurikhin, V. Fomin, A. Mashkin, S. Ferin, in *2 kW CW Ytterbium Fiber Laser with Record Diffraction-Limited Brightness.* Conference on Lasers and Electro-Optics Europe (2005), p. 508
21. V. Fomin, A. Mashkin, M. Abramov, A. Ferin, V. Gapontsev, in *3 kW Yb Fibre Lasers with a Single-Mode Output.* Symposium on High Power Fiber Lasers and Their Applications, St. Petersburg, Russia (2006)
22. V.P. Gapontsev, in *New Milestones in the Development of Super High Power Fiber Laser.* Photonics West (2006), pp. 21–26
23. J. Limpert, F. Roser, S. Klingebiel, T. Schreiber, C. Wirth, T. Peschel, A. Tiinnermann, The rising power of fiber lasers and amplifiers. IEEE J. Sel. Top. Quantum Electron. **13**(3), 537–545 (2007)
24. Y. Jeong, A.J. Boyland, J.K. Sahu, S. Chung, J. Nilsson, D.N. Payne, Multi-kilowatt single-mode ytterbium-doped large-core fiber laser. J. Opt. Soc. Korea **13**(4), 416–422 (2009)
25. N.S. Platonov, D.V. Gapontsev, V.P. Gapontsev, V. Shumilin, 135W CW fiber laser with perfect single mode output. IEEE Laser and Electro-Optics, vol. 2 (2002), p. CPDC3-1-4
26. J. Boullet, Y. Zaouter, R. Desmarchelier, M. Cazaux, F. Salin, J. Saby, R. Bello-Doua, E. Cormier, High power ytterbium-doped rod-type threelevel photonic crystal fiber laser. Opt. Exp. **16**(22), 17891–17902 (2008)
27. J.K. Kim, C. Hagemann, T. Schreiber, T. Peschel, S. Böhme, R. Eberhardt, A. Tünnermann, Monolithic all-glass pump combiner scheme for high-power fiber laser systems. Opt. Exp. **18**(12), 13194–13203 (2010)
28. http://www.itflabs.com/data/File/Tech/CleoEu-Gonthier.pdf
29. W. Li, X. Chen, Z.C. Wu, Combining power of high power fiber laser exceeds 3kW. High Power Laser Part. Beams **22**(2), 242 (2010)
30. Goldberg, M.L. Flohic, Optical fiber amplifiers and lasers and optical pumping devices there for and methods of fabricating same, U.S. Patent 6,731,837 (2004)
31. P. Loosen, G. Treusch, C.R. Haas, U. Gardenier, M. Weck, V. Sinnhoff, S. Kasperowski, R. vor dem Esche, High-power diodelasers and their direct industrial applications. Proc. SPIE **2382**, 79–88 (1995)
32. L. Golderg, B. Cole, E. Snitzer, A review of the use of high power diode lasers in surface hardening. Electron. Lett. **33**(25), 2127–2129 (1997)
33. E. Kennedy, G. Byrne, D.N. Collins, W.L. Wei, Multi-coupler side-pumped Yb-doped double clad fiber laser. J. Mater. Process. Technol. **155–156**, 1855–1860 (2004)
34. J.Q. Xu, J.H. Lu, G. Kumar, J.R. Lu, K. Ueda, A non-fused fiber coupler for side-pumping of double-clad fiber lasers. Opt. Commun. **220**, 389–395 (2003)
35. http://www.auniontech.com/uploadfile/2011/35/13149580538536.pdf
36. http://www.dilas.com
37. T.J. Lienert, S.S. Babu, T.A. Siewert, V.L. Acoff, *Welding Fundamentals and Processes.* ASM Handbook, vol. 6A. (ASM International, 2011)

38. F. Lusquiños, J.C. Conde, S. Bonss, A. Riveiro, F. Quintero, R. Comesaña, J. Pou, Theoretical and experimental analysis of high power diode laser (HPDL) hardening of AISI 1045 steel. Appl. Surf. Sci. **254**, 948–954 (2007)
39. N.S. Bailey, W.T. Yung, C. Shin, Predictive modeling and experimental results for residual stresses in laser hardening of AISI 4140 steel by a high power diode laser. Surf. Coat. Technol. **203**, 2003–2012 (2009)
40. M.J. Balart, A. Bouzina, L. Edwards, M.E. Fitzpatrick, The onset of tensile residual stresses in grinding of hardened steels. Mater. Sci. Eng. **367**, 132–142 (2004)
41. P.M. Beckett, A.R. Fleming, R.J. Foster, J.M. Gilbert, D.G. Whitehead, The application of semiconductor diode lasers to the soldering of electronic components. Opt. Quantum Electron. **27**, 1303–1311 (1995)
42. http://www.coherent.com/Products/index.cfm?807/Diode-Laser-Systems
43. L. Li, The advances and characteristics of high power diode laser materials processing. Opt. Lasers Eng. **34**, 231–253 (2001)
44. http://www.dilas.com/gdresources/downloads/DILAS_ThinMetal
45. E. Kennedy, G. Byrne, D.N. Collins, A review of the use of high power diode lasers in surface hardening. J. Mater. Process. Technol. **155–156**, 1855–1860 (2004)
46. https://www.coherent.com/download/6698/An-Introduction-to-Diode-Lasers-for-Materials-Processing.pdf
47. J.F. Ready, *Industrial Applications of Lasers*, 2nd edn. (Academic Press, New York, 1997)
48. F. Bachmann, High-power diode lasers technology and applications. Proc. SPIE **3888**, 394–403 (2000)
49. S.H. Ghasemi, M. Lafouti, A beam shaping design for coupling high power diode laser stack to fiber with capability of spectral narrowing and stabilizing. Opt. Commun. **285**(12), 2879–2882 (2012)
50. L.L. Xiong, M. Wang, X.B. Wang, Y.F. Zheng, D. Wu, P. Zhang, X.N. Li, Z.F. Wang, X.S. Liu, 3000W CW diode laser cladding system. Proc. SPIE **8241**, 824106(1–5) (2012)
51. W.B. Qin, Y.Q. Liu, Y.H. Cao, J. Gao, F. Pan, Z.Y. Wang, 2000W high beam quality diode laser for direct materials processing. Proc. SPIE **8197**, 81971J-4 (2011)
52. http://www.dilas-inc.com/gdresources/downloads/whitepapers/DILAS_High-brightness-high-power-kW-system.pdf
53. http://www.yumpu.com/en/document/view/5684664/new-innovations-in-diode-laser-cladding-a-fraunhofer-usa
54. http://usuarios.fceia.unr.edu.ar/~adruker/ASM%20Metals%20HandBook%20Volume%204%20-%20Heat%20Treating.pdf
55. J.R. Davis, Surface Engineering for Corrosion and Wear Resistance, (ASM International, 2001)
56. L.A. Dobrzañski, S. Malara, T. Tañski, J. Konieczny, Effect of high power diode laser surface alloying on structure of MCMgAl12Zn1 alloy. Arch. Mater. Sci. Eng. **43**(1), 54–61 (2010)
57. A. Sadighha, G. Mohaghegh Zahed, Meta-analysis of hair removal laser trials. Lasers Med. Sci. **24**, 21–25 (2009)
58. C.C. Dierickx. *Laser Hair Removal: Scientific Principles and Practical Aspects*. Coherent Medical (1999), pp. 1–8
59. J.L. Boulnois, Photophysical processes in recent medical laser developments: a review. Lasers Med. Sci. **1**(1), 47–66 (1986)
60. http://www.complexionsrx.com/laser-hair-removal-san-diego-ca.html
61. http://www.aesthetic.lumenis.com/pdf/400ms.pdf
62. http://www.aesthetic.lumenis.com/pdf/laser_hr_all_skintypes.pdf
63. http://www.laserplusmed.com/pdf/Doc1.pdf
64. M.C. Grossman, C. Dierickx, W. Farinelli et al., Damage to hair follicles by normal mode ruby laser pulses. J. Am. Acad. Dermatol. **35**(6), 889–894 (1996)

65. B. Anvari, T.E. Milner, B.S. Tanenbaum, J.S. Nelson, A comparative study of human skin thermal response to sapphire contact and cryogen spray cooling. IEEE Trans. Biomed. Eng. **45**, 934–941 (1998)
66. Y. Sun, Y. Dai, T. Song, *The Design and Packaging of Laser Heads of Hair Removal*. Internal Talk from Focuslight Technologies Co., Ltd. (2012), pp. 35–41
67. http://www.coeoptics.com/products.Asp?act=xx&ID=310
68. http://www.focuslight.com.cn/products.asp

Chapter 11
Development Trend and Challenges in High Power Semiconductor Laser Packaging

Driven by lower cost, longer lifetime, and new applications, the requirements of high power semiconductor lasers have been changing and the demand for new products has been accelerated in recent years. As a result, the packaging technologies for high power semiconductor lasers have been advanced rapidly and have become more sophisticated. In this chapter, we review and discuss the technology development trend of high power semiconductor lasers, including single emitters, bars, horizontal bar arrays, and vertical bar stacks. The packaging technology is still one of the bottlenecks of the advancement of high power semiconductor lasers. We will discuss the challenges and issues in high power laser packaging and some approaches and strategies in addressing the challenges and issues will be presented. The application and price trend are also discussed briefly in this chapter.

11.1 Introduction

High power semiconductor lasers, including single emitters, laser bars, horizontal bar arrays, and vertical bar stacks, have found increased applications in pumping of solid state laser systems for industrial, military, and medical applications as well as direct material processing applications such as welding, cutting, and surface treatment [1, 2]. As power, efficiency, reliability, manufacturability, and cost of high power semiconductor laser technology continuing to improve, many new applications are being enabled. Commercially, there are numerous types of packaged high power semiconductor lasers, either with or without fiber coupling. However, the basic unit of a packaged high power semiconductor laser is either a single emitter or a bar, which consists of an array of single emitters. Figure 11.1 is an illustration of a single emitter semiconductor laser [3]. The laser chip is epi-down mounted on a heat sink, which is typically a Cu material or CuW material. The laser chip is normally of 500–600 μm wide and 100–150 μm thick. The cavity length is generally 1–4 mm depending on the chip design and the stripe width typically ranges from 100 to 200 μm. The light output from a semiconductor laser is

X. Liu et al., *Packaging of High Power Semiconductor Lasers*, Micro- and Opto-Electronic Materials, Structures, and Systems, DOI 10.1007/978-1-4614-9263-4_11,

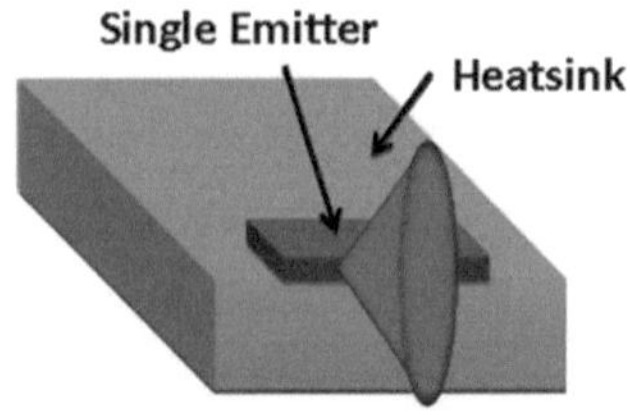

Fig. 11.1 An illustration of a single emitter semiconductor laser [3]

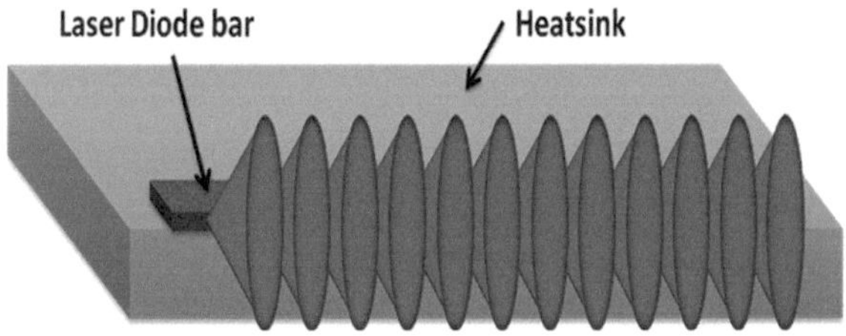

Fig. 11.2 An illustration of a high power semiconductor laser bar [3]

highly diverged. For an 808 nm laser, the fast axis (vertical direction) divergence (full width at half maximum, FWHM) is normally about 40° and the slow axis (horizontal direction) divergence (FWHM) at operation power is about 8°. Figure 11.2 illustrates a laser bar, which is an array of single emitters [3]. The laser bar is also epi-down bonded to a heat sink. A standard laser bar is of 10 mm wide and generally has 19–75 emitters depending on the fill factor and stripe width. The far-field performance is similar to that of a single emitter laser.

Traditionally, output power, conversion efficiency, and reliability are the three key performance measures of high power semiconductor lasers. However, with the advancement of chip technology and driven by lower cost, longer lifetime, new applications, high pumping efficiency of diode-pumped solid state lasers and compactness of laser systems, there emerges new technology development trend for high power semiconductor lasers, namely, output power scaling, high brightness, indium free die/bar bonding, narrow spectrum, and low "smile". With these development trends, there also exist challenges and issues.

11.2 Output Power Scaling

Many new applications require ever increasing output power of a semiconductor laser. The output power of a single emitter laser has been significantly improved with the advancement of laser chip technology. Laser arrays (or laser bars) provide a magnitude of output power of single emitter by integration the single emitters at the wafer level. The output power evolution of a diode laser bar is shown in Fig. 11.3 [4, 5]. The output power has improved largely due to the development

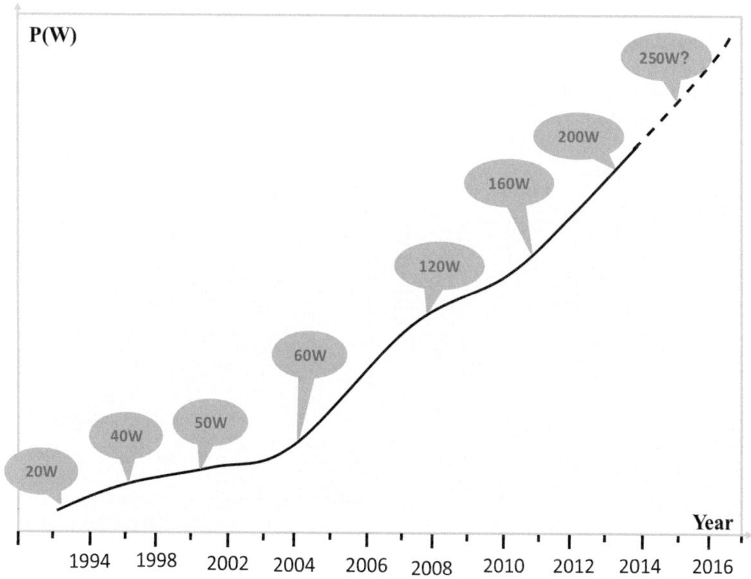

Fig. 11.3 The output power evolution of a diode laser bar [4, 5]

of chip and packaging technologies. To further scaling the output power, several packaging technologies have been developed including multiple single-emitter modules, horizontal bar arrays, vertical bar stacks, and stack arrays.

11.2.1 Single Emitter and Bar

Single emitter: The output power of a single emitter laser is limited either by catastrophic optical mirror damage (COMD) or by thermal rollover, as shown in Fig. 11.4 [3, 5]. The COMD is caused mainly by optical absorption and non-radiative recombination at the front facet which lead to localized overheating. Some advanced technologies, such as facet passivation, non-absorbing-mirror, and unpumped window, have been developed to significantly improve the COMD level of lasers [6–8]. Thermal rollover happens when more heat is generated than dissipated by the device. Typically, it is caused by excessive heat generation which leads to heat accumulation and significant temperature rise. As the conversion efficiency of a laser is fixed, with increased output power, more heat is generated. In order to avoid thermal rollover, thermal resistance of the device has to be reduced. Longer cavity length and wider stripe width can significantly reduce thermal resistance and thus generally higher output power single emitter lasers have longer cavity length. With the improved COMD and thermal rollover levels, 5–10 W 808 nm and 8–15 W 9xx nm single emitter lasers with stripe width of 200 μm

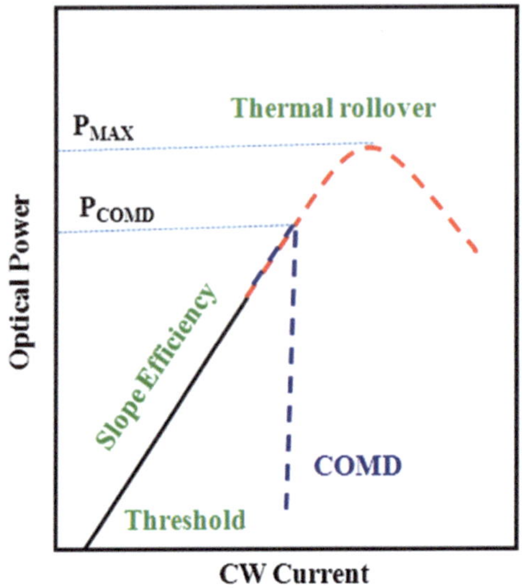

Fig. 11.4 An illustration of the output optical power as a function of driving current of a single emitter laser (*red dot line* and *blue dot line* present the laser with thermal-over and COMD, respectively) [3, 5]

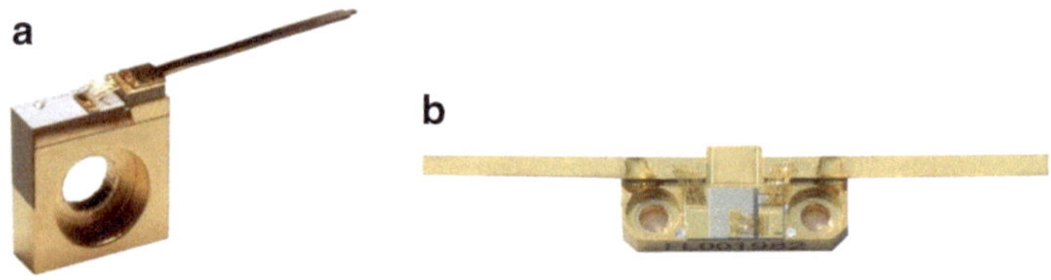

Fig. 11.5 Physical structure of the single emitter laser. (**a**) C-mount laser. (**b**) F-mount laser [9]

and 100 μm, respectively, are commonly available commercially. Figures 11.5 and 11.6 show two typical structures laser devices of C-mount and F-mount and the power-voltage-injection current (PVI) curves of an 808 nm F-mount diode laser, respectively [9].

Single bar: To increase the output power at the chip wafer level, single emitters are integrated at the wafer level to form a one-dimensional array, which is more commonly called a laser bar, as shown in Fig. 11.2. In a bar, the single emitters are electrically connected in parallel. The most common package format is a conduction-cooled CS package and a micro-channel liquid-cooled package, as shown in Fig. 11.7a, b [3, 9]. Depending on the fill factor and cavity length, the CW output power of a laser bar can be as high as hundreds of Watts [1, 2, 10]. For reliable operation and commercial products, the continuous wave (CW) output power of a single bar is normally not higher than 100 and 150 W for 808 nm and 9xx nm (e.g., 976 nm), respectively. For a state-of-the-art commercial 20 or 30 % fill factor 808 nm and 9xx nm conduction-cooled laser bar, the operation power is 60–80 W and 80–100 W, respectively. For higher fill factors, the operation power

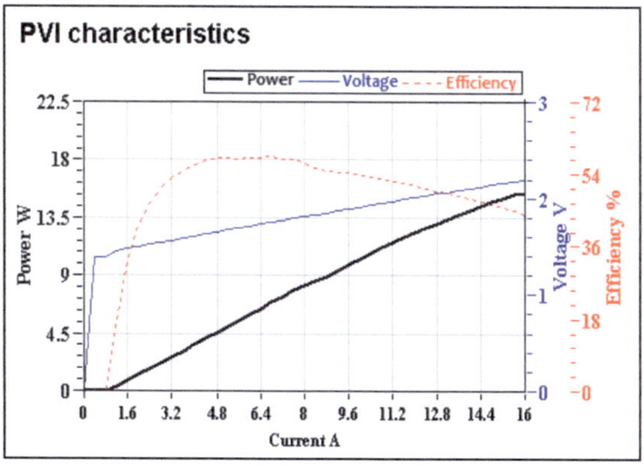

Pop(W)	15.00
Iop(A)	15.21
Ith(A)	0.55
Slope Eff.(W/A)	1.09
Eff.@Iop(%)	45.63
Max Eff.(%)	58.58
Rs(mOhms)	48.96
V @Iop(V)	2.16
TC	25.0

Fig. 11.6 PVI curve of F-mount [9]

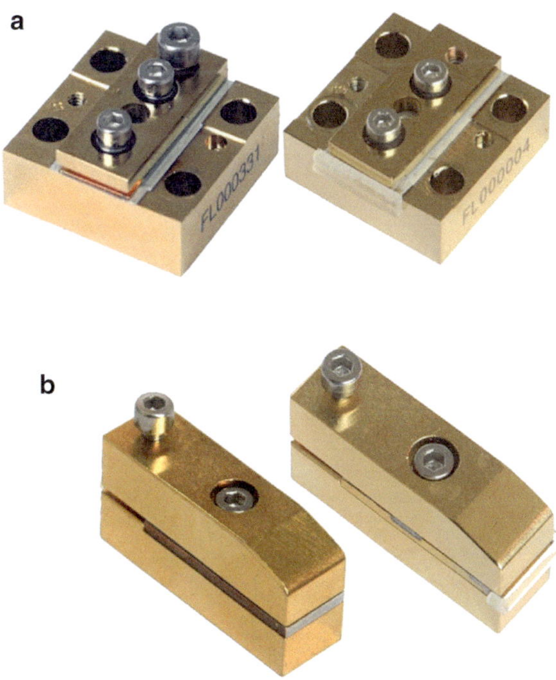

Fig. 11.7 An example of a (**a**) conduction-cooled package and (**b**) micro-channel liquid-cooled single bar package with (*right*) and without (*left*) collimating lens [3, 9]

can be as high as 100 and 150 W. At such power levels, the package is normally liquid cooled. Figure 11.8 shows typical PVI plots and spectrum data of a commercial conduction-cooled package and micro-channel liquid-cooled single bar packages of different wavelength [9].

a

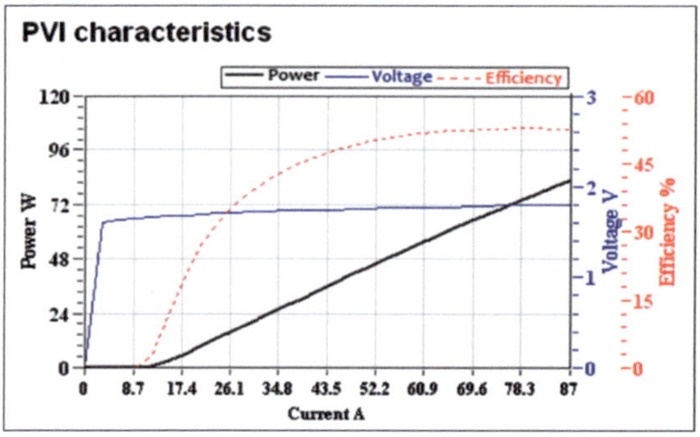

Pop(W)	80.00
Iop(A)	84.13
Ith(A)	12.32
Slope Eff.(W/A)	1.15
Eff.@Iop(%)	52.83
Max Eff.(%)	53.06
Rs(mOhms)	1.58
V @Iop(V)	1.80
TC	25.0

b

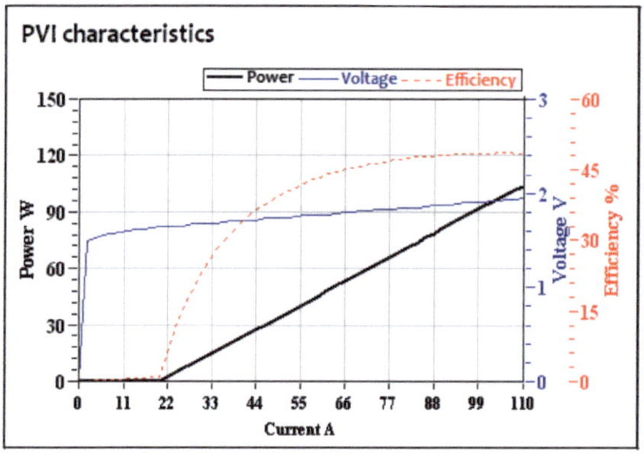

Pop(W)	100.00
Iop(A)	106.41
Ith(A)	20.53
Slope Eff.(W/A)	1.17
Eff.@Iop(%)	48.62
Max Eff.(%)	48.69
Rs(mOhms)	3.39
V @Iop(V)	1.93
TC	

c

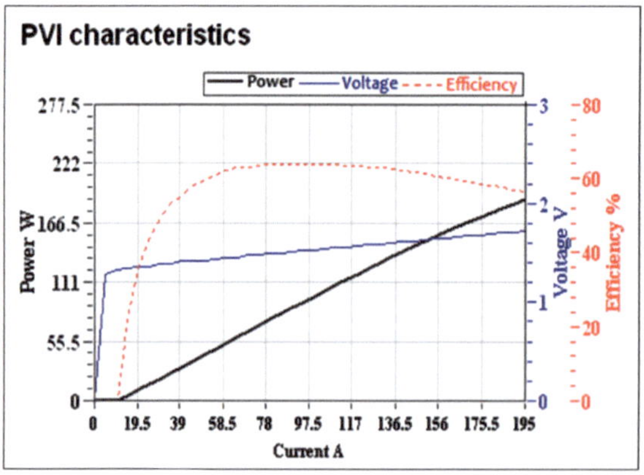

Pop(W)	185.00
Iop(A)	189.86
Ith(A)	7.59
Slope Eff.(W/A)	1.03
Eff.@Iop(%)	56.99
Max Eff.(%)	63.95
Rs(mOhms)	2.08
V @Iop(V)	1.71
TC	25.0

Fig. 11.8 A typical PVI and spectrum data of a commercial (**a**) conduction-cooled package. (**b**) micro-channel liquid-cooled single bar of 808 nm. (**c**) micro-channel liquid-cooled single bar of 980 nm [9]

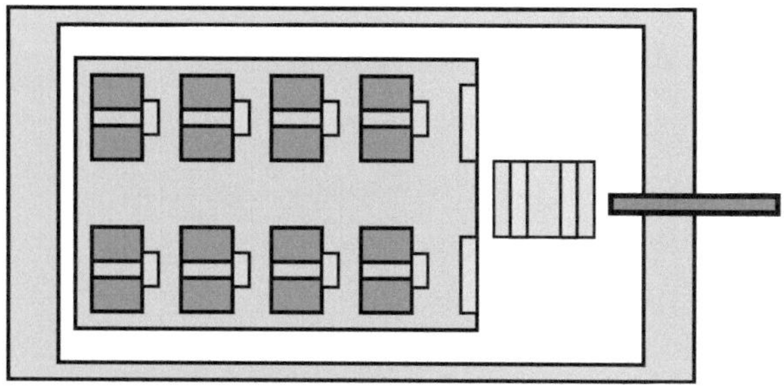

Fig. 11.9 A schematic drawing of a multiple single-emitter module [3]

For a single bar, the total output power is significantly increased compared to a single emitter. However, the output power from each individual emitter of the laser bar is not as high as if it were a single emitter for the same laser design due to thermal crosstalk among the individual emitters and the overall thermal management. The major challenges in single bar packaging are thermal management and thermal stress management. Thermal management includes thermal design and process control to achieve voids "free" in bar bonding interface. Although numerous thermal management approaches have been proposed, including using diamond heat sinks and micro-thermal management, efficient cooling of laser bars is still one of the major obstacles high power laser bar development [6, 11–13]. Voids in bar bonding interface can affect the performance of a laser bar including power and reliability significantly [6, 14]. Two process approaches are used to reduce solder voids in the bonding interface. One is to bond a laser bar using die bonders with controlled pressure, and protected environment under a designed temperature profile. The other approach is to use a vacuum solder reflow system. Thermal stress is mainly caused by the coefficient of thermal expansion (CTE) mismatch between the laser bar and the mounting substrate. Thermal stress not only limits the choices of mounting substrates/heat sinks but also influences reliability, spectrum, and smile of a laser bar. To reduce thermal stress, high thermal conductive and more closely CTE-matched mounting substrates/heat sinks are being developed [11, 13, 14].

11.2.2 Multiple Single-Emitter and Bar Modules

Multiple single-emitter modules: Although the output power of a single emitter has improved significantly in recent years, the power is still generally limited to several watts for 808 nm laser and in low tens of watts for a 9xx nm laser. Naturally, integration of single emitters at the module level is another choice to further increase the output power. Figure 11.9 shows a schematic drawing of a multiple single-emitter module [3]. The individual single emitters are normally serially connected and thus the driving current of such a module is low and the voltage of the module is the sum of the multiple single emitters. For such a module, the output

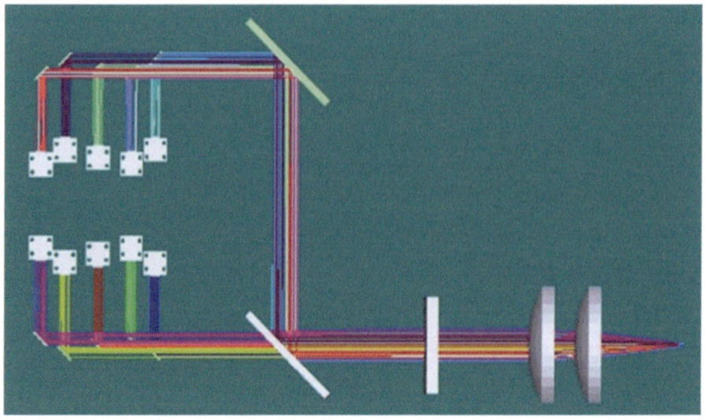

Fig. 11.10 A schematic drawing of a multiple bar-emitter module [5]

power from the single emitter is optically combined and is typically coupled into a delivery fiber. As there is virtually no thermal crosstalk among the single emitters, the output power of the single emitters is not affected in a multiple single-emitter module. However, there do have optical loss in the beam combining and fiber coupling. Today, multiple single-emitter modules with output powers of tens of Watts to over one hundred of watts are commercially available.

Due to the output power limit of a single emitter and the complexity of beam combining, the challenges of the multiple single-emitter module is to push the output power to even higher level such as hundreds of watts. When more and more single emitters are used, the optical design has become more difficult, the cost of the micro-optics, and the footprint of the module make this approach less attractive. The other challenge of the multiple single-emitter module approach is the wavelength uniformity of the single emitters. The single emitters have to be carefully selected to match the wavelength so that the module would have a narrow spectrum.

Multiple single-bar modules: Output power can be achieved by laser bars combination technologies, such as spatial beam combining, the polarization, and multiple wavelength beam combining. Figure 11.10 shows a schematic drawing of a multiple bar-emitter module [5]. The individual single bars are serially connected. The output power is delivered by high energy fiber. Same as multiple single-emitter modules, the output power of the single bars is not affected significantly in a multiple single-bar module as there is virtually no thermal crosstalk among the single bars. Today, multiple single-bar modules with output powers of hundreds of Watts to a few thousands of watts are commercially available.

However, due to the output power limit of a single bar and the complexity of beam combining system, it is difficult to achieve the output power of several thousand watts by beam combination of multiple single laser bars. Just as the same of the multiple single-emitter modules, the challenge of the multiple single-bar module approach is the complexity of the optical structure. The more micro-optics device is used, the more cross interruption among the beams happens. Also, the fabrication of the module becomes difficult, and the optical loss increases.

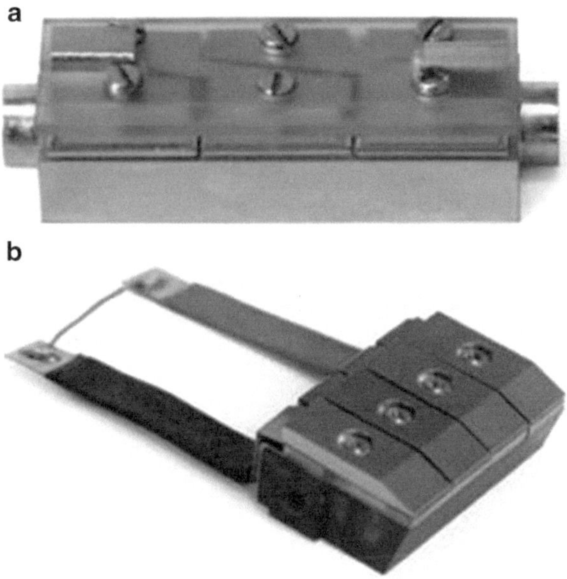

Fig. 11.11 Examples of horizontal bar arrays; (**a**) 1 × 3 and (**b**) 1 × 4 [3]

11.2.3 Horizontal Arrays

For certain applications, such as side pumping of a solid state laser, when higher optical power are required and the laser beams are not needed to be focused, an array of laser bars are packaged horizontally. Figure 11.11 shows two examples of horizontal bar arrays [3]. For Fig. 11.11a, three bars are packaged in series. The bars are water cooled and they are isolated from the cooler by a thermally conductive and electrically insulating material. For Fig. 11.11b, the bars are electrically connected in series as well, but the bars are liquid cooled individually. For these packages, the output power can range from tens of watts to hundreds of Watts to even thousands of Watts depending on how many bars are packaged together and the power of each bar. Figure 11.12 is an example of a LIV (Light power-Injection current–voltage) and spectrum testing data of a quasi-continuous wave (QCW) horizontal bar array [3]. And Fig. 11.13 is an example of a LIV and spectrum testing data of a CW horizontal bar array [5].

For a horizontal bar array shown in Fig. 11.11a, when the bars are electrically isolated from the cooling water, industrial water can be used. However, when the bars are cooled individually with the water flow through the mounting substrate/heat sink which is generally a metallic material for electrical conduction and heat conduction, deionized water is required which causes inconvenience for applications. On the other hand, when laser bars are electrically isolated from the cooler, they are thermally in series, which means the heat from the previous bars will be transferred to the next bars and causes temperature rise in the last bars and temperature nonuniformity among the bars. This poses reliability and wavelength

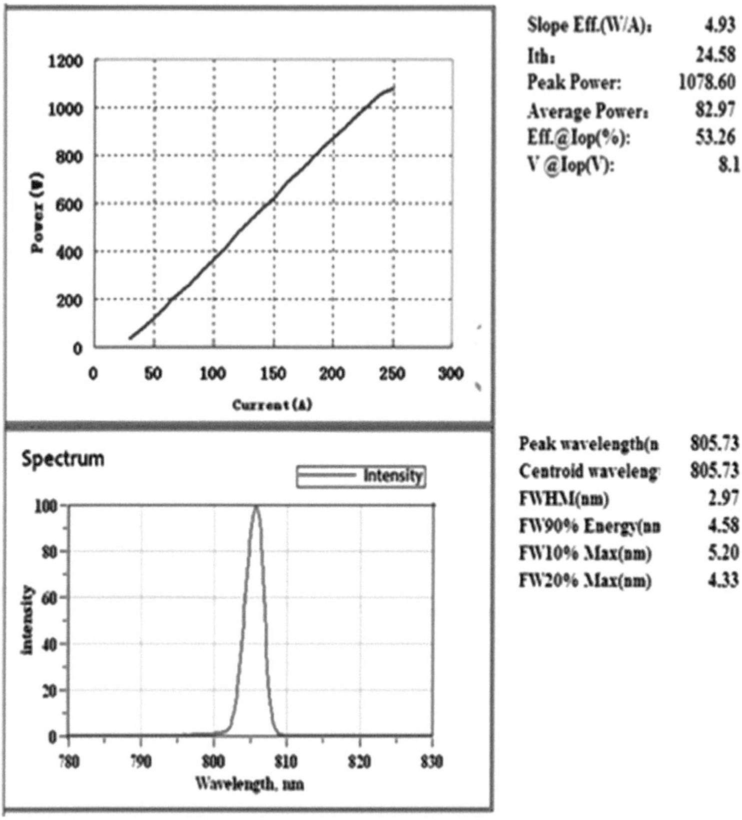

Fig. 11.12 An example of a PI and spectrum testing data of a QCW horizontal bar array [3]

uniformity challenge. Therefore, for a horizontal bar array similar to that shown in Fig. 11.11a, the number of bars and the output power are limited.

11.2.4 Vertical Stacks

Vertical stacks: With the demand of higher output power, vertical bar stacks have become the preferred choice. Figure 11.14a, b shows a typical conduction-cooled G-Stack for QCW application and the typical micro-channel liquid-cooled vertical stacks (V-Stack) with (right) and without (left) collimating lens, respectively. For both configurations, the bars are electrically connected in series [5]. For Fig. 11.14a, the bars are conduction cooled and they are isolated from the heat sink by a thermally conductive and electrically insulating material. For the configuration as shown in Fig. 11.14b, the bars are liquid cooled individually.

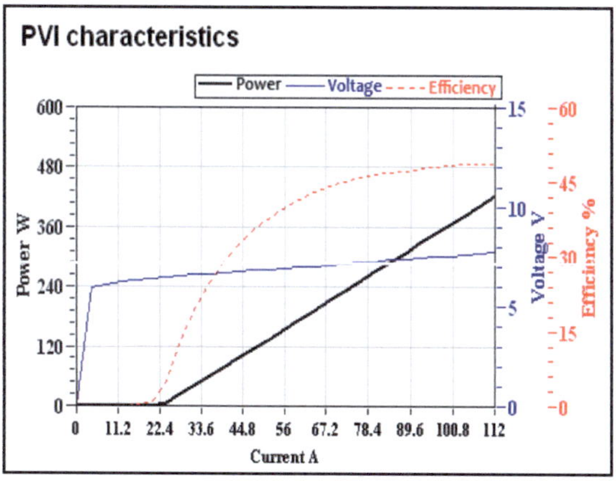

Pop(W)	400.00
Iop(A)	107.14
Ith(A)	23.82
Slope Eff.(W/A)	4.81
Eff.@Iop(%)	48.68
Max Eff.(%)	48.81
Rs(mOhms)	14.18
V @Iop(V)	7.67
TC	25.0

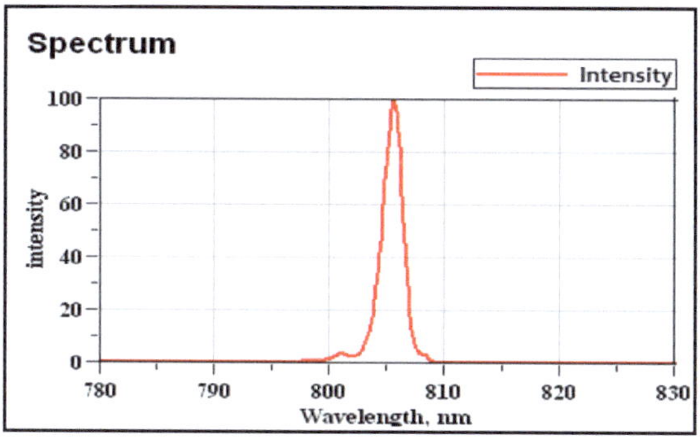

Peak wavelength(nm)	805.79
Centroid wavelength(nm)	805.66
FWHM(nm)	1.98
FW90% Energy(nm)	4.58

Fig. 11.13 An example of a PI and spectrum testing data of a CW horizontal bar array [5]

For a G-Stack, it can be only used at QCW mode due to poor cooling method. The upper limit of the duty cycle is related to the pitch of the bars (the thickness of the mounting substrates). The state-of-the-art commercial product can deliver 300 W out of a bar and as many as 40 bars are packaged in a G-stack. For a vertical bar stack, the stack shown in configuration in Fig. 11.14b is composed by 25 bars, and the number of laser bars can be extended to 60 bars as shown in Fig. 11.14c. The output power of the 60 bars packaged V-Stack can be as high as 300 W per bar, and the total power of the stack reaches to 18 kW at the duty cycle up to 10 %, which is shown in Fig. 11.15 [9].

Multiple stack laser head: At the vertical stack level, the output power also can be improved much by beam combining technologies. The internal optical design in multiple stack laser head and the assembled multiple stack laser head are shown in Fig. 11.16a, b, respectively [15, 16]. In multiple stack laser head, two laser stacks with 25 laser bars are combined together by polarization beam combining, and the output power can reach up to 4,000 W in CW operation mode as shown in Fig. 11.17 [5].

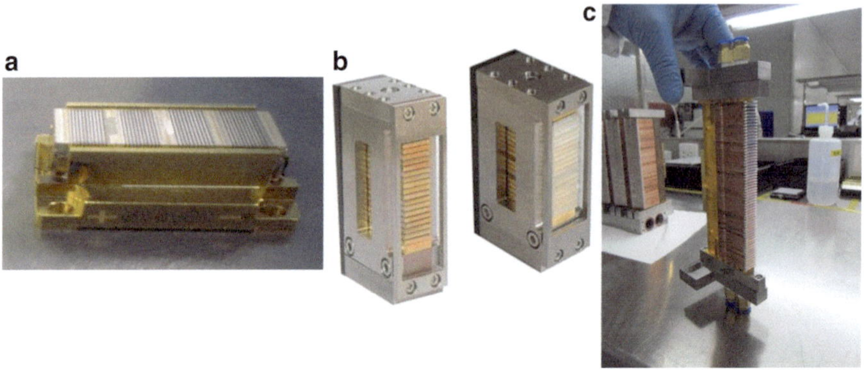

Fig. 11.14 Examples of vertical bar stacks: (**a**) a conduction-cooled G-Stack for QCW application. (**b**) micro-channel liquid-cooled stacks with (*right*) and without (*left*) fast axis collimating lens. (**c**) 60 bars micro-channel liquid-cooled stacks with fast axis collimating lens [5]

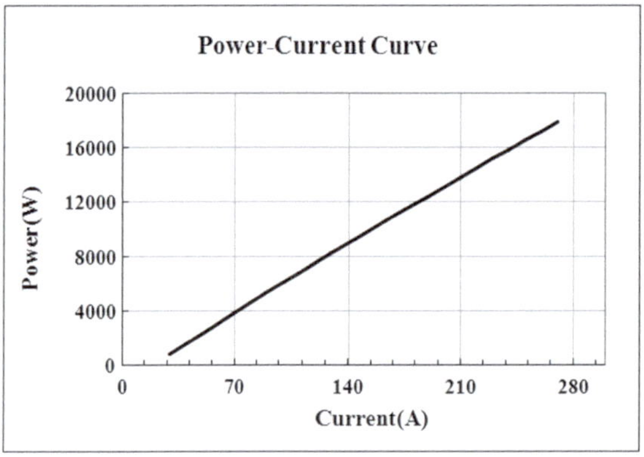

Peak Pop(W)	18522.00
Iop(A)	280.00
Ith(A)	15.92
Slope Eff.(W/A)	71.66
V @Iop(V)	121.00
TC	22.00

Fig. 11.15 PI testing data of a QCW (Quasi CW) 60 bar V-stack [9]

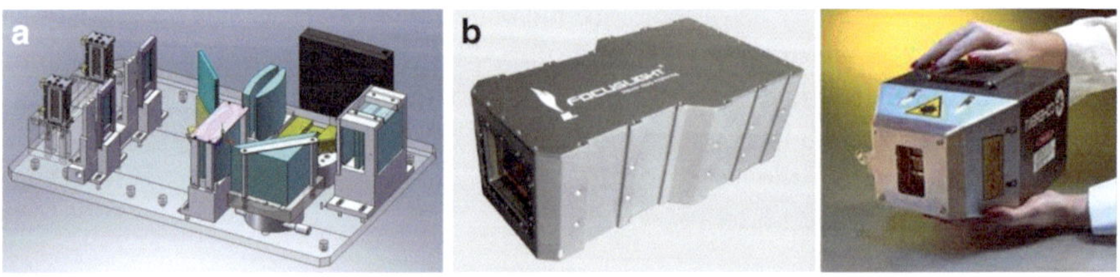

Fig. 11.16 Beam combining of multiple vertical stacks. (**a**) Optical design of the multiple stack laser head. (**b**) Multiple stack laser heads used for material surface processing [15, 16]

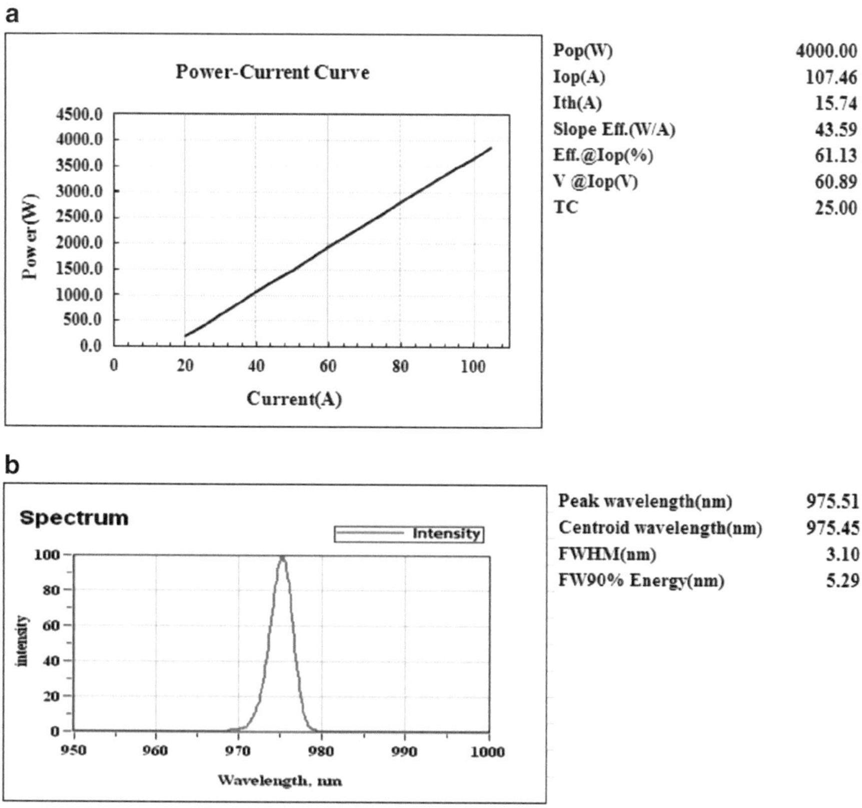

Fig. 11.17 Testing results of the multiple stack laser head. (**a**) power-current curve. (**b**) spectrum [5]

11.2.5 Stack Arrays

The stack array is a further step to realize power scaling. A stack array is composed of several vertical stacks, and the output power can be improved substantially. The output power of a stack array can be as high as tens thousand watts. Figure 11.18 shows an example of micro-channel-cooled stack array [9]. For the configuration, the vertical stacks are electrically connected in series. For Fig. 11.18, laser bars of the stacks are micro-channel cooled individually. For the stack array, each vertical stack consists of 60 bars. The output power of the laser array with eight vertical stacks of 60 bars can be as high as 38.4 kW CW [9].

The major challenges in stack array packaging are the spectrum control and beam control. Although the laser bars in the vertical stack are cooled in parallel in both conduction cooled and micro-channel liquid-cooled configurations, there remains temperature nonuniformity among the bars and among the stacks due to thermal crosstalk and/or liquid flow nonuniformity. This would alter the

Fig. 11.18 An example of micro-channel-cooled stack array [9]

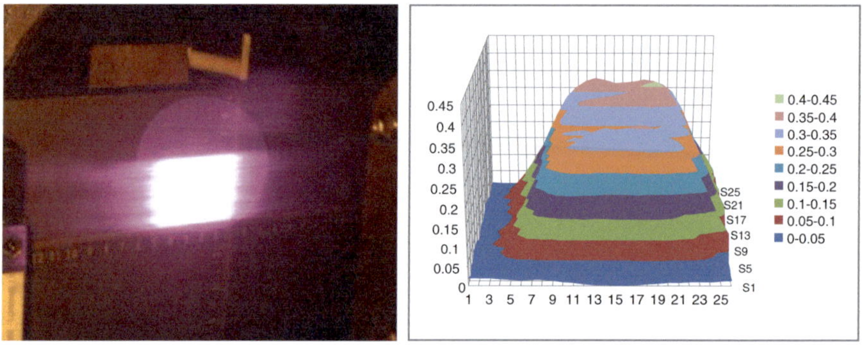

Fig. 11.19 A beam-shaped rectangular beam output from a micro-channel liquid-cooled stack array; *left*, picture of the beam; *right*, uniformity of the beam [3, 5]

wavelength of the bars and broadened the spectrum of the stack array. Also, as the power of a stack array increases, the thermal management should be improved to enhance the reliability. Beam control includes beam size, intensity uniformity, and pointing direction control. Beam shaping optical systems need to be designed and installed to achieve beam control. Figure 11.19 shows a beam-shaped rectangular beam output from a micro-channel liquid-cooled stack array [5].

11.3 High Brightness

For most applications, the output power from a semiconductor laser system (no matter if it is a multiple single-emitter module, a laser bar, a stack or multiple stacks) is most useful when it comes from a relatively small area. This is quantified by the parameter known as brightness, which is defined as the total power emitted by the source per unit solid angle, per unit area. The higher the source brightness,

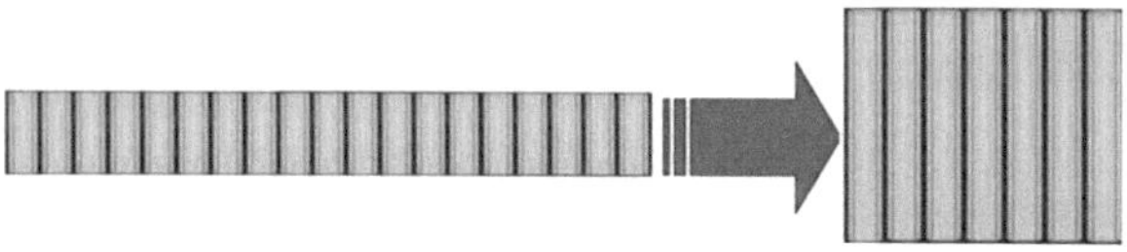

Fig. 11.20 Schematic chip structure of a mini-bar [5]

the more easily and efficiently it can be concentrated into a small focused line, spot, or area. In other words, a higher brightness source can deliver greater power density at the work surface [17]. Even for pumping applications, increasing the spatial brightness of the pump diode enables the laser system to improve the volume, efficiency, power, and beam quality while at the same time reducing thermal management cost in the system [1, 2, 10, 17, 18].

11.3.1 Chip Design

To develop high power laser chip is a classic method to improve the beam brightness of semiconductor lasers. On one hand, an important method to achieve higher output is to increase the length of cavity of laser chips. Up to now, it is no longer uncommon to obtain a single emitter laser chip with the cavity length of 4 mm commercially. Single emitter laser chips with the cavity length of 6 mm are being developed and available as engineering samples. The typical output power of these chips are up to 10 and 15 W for 808 nm and 9xx nm (e.g., 980 nm), respectively.

On the other hand, the new structures of semiconductor laser chips are proposed to improve the beam brightness. A very promising new design of single emitters is the tapered laser diode. This design uses a mono-mode laser diode (ridge section) of low power but high beam quality which emits in an amplifier, which is adapted to the lateral divergence of the beam while still being a waveguide for the vertical direction (tapered section). This leads to high values for the brightness of tapered diode lasers. High brightness of 660 MW/cm^{2}str has already been achieved for a tapered laser emitting a diffraction limited output power of more than 8 W at a wavelength of 980 nm [19]. Furthermore, the optimized tapered lasers at 978 nm with improved layer structures can achieve brightness of 1,100 MW/cm^{2}str [20].

The beam quality and brightness of semiconductor lasers can be improved observably based on the vertical-cavity surface emitting laser (VCSEL). The thermal management technology is a block bottleneck to high power VCSEL. However, it is one of the trends for high power and high brightness semiconductor lasers.

The beam shaping of full laser bars which has the large beam size of 10 mm $\times$ 1 μm are difficult for beam shaping and fiber coupling. In recent years, mini-bars have been designed and are beneficial to obtain high brightness. Figure 11.20 shows the schematic chip structure of a mini-bar and the typical

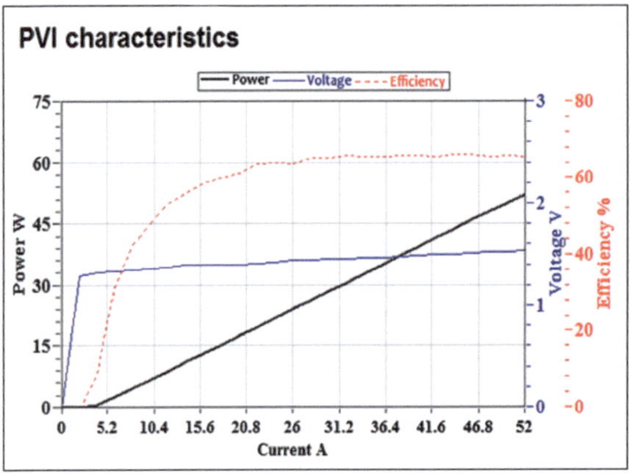

Pop(W)	50.00
Iop(A)	50.01
Ith(A)	4.29
Slope Eff.(W/A)	1.10
Eff.@Iop(%)	65.77
Max Eff.(%)	66.16
Rs(mOhms)	4.05
V @Iop(V)	1.52
TC	25.0

Fig. 11.21 Typical PVI testing result of a mini-bar [5]

testing result is shown in Fig. 11.21 [5]. A typical mini-bar has seven emitters and the lateral size is 3.5 mm. Some other designs have even less number of emitters and the pitch is larger, e.g., five emitters and a pitch of 1,000 μm, that is, the total lateral size is 5 mm. The typical cavity length of mini-bar is 4 mm, but longer cavity length such as 6 mm is being developed. The mini-bar can generate higher power for each emitter as a result of the longer cavity, and the higher brightness can be achieved in mini-bar. However, the package for the mini-bar presents more difficult as a result of the longer cavity.

11.3.2 Beam Shaping and Fiber Coupling Technologies

As analyzed in Chap. 5, the beam brightness of semiconductor lasers can be improved substantially by beam shaping and fiber coupling technologies. For single semiconductor laser device, such as single emitter, single mini-bar, single bar, and single stack, the spatial beam shaping technology based on optical system is applied. The laser device is to collimate first, and then to improve beam parameter product (BPP) by beam shaping, and to fiber coupling at last step. By using this approach, the beam brightness of laser devices can be improved to 10^5 W/cm^2. The beam brightness can be improved further by fiber coupling of multiple semiconductor laser devices.

Single emitter-based technologies: Due to small sizes of active region of a single emitter chip, a single emitter laser is directly coupled into a fiber after beam collimation at fast axis. Based on this method, the state-of-the-art commercial product is a 10 W single emitter laser with 808 nm coupled into fiber with

100 μm diameter and the output power from the fiber is 8 W [21]. Commercially, the optical power of 7 W is offered from a 200 μm core fiber [22].

In order to improve the beam brightness, multiple single-emitters are normally serially connected by beam combining technologies. The brightness can be improved several times to tens of times. The multiple single-emitters products of 50–100 W from 100 μm fiber have been produced by many companies. Optical output power in excess of 100 W from a 105 μm core, 0.15 numerical aperture (NA) fiber is demonstrated with high coupling efficiency [23]. Furthermore, it has been reported an architectural framework and prototype results for kW-class laser tools based on single emitters that addresses a range of output powers (500 W to multiple kW) and BPPs (20–100 mm × mrad) in a system with an operating efficiency near 50 % [24].

Single bar-based technologies: At the single bar level, the fiber coupling architecture using fiber array bundle offers limited brightness. The state-of-the-art commercial product offers 60 W CW power out of a 400 μm fiber [25]. In order to improve the brightness, the beam shaping elements, such as beam transformation system (BTS), V-step and micro-prism, have been used, and the higher brightness fiber module of 110 W CW power out of a 200 μm also has been produced [25].

Technologies also have been developed to coupling the light output from multiple mini-bars into a small fiber using micro-optics and 500 W output from a 200 μm fiber has been reported [18]. Within the German national research project "Briolas", different fiber (400 μm)-coupled laser sources have been built up: fiber-coupled diode laser beam source with BPP of 40 mm × mrad built from diode laser bars with 10 mm bar width, in which the output power is 1,350 W [26].

Mini-bar-based technologies: The fiber coupling technology for mini-bars has been proposed in recent years. Due to good beam quality, the higher brightness can be obtained based on simpler beam shaping system. It has been reported 40 W mini-bar has been coupled into 200 μm fiber just with collimation and focus optical system [25]. Many other fiber coupling technologies for mini-bars also have been developed. Fiber-coupled diode laser beam source with a BPP of 20 mm × mrad built from 3.2 mm mini-bars with a bar width of 3.0 mm and eight emitters, in which two wavelength 940/980 nm wavelength are combined at an output power of 1,050 W [26]. It is demonstrated that fiber-coupled diode laser systems in the 10 kW class output with fiber core diameter here 1,000 μm (0.22 NA) has been achieved [27].

Technologies are also being developed to couple the light output from multiple mini-bars into a small fiber using micro-optics and 500 W output from a 200 μm fiber has been reported [18]. By the micro-optic beam shaping, more than 400 W out of 200 μm fiber based on broad area laser bars of a single wavelength has been demonstrated, even higher power levels with up to 1.2 kW out of 200 μm fiber with wavelength coupling technology are achieved [7]. The BPP is still 22 mm × mrad with a power density of 3,800 kW/cm^2 if focused to a 200 μm fiber spot. Furthermore, each of the four wavelength modules is separately exchangeable and checkable.

The mini-bar with high power, large pitch, and good beam collimation are called tailored bars (T-bar) have been developed [28]. It was also reported that the optical output power is scaled from 180 W coupled into a 100 μm NA 0.22 fiber up to 1.7 kW coupled into a 400 μm NA 0.22 fiber [28].

Vertical stack-based technologies: At the vertical stack level, spatial coupling architecture to fill the "gaps" between the bar emission lines to increase the fill factor, there are two methods. For a single vertical stack, the beam shaping technology of beam cutting, rotation, and rearrangement has been used. It was reported that the highest output power of the fiber module of a laser stack is 400 W from a fiber with core diameter of 400 μm [29].

The new beam shaping element designed by diffraction theory, such as optical grating, holographic phase plates, and amplitude phase plates, is an important development trend of beam shaping technology for high power semiconductor lasers. By a new beam shaping technology based on optical grating, a fiber-coupled direct diode laser with a power level of 1,040 W from a 200 μm core diameter, 0.18 NA output fiber at a single center wavelength has been produced [30]. A fiber-coupled direct diode laser with a power level of 2,030 W from a 50 μm core diameter, 0.15 NA output fiber at a single center wavelength was demonstrated [31].

On other hand, the beam combined technology by multiple vertical stacks also can be filled the "gaps" between the bar emission lines to increase the output power and brightness. The beam combined technologies are polarization coupling architecture to combine the light output from two stacks by a polarization filter or a special prism, and wavelength coupling architecture to use wavelength selective edge filters to combine the output of several stacks with different emission wavelengths on the same optical path. Based on beam shaping technology, fiber-coupled diode laser systems in the 10 kW class output with fiber core diameter here 1,000 μm (0.22 NA) has been achieved [17].

The challenges of achieving high brightness are innovative optical design and beam combining and optical coupling process. All the beaming shaping optical components and fibers have to have high power handling capability. High-power handling requires high-quality end-face finish and special connectors that can withstand back reflections [30].

11.4 Narrow Spectrum

11.4.1 Spectral Control for Laser Bars

Semiconductor laser bar products are required to have narrow spectral width for applications. Increasing the spectral accuracy by reducing the spectral width of the pump diode enables the laser system design engineer to improve the laser system compactness, efficiency, power, and beam quality while at the same time reducing thermal management cost in the system because the pumping source energy out of

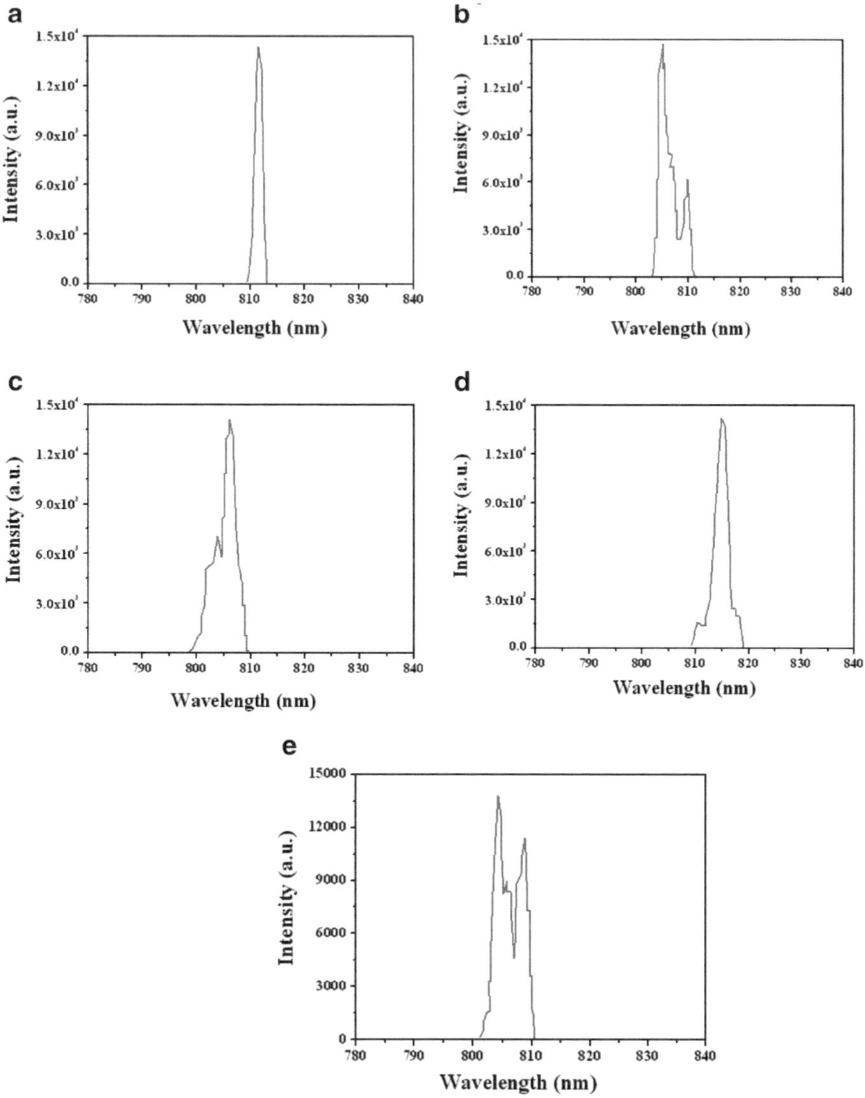

Fig. 11.22 Typical examples of spectra of laser bars [9, 33]

the absorption bandwidth of the active laser media is wasted and becomes heat. Spectral width is one of the key specifications of laser bar products and it is very important to improve the spectral performance to improve production yield, reduce cost, and gain competitiveness.

The spectral broadening of laser arrays is a result of nonuniform emitting wavelength from individual emitters. The broadened spectrum of a laser array/bar can have double or even multiple peaks; some may have shoulders or tails on either or both sides of the spectrum. Figure 11.22 shows some typical spectra of laser bars [9, 33].

The emitting wavelength from individual emitters is affected by wafer uniformity as well as packaging-related thermal and thermal stress effects, with the latter being the major factor. The wavelength of an 808 nm laser shifts to the longer direction at a rate of ~0.28 nm/°C. When the junction temperature of the emitters across the whole array is not uniform, the emitting wavelength varies. The CTE mismatch between the laser array and the mounting substrate material would cause thermal stress in the package structure, which could be imposed on the laser array. Tensile or compressive stress in the epitaxial material of a laser affects the emitting wavelength with a coefficient on the order of ~1×10^{-5} eV/bar (or ~0.005 nm/bar), with tensile stress causing red-shift and compressive stress having blue-shift [32]. Nonuniform thermal stress experienced by the emitters across the width of the array would cause uneven wavelength and thus broad spectrum.

Conduction-cooled 808 nm laser bars with different types of broadened spectral shapes were characterized using spatial spectral mapping, scanning acoustic microscopy, and material analysis techniques to study the emitting wavelength of each emitter, wavelength distribution, and package structure, especially die bonding [33]. It was concluded that while solder voiding causes local heating and therefore there could have a "shoulder" or "tail" on the right side of the spectrum, nonuniform thermal stress on the emitters could be the dominant factor in causing double or multiple peaks in spectrum. Intermetallic formation at some local die-attach solder interface causes additional local compressive stress on some emitters, which lead to a "shoulder" or "tail" on the left side of the spectrum.

Knowing the mechanism of the spectral broadening, it is important to know the shape of the broadened spectrum and determine if it is mainly due to thermal effect or stress effect. The challenge remains how to achieve the temperature uniformity and stress uniformity across the laser bar to eliminate the thermal effect and stress effect. Local temperature rise is mainly caused by solder voiding in bar bonding interface. So the approach to achieve temperature uniformity across the bar is to minimize solder voiding or even achieve void "free" bonding. Furthermore, proper epi-side metallization design of the laser bar can also minimize the impact of solder voiding and local temperature rise [34]. The stress in a laser bar can be induced by CTE mismatch between the laser bar and the mounting substrate, the bonding process and intermetalic formation in the solder bonding interface [33]. New package structures should be designed to properly accommodate the stress caused by CTE mismatch between the laser bar and the mounting substrate as well as the bonding process. Also the packaging process should be optimized to minimize the thermal stress. Furthermore, a proper metallization structure of the laser bar as well as the mounting substrate should be used to avoid intermetalic formation in the solder bonding interface.

11.4.2 Spectral Control for Vertical Stacks

Although the laser bars in the stack lasers are cooled in parallel in both conduction-cooled and micro-channel liquid-cooled configurations, there

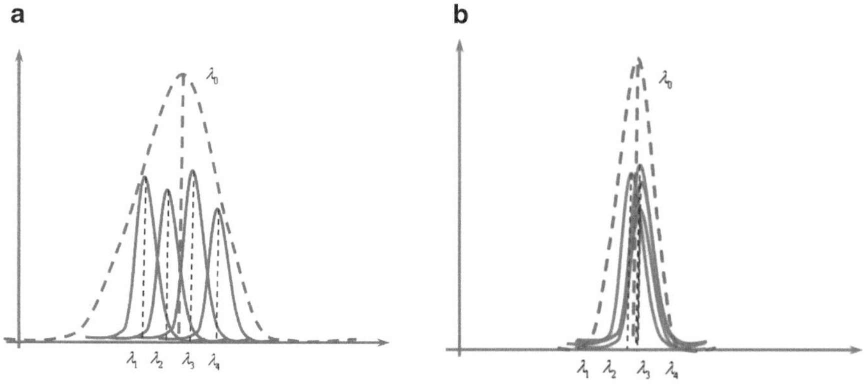

Fig. 11.23 Illustration of output spectrum of a vertical stack (a) without and (b) with spectral control [35]

remains temperature nonuniformity among the bars due to thermal crosstalk and/or liquid flow nonuniformity. This would alter the wavelength of the bars and broaden the spectrum of the stack, as illustrated in Fig. 11.23a [35]. With proper spectral control technology, the spectral of the stack can be improved significantly, as illustrated in Fig. 11.23b [35].

There are mainly two steps to carry out the control of the spectrum width for the stack: thermal analysis of the stack; wavelength selection of the bars based on the temperature distribution in the vertical stacks. The principle of the wavelength selection for the vertical stacks is shown in Fig. 11.24 [5]. First, the temperature distribution is simulated and calculated in thermal analysis of the stack with experimental verification. According to the temperature distribution in stack, the relationship between wavelength shift and the position can be deduced. Second, the appropriate bar is selected, the wavelength of which should be matched with the temperature distribution. After the above two steps, the bars are assembled into vertical stacks to achieve optimal and uniform output wavelength.

Using this method the spectrum broadening of the stack lasers can be controlled effectively as shown in Fig. 11.25 [5]. Figure 11.25 shows the testing results of the wavelength selection in the 60 bar micro-channel coolers V-Stack laser. Based on the wavelength selection technique, the range of the peak wavelength for each bar of the 60 bar MCC V-Stack laser can be controlled within 0.5 nm.

Figure 11.26 shows the typical spectrum of a laser stack with 60 bars. Based on the narrow spectrum technology, the FWHM and FW90 % of spectrum in a laser stack with 60 bars are 2.68 nm and 3.9 nm, respectively [5]. And the FWHM of the area array assembled by eight stacks can be controlled within 3 nm shown in Fig. 11.27 [5].

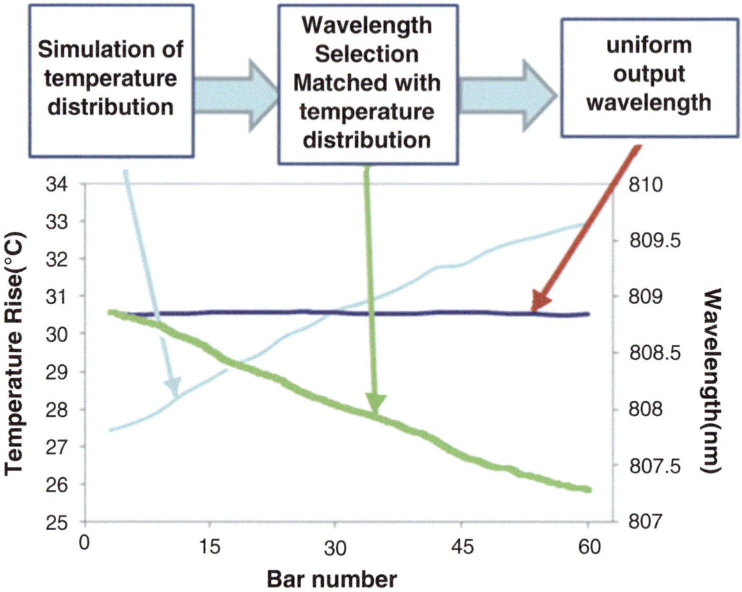

Fig. 11.24 Principle of spectral control for the vertical stacks [5]

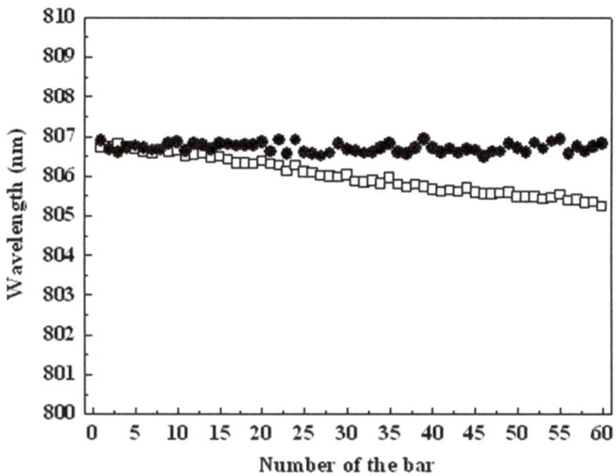

Fig. 11.25 Wavelength distribution of the 60 bar MCC before (*open squares*) and after (*filled circles*) constituting V-stack laser [5]

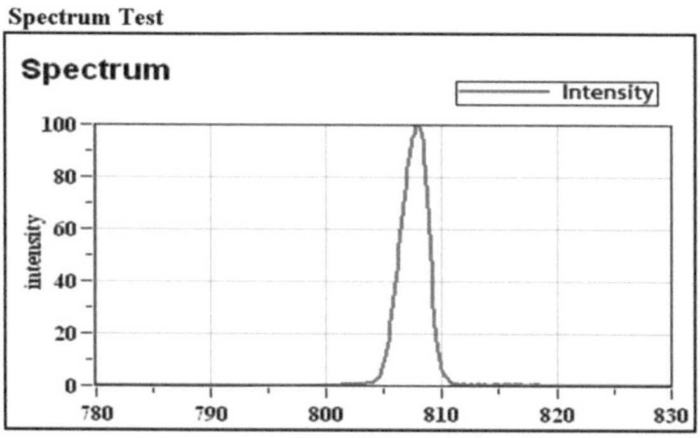

Peak wavelength(nm)	808.12
Centroid wavelength(nm	807.75
FWHM(nm)	2.68
FW90% Energy(nm)	3.90

Fig. 11.26 Spectrum of the laser stack with 60 bars [5]

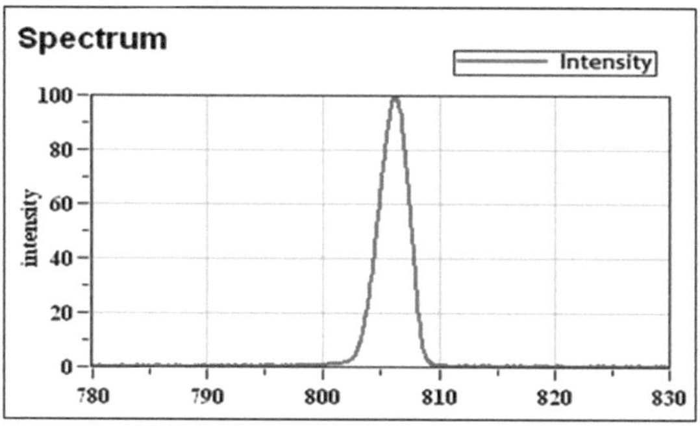

Peak wavelength(nm)	806.83
Centroid wavelength(nm)	806.65
FWHM(nm)	2.98
FW90% Energy(nm)	4.65

Fig. 11.27 Output spectrum of the area array assembled by eight stacks [5]

11.5 Low "Smile"

The nonlinearity of the near-field of emitters (or the so called "smile") in a laser diode array poses significant challenges in optical coupling and beam shaping and has become one of the major roadblocks in broader applications of laser arrays. If near-field linearity of a laser array is poor, in other words, if the smile is very large, the coupling efficiency of the laser array to a fiber array or micro-optics such as a fast axis collimation lens is very low. Increasing the near-field linearity of a pumping laser diode array enables the laser system manufacturer to improve the laser system compactness, optical coupling efficiency, power, and beam quality while at the same time reducing manufacturing cost in the laser system, such as diode-pumped solid state lasers and fiber lasers. Therefore, the near-field linearity of a laser bar is one of the key specifications of laser array products and improving the near-field performance is especially important in order to increase production yield, reduce cost, and differential product performance.

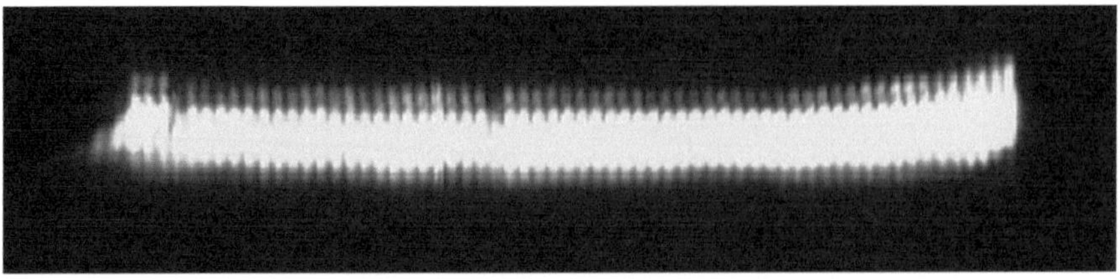

Fig. 11.28 Enlarged "smile" image of a typical good laser diode array [36]

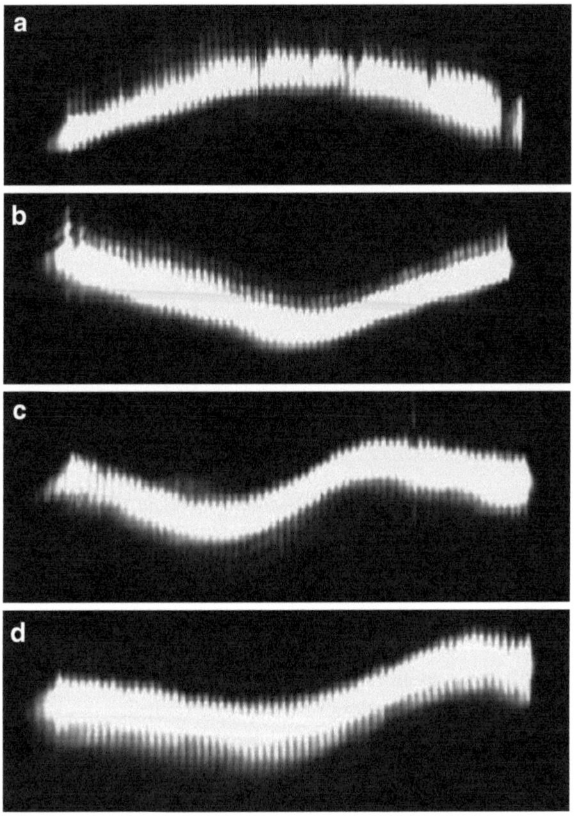

Fig. 11.29 Magnified images of a diode laser bar with various "smile". (**a**)"cry" of approx. 2 μm, (**b**) "smile" of approx. 2.5 μm, or others (**c**), (**d**) [36]

Figure 11.28 shows a magnified "smile" image of a typical good diode laser array [36]. As can be seen from Fig. 11.28, a typical good laser diode array is nearly linear. This is a better emitting light source for beam coupling. The individual emitters form a curvature, which can be concave or convex, corresponding to "cry" as can be seen from Fig. 11.29. Both "smile", "cry" or others are commonly called "smile" due to historical reason [36].

The near-field nonlinearity of laser diode array is caused by CTE mismatch among the different layers of a bare bar, the packaging process, and CTE mismatch between the laser bar and the bonding heat sink. The factors that affect the "smile" are solder layer material, mounting substrate/heat-sink material and thickness, pick-up tool of the die bonding process, and die bonding temperature profile.

11.6 Indium Free Bonding

Indium solder is one of the most widely used solders in high power laser die bonding. Indium solder has some advantages in laser die bonding. It also has some concerns, however, especially in terms of reliability. It was known that indium solder-bonded lasers have much shorter lifetime than AuSn solder-bonded devices due to electromigration, thermal-electro-migration, thermal fatigue, and oxidation [37–39]. Catastrophic degradation was observed in indium solder-bonded lasers. Nondestructive optical and acoustic microscopy was conducted during the lifetime testing to monitor the failure process and destructive failure analysis was performed after the lasers failed revealing that the sudden failure was caused by electromigration of indium solder [37]. Figure 11.30 shows the schematic illustration of the electromigration mechanism in an indium-bonded laser assembly [37]. Indium solder voids were created and gradually enlarged by indium solder electromigration, which caused local heating near the facets of the laser. The local heating induced COMD of the lasers. Current crowding, localized high temperature, and large temperature gradient contributed to the fast indium solder electromigration. To overcome the indium solder problems, indium-free bonding technology is being developed and some technologies have been used in commercial products [7, 38–40].

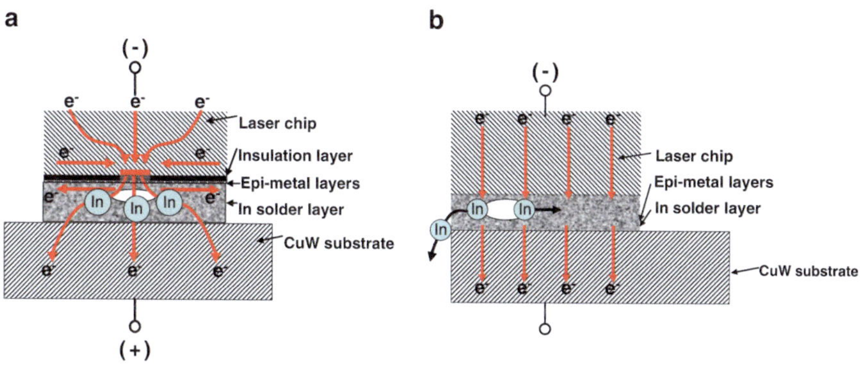

Fig. 11.30 Schematic illustration of the electromigration mechanism in an indium-bonded laser assembly: (**a**) cross-sectional view and (**b**) side view [37]

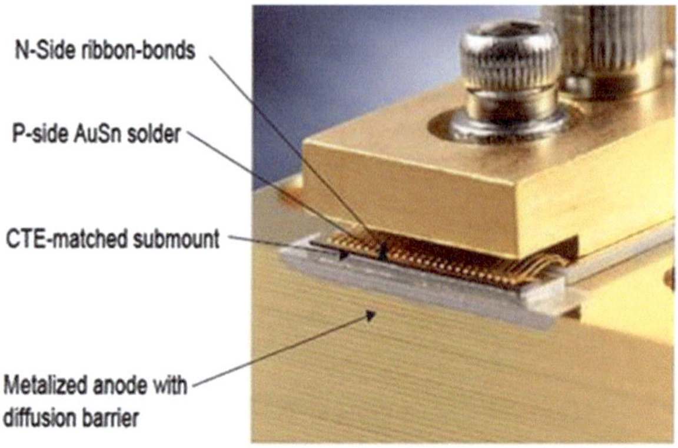

Fig. 11.31 An example of AuSn bonding structure [39]

11.6.1 AuSn Solder

AuSn solder become the choice of indium free bonding due to its stable performance and properties. Figure 11.31 is an example of AuSn bonding structure [39]. As AuSn is a hard solder, a CTE-matched submount has to be added between the laser bar and the heat sink to serve as a buffer layer.

For AuSn bonding, the challenges are thermal management and stress management. As a CTE-matched buffer layer is needed between the laser bar and the heat sink, there are more interfaces and interfaces generally have voids which could lead to overheating of the device. Also, the CTE-matched buffer lay normally has lower thermal conductivity than the heat-sink material which in most cases is a copper material. On the other hand, although the CTE-matched buffer layer has a closer CTE to the laser bar than the heat sink, there remains CTE mismatch and AuSn solder could not release the thermal stress as indium solder does.

11.6.2 Nano-Scale Silver Paste

Nano-scale silver paste as a diode laser bonding material has been investigated [41]. The pure silver joint shows feature of high melting point, high electric conductivity, and high thermal conductivity which are good at heat dissipation. More importantly, the nano-scale silver paste possesses a uniform micro-porous structure that makes its elastic modulus equivalent to that of indium solder and much lower than that of gold-tin solder. Accordingly, low thermal stress is introduced using nano-scale silver paste for soldering. Also, the nano-scale silver paste belongs to green electrical die-attach material. All of these characteristics are beneficial to die bonding the laser diodes.

The nano-scale silver paste has been used for die bonding the single emitter laser diode [41]. The die shear test, function test, and far-field characteristic test were conducted on the device. The thermal resistance of the device was also calculated. All of the performance tests show that the new interconnecting material has the potential to be applied to the field of semiconductor laser. In addition, the new die-attach material can avoid producing voids and have low thermal resistance and high conversion efficiency. However, the packaging process has yet to be improved to achieve uniform bonding thickness. This technology is not matured and the manufacturability, cost effectiveness, and reliability of the devices have yet to be proven.

11.7 Application Trend

Previously, the poor beam quality and low power of diode lasers had prevented them from being used successfully in direct diode laser applications where high brightness is needed. With the advancement of laser technology, the performance of the semiconductor laser is significantly improved exhibiting higher brightness and higher power as well as higher reliability. With these advantages, semiconductor lasers are more widely used as light sources for direct applications. In the following, we introduce the applications in pumping solid state and fiber lasers, material processing as well as medical and cosmetic fields.

11.7.1 Pumping Applications

The most common application of semiconductor lasers is pumping the gain medium of solid state and fiber laser systems. In diode-pumped laser architectures, the semiconductor laser light "pumps" or excites the gain medium which is configured inside its own laser cavity. The pump light is absorbed by the crystal or fiber, and a solid state or fiber laser is created. Diode pumping has significant advantages over traditional approaches that utilize lamps as pumps. In fact, nearly every important parameter of a laser system is improved with diode-pumped architectures including system size, weight, performance, and cost of ownership.

In recent years, high power semiconductor laser with the output power of 100 kW or even higher have been used in pumping solid state laser and fiber laser. The most dramatic advantage to diode pumping is the improved efficiency of the laser system. The center wavelength and bandwidth of the semiconductor laser are important factors to enhance the pumping efficiency. Further in ultrahigh power pumping solid state laser, the wavelength and intensity distribution uniformity should be concerned in improving the pumping efficiency.

11.7.2 Material Processing Applications

Recently, semiconductor laser with high power and high brightness are being used to directly process materials, such as cladding, sheet welding, metal and plastics soldering. For example, the semiconductor laser with output power from 500 to 1,000 W fits for laser quencher, and the power above 2,500 W can be used in laser harden and cladding. Further, deep-penetration welding using direct diode lasers with high beam quality has become a reality. This can be considered as a first step into the high-power laser market, where laser cutting is dominant. The beam quality needed to achieve laser cutting of 5 mm $\times$ mrad out of a 100 μm of fiber in the multi-kW-range is a challenge; it would result in an intensity that is 36 times higher compared to a current 600 μm coupled high-brilliance system. With the new innovation in beam shaping and fiber coupling, it is possible to achieve in the future.

11.7.3 Medical and Cosmetic Applications

Wavelength is the key parameter of the semiconductor laser for Medical and cosmetic. Different virus or human tissues need to absorb different wavelength, and hence the choice of adapted wavelength is very important, and furthermore the output power and operation mode also influence therapeutic efficiency. Semiconductor lasers have diversified wavelength to choose ranging from ultraviolet to infrared wave band and can meet the requirements in medical and cosmetic applications, such as surgery gasification, laser scalpel, and hair removal. Semiconductor laser can extend the application in cancer diagnosing, eye disease detecting, cholesterol testing, tattoo removal, etc. For example in tattoo removal, laser therapy works by breaking down this fibrous barrier, allowing the pigment to disperse into smaller pieces, which the body's natural defenses can slowly remove.

11.7.4 Price Trend

Figure 11.32 shows the price evolution and prediction of the semiconductor laser operated at QCW and CW operation mode [5]. When the accumulated volume is low, e.g., with a total power up to 1 kW, in 2012 the price is around US$3.00 and 5.80 per watt for QCW and CW operation mode, respectively. As the accumulated volume is increased to 10^7 kW, the price is reduced to only US$0.10 and 0.50 per watt for the QCW and CW operation mode, respectively, in the year 2012. In the coming years, with the development of the semiconductor laser science and technology one may predict that the price be reduced by about 15–20 % for every period of 2 years.

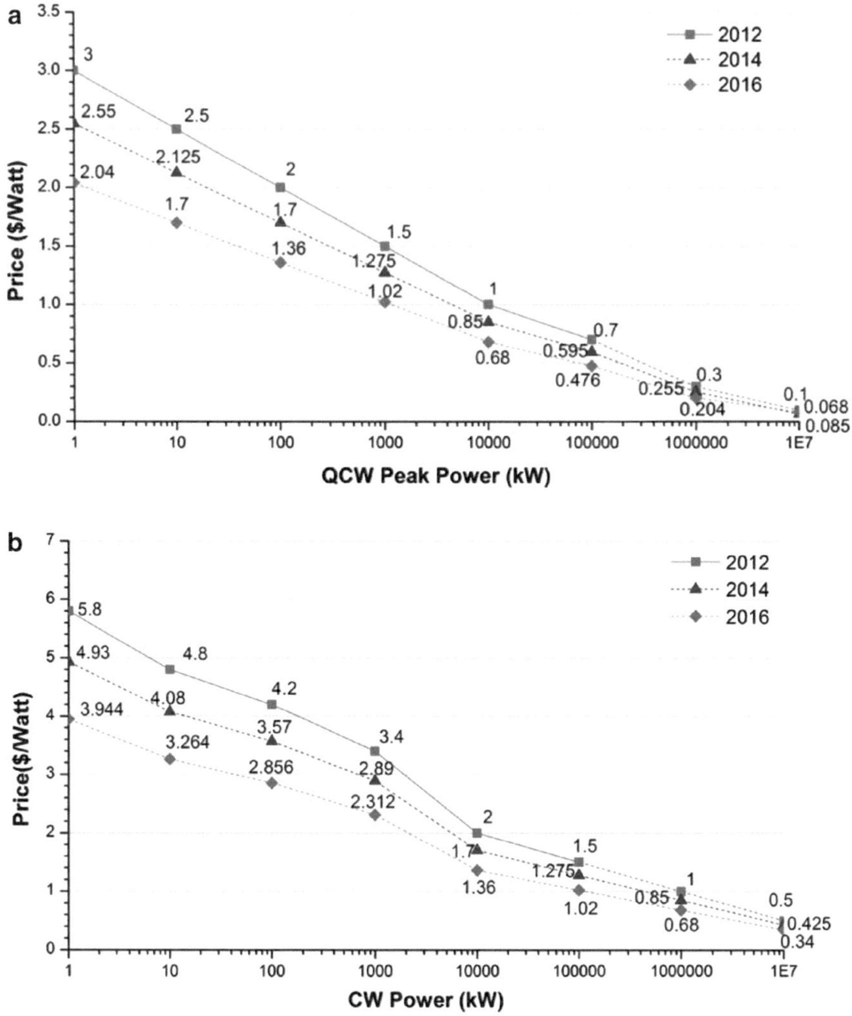

Fig. 11.32 Price evolution of the semiconductor laser operated at: (**a**) QCW and (**b**) CW mode [5]

References

1. B. Faircloth, High-brightness high-power fiber coupled diode laser system for material processing and laser pumping. Proc. SPIE **4973**, 34–41 (2003)
2. N. Lichtenstein, B. Schmidt, A. Fily, S. Weiß, S. Arlt, S. Pawlik, B. Sverdlov, J. Müller, C. Harder, DPSSL and FL pumps based on 980 nm-Telecom Pump Laser Technology: changing the industry. Proc. SPIE **5336**, 77–83 (2004)
3. X.S. Liu, W. Zhao, *Technology Trend and Challenges in High Power Semiconductor Laser Packaging*. 2009 Electronic Components and Technology Conference (2009), pp. 2106–2113.

4. http://www.google.com.hk/url?q=http://www.nlight.net/nlight-files/file/articles/HP_May05_OLE.pdf&sa=U&ei=YPHUUtX-Cu2ciAe7wIDYDg&ved=0CE8QFjAI&usg=AFQjCNEMghYrnpzlYrRZgiKDyEXJYDQ2ig
5. J.W. Wang, D. Hou, B. Hao, G.D. Xu, X.S. Liu, *Performance on Semiconductor Laser With Different Packaging Structure and Development Trend*. (Internal Talk from Focuslight Technologies Co., Ltd., 2011), pp. 13–29
6. X.S. Liu, M.H. Hu, C.G. Caneau, R. Bhat, L.C. Hughes, C.E. Zah, Thermal management strategies for high power semiconductor pump lasers. IEEE Trans. Compon. Pack. Technol. **29** (2), 268–276 (2006)
7. A. Hodges, J. Wang, M. DeFranza, X.S. Liu, B. Vivian, C. Johnson, P. Crump, P. Leisher, M. DeVito, R. Martinsen, J. Bell, A CTE matched, hard solder, passively cooled laser diode package combined with nXLT™ facet passivation enables high power, high reliability operation. Proc. SPIE **6552**, 65521E(1–9) (2007)
8. K.C. Song, Y.B. Li, N. Visovsky, H.K. Nguyen, M. Hu, X.S. Liu, S. Coleman, B. Paddock, M. Turner, C. Catherine, R. Bhat, C.E. Zah, *High Power 1060 nm DBR Lasers with Impurity Free Quantum Well Intermixing Passive Section*. Proceedings of LEOS Annual Conference (2005), ThZ6
9. D. Wu, L.L. Zhu, G.F. Zhou, G.D. Xu, X.S. Liu, *Performance of High Power Semiconductor Lasers*. (Internal Talk from Focuslight Technologies Co., Ltd., 2012), pp. 18–22
10. R. Szweda, Diode lasers now pushing the power limits. III-Vs Review **18**(5), 42–44 (2005)
11. S. Weib, E. Zakel, H. Reichl, Mounting of high power laser diodes on diamond heatsinks. IEEE Trans. Compon. Packag. Manuf. Tech. A **19**(1), 46–53 (1996)
12. D. Lorenzen, J. Bonhaus, W.R. Fahrner, E. Kaulfersch, E. Wörner, P. Koidl, K. Unger, D. Müller, S. Rölke, H. Schmidt, M. Grellmann, Micro thermal management of high-power diode laser bars. IEEE Trans. Ind. Electron. **48**(2), 286–297 (2001)
13. M. Leers, K. Boucke, *Cooling Approaches for High Power Diode Laser Bars*. Electronic Components and Technology Conference (2008), pp. 1011–1016
14. W. Pittroff, G. Erbert, G. Beister, F. Bugge, A. Klein, A. Knauer, J. Maege, P. Ressel, J. Sebastian, R. Staske, G. Traenkle, *Mounting of High Power Laser Diodes on Boron Nitride Heat Sinks using an Optimized AuSn Metallurgy*. 2000 Electronic Components and Technology Conference (2000), pp. 119–124
15. http://www.focuslight.com.cn/product.asp?tid=9
16. http://www.coherent.com/products/?1538/HighLight-4000L
17. F. Leibreich, H.G. Treusch, *Powering Brightness*. SPIE's OE Magazine (2001), pp. 18–19
18. M. Haag, B. Köhler, J. Biesenbach, T. Brand, Novel high-brightness fiber coupled diode laser device. Proc. SPIE **6456**, 64560T(1–8) (2007)
19. M.T. Kelemen, J. Weber, G. Kaufel, R. Moritz, M. Mikulla, G. Weimann, 8 W high-efficiency high-brightness tapered diode lasers at 976 nm. Proc. SPIE **6104**, 61040D(1–9) (2006)
20. B. Sumpf, K. Hasler, P. Adamiec, F. Bugge, F. Dittmar, J. Fricke, H. Wenzel, M. Zorn, G. Erbert, High-brightness quantum well tapered lasers. IEEE J. Sel. Top. Quantum Electron. **15**(3), 1009–1020 (2009)
21. http://www.focuslight.com.cn/product.asp?tid=8
22. http://www.lumics.com/fileadmin/user_upload/Datasheets_Lumics_Products/Multimode/TO_package/LU0808T070.pdf
23. R. Duesterberg, L. Xu, J.A. Skidmore, J. Guo, J. Cheng, J.H. Du, B. Johnson, D.L. Vecht, N. Guerin, B. Huang, D.L. Yin, P. Cheng, R. Raju, K.W. Lee, J. Cai, V. Rossin, E.P. Zucker, 100 W high-brightness multi-emitter laser pump. Proc. SPIE **7918**, 79180V(1–8) (2011)
24. K. Price, F. Pfeffer, P. Leisher, S. Karlsen, R. Martinsen, KW-class industrial diode lasers comprised of single emitters. Proc. SPIE **7583**, 75830E(1–9) (2010)
25. http://www.dilas.de/pages/products.php?category=22&series=40
26. V. Krause, A. Koesters, H. Koenig, U. Strauss, Brilliant high-power diode lasers based on broad area lasers. Proc. SPIE **6876**, 687615(1–13) (2008)

27. M. Baumann, V. Krausea, G. Bergweilerb, M. Flaischerowitzc, J. Banikd, Local heat treatment of high strength steels with zoom-optics and 10 kW-diode laser. Proc. SPIE **8239**, 82390J(1–9) (2012)
28. B. Köhler, S. Ahlert, A. Bayer, H. Kissel, H. Müntz, A. Noeske, K. Rotter, A. Segref, M. Stoiber, A. Unger, P. Wolf, J. Biesenbach, Scalable high-power and high-brightness fiber coupled diode laser devices. Proc. SPIE **8241**, 824108(1–9) (2012)
29. X. Gao, H. Ohashi, H. Okamoto, M. Takasaka, K. Shinoda, Beam-shaping technique for improving the beam quality of a high-power laser-diode stack. Opt. Lett. **31**(11), 1654–1656 (2006)
30. R.K. Huang, B. Chann, J.D. Glenn, Ultra-high brightness, wavelength-stabilized, kW-class fiber coupled diode laser. Proc. SPIE **7918**, 791810(1–9) (2011)
31. R.K. Huang, B. Chann, J. Burgess, M. Kaiman, R. Overman, J.D. Glenn, P. Tayebati, Direct diode lasers with comparable beam quality to fiber, CO2, and solid state lasers. Proc. SPIE **8241**, 824102(1–6) (2012)
32. R. Stakse, J. Sebastien, J. Wenzel, G. Erbert, H.G. Hansel, *Influence of Mounting Stress on Polarization Degree of Electroluminescence of Laser Diode Bars*. IEEE Lasers and Electro—Optics Society13th Annual Meeting (2000), pp. 10–15
33. X.S. Liu, J.W. Wang, P.Y. Wei, *Study of the Mechanisms of Spectral Broadening in High Power Semiconductor Laser Arrays*. Electronic Components and Technology Conference, IEEE (2008), pp. 1005–1010
34. X.S. Liu, K.C. Song, R.W. Davis, L.C. Hughes, M.H. Hu, C.E. Zah, A metallization scheme for junction-down bonding of high power semiconductor lasers. IEEE Trans. Adv. Packag. **29** (3), 533–541 (2006)
35. X.N. Li, C.H. Peng, Y.X. Zhang, J.W. Wang, L.L. Xiong, P. Zhang, X. S. Liu, *A New Continuous Wave 2500 W Semiconductor Laser Vertical Stack*. IEEE 2010 11th International Conference on Electronic Packaging Technology & High Density Packaging (2010), pp. 1350–1354
36. J.W. Wang, Z.B. Yuan, L.J. Kang, K. Yang, Y.X. Zhang, X.S. Liu, *Study of the Mechanism of "Smile" in High Power Diode Laser Arrays and Strategies in Improving Near-Field Linearity*. 2009 Electronic Components and Technology Conference (2009), pp. 837–842
37. X.S. Liu, R.W. Davis, L.C. Hughes, M.H. Rasmussen, C.E. Zah, A study on the reliability of indium solder die bonding of high power semiconductor lasers. J. Appl. Phys. **100**(1), 013104 (1–11) (2006)
38. D. Lorenzen, M. Schroder, J. Meusel, P. Hennig, H. Konig, M. Philippens, J. Sebastian, R. Hulsewede, Comparative performance studies of indium and gold-tin packaged diode laser bars, High-power diode laser technology and applications IV. Proc. SPIE **6104**, 610404 (1–12) (2006)
39. D. Schleuning, M. Griffin, P. James, J. McNulty, D. Mendoza, J. Morales, D. Nabors, M. Peters, H.L. Zhou, M. Reed, Robust hard-solder packaging of conduction cooled laser diode bars. Proc. SPIE **6456**, 645604(1–11) (2007)
40. X.N. Li, L.J. Kang, J.W. Wang, P. Zhang, L.L. Xiong, X.S. Liu, Hard solder 20 kW QCW stack array diode laser. Proc. SPIE **8241**, 82410C(1–8) (2012)
41. Y. Yan, X. Chen, X.S. Liu, G.-Q. Lu, *Die Bonding of Single Emitter Semiconductor Laser with Nano-Scale Silver Paste*. 2011 International Conference on Electronic Packaging Technology & High Density Packaging (2011), pp. 1143–1147

Index

X. Liu et al., *Packaging of High Power Semiconductor Lasers*, Micro- and Opto-Electronic Materials, Structures, and Systems, DOI 10.1007/978-1-4614-9263-4,

D

E

F

MIX
Papier aus verantwortungsvollen Quellen
Paper from responsible sources
FSC® C105338

If you have any concerns about our products,
you can contact us on
ProductSafety@springernature.com

In case Publisher is established outside the EU,
the EU authorized representative is:
Springer Nature Customer Service Center GmbH
Europaplatz 3, 69115 Heidelberg, Germany

Printed by Libri Plureos GmbH
in Hamburg, Germany